GEOMETRIC AND TOPOLOGICAL METHODS
FOR QUANTUM FIELD THEORY

Based on lectures given at the renowned Villa de Leyva summer school, this book provides a unique presentation of modern geometric methods in quantum field theory. Written by experts, it enables readers to enter some of the most fascinating research topics in this subject.

Covering a series of topics on geometry, topology, algebra, number theory methods, and their applications to quantum field theory, the book covers topics such as Dirac structures, holomorphic bundles and stability, Feynman integrals, geometric aspects of quantum field theory and the standard model, spectral and Riemannian geometry, and index theory. This is a valuable guide for graduate students and researchers in physics and mathematics wanting to enter this interesting research field at the borderline between mathematics and physics.

ALEXANDER CARDONA is Associate Professor in Mathematics at the Universidad de los Andes, Bogotá, where he is part of the research group in geometry, topology, and global analysis. His research interest includes a wide range of applications of mathematics in theoretical physics.

IVÁN CONTRERAS is a Ph.D. student at the Institute of Mathematics, University of Zurich, working in the mathematical physics group. His areas of interest cover the connection between geometry, topology, and field theories.

ANDRÉS F. REYES-LEGA is Associate Professor in the Physics Department at the Universidad de los Andes, Bogotá, and is a member of the theoretical physics group. His recent research work has been in quantum field theory and quantum information theory.

GEOMETRIC AND TOPOLOGICAL METHODS FOR QUANTUM FIELD THEORY

Proceedings of the 2009
Villa de Leyva Summer School

Edited by

ALEXANDER CARDONA
Universidad de los Andes

IVÁN CONTRERAS
University of Zurich

ANDRÉS F. REYES-LEGA
Universidad de los Andes

CAMBRIDGE
UNIVERSITY PRESS

Shaftesbury Road, Cambridge CB2 8EA, United Kingdom

One Liberty Plaza, 20th Floor, New York, NY 10006, USA

477 Williamstown Road, Port Melbourne, VIC 3207, Australia

314–321, 3rd Floor, Plot 3, Splendor Forum, Jasola District Centre, New Delhi – 110025, India

103 Penang Road, #05–06/07, Visioncrest Commercial, Singapore 238467

Cambridge University Press is part of Cambridge University Press & Assessment,
a department of the University of Cambridge.

We share the University's mission to contribute to society through the pursuit of
education, learning and research at the highest international levels of excellence.

www.cambridge.org
Information on this title: www.cambridge.org/9781107026834

© Cambridge University Press & Assessment 2013

First published 2013

A catalogue record for this publication is available from the British Library

Library of Congress Cataloging-in-Publication data
Geometric and topological methods for quantum field theory : proceedings of the 2009
Villa de Leyva summer school / edited by Alexander Cardona,
Iván Contreras, Andrés F. Reyes-Lega.
pages   cm
Includes bibliographical references and index.
ISBN 978-1-107-02683-4 (hardback)
1. Geometric quantization.   2. Quantum field theory – Mathematics.
I. Cardona, Alexander, editor of compilation.   II. Contreras, Iván, 1985– editor of compilation.
III. Reyes-Lega, Andrés F., 1973– editor of compilation.
QC174.17.G46G46   2013
530.14′301516 – dc23   2012048560

ISBN   978-1-107-02683-4   Hardback

Contents

Contributors

Luis J. Boya
Departamento de Física Teórica, Universidad de Zaragoza, España.

Francis Brown
CNRS – Institut de Mathématiques de Jussieu, Paris, France.

Henrique Bursztyn
Instituto de Matemática Pura e Aplicada, Rio de Janeiro, Brasil.

Leonardo A. Cano García
Mathematics Department, Universidad de los Andes, Bogotá, Colombia.

Alexander Cardona
Mathematics Department, Universidad de los Andes, Bogotá, Colombia.

Iván Contreras
Institüt für Mathematik, Universität Zürich, Switzerland.

César Del Corral
Mathematics Department, Universidad de los Andes, Bogotá, Colombia.

Sylvie Paycha
Universität Potsdam, Germany (on leave from: Université Blaise Pascal, Clermont-Ferrand, France).

Florent Schaffhauser
Mathematics Department, Universidad de los Andes, Bogotá, Colombia.

Florian Scheck
Institute of Physics, Theoretical Particle Physics, Johannes Gutenberg–University, Mainz, Germany.

Andrés Vargas
Departamento de Matemáticas, Pontificia Universidad Javeriana, Bogotá, Colombia.

Stefan Weinzierl
Institut für Physik, Universität Mainz, Germany.

Introduction

This volume offers an introduction to some recent developments in several active topics at the interface between geometry, topology, number theory and quantum field theory:

- new geometric structures, Poisson algebras and quantization,
- multizeta, polylogarithms and periods in quantum field theory,
- geometry of quantum fields and the standard model.

It is based on lectures and short communications delivered during a summer school on "Geometric and Topological Methods for Quantum Field Theory" held in Villa de Leyva, Colombia, in July 2009. This school was the sixth of a series of summer schools to take place in Colombia, which have taken place every other year since July 1999. The invited lectures, aimed at graduate students in physics or mathematics, start with introductory material before presenting more advanced results. Each lecture is self-contained and can be read independently of the others.

The volume begins with the introductory lectures on the geometry of Dirac structures by Henrique Bursztyn, in which the author provides the motivation, main features and examples of these new geometric structures in theoretical physics and their applications in Poisson geometry. These lectures are followed by an introduction to the geometry of holomorphic vector bundles over Riemann surfaces by Florent Schaffhauser, in which the author discusses the structure of spaces of connections, the notion of stability and takes us to the celebrated classification theorem of Donaldson for stable bundles. The third lecture, by Sylvie Paycha, explores possible extensions of the theory of characteristic classes and Chern–Weil theory to a class of infinite-dimensional bundles by means of pseudo-differential techniques. After some geometric preliminaries, the author presents the analytic tools (regularized traces and their properties) which are then used to extend the

1

finite-dimensional Chern–Weil calculus to certain infinite-rank vector bundles, with a brief incursion in the Hamiltonian formalism in gauge theory.

The reader is led into the realm of perturbative quantum field theory with an introductory lecture by Stefan Weinzierl on the theory of Feynman integrals. Together with practical algorithms for evaluating Feynman integrals, the author discusses mathematical aspects of loop integrals related to periods, shuffle algebras and multiple polylogarithms. A further lecture by Francis Brown provides an introduction to recent work on iterated integrals and polylogarithms, with emphasis on the case of the thrice punctured Riemann sphere. The author also gives an overview of some recent results connecting such iterated integrals and polylogarithms with Feynman diagrams in perturbative quantum field theory.

In a subsequent lecture, Luis Boya discusses geometric structures that are relevant in quantum field theory and string theory. After an introduction to the basics of differential geometry aimed at physicists, the author discusses holonomy groups, higher-dimensional models relevant for string theory, M-theory and F-theory, as well as geometric aspects of compactification. The last lecture, by Florian Scheck, presents a critical account of some of the more puzzling aspects of the standard model, emphasizing phenomenological as well as geometric aspects. This includes a presentation of the basic geometric structures underlying gauge theories, a discussion of mass matrices and state mixing, a geometric account of anomalies and a review of the noncommutative geometry approach to the standard model. The lecture finishes with a discussion of spontaneous symmetry breaking based on causal gauge invariance.

The invited lectures are followed by four short communications on a wide spectrum of topics. In the first contribution, Leonardo Cano adapts some well-known techniques of spectral analysis for Schrödinger operators to the study of Laplacians on complete manifolds with corners of codimension 2. The author presents results on the absence of a singular continuous spectrum for such operators, as well as a description of the behavior of its pure point spectrum in terms of the underlying geometry. The chapter by Iván Contreras gives a categorical overview of the so-called formal groupoids and studies their associated Hopf algebroids, mentioning their relevance in the field of Poisson geometry as formal realizations of Poisson manifolds. Andrés Vargas presents in his contribution a detailed study of the Einstein condition on Riemannian manifolds with metrics of Hölder regularity, introducing the use of harmonic coordinates and considering the smoothness of the differentiable structure of the underlying manifold. Finally, in the last contribution, Alexander Cardona and César Del Corral study the index of Dirac-type operators associated to Atiyah–Patodi–Singer type boundary conditions from the point of view of weighted (super-)traces. The authors show that both the index of such an operator and the reduced eta-invariant term can be expressed in terms of weighted (super-) traces of identity operators determined by the boundary conditions.

We hope that these contributions will give – as much as the school itself seems to have given – young students the desire to pursue what might be their first acquaintance with some of the problems on the edge of mathematics and physics presented here. On the other hand, we hope that the more advanced reader will find some pleasure in reading about different outlooks on related topics and seeing how the well-known mathematical tools prove to be very useful in some areas of quantum field theory.

We are indebted to various organizations for their financial support for this school. Let us first of all thank Universidad de los Andes, which was our main source of financial support in Colombia. Other organizations such as CLAF (Centro Latino Americano de Fisica) in Brazil, and the International Mathematical Union, through the CDE (Commission on Development and Exchanges) program, also contributed in a substantial way to the financial support needed for this school.

Special thanks to Sergio Adarve (Universidad de los Andes) and Hernán Ocampo (Universidad del Valle) and Sylvie Paycha (Universität Potsdam), coorganizers of the school, who dedicated time and energy to make this school possible. We are also very grateful to Marta Kovacsics who did a great job for the practical organization of the school, the quality of which was very much appreciated by participants and lecturers. We are also very indebted to Diana Tarrifa, Julie Pinzón, Luz Malely Gutiérrez, Mauricio Morales and Alexandra Parra for their help in various essential tasks needed for the successful development of the school.

We also would like to thank the administrative staff at Universidad de los Andes, particularly José Luis Villaveces, Vice-rector; Carlos Montenegro, Dean of the School of Sciences; René Meziat, Director of the Mathematics Department; and Ferney Rodríguez, Director of the Physics Department, for their constant encouragement and support.

Without the people named here, all of whom helped in the organization in some way or another, before, during and after the school, this scientific event would not have left such vivid memories in the lecturers' and participants' minds. Last but not least, thanks to all the participants who gave us all, lecturers and editors, the impulse to prepare this volume through the enthusiasm they showed during the school, and thank you to all the contributors and referees for their participation in the realization of these proceedings.

The editors:

Alexander Cardona, Iván Contreras and Andrés Reyes-Lega

1
A brief introduction to Dirac manifolds

HENRIQUE BURSZTYN

Abstract

These lecture notes are based on a series of lectures given at the school
on "Geometric and Topological Methods for Quantum Field Theory",
in Villa de Leyva, Colombia. We present a basic introduction to Dirac
manifolds, recalling the original context in which they were defined,
their main features, and briefly mentioning more recent developments.

1.1 Introduction

Phase spaces of classical mechanical systems are commonly modeled by symplectic
manifolds. It often happens that the dynamics governing the system's evolution
are constrained to particular submanifolds of the phase space, e.g. level sets of
conserved quantities (typically associated with symmetries of the system, such
as momentum maps), or submanifolds resulting from constraints in the possible
configurations of the system, etc. Any submanifold C of a symplectic manifold M
inherits a *presymplectic* form (i.e. a closed 2-form, possibly degenerate), given by
the pullback of the ambient symplectic form to C. It may be desirable to treat C in
its own right, which makes presymplectic geometry the natural arena for the study
of constrained systems; see e.g. [23, 25].

In many situations, however, phase spaces are modeled by more general objects:
Poisson manifolds (see e.g. [35]). A Poisson structure on a manifold M is a bivector field $\pi \in \Gamma(\wedge^2 TM)$ such that the skew-symmetric bracket $\{f, g\} := \pi(df, dg)$
on $C^\infty(M)$ satisfies the Jacobi identity. Just as for symplectic phase spaces, there
are natural examples of systems on Poisson phase spaces which are constrained

---

Geometric and Topological Methods for Quantum Field Theory, ed. Alexander Cardona, Iván Contreras and
Andrés F. Reyes-Lega. Published by Cambridge University Press. © Cambridge University Press 2013.

to submanifolds. The present notes address the following motivating questions: what kind of geometric structure is inherited by a submanifold C of a Poisson manifold M? Can one "pullback" the ambient Poisson structure on M to C, in a similar way to what one does when M is symplectic? From another viewpoint, recall that M carries a (possibly singular) symplectic foliation, which completely characterizes its Poisson structure. Let us assume, for simplicity, that the intersection of C with each leaf $\mathcal{O}$ of M is a submanifold of C. Then $\mathcal{O} \cap C$ carries a presymplectic form, given by the pullback of the symplectic form on $\mathcal{O}$. So the Poisson structure on M induces a decomposition of C into presymplectic leaves. Just as Poisson structures define symplectic foliations, we can ask whether there is a more general geometric object underlying foliations with presymplectic leaves.

The questions posed in the previous paragraph naturally lead to *Dirac structures* [16, 17], a notion that encompasses presymplectic and Poisson structures. A key ingredient in the definition of a Dirac structure on a manifold M is the so-called *Courant bracket* [16] (see also [21]), a bilinear operation on the space of sections of $TM \oplus T^*M$ used to formulate a general integrability condition unifying the requirements that a 2-form is closed and that a bivector field is Poisson. These notes present the basics of Dirac structures, including their main geometric features and key examples. Most of the material presented here goes back to Courant's original paper [16], perhaps the only exception being the discussion about morphisms in the category of Dirac manifolds in Section 1.5.

Despite its original motivation in constrained mechanics,[1] recent developments in the theory of Dirac structures are related to a broad range of topics in mathematics and mathematical physics. Owing to space and time limitations, this chapter is *not* intended as a comprehensive survey of this fast growing subject (which justifies the omission of many worthy contributions from the references). A (biased) selection of recent aspects of Dirac structures is briefly sketched at the end of the chapter.

This chapter is structured as follows. In Section 1.2, we recall the main geometric properties of presymplectic and Poisson manifolds. Section 1.3 presents the definition of Dirac structures and their first examples. The main properties of Dirac structures are presented in Section 1.4. Section 1.5 discusses morphisms between Dirac manifolds. Section 1.6 explains how Dirac structures are inherited by submanifolds of Poisson manifolds. Section 1.7 briefly mentions some more recent developments and applications of Dirac structures.

---

[1] Dirac structures are named after Dirac's work on the theory of constraints in classical mechanics (see e.g. [20, 41]), which included a classification of constraint surfaces (first class, second class...), the celebrated Dirac bracket formula, as well as applications to quantization and field theory.

1.1.1 Notation, conventions, terminology

All manifolds, maps, vector bundles, etc. are smooth, i.e. in the C^∞ category. Given a smooth map $\varphi : M \to N$ and a vector bundle $A \to N$, we denote the pullback of A to M by $\varphi^* A \to M$.

For a vector bundle $E \to M$, a *distribution* D in E assigns to each $x \in M$ a vector subspace $D_x \subseteq E_x$. If the dimension of D_x, called the *rank* of D at x, is independent of x, we call the distribution *regular*. A distribution D in E is *smooth* if, for any $x \in M$ and $v_0 \in D_x$, there is a smooth local section v of E (defined on a neighborhood of x) such that $v(y) \in D_y$ and $v(x) = v_0$. A distribution that is smooth and regular is a subbundle. The rank of a smooth distribution is a lower semi-continuous function on M. For a vector bundle map $\Phi : E \to A$ covering the identity, the image $\Phi(E)$ is a smooth distribution of A; the kernel ker(Φ) is a distribution of E whose rank is an upper semi-continuous function, so it is smooth if and only if it has locally constant rank. A smooth distribution D in TM is *integrable* if any $x \in M$ is contained in an *integral submanifold*, i.e. a connected immersed submanifold $\mathcal{O}$ so that $D|_{\mathcal{O}} = T\mathcal{O}$. An integrable distribution defines a decomposition of M into *leaves* (which are the maximal integral submanifolds); we generally refer to this decomposition of M as a *singular foliation*, or simply a *foliation*; see e.g. [22, Sec. 1.5] for details. When D is smooth and has constant rank, the classical Frobenius theorem asserts that D is integrable if and only if it is involutive. We refer to the resulting foliation in this case as *regular*.

Throughout the chapter, the Einstein summation convention is consistently used.

1.2 Presymplectic and Poisson structures

A symplectic structure on a manifold can be defined in two equivalent ways: either by a nondegenerate closed 2-form or by a nondegenerate Poisson bivector field. If one drops the nondegeneracy assumption, the first viewpoint leads to the notion of a *presymplectic* structure, while the second leads to *Poisson* structures. These two types of "degenerate" symplectic structures have distinct features that will be recalled in this section.

1.2.1 Two viewpoints on symplectic geometry

Let M be a smooth manifold. A 2-form $\omega \in \Omega^2(M)$ is called *symplectic* if it is nondegenerate and $d\omega = 0$. The nondegeneracy assumption means that the bundle map

$$\omega^\sharp : TM \to T^*M, \quad X \mapsto i_X \omega, \tag{1.1}$$

is an isomorphism; in local coordinates, writing $\omega = \frac{1}{2}\omega_{ij}dx^i \wedge dx^j$, this amounts to the pointwise invertibility of the matrix (ω_{ij}). The pair (M, ω), where ω is a symplectic 2-form, is called a *symplectic manifold*.

The basic ingredients of the Hamiltonian formalism on a symplectic manifold (M, ω) are as follows. For any function $f \in C^\infty(M)$, there is an associated *Hamiltonian vector field* $X_f \in \mathcal{X}(M)$, uniquely defined by the condition

$$i_{X_f}\omega = df. \tag{1.2}$$

In other words, $X_f = (\omega^\sharp)^{-1}(df)$. There is an induced bilinear operation

$$\{\cdot, \cdot\} : C^\infty(M) \times C^\infty(M) \to C^\infty(M),$$

known as the *Poisson bracket*, that measures the rate of change of a function g along the Hamiltonian flow of a function f,

$$\{f, g\} := \omega(X_g, X_f) = \mathcal{L}_{X_f}g. \tag{1.3}$$

The Poisson bracket is skew-symmetric, and one verifies from its definition that

$$d\omega(X_f, X_g, X_h) = \{f, \{g, h\}\} + \{h, \{f, g\}\} + \{g, \{h, f\}\}; \tag{1.4}$$

it follows that the Poisson bracket satisfies the Jacobi identity, since ω is closed. The pair $(C^\infty(M), \{\cdot, \cdot\})$ is a *Poisson algebra*, i.e. $\{\cdot, \cdot\}$ is a Lie bracket on $C^\infty(M)$ that is compatible with the associative commutative product on $C^\infty(M)$ via the Leibniz rule:

$$\{f, gh\} = \{f, g\}h + \{f, h\}g.$$

It follows from the Leibniz rule that the Poisson bracket is defined by a bivector field $\pi \in \Gamma(\wedge^2 TM)$, uniquely determined by

$$\pi(df, dg) = \{f, g\} = \omega(X_g, X_f); \tag{1.5}$$

we write this locally as

$$\pi = \frac{1}{2}\pi^{ij}\frac{\partial}{\partial x^i} \wedge \frac{\partial}{\partial x^j}. \tag{1.6}$$

The bivector field π defines a bundle map

$$\pi^\sharp : T^*M \to TM, \quad \alpha \mapsto i_\alpha\pi, \tag{1.7}$$

in such a way that $X_f = \pi^\sharp(df)$. Since $df = \omega^\sharp(X_f) = \omega^\sharp(\pi^\sharp(df))$, we see that ω and π are related by

$$\omega^\sharp = (\pi^\sharp)^{-1} \quad \text{and} \quad (\omega_{ij}) = (\pi^{ij})^{-1}. \tag{1.8}$$

The whole discussion so far can be turned around, in that one can take the bivector field $\pi \in \Gamma(\wedge^2 TM)$, rather than the 2-form ω, as the starting point to define a symplectic structure. Given a bivector field $\pi \in \Gamma(\wedge^2 TM)$, we call it *nondegenerate* if the bundle map (1.7) is an isomorphism or, equivalently, if the local matrices (π^{ij}) in (1.6) are invertible at each point. We say that π is *Poisson* if the skew-symmetric bilinear bracket $\{f, g\} = \pi(df, dg)$, $f, g \in C^\infty(M)$, satisfies the Jacobi identity:

$$\text{Jac}_\pi(f, g, h) := \{f, \{g, h\}\} + \{h, \{f, g\}\} + \{g, \{h, f\}\} = 0, \qquad (1.9)$$

for all $f, g, h \in C^\infty(M)$.

The relation

$$\pi(df, dg) = \omega(X_g, X_f)$$

establishes a 1–1 correspondence between nondegenerate bivector fields and nondegenerate 2-forms on M, in such a way that the bivector field is Poisson if and only if the corresponding 2-form is closed (see (1.4)). So a symplectic manifold can be equivalently defined as a manifold M equipped with a nondegenerate bivector field π that is Poisson.

The two alternative viewpoints to symplectic structures are summarized in the following table:

Nondegenerate $\pi \in \Gamma(\wedge^2 TM)$	Nondegenerate $\omega \in \Omega^2(M)$
$\text{Jac}_\pi = 0$	$d\omega = 0$
$X_f = \pi^\sharp(df)$	$i_{X_f}\omega = df$
$\{f, g\} = \pi(df, dg)$	$\{f, g\} = \omega(X_g, X_f)$

Although the viewpoints are interchangeable, one may turn out to be more convenient than the other in specific situations, as illustrated next.

1.2.2 Going degenerate

There are natural geometric constructions in symplectic geometry that may spoil the nondegeneracy condition of the symplectic structure, and hence take us out of the symplectic world. We mention two examples.

Consider the problem of passing from a symplectic manifold M to a submanifold $\iota : C \hookrightarrow M$. To describe the geometry that C inherits from M, it is more natural to represent the symplectic structure on M by a 2-form ω, which can then be pulled back to C. The resulting 2-form $\iota^*\omega$ on C is always closed, but generally fails to be nondegenerate.

As a second example, suppose that a Lie group G acts on a symplectic manifold M by symmetries, i.e. preserving the symplectic structure, and consider the geometry inherited by the quotient M/G (we assume, for simplicity, that the action is free and proper, so the orbit space M/G is a smooth manifold). In this case, it is more convenient to think of the symplectic structure on M as a Poisson bivector field π, which can then be projected, or pushed forward, to M/G since π is assumed to be G-invariant. The resulting bivector field on M/G always satisfies (1.9), but generally fails to be nondegenerate.

These two situations illustrate why one may be led to generalize the notion of a symplectic structure by dropping the nondegeneracy condition, and how there are two natural ways to do it. Each way leads to a different kind of geometry: a manifold equipped with a closed 2-form, possibly degenerate, is referred to as *presymplectic*, while a *Poisson manifold* is a manifold equipped with a Poisson bivector field, not necessarily nondegenerate. The main features of presymplectic and Poisson manifolds are summarized below.

Presymplectic manifolds

On a presymplectic manifold (M, ω), there is a natural *null distribution* $K \subseteq TM$, defined at each point $x \in M$ by the kernel of ω:

$$K_x := \ker(\omega)_x = \{X \in T_x M \mid \omega(X, Y) = 0 \ \forall \ Y \in T_x M\}.$$

This distribution is not necessarily regular or smooth. In fact, K is a smooth distribution if and only if it has locally constant rank (see Section 1.1.1). For $X, Y \in \Gamma(K)$, note that

$$i_{[X,Y]}\omega = \mathcal{L}_X i_Y \omega - i_Y \mathcal{L}_X \omega = \mathcal{L}_X i_Y \omega - i_Y (i_X d + d i_X)\omega = 0;$$

it follows that, when K is regular, it is integrable by Frobenius' theorem. We refer to the resulting regular foliation tangent to K as the *null foliation* of M.

One may still define Hamiltonian vector fields on (M, ω) via (1.2), but, without the nondegeneracy assumption on ω, there might be functions admitting no Hamiltonian vector fields (e.g. if df lies outside the image of (1.1) at some point). We say that a function $f \in C^\infty(M)$ is *admissible* if there exists a vector field X_f such that (1.2) holds. In this case, X_f is generally not uniquely defined, as we may change it by the addition of any vector field tangent to K. Still, the Poisson bracket formula

$$\{f, g\} = \mathcal{L}_{X_f} g \tag{1.10}$$

is well defined (i.e. independent of the choice of X_f) when f and g are admissible. Hence the space of admissible functions, denoted by

$$C^\infty_{\mathrm{adm}}(M) \subseteq C^\infty(M),$$

is a Poisson algebra.

When K is regular, a function is admissible if and only if $df(K) = 0$, i.e. f is constant along the leaves of the null foliation; in particular, depending on how complicated this foliation is, there may be very few admissible functions (e.g. if there is a dense leaf, only the constant functions are admissible). When K is regular and the associated null foliation is simple, i.e. the leaf space M/K is smooth and the quotient map $q : M \to M/K$ is a submersion, then M/K inherits a symplectic form ω_{red}, uniquely characterized by the property that $q^*\omega_{red} = \omega$; in this case, the Poisson algebra of admissible functions on M is naturally identified with the Poisson algebra of the symplectic manifold $(M/K, \omega_{\text{red}})$ via

$$q^* : C^\infty(M/K) \xrightarrow{\sim} C^\infty_{\text{adm}}(M)$$

(see e.g. [37, Sec. 6.1] and references therein).

Poisson manifolds

If (M, π) is a Poisson manifold, then any function $f \in C^\infty(M)$ defines a (unique) Hamiltonian vector field $X_f = \pi^\sharp(df)$, and the whole algebra of smooth functions $C^\infty(M)$ is a Poisson algebra with bracket $\{f, g\} = \pi(df, dg)$.

The image of the bundle map $\pi^\sharp$ in (1.7) defines a distribution on M,

$$R := \pi^\sharp(T^*M) \subseteq TM, \tag{1.11}$$

not necessarily regular, but always smooth and integrable. (The integrability of the distribution R may be seen as a consequence of Weinstein's splitting theorem [43].) So it determines a singular foliation of M, in such a way that two points in M lie in the same leaf if and only if one is accessible from the other through a composition of local Hamiltonian flows. One may verify that the bivector field π is "tangent to the leaves", in the sense that, if $f \in C^\infty(M)$ satisfies $\iota^* f \equiv 0$ for a leaf $\iota : \mathcal{O} \hookrightarrow M$, then $X_f \circ \iota \equiv 0$. So there is an induced Poisson bracket $\{\cdot, \cdot\}_\mathcal{O}$ on $\mathcal{O}$ determined by

$$\{f \circ \iota, g \circ \iota\}_\mathcal{O} := \{f, g\} \circ \iota, \quad f, g \in C^\infty(M),$$

which is nondegenerate; in particular, each leaf carries a symplectic form, and one refers to this foliation as the *symplectic foliation* of π. The symplectic foliation of a Poisson manifold uniquely characterizes the Poisson structure. For more details and examples, see e.g. [11, 22, 35].

Remark 1   The integrability of the distribution (1.11) may be also seen as resulting from the existence of a Lie algebroid structure on T^*M, with anchor $\pi^\sharp : T^*M \to TM$ and Lie bracket on $\Gamma(T^*M) = \Omega^1(M)$ uniquely characterized by

$$[df, dg] = d\{f, g\},$$

see e.g. [11, 15]; we will return to Lie algebroids in Section 1.4.

1.3 Dirac structures

Dirac structures were introduced in [16, 17] as a way to treat both types of "degenerate" symplectic structures, namely presymplectic and Poisson, in a unified manner. This common framework relies on viewing presymplectic and Poisson structures as subbundles of

$$\mathbb{T}M := TM \oplus T^*M,$$

defined by the graphs of the bundle maps (1.1) and (1.7). The precise definition of a Dirac structure resorts to additional geometrical structures canonically present on $\mathbb{T}M$, as we now recall.

Let us consider $\mathbb{T}M$ equipped with the natural projections

$$\mathrm{pr}_T : \mathbb{T}M \to TM \quad \text{and} \quad \mathrm{pr}_{T^*} : \mathbb{T}M \to T^*M,$$

as well as two extra structures: the nondegenerate, symmetric fibrewise bilinear form $\langle \cdot, \cdot \rangle$ on $\mathbb{T}M$, given at each $x \in M$ by

$$\langle (X, \alpha), (Y, \beta) \rangle = \beta(X) + \alpha(Y), \quad X, Y \in T_x M, \ \alpha, \beta \in T_x^* M, \tag{1.12}$$

and the *Courant bracket* $[\![\cdot, \cdot]\!] : \Gamma(\mathbb{T}M) \times \Gamma(\mathbb{T}M) \to \Gamma(\mathbb{T}M)$,

$$[\![(X, \alpha), (Y, \beta)]\!] = \left([X, Y], \mathcal{L}_X \beta - \mathcal{L}_Y \alpha + \frac{1}{2} d(\alpha(Y) - \beta(X)) \right). \tag{1.13}$$

A *Dirac structure* on M is a vector subbundle $L \subset \mathbb{T}M$ satisfying:

(i) $L = L^\perp$, where the orthogonal is with respect to $\langle \cdot, \cdot \rangle$,
(ii) $[\![\Gamma(L), \Gamma(L)]\!] \subseteq \Gamma(L)$, i.e. L is involutive with respect to $[\![\cdot, \cdot]\!]$.

Remarks

- Since the pairing $\langle \cdot, \cdot \rangle$ has split signature, condition (i) is equivalent to $\langle \cdot, \cdot \rangle|_L = 0$ and rank$(L) = \dim(M)$.
- The Courant bracket satisfies

$$[\![[\![a_1, a_2]\!], a_3]\!] + c.p. = \frac{1}{3} d(\langle [\![a_1, a_2]\!], a_3 \rangle + c.p.), \tag{1.14}$$

for $a_1, a_2, a_3 \in \Gamma(\mathbb{T}M)$, where c.p. stands for "cyclic permutations". So it fails to satisfy the Jacobi identity, and it is not a Lie bracket.[2]
- One may alternatively use, instead of (1.13), the non-skew-symmetric bracket

$$((X, \alpha), (Y, \beta)) \mapsto ([X, Y], \mathcal{L}_X \beta - i_Y d\alpha)$$

---

[2] The properties of the Courant bracket are axiomatized in the notion of a *Courant algebroid* [33]; see Section 1.7.

for condition (ii); (1.13) is the skew-symmetrization of this bracket, and a simple computation shows that both brackets agree on sections of subbundles satisfying (i).

A subbundle $L \subset \mathbb{T}M$ satisfying (i) is sometimes referred to as a *Lagrangian subbundle* of $\mathbb{T}M$ (in analogy with the terminology in symplectic geometry), or as an *almost Dirac structure* on M. Condition (ii) is referred to as the *integrability condition*; by (i), condition (ii) can be equivalently written as

$$\langle [\![a_1, a_2]\!], a_3 \rangle \equiv 0, \quad \forall a_1, a_2, a_3 \in \Gamma(L). \tag{1.15}$$

For any Lagrangian subbundle $L \subset \mathbb{T}M$, the expression

$$\Upsilon_L(a_1, a_2, a_3) := \langle [\![a_1, a_2]\!], a_3 \rangle, \tag{1.16}$$

for $a_1, a_2, a_3 \in \Gamma(L)$, defines an element $\Upsilon_L \in \Gamma(\wedge^3 L^*)$ that we call the *Courant tensor* of L. Hence, for a Lagrangian subbundle L, the integrability condition (ii) is equivalent to the vanishing condition $\Upsilon_L \equiv 0$.

Example 1.1 Any bivector field $\pi \in \Gamma(\wedge^2 TM)$ defines a Lagrangian subbundle of $\mathbb{T}M$ given by the graph of (1.7),

$$L_\pi = \{ (\pi^\sharp(\alpha), \alpha) \mid \alpha \in T^*M \}. \tag{1.17}$$

One may verify that, for $a_i = (\pi^\sharp(df_i), df_i)$, $i = 1, 2, 3$,

$$\Upsilon_{L_\pi}(a_1, a_2, a_3) = \langle [\![a_1, a_2]\!], a_3 \rangle = \{ f_1, \{ f_2, f_3 \} \} + \text{c.p.}$$

So (1.15) holds, i.e. (1.17) is a Dirac structure if and only if π is a Poisson bivector field. In fact, Poisson structures can be identified with Dirac structures $L \subset \mathbb{T}M$ with the additional property that

$$L \cap TM = \{ 0 \}. \tag{1.18}$$

Similarly, any 2-form $\omega \in \Omega^2(M)$ defines the Lagrangian subbundle

$$L_\omega = \text{graph}(\omega^\sharp) = \{ (X, \omega^\sharp(X)) \mid X \in TM \} \subset \mathbb{T}M. \tag{1.19}$$

In this case, for $a_i = (X_i, \omega^\sharp(X_i))$, $i = 1, 2, 3$,

$$\Upsilon_{L_\omega}(a_1, a_2, a_3) = \langle [\![a_1, a_2]\!], a_3 \rangle = d\omega(X_1, X_2, X_3).$$

So (1.19) is a Dirac structure if and only if ω is presymplectic, and presymplectic structures are identified with Dirac structures L satisfying

$$L \cap T^*M = \{ 0 \}. \tag{1.20}$$

Example 1.2 Let $F \subseteq TM$ be a regular distribution, i.e. a vector subbundle, and let $L = F \oplus F^\circ$, where $F^\circ \subseteq T^*M$ is the annihilator of F. Clearly $L = L^\perp$, and one may check that L satisfies the integrability condition (ii) if and only if F is involutive, i.e. it is tangent to the leaves of a regular foliation (by Frobenius' theorem). So one may view regular foliations as particular cases of Dirac structures.

Remark 2 The notion of Dirac structure may be naturally extended to the complexification $\mathbb{T}M \otimes \mathbb{C}$. A natural example arises as follows: consider an almost complex structure J on M, i.e. an endomorphism $J : TM \to TM$ satisfying $J^2 = -1$, and let $T_{1,0} \subset TM \otimes \mathbb{C}$ be its $+i$-eigenbundle. Following the previous example, $L = T_{1,0} \oplus (T_{1,0})^{\circ} = T_{1,0} \oplus T^{0,1}$ is a (complex) Dirac structure provided $T_{1,0}$ is involutive, i.e. J is integrable as a complex structure. An interesting class of complex Dirac structures that includes this example is studied in [26] (see Section 1.7).

Example 1.3 Let P be a manifold, and let ω_t, $t \in \mathbb{R}$, be a smooth family of closed 2-forms on P. This family determines a Dirac structure L on $M = P \times \mathbb{R}$ by

$$L_{(x,t)} = \{((X, 0), (i_X \omega_t, \gamma)) \mid X \in T_x P, \ \gamma \in \mathbb{R}\} \subset (T_x P \times \mathbb{R}) \oplus (T_x^* P \times \mathbb{R}),$$

where $(x, t) \in P \times \mathbb{R}$.

Example 1.4 Consider $M = \mathbb{R}^3$, with coordinates (x, y, z), and let

$$L = \mathrm{span}\left\langle \left(\frac{\partial}{\partial y}, z dx \right), \left(\frac{\partial}{\partial x}, -z dy \right), (0, dz) \right\rangle.$$

For $z \neq 0$, condition (1.18) is satisfied, and we see that we can write L as the graph of the Poisson structure $\pi = \frac{1}{z} \frac{\partial}{\partial x} \wedge \frac{\partial}{\partial y}$, with brackets

$$\{x, y\} = \frac{1}{z}, \quad \{x, z\} = 0, \quad \{y, z\} = 0.$$

Note that L is a *smooth Dirac structure*, despite the fact that this Poisson structure is singular at $z = 0$; the singularity just reflects the fact that L ceases to be the graph of a bivector field (i.e. to satisfy (1.18)) when $z = 0$.

Example 1.5 Other examples of Dirac structure arise from the observation [40] that the Abelian group of closed 2-forms $\Omega_{cl}^2(M)$ acts on $\mathbb{T}M$ preserving both $\langle \cdot, \cdot \rangle$ and $[\![\cdot, \cdot]\!]$ (more generally, the group of vector-bundle automorphisms of $\mathbb{T}M$ preserving $\langle \cdot, \cdot \rangle$ and $[\![\cdot, \cdot]\!]$ is $\mathrm{Diff}(M) \ltimes \Omega_{cl}^2(M)$, see e.g. [26, Sec. 3]); the action is given by

$$(X, \alpha) \overset{\tau_B}{\mapsto} (X, \alpha + i_X B), \quad \text{for } B \in \Omega_{cl}^2(M).$$

As a result, this operation, called *gauge transformation* in [40], sends Dirac structures to Dirac structures. So one may construct new Dirac structures from old with the aid of closed 2-forms. For example, given an involutive distribution $F \subseteq TM$, the associated Dirac structure of Example 1.2 may be modified by $B \in \Omega_{cl}^2(M)$ to yield a new Dirac structure

$$\{(X, \alpha) \mid X \in F, \ (\alpha - i_X B)|_F = 0\}.$$

More examples of Dirac structures coming from "constraints" in Poisson manifolds will be discussed in Section 1.6.

1.4 Properties of Dirac structures

We now describe the main geometric features of Dirac structures, generalizing those of presymplectic and Poisson manifolds recalled in Section 1.2. Specifically, we shall see how the null distribution and the Poisson algebra of admissible functions of presymplectic manifolds, and the symplectic foliation of Poisson manifolds, generalize to Dirac manifolds.

1.4.1 Lie algebroid

A *Lie algebroid* is a vector bundle $A \to M$ equipped with a Lie bracket $[\cdot, \cdot]$ on $\Gamma(A)$ and a bundle map (called the *anchor*) $\rho : A \to TM$ so that

$$[u, fv] = (\mathcal{L}_{\rho(u)} f)v + f[u, v],$$

for $u, v \in \Gamma(A)$ and $f \in C^\infty(M)$. A key property of a Lie algebroid A is that $\rho(A) \subseteq TM$ is an integrable smooth distribution (possibly of non-constant rank); one refers to the leaves of the associated foliation as *leaves* or *orbits* of the Lie algebroid. More details on Lie algebroids can be found e.g. in [11, 22].

For a Dirac structure L on M, the vector bundle $L \to M$ inherits a Lie algebroid structure, with bracket on $\Gamma(L)$ given by the restriction of $[\![\cdot, \cdot]\!]$ and anchor given by the restriction of pr_T to L:

$$[\cdot, \cdot]_L := [\![\cdot, \cdot]\!]|_{\Gamma(L)}, \qquad \rho_L := \mathrm{pr}_T|_L : L \to TM. \tag{1.21}$$

Note that the Jacobi identity of $[\cdot, \cdot]_L$ follows from (1.14) and (1.15).

On a Poisson manifold (M, π), viewed as a Dirac structure with L given by (1.17), we have a vector-bundle isomorphism $\mathrm{pr}_{T^*}|_L : L \to T^*M$. The Lie algebroid structure on L can be carried over to T^*M, and it coincides with the one mentioned in Remark 1.

For a presymplectic manifold (M, ω), the anchor $\rho_L : L \to TM$ is an isomorphism between the Lie algebroid structure of L and the canonical one on TM (defined by the usual Lie bracket of vector fields and anchor given by the identity).

1.4.2 Presymplectic leaves and null distribution

Since (1.21) endows any Dirac structure L on M with a Lie algebroid structure, the distribution

$$R = \mathrm{pr}_T(L) \subseteq TM \qquad (1.22)$$

is integrable and defines a (singular) foliation on M. Each leaf $\mathcal{O}$ of this foliation naturally inherits a closed 2-form $\Omega_L \in \Omega^2(\mathcal{O})$, defined at each $x \in \mathcal{O}$ by

$$\Omega_L(X, Y) = \alpha(Y), \quad \text{where } X, Y \in T_x\mathcal{O} = R|_x \text{ and } (X, \alpha) \in L|_x.$$

The fact that the formula for Ω_L does not depend on the choice of α results from the observation that

$$R^\circ = L \cap T^*M,$$

while the smoothness of Ω_L follows from the existence of smooth splittings for the surjective bundle map $\mathrm{pr}_T|_L : L|_\mathcal{O} \to T\mathcal{O}$; the condition $d\Omega_L = 0$ follows from the integrability condition of L. One refers to the foliation defined by (1.22) as the *presymplectic foliation* of L. In the case of a Poisson manifold, this is just its symplectic foliation, while for a presymplectic manifold the leaves are its connected components.

The distribution

$$K := L \cap TM \subseteq TM \qquad (1.23)$$

agrees, at each point, with the kernel of the leafwise 2-form Ω_L, so we refer to it as the *null distribution* (or *kernel*) of the Dirac structure; as we saw for presymplectic manifolds, this distribution is not smooth unless it is regular, in which case it is automatically integrable, giving rise to a regular foliation – the *null foliation*. Note that the null distribution is zero at all points if and only if L is defined by a Poisson structure (cf. (1.18)).

In analogy to what happens for Poisson manifolds, a Dirac structure L is completely determined by its presymplectic foliation; indeed, at each point,

$$L = \{(X, \alpha) \mid X \in \mathrm{pr}_T(L), \ \alpha|_{\mathrm{pr}_T(L)} = i_X \Omega_L\}.$$

The reader should have no problem identifying the distributions K, R, and the presymplectic foliation in each example of Section 1.3.

In the spirit of Section 1.2 we have, schematically, the following equivalences:

$$\text{nondegenerate Poisson structure} \rightleftharpoons \text{symplectic structure}$$

$$\text{Poisson structure} \rightleftharpoons \text{symplectic foliation}$$

$$\text{Dirac structure} \rightleftharpoons \text{presymplectic foliation.}$$

1.4.3 Hamiltonian vector fields and Poisson algebra

The notion of Hamiltonian vector field also extends to Dirac structures. Let L be a Dirac structure on M. Following the discussion for presymplectic manifolds in Section 1.2.2, a function $f \in C^\infty(M)$ is called *admissible* if there is a vector field X such that

$$(X, df) \in L,$$

in which case X is called *Hamiltonian* relative to f. Just as for presymplectic manifolds, X is not uniquely determined, as it may be modified by the addition of any vector field tangent to the null distribution $K = L \cap TM$. The space of admissible functions, denoted by $C^\infty_{\mathrm{adm}}(M)$, is always a Poisson algebra, with Poisson bracket defined as in (1.10).

When K is regular, a function f is admissible if and only if its differential annihilates K, i.e. if f is constant along the leaves of the null foliation. In particular, if the null foliation is simple, its leaf space acquires a Poisson structure (generalizing the symplectic structure ω_{red} of Section 1.2.2) through the identification of its functions with admissible functions on M.

In conclusion, Dirac structures naturally mix presymplectic and Poisson features (controlled by the distributions R and K): presymplectic structures arise on their leaves, while Poisson structures arise on their algebras of admissible functions.

1.5 Morphisms of Dirac manifolds

As we saw in Section 1.2, symplectic structures are equivalently described in terms of (nondegenerate) 2-forms or bivector fields. As much as the two approaches are equivalent, they naturally lead to different notions of morphism, reflecting the fact that covariant and contravariant tensors have different functorial properties with respect to maps.

Let M_i be equipped with a symplectic form ω_i, and corresponding Poisson bivector field π_i, $i = 1, 2$. There are two natural ways in which a map $\varphi : M_1 \to M_2$ can preserve symplectic structures: either by preserving symplectic 2-forms,

$$\varphi^* \omega_2 = \omega_1 \tag{1.24}$$

or by preserving Poisson bivector fields,

$$\varphi_* \pi_1 = \pi_2; \tag{1.25}$$

by (1.25) we mean that π_1 and π_2 are φ-related: for all $x \in M_1$,

$$(\pi_2)_{\varphi(x)}(\alpha, \beta) = (\pi_1)_x(\varphi^*\alpha, \varphi^*\beta), \quad \forall \alpha, \beta \in T^*_{\varphi(x)}M_2.$$

Conditions (1.24) and (1.25) are not equivalent. Consider $\mathbb{R}^2$ with coordinates (q^1, p_1) and symplectic structure (written both as a 2-form and as a bivector field)

$$\omega_{\mathbb{R}^2} = dq^1 \wedge dp_1, \quad \pi_{\mathbb{R}^2} = \frac{\partial}{\partial p_1} \wedge \frac{\partial}{\partial q^1},$$

and $\mathbb{R}^4$ with coordinates (q^1, p_1, q^2, p_2) and symplectic structure

$$\omega_{\mathbb{R}^4} = dq^1 \wedge dp_1 + dq^2 \wedge dp_2, \quad \pi_{\mathbb{R}^4} = \frac{\partial}{\partial p_1} \wedge \frac{\partial}{\partial q^1} + \frac{\partial}{\partial p_2} \wedge \frac{\partial}{\partial q^2}.$$

One readily verifies that the projection $\mathbb{R}^4 \to \mathbb{R}^2, (q^1, p_1, q^2, p_2) \mapsto (q^1, p_1)$ satisfies (1.25), but not (1.24), while the inclusion $\mathbb{R}^2 \hookrightarrow \mathbb{R}^4, (q^1, p_1) \mapsto (q^1, p_1, 0, 0)$, satisfies (1.24) but not (1.25). More generally, note that the nondegeneracy of ω_1 in (1.24) forces φ to be an immersion, while the nondegeneracy of π_2 in (1.25) forces φ to be a submersion. The two conditions (1.24) and (1.25) become equivalent when φ is a local diffeomorphism. It is also clear that (1.24) naturally extends to a notion of morphism between presymplectic manifolds, whereas (1.25) leads to the usual notion of *Poisson map* between Poisson manifolds.

As first noticed by A. Weinstein, the pullback and pushforward relations carry over to Dirac structures, leading to two distinct notions of morphism for Dirac manifolds, generalizing (1.24) and (1.25).

1.5.1 Pulling back and pushing forward

In order to extend the notions of pullback and pushforward to the realm of Dirac structures, it is convenient to start at the level of linear algebra.

Lagrangian relations

In the linear-algebra context, by a *Dirac structure on a vector space V* we will simply mean a subspace $L \subset V \oplus V^*$ that is Lagrangian, i.e. such that $L = L^\perp$ with respect to the natural symmetric pairing

$$\langle (v_1, \alpha_1), (v_2, \alpha_2) \rangle = \alpha_2(v_1) + \alpha_1(v_2).$$

We also simplify the notation by writing $\mathbb{V} = V \oplus V^*$; we denote by $\mathrm{pr}_V : \mathbb{V} \to V$ and $\mathrm{pr}_{V^*} : \mathbb{V} \to V^*$ the natural projections.

Let $\varphi : V \to W$ be a linear map. We know that 2-forms $\omega \in \wedge^2 W^*$ can be pulled back to V, while bivectors $\pi \in \wedge^2 V$ can be pushed forward to W:

$$\varphi^* \omega(v_1, v_2) = \omega(\varphi(v_1), \varphi(v_2)), \quad \varphi_* \pi(\beta_1, \beta_2) = \pi(\varphi^* \beta_1, \varphi^* \beta_2),$$

for $v_1, v_2 \in V$, $\beta_1, \beta_2 \in W^*$. These notions are generalized to Dirac structures as follows. The *backward image* of a Dirac structure $L_W \subset \mathbb{W}$ under φ is defined by

$$\mathfrak{B}_\varphi(L_W) = \{(v, \varphi^*\beta) \mid (\varphi(v), \beta) \in L_W\} \subset V \oplus V^*, \tag{1.26}$$

while the *forward image* of $L_V \in V \oplus V^*$ is

$$\mathfrak{F}_\varphi(L_V) = \{(\varphi(v), \beta) \mid (v, \varphi^*\beta) \in L_V\} \subset W \oplus W^*. \tag{1.27}$$

Proposition 1.6 Both $\mathfrak{B}_\varphi(L_W)$ and $\mathfrak{F}_\varphi(L_V)$ are Dirac structures.

The proof will follow from Lemma 1.7 below. Note that when L_W is defined by $\omega \in \wedge^2 W^*$, then $\mathfrak{B}_\varphi(L_W)$ is the Dirac structure associated with $\varphi^*\omega$ and, similarly, if L_V is defined by $\pi \in \wedge^2 V$, then $\mathfrak{F}_\varphi(L_V)$ is the Dirac structure corresponding to $\varphi_*\pi$. So we also refer to the Dirac structures (1.26) and (1.27) as the *pullback* of L_W and the *pushforward* of L_V, respectively.

The pullback and pushforward operations (1.26) and (1.27) are not inverses of one another; one may verify that

$$\mathfrak{F}_\varphi(\mathfrak{B}_\varphi(L_W)) = L_W \text{ if and only if } \mathrm{pr}_W(L_W) \subseteq \varphi(V), \tag{1.28}$$

$$\mathfrak{B}_\varphi(\mathfrak{F}_\varphi(L_V)) = L_V \text{ if and only if } \ker(\varphi) \subseteq L_V \cap V. \tag{1.29}$$

As a result, $\mathfrak{F}_\varphi \circ \mathfrak{B}_\varphi = \mathrm{Id}$ if and only if φ is surjective, while $\mathfrak{B}_\varphi \circ \mathfrak{F}_\varphi = \mathrm{Id}$ if and only if φ is injective.

Pullback and pushforward of Dirac structures can be understood in the broader context of the composition of Lagrangian relations (which is totally analogous to the calculus of canonical relations in the linear symplectic category; see [27], [5, Sec. 5]; see also [45] and references therein), that we now recall. Let us denote by $\overline{\mathbb{V}}$ the vector space $\mathbb{V} = V \oplus V^*$ equipped with the pairing $-\langle \cdot, \cdot \rangle$. The product $\mathbb{V} \times \mathbb{W}$ inherits a natural pairing from $\mathbb{V}$ and $\mathbb{W}$. We call a Lagrangian subspace $L \subset \mathbb{V} \times \overline{\mathbb{W}}$ a *Lagrangian relation* (from W to V). Note that any Dirac structure $L \subset \mathbb{V}$ may be seen as a Lagrangian relation (either from V to the trivial vector space $\{0\}$, or the other way around), and any linear map $\varphi : V \to W$ defines a Lagrangian relation

$$\Gamma_\varphi = \{((w, \beta), (v, \alpha)) \mid w = \varphi(v), \ \alpha = \varphi^*(\beta)\} \subset \mathbb{W} \times \overline{\mathbb{V}}. \tag{1.30}$$

Lagrangian relations can be composed much in the same way as maps. Consider vector spaces U, V, W and Lagrangian relations $L_1 \subseteq \mathbb{U} \times \overline{\mathbb{V}}$ and $L_2 \subset \mathbb{V} \times \overline{\mathbb{W}}$. The *composition* of L_1 and L_2, denoted by $L_1 \circ L_2 \subset \mathbb{U} \times \overline{\mathbb{W}}$, is defined by

$$L_1 \circ L_2 := \{((u, \alpha), (w, \gamma)) \mid \exists (v, \beta) \in \mathbb{V} \ s.t. \ (u, \alpha, v, \beta) \in L_1,$$

$$(v, \beta, w, \gamma) \in L_2\}.$$

Lemma 1.7 *The composition $L_1 \circ L_2$ is a Lagrangian relation.*

Proof To verify that $(L_1 \circ L_2)^\perp = L_1 \circ L_2$, consider the subspace

$$C := \mathbb{U} \times \Delta \times \overline{\mathbb{W}} \subset \mathbb{U} \times \overline{\mathbb{V}} \times \mathbb{V} \times \overline{\mathbb{W}}, \qquad (1.31)$$

where Δ is the diagonal in $\overline{\mathbb{V}} \times \mathbb{V}$. Then

$$C^\perp = \{0\} \times \Delta \times \{0\} \subseteq C,$$

and $L_1 \circ L_2$ is the image of $L_1 \times L_2 \subset \mathbb{U} \times \overline{\mathbb{V}} \times \mathbb{V} \times \overline{\mathbb{W}}$ in the quotient $C/C^\perp = \mathbb{U} \times \overline{\mathbb{W}}$:

$$L_1 \circ L_2 = \frac{(L_1 \times L_2) \cap C + C^\perp}{C^\perp}. \qquad (1.32)$$

The projection $p : C \to C/C^\perp$ satisfies $\langle p(c_1), p(c_2) \rangle = \langle c_1, c_2 \rangle$ for all $c_1, c_2 \in C$. So $p(c_1) \in (L_1 \circ L_2)^\perp$, i.e. $\langle p(c_1), p(c_2) \rangle = 0$ for all $p(c_2) \in L_1 \circ L_2$, if and only if $\langle c_1, c_2 \rangle = 0$ for all $c_2 \in (L_1 \times L_2) \cap C + C^\perp$, i.e.

$$c_1 \in ((L_1 \times L_2) \cap C + C^\perp)^\perp.$$

Now note that[3]

$$((L_1 \times L_2) \cap C + C^\perp)^\perp = (L_1 \times L_2 + C^\perp) \cap C$$

$$= (L_1 \times L_2) \cap C + C^\perp,$$

from which we conclude that $(L_1 \circ L_2)^\perp = L_1 \circ L_2$. □

We can now conclude the proof of Proposition 1.6.

Proof (of Proposition 1.6) Let us view a Dirac structure L_V on V as a Lagrangian relation $L_V \subset V \times \{0\}$; similarly, we regard a Dirac structure L_W on W as a Lagrangian relation $L_W \subset \{0\} \times \overline{W}$. Then one immediately verifies that the pushforward and pullback with respect to a linear map $\varphi : V \to W$ are given by compositions of relations with respect to the Lagrangian relation Γ_φ in (1.30):

$$\mathfrak{B}_\varphi(L_W) = L_W \circ \Gamma_\varphi, \qquad \mathfrak{F}_\varphi(L_V) = \Gamma_\varphi \circ L_V. \qquad (1.33)$$

If follows from Lemma 1.7 that (1.26) and (1.27) are Dirac structures. □

Before moving on to Dirac structures on manifolds, we discuss for later use how the composition expression in (1.32) may be simplified for the particular cases in

---

[3] Here we use that, for subspaces A, B of any vector space equipped with a symmetric pairing with split signature, we have $(A + B)^\perp = A^\perp \cap B^\perp$ and $(A \cap B)^\perp = A^\perp + B^\perp$.

(1.33). For the composition $\mathfrak{B}_\varphi(L_W) = L_W \circ \Gamma_\varphi$, we have (cf. (1.31))

$$C = \Delta_W \times \overline{\mathbb{V}} \subset \overline{W} \times W \times \overline{\mathbb{V}},$$

where Δ_W is the diagonal subspace in $\overline{W} \times W$, so that $C^\perp = \Delta_W \times \{0\}$; then, by (1.32), $\mathfrak{B}_\varphi(L_W)$ is the image of $(L_W \times \Gamma_\varphi) \cap C$ under the projection $p : C \to C/C^\perp = \overline{\mathbb{V}}$. We may write p as a composition of two maps: the isomorphism $p_1 : C \xrightarrow{\sim} W \times \mathbb{V}, (w, w, v) \mapsto (w, v)$, followed by the projection $p_2 : W \times \mathbb{V} \to \mathbb{V}$. Noticing that $p_1((L_W \times \Gamma_\varphi) \cap C) = (L_W \times \mathbb{V}) \cap \Gamma_\varphi$, we can write $\mathfrak{B}_\varphi(L_W) = p_2((L_W \times \mathbb{V}) \cap \Gamma_\varphi)$. One also checks that

$$\ker(p_2|_{(L_W \times \mathbb{V}) \cap \Gamma_\varphi}) = \{((0, \beta), (0, 0)) \mid (0, \beta) \in L_W, \ \varphi^*\beta = 0\}$$

$$= (\ker(\varphi^*) \cap L_W) \times \{0\}.$$

As a result, we obtain:

Proposition 1.8 The following is a short exact sequence:

$$(\ker(\varphi^*) \cap L_W) \times \{0\} \hookrightarrow (L_W \times \mathbb{V}) \cap \Gamma_\varphi \twoheadrightarrow \mathfrak{B}_\varphi(L_W), \qquad (1.34)$$

where the second map is the projection $W \times \mathbb{V} \to \mathbb{V}$. Similarly, we have the short exact sequence

$$\{0\} \times (\ker(\varphi) \cap L_V) \hookrightarrow \Gamma_\varphi \cap (W \times L_V) \twoheadrightarrow \mathfrak{F}_\varphi(L_V), \qquad (1.35)$$

where the second map is the projection $W \times \mathbb{V} \to W$.

Morphisms on Dirac manifolds

We now transfer the notions of pullback and pushforward of Dirac structures to the manifold setting, i.e. the role of $\mathbb{V}$ will be played by $\mathbb{T}M$.

Let $L_M \subset \mathbb{T}M$ and $L_N \subset \mathbb{T}N$ be almost Dirac structures, and $\varphi : M \to N$ be a smooth map. Analogously to (1.30), we consider the Lagrangian relation

$$\Gamma_\varphi = \{((Y, \beta), (X, \alpha)) \mid Y = d\varphi(X), \ \alpha = \varphi^*\beta\} \subset \varphi^*\mathbb{T}N \oplus \overline{\mathbb{T}M}$$

where $\varphi^*\mathbb{T}N \to M$ is the pullback bundle of $\mathbb{T}N$ by φ, and we define, as in (1.33), the *backward image* of L_N by

$$\mathfrak{B}_\varphi(L_N) = (\varphi^*L_N) \circ \Gamma_\varphi \subset \mathbb{T}M, \qquad (1.36)$$

and the *forward image* of L_M by

$$\mathfrak{F}_\varphi(L_M) = \Gamma_\varphi \circ L_M \subset \varphi^*\mathbb{T}N. \qquad (1.37)$$

By what we saw in Section 1.5.1, $\mathfrak{B}_\varphi(L_N)$ and $\mathfrak{F}_\varphi(L_M)$ define Lagrangian distributions in $\mathbb{T}M$ and $\varphi^*\mathbb{T}N$, respectively; the issue of whether they are smooth will be addressed in the next section.

We call $\varphi : M \to N$ a *backward Dirac map* (or simply *b-Dirac*) when L_M coincides with $\mathfrak{B}_\varphi(L_N)$,

$$(L_M)_x = \mathfrak{B}_\varphi(L_N)_x = \{(X, d\varphi^*(\beta)) \mid (d\varphi(X), \beta) \in (L_N)_{\varphi(x)}\}, \quad \forall x \in M, \tag{1.38}$$

and we call it a *forward Dirac map* (or simply *f-Dirac*) when φ^*L_N (the pullback of L_N to N as a vector bundle) coincides with $\mathfrak{F}_\varphi(L_M)$, i.e. $\forall x \in M$,

$$(L_N)_{\varphi(x)} = \mathfrak{F}_\varphi(L_M)_{\varphi(x)} = \{(d\varphi(X), \beta) \mid (X, d\varphi^*(\beta)) \in (L_M)_x\}. \tag{1.39}$$

As already observed, (1.38) generalizes the pullback of 2-forms, while (1.39) extends the notion of two bivector fields being φ-related.

If $\varphi : M \to M$ is a diffeomorphism, then it is f-Dirac if and only if it is b-Dirac, and these are both equivalent to the condition that the canonical lift of φ to $\mathbb{T}M$ preserves the subbundle $L_M \subset \mathbb{T}M$:

$$\Phi := (d\varphi, (d\varphi^{-1})^*) : \mathbb{T}M \to \mathbb{T}M, \quad \Phi(L_M) = L_M. \tag{1.40}$$

We refer to φ as a *Dirac diffeomorphism*.

Remark 3 Forward Dirac maps $\varphi : M \to N$ satisfying the additional transversality condition $\ker(d\varphi) \cap L_M = \{0\}$ are studied in [1, 7, 8] in the context of actions and momentum maps.

1.5.2 Clean intersection and smoothness issues

In the previous subsection, we used pullback and pushforward to obtain two ways to relate Dirac structures on M and N by a map $\varphi : M \to N$: via (1.38) and (1.39). We now discuss the possibility of using pullback and pushforward to *transport* Dirac structures from one manifold to the other. In other words, we know that the backward (resp. forward) image of a Dirac structure (1.36) (resp. (1.37)) gives rise to a Lagrangian distribution in $\mathbb{T}M$ (resp. $\varphi^*\mathbb{T}N$), and the issue is whether this distribution fits into a smooth bundle defining a Dirac structure. Note that, while the Lagrangian distribution generated by the backward image (1.36) is defined over all of M, the forward image (1.37) only defines the distribution over points in the image of φ. (This reflects the well-known fact that, while any map $\varphi : M \to N$ induces a map $\Omega^\bullet(N) \to \Omega^\bullet(M)$ by pullback, there is in general no induced map from $\mathcal{X}^\bullet(M)$ to $\mathcal{X}^\bullet(N)$.) This makes the discussion for backward images a bit simpler, so we treat them first.

Backward images

Let L_N be an almost Dirac structure on N, and consider a map $\varphi : M \to N$. As we already remarked, (1.38) defines, at each point $x \in M$, a Lagrangian subspace of $T_x M \oplus T_x^* M$. This Lagrangian distribution does *not* necessarily fit into a smooth subbundle of $\mathbb{T}M$, as illustrated by the following example.

Example 1.9 Let $\varphi : \mathbb{R} \to \mathbb{R}^2$ be the inclusion $x \mapsto (x, 0)$, and consider the Dirac structure on $\mathbb{R}^2$ defined by the bivector field

$$\pi = x \frac{\partial}{\partial x} \wedge \frac{\partial}{\partial y}.$$

We consider its backward image to the x-axis via φ. Whenever $x \neq 0$, π is non-degenerate, and hence corresponds to a 2-form; its pullback is then a 2-form on $\mathbb{R} - \{0\}$, which must therefore vanish identically (by dimension reasons). So the backward image of π on $\mathbb{R}$ is $T_x \mathbb{R} \subset T_x \mathbb{R} \oplus T_x^* \mathbb{R}$ for $x \neq 0$. For $x = 0$, one can readily compute the backward image to be $T_x^* \mathbb{R} \subset T_x \mathbb{R} \oplus T_x^* \mathbb{R}$, so the resulting family of Lagrangian subspaces is not a smooth vector bundle over $\mathbb{R}$.

As we now see, a suitable clean-intersection assumption guarantees the smoothness of backward images. For a smooth map $\varphi : M \to N$, let $(d\varphi)^* : \varphi^* T^* N \to T^* M$ be the dual of the tangent map $d\varphi : TM \to \varphi^* TN$. For an almost Dirac structure L_N on N, we denote by $\varphi^* L_N$ the vector bundle over M obtained by the pullback of L_N *as a vector bundle*.

Proposition 1.10 Let $\varphi : M \to N$ be a smooth map, and let L_N be an almost Dirac structure on N. If $\ker((d\varphi)^*) \cap \varphi^* L_N$ has constant rank, then the backward image $\mathfrak{B}_\varphi(L_N) \subset \mathbb{T}M$ defines a smooth Lagrangian subbundle, i.e. it is an almost Dirac structure on M. If $\mathfrak{B}_\varphi(L_N)$ is a smooth bundle and L_N is integrable, then $\mathfrak{B}_\varphi(L_N)$ is integrable, i.e. it is a Dirac structure.

Proof The backward image of L_N fits into a short exact sequence that is a pointwise version of (1.34):

$$\ker((d\varphi)^*) \cap \varphi^* L_N \to (\varphi^* L_N \oplus \mathbb{T}M) \cap \Gamma_\varphi \to \mathfrak{B}_\varphi(L_N), \qquad (1.41)$$

where the first map is the inclusion $\varphi^* \mathbb{T}N \to \varphi^* \mathbb{T}N \oplus \mathbb{T}M$, and the second is the projection of $\varphi^* \mathbb{T}N \oplus \mathbb{T}M$ onto $\mathbb{T}M$. Since $\mathfrak{B}_\varphi(L_N)$ has constant rank, it is clear that $(\varphi^* \mathbb{T}N \oplus \mathbb{T}M) \cap \Gamma_\varphi$ has constant rank if and only if $\ker((d\varphi)^*) \cap \varphi^* L_N$ does too; in this case both are smooth vector bundles, and hence so is $\mathfrak{B}_\varphi(L_N)$.

For the assertion about integrability, we compare the tensors Υ_M and Υ_N associated with the almost Dirac structures $\mathcal{B}_\varphi(L_N)$ and L_N, respectively (see (1.16)). Let us call (local) sections $a = (X, \alpha)$ of $\mathbb{T}M$ and $b = (Y, \beta)$ of $\mathbb{T}N$ φ-*related* if the vector fields X and Y are φ-related and $\alpha = \varphi^*\beta$. If now $a_i = (X_i, \alpha_i)$, $b_i = (Y_i, \beta_i)$, $i = 1, 2, 3$, are sections of $\mathbb{T}M$ and $\mathbb{T}N$, respectively, such that a_i and b_i are φ-related, then a direct computation shows that

$$\langle a_1, [\![a_2, a_3]\!]\rangle = \varphi^*\langle b_1, [\![b_2, b_3]\!]\rangle. \tag{1.42}$$

Suppose that $x_0 \in M$ is such that

(i) $d\varphi$ has constant rank around x_0, and
(ii) $\ker((d\varphi)^*) \cap \varphi^* L_N$ has constant rank around x_0.

By (i), we know that locally, around x_0 and $\varphi(x_0)$, the map φ looks like

$$(x^1, \ldots, x^m) \overset{\varphi}{\mapsto} (x^1, \ldots, x^k, 0, \ldots, 0), \tag{1.43}$$

where k is the rank of $d\varphi$ around x. In this local model, $M = \{(x^1, \ldots, x^m)\}$, $N = \{(x^1, \ldots, x^n)\}$, and we identify the submanifolds of M and N defined by $x^i = 0$ for $i > k$ with $S = \{(x^1, \ldots, x^k)\}$.

Claim: *For any $a_{x_0} \in \mathcal{B}_\varphi(L_N)|_{x_0}$, one can find local sections a of $\mathcal{B}_\varphi(L_N)$ and b of L_N so that $a|_{x_0} = a_{x_0}$, and a and b are φ-related.*

To verify the claim, first extend a_{x_0} to a local section $a_S = (X_S, \alpha_S)$ of $\mathcal{B}_\varphi(L_N)|_S$; using (ii) and (1.41), one can then find a local section $b_S = (Y_S, \beta_S)$ of $L_N|_S$ so that (b_S, a_S) is a section of $\Gamma_\varphi|_S$, i.e. $d\varphi(X_S) = Y_S$ and $\alpha_S = \varphi^*\beta_S$. Using the local form (1.43), one may naturally extend X_S to a vector field X on M satisfying $d\varphi(X) = Y_S$. It follows that $a = (X, \varphi^*\beta_S)$ is a local section of $\mathcal{B}_\varphi(L_N)$ and, if b is any local section of L_N extending b_S, a and b are φ-related. This proves the claim.

Given arbitrary $(a_i)_{x_0} \in \mathcal{B}_\varphi(L_N)|_{x_0}$, $i = 1, 2, 3$, by the previous claim we can extend them to local sections a_i of $\mathcal{B}_\varphi(L_N)$ and find local sections b_i of L_N, so that a_i and b_i are φ-related. From (1.42), we have

$$\Upsilon_M((a_1)_{x_0}, (a_2)_{x_0}, (a_3)_{x_0}) = \Upsilon_M(a_1, a_2, a_3)(x_0) = \Upsilon_N(b_1, b_2, b_3)(\varphi(x_0)) = 0,$$

since L_N is integrable. It directly follows that Υ_M vanishes at all points $x \in M$ satisfying (i) and (ii). Since these points form an open dense subset in M, we conclude that $\Upsilon_M \equiv 0$, as desired. $\quad\square$

We refer to the condition that $\ker((d\varphi)^*) \cap \varphi^* L_N$ has constant rank as the *clean-intersection condition* (cf. [5, Sec. 5]). By (1.20), it always holds when L_N is defined by a 2-form on N.

Example 1.11 Let L be a Dirac structure on M, and let $\varphi : C \hookrightarrow M$ be the inclusion of a submanifold. In this case, $\ker(d\varphi)^* = TC^\circ$, so the clean-intersection condition is that $L \cap TC^\circ$ has constant rank. This guarantees that C inherits a Dirac structure by pullback, explicitly given (noticing that $(L|_C \oplus \mathbb{T}C) \cap \Gamma_\varphi$ projects isomorphically onto $L \cap (TC \oplus T^*M|_C)$ in (1.41)) by

$$\mathcal{B}_\varphi(L) = \{(X, \varphi^*\beta) \mid (X, \beta) \in L\} \cong \frac{L \cap (TC \oplus T^*M|_C)}{L \cap TC^\circ} \subset TC \oplus T^*C,$$

as originally noticed in [16].

Forward images

Let L_M be an almost Dirac structure on M and $\varphi : M \to N$ be a smooth map. The forward image (1.37) of L_M by φ defines a Lagrangian distribution in $\varphi^*\mathbb{T}N$. The question of whether it is a Dirac structure on N involves two issues: first, as in the discussion of backward images, whether this distribution fits into a smooth subbundle of $\varphi^*\mathbb{T}N$; second, if this subbundle determines a subbundle of $\mathbb{T}N$. The first issue leads to a clean-intersection condition, analogous to the one for backward images, while the second issue involves an additional invariance condition with respect to the map φ. This invariance condition is of course already necessary when one considers the pushforward of bivector fields: if $\varphi : M \to N$ is a surjective submersion, for the pushforward of a bivector field $\pi \in \Gamma(\wedge^2 TM)$ to be well defined, one needs that π satisfies, for all $\alpha, \beta \in T_y^*N$,

$$\pi_x(\varphi^*\alpha, \varphi^*\beta) = \pi_{x'}(\varphi^*\alpha, \varphi^*\beta)$$

whenever $\varphi(x) = \varphi(x') = y$. We extend this invariance to the context of Dirac structures as follows: we say that L_M is φ-*invariant* if the right-hand side of (1.39) is invariant for all x on the same φ-fibre.

The next example illustrates a situation where a Dirac structure satisfies the invariance condition, but its forward image is not a smooth bundle.

Example 1.12 Consider the 2-form

$$\omega = x dx \wedge dy$$

on $\mathbb{R}^2$, independent of y, and let $\varphi : \mathbb{R}^2 \to \mathbb{R}$ be the projection $(x, y) \mapsto x$. For $x \neq 0$, its forward image is $T_x^*\mathbb{R} \subset T_x\mathbb{R} \oplus T_x^*\mathbb{R}$ (since the 2-form is nondegenerate at these points, so we may consider the pushforward of the corresponding bivector field), while for $x = 0$ it is $T_x\mathbb{R} \subset T_x\mathbb{R} \oplus T_x^*\mathbb{R}$.

The following result is parallel to Proposition 1.10.

Proposition 1.13 Let $\varphi : M \to N$ be a surjective submersion and L_M be an almost Dirac structure on M. If $\ker(d\varphi) \cap L_M$ has constant rank, then the forward image of L_M by φ is a smooth Lagrangian subbundle of $\varphi^*\mathbb{T}N$. If the forward image of L_M is a smooth bundle in $\varphi^*\mathbb{T}N$ and L_M is φ-invariant, then it defines an almost Dirac structure on M, which is integrable provided L_M is.

Proof Similarly to the proof of Proposition 1.10, one has the short exact sequence (cf. (1.35))

$$\ker(d\varphi) \cap L_M \rightarrow \Gamma_\varphi \cap (\varphi^*\mathbb{T}N \oplus L_M) \rightarrow \mathfrak{F}_\varphi(L_M), \qquad (1.44)$$

where the first map is the inclusion $\mathbb{T}M \rightarrow \varphi^*\mathbb{T}N \oplus \mathbb{T}M$ and the second is the projection $\varphi^*\mathbb{T}N \oplus \mathbb{T}M \rightarrow \varphi^*\mathbb{T}N$. Since the rank of $\mathfrak{F}_\varphi(L_M)$ is constant, one argues as in the proof of Proposition 1.10: the constant-rank condition for $\ker(d\varphi) \cap L_M$ guarantees that $\Gamma_\varphi \cap (\varphi^*\mathbb{T}N \oplus L_M)$ has constant rank, hence $\mathfrak{F}_\varphi(L_M) \subset \varphi^*\mathbb{T}N$ is a (Lagrangian) vector subbundle.

The φ-invariance of L_M guarantees that $\mathfrak{F}_\varphi(L_M)$ determines a Lagrangian subbundle of $\mathbb{T}N$, which we denote by L_N, given by the image of $\mathfrak{F}_\varphi(L_M)$ under the natural projection $\varphi^*\mathbb{T}N \rightarrow \mathbb{T}N$; one may also verify that $\mathfrak{F}_\varphi(L_M) = \varphi^*L_N$ (the pullback of L_N as a vector bundle).

The assertion about integrability is verified as in Proposition 1.10. If $x \in M$ is such that $\ker(d\varphi) \cap L_M$ has constant rank around it, for any local section b of L_N around $\varphi(x)$ one can choose a local section a of L_M around x (by splitting (1.44)) which is φ-related to b. Given arbitrary $(b_y)_i \in L_N|_y$, $i = 1, 2, 3$, where $y = \varphi(x)$, we extend them to local sections b_i of L_N, and take local sections a_i of L_M so that a_i and b_i are φ-related. Denoting by Υ_N and Υ_M the Courant tensors of L_N and L_M, respectively, we have

$$\Upsilon_N((b_y)_1, (b_y)_2, (b_y)_3) = \Upsilon_N(b_1, b_2, b_3)(y) = \Upsilon_M(a_1, a_2, a_3)(x) = 0,$$

assuming that L_N is integrable. Using the fact that points $x \in M$ around which $\ker(d\varphi) \cap L_M$ has constant rank form an open dense subset of M, we conclude that $\Upsilon_N \equiv 0$. $\square$

Example 1.14 Let M be equipped with a Dirac structure L, and suppose that $\varphi : M \rightarrow N$ is a surjective submersion whose fibres are connected and such that

$$\ker(d\varphi) \subseteq L \cap \mathbb{T}M = K.$$

Then $\ker(d\varphi) \cap L = \ker(d\varphi)$ has constant rank. We claim that L is automatically φ-invariant. To verify this fact, the following is the key observation: given a vector field Z (assume it to be complete, for simplicity), its flow ψ_Z^t preserves L (i.e. it is a family of Dirac diffeomorphisms, see (1.40)), if and only if

$$\mathcal{L}_Z(X, \alpha) = ([Z, X], \mathcal{L}_Z\alpha) \in \Gamma(L), \qquad \text{for all } (X, \alpha) \in \Gamma(L). \qquad (1.45)$$

Since φ is a surjective submersion, it is locally a projection, so any $x \in M$ admits a neighborhood in which, for any other x' on the same φ-fibre, one can find a (compactly supported) vector field Z, tangent to $\ker(d\varphi)$, whose flow (for some time t) takes x to x'. Note that (1.45) holds, as the right-hand side of this equation is $[\![(Z, 0), (X, \alpha)]\!]$, which must lie in $\Gamma(L)$ as a result of the integrability of L (since Z is a section of $\ker(d\varphi) \subseteq L \cap TM$). So the (time-$t$) flow of Z defines a Dirac diffeomorphism taking x to x'. Since φ has connected fibres, it follows that one can find a Dirac diffeomorphism (by composing finitely many flows) ψ taking x to any x' on the same φ-fibre. Using ψ, we see that

$$\mathfrak{F}_\varphi(L|_{x'}) = \mathfrak{F}_\varphi(\mathfrak{F}_\psi(L|_x)) = \mathfrak{F}_{\varphi \circ \psi}(L|_{x'}) = \mathfrak{F}_\varphi(L|_x),$$

so L is φ-invariant.

It follows that the forward image of L, denoted by L_N, defines a Dirac structure on N. Moreover, $L_N \cap TN = d\varphi(K)$. A particular case of this situation is when the null foliation of L is simple, N is the leaf space, $\varphi : M \to N$ is the natural projection, and $K = \ker(d\varphi)$. In this case, $L_N \cap TN = \{0\}$, so L_N is given by a bivector field π_N, determined by

$$\{f, g\}_N(\varphi(x)) = dg(\pi_N^\sharp(df))(\varphi(x)) = dg(d\varphi(X_{\varphi^* f}))(\varphi(x)) = \{\varphi^* f, \varphi^* g\}(x).$$

So π_N is the Poisson structure on N arising from the Poisson bracket on admissible functions (in this case, these agree with the basic functions relative to the null foliation) on M via the identification

$$\varphi^* : C^\infty(N) \to C^\infty(M)_{\mathrm{adm}}.$$

Note that φ is f-Dirac by construction, but it is also b-Dirac by (1.29).

Remark 4 Consider a Lie group G acting on M preserving the Dirac structure L, i.e. an action by Dirac diffeomorphisms. If the action is free and proper, the quotient map $\varphi : M \to N = M/G$ is a surjective submersion and L is φ-invariant. In this case the clean-intersection assumption, guaranteeing that the forward image of L on N is a Dirac structure, is that the distribution $L \cap D$ has constant rank, where D is the distribution tangent to the G-orbits. Extensions of this discussion to non-free actions have been considered e.g. in [29].

1.6 Submanifolds of Poisson manifolds and constraints

We now discuss Dirac manifolds arising as submanifolds of Poisson manifolds, analogous to the presymplectic structures inherited by submanifolds of symplectic manifolds. The discussion illustrates how Dirac structures provide a convenient setting for the description of the intrinsic geometry of constraints in Poisson phase spaces.

As pointed out in Example 1.11, submanifolds of Dirac manifolds inherit Dirac structures (modulo a cleanness issue) via pullback. We will focus on the Dirac structures inherited by submanifolds $\iota : C \hookrightarrow M$ of a given *Poisson manifold* (M, π). We denote by L the Dirac structure on M defined by π. Then

$$\mathfrak{B}_\iota(L) = \{(X, \iota^*\beta) \in \mathbb{T}C \mid X = \pi^\sharp(\beta) \in TC, \ \beta \in T^*M\}. \tag{1.46}$$

We will focus on the case where this is a smooth bundle, hence a Dirac structure on C. Note that the clean-intersection condition for this pullback is that $\ker(\pi^\sharp) \cap TC^\circ$ has constant rank, which is equivalent to $\pi^\sharp(TC^\circ) \subset TM|_C$ having constant rank, as a result of the short exact sequence $\ker(\pi^\sharp) \cap TC^\circ \to TC^\circ \to \pi^\sharp(TC^\circ)$.

The null distribution of $\mathfrak{B}_\iota(L)$ is

$$TC \cap \mathfrak{B}_\iota(L) = TC \cap \pi^\sharp(TC^\circ). \tag{1.47}$$

We present a description of the Poisson bracket on admissible functions on C in terms of the Poisson structure on M.

1.6.1 The induced Poisson bracket on admissible functions

Assume that a function f on C admits a local extension $\hat{f}$ to M so that

$$d\hat{f}(\pi^\sharp(TC^\circ)) = 0. \tag{1.48}$$

Then $X_{\hat{f}} = \pi^\sharp(d\hat{f})$ satisfies $\beta(X_{\hat{f}}) = 0$ for all $\beta \in TC^\circ$; i.e. $X_{\hat{f}}|_C \in TC$. It is then clear from (1.46) that

$$(X_{\hat{f}}|_C, df) \in \mathfrak{B}_\iota(L),$$

i.e. $X_{\hat{f}}|_C$ is a Hamiltonian vector field for f.

Recall that, if the null distribution (1.47) has constant rank, then a function f on C is admissible for $\mathfrak{B}_\iota(L)$ if and only if df vanishes on $TC \cap \pi^\sharp(TC^\circ)$. If we assume additionally that $\pi^\sharp(TC^\circ)$ has constant rank, then any admissible function f on C can be extended to a function $\hat{f}$ on M satisfying (1.48). Denoting by $\{\cdot, \cdot\}_C$ the Poisson bracket on admissible functions, we have

$$\{f, g\}_C = X_f g = X_{\hat{f}}|_C g = X_{\hat{f}} \hat{g}|_C, \tag{1.49}$$

i.e. the Poisson brackets on C and M are related by

$$\{f, g\}_C = \{\hat{f}, \hat{g}\}|_C, \tag{1.50}$$

where the extension $\hat{f}$ (resp. $\hat{g}$) satisfies (1.48).

1.6.2 A word on coisotropic submanifolds (or first-class constraints)

A submanifold C of M is called *coisotropic* if

$$\pi^\sharp(TC^\circ) \subseteq TC. \tag{1.51}$$

These submanifolds play a key role in Poisson geometry (see e.g. [11]) and general-ize the notion of first-class constraints in physics [20, 23]: when C is defined as the zero set of k independent functions ψ^i, condition (1.51) amounts to $\{\psi^i, \psi^j\}|_C = 0$.

Note that (1.51) does not guarantee that the pullback image of π to C is smooth; indeed, Example 1.9 illustrates a pullback image to a coisotropic submanifold that is not smooth. Despite this fact, Dirac structures play an important role in the study of coisotropic submanifolds: they provide a way to extend the formulation of coisotropic embedding problems in symplectic geometry (treated in [24, 34]) to the realm of Poisson manifolds.

More specifically, assume that the pullback image of L to C is smooth, so that it defines a Dirac structure $\mathfrak{B}_l(L)$. Condition (1.51) implies that this Dirac structure has null distribution (cf. (1.47))

$$TC \cap \mathfrak{B}_l(L) = \pi^\sharp(TC^\circ), \tag{1.52}$$

which necessarily has (locally) constant rank (since the rank of the left-hand side of (1.52) is an upper semi-continuous function on C, while the rank of the right-hand side is lower semi-continuous; cf. Section 1.1.1).

Using Dirac structures, one can then pose the converse question: can any Dirac manifold with constant-rank null distribution be coisotropically embedded into a Poisson manifold? It is proven in [13] that this is always possible, extending a result of Gotay [24] concerning coisotropic embeddings of constant-rank presymplectic manifolds into symplectic manifolds. The reader can find much more on coisotropic embeddings into Poisson manifolds in [13].

1.6.3 Poisson–Dirac submanifolds and the Dirac bracket

Let us assume that the pullback image (1.46) is a smooth bundle, hence defining a Dirac structure $\mathfrak{B}_l(L)$ on C. As we have seen, this is guaranteed by the condition that $\ker(\pi^\sharp) \cap TC^\circ$, or $\pi^\sharp(TC^\circ)$, has constant rank. The Dirac structure $\mathfrak{B}_l(L)$ on C is defined by a *Poisson* structure if and only if its null distribution is zero, i.e. by (1.47), if and only if

$$TC \cap \pi^\sharp(TC^\circ) = 0. \tag{1.53}$$

In this case C is called a *Poisson–Dirac submanifold* [18] (cf. [43, Prop. 1.4]). When M is symplectic, a Poisson–Dirac submanifold amounts to a symplectic submanifold (i.e. a submanifold whose natural presymplectic structure is nonde-generate). In general, Poisson–Dirac submanifolds are "leafwise symplectic sub-manifolds", thus generalizing symplectic submanifolds to the Poisson world.

On Poisson–Dirac submanifolds, all smooth functions are admissible. Supposing that $\pi^\sharp(TC^\circ)$ has constant rank, the Poisson bracket $\{\cdot, \cdot\}_C$ on C may be computed as in (1.50):

$$\{f, g\}_C = \{\hat{f}, \hat{g}\}|_C, \tag{1.54}$$

where $\hat{f}, \hat{g}$ are local extensions of f, g to M, satisfying (1.48).

Amongst Poisson–Dirac submanifolds, there are two special cases of interest. The first one is when

$$\pi^\sharp(TC^\circ) = 0,$$

which is equivalent to $TC^\circ \subseteq \ker(\pi^\sharp)$, or $\mathrm{im}(\pi^\sharp) \subseteq TC$; this is precisely the con-dition (1.28) for $\mathfrak{F}_\iota(\mathfrak{B}_\iota(L)) = L$, i.e. for the inclusion $\iota : C \hookrightarrow M$ to be a Poisson map, and hence for C to be a Poisson submanifold of M.

The second special case is when

$$TM|_C = TC \oplus \pi^\sharp(TC^\circ),$$

and C is called a *cosymplectic* submanifold. Such submanifolds extend the notion of *second-class constraints* to the realm of Poisson manifolds; this is the context in which the Dirac-bracket formula for the pullback Poisson structure $\{\cdot, \cdot\}_C$ on C can be derived (see e.g. [37, Ch. 10]), providing an alternative expression to (1.54) (see (1.56) below).

When C is a cosymplectic submanifold, one has natural projections

$$\mathrm{pr}_{TC} : TM|_C \to TC, \quad \mathrm{pr}_{\pi^\sharp(TC^\circ)} = \mathrm{Id} - \mathrm{pr}_{TC} : TM|_C \to \pi^\sharp(TC^\circ),$$

and it is clear from (1.54) that the Poisson structure on C is given by the bivector field $\pi_C = (\mathrm{pr}_{TC})_*\pi|_C$. Note also that, for f, g smooth functions on C and F an arbitrary local extensions of f to M, we have (see (1.49), recalling the extensions $\hat{f}, \hat{g}$ in (1.48))

$$dg(X_f) = \{f, g\}_C = -df(X_{\hat{g}}|_C) = -dF(X_{\hat{g}})|_C$$
$$= d\hat{g}(X_F)|_C = dg(\mathrm{pr}_{TC}(X_F|_C));$$

we conclude that the Hamiltonian vector fields X_f and X_F (with respect to π_C and π, respectively) are related by

$$X_f = \mathrm{pr}_{TC}(X_F|_C). \tag{1.55}$$

Let us suppose that the "constraint" submanifold C is defined by k independent functions ψ^i, i.e. $C = \Psi^{-1}(0)$ for a submersion $\Psi = (\psi^1, \ldots, \psi^k) : M \to \mathbb{R}^k$. Then $d\psi^i$, $i = 1, \ldots, k$, form a basis for TC°, and $TC = \ker(d\Psi)$ along C. Consider the matrix (c^{ij}), where

$$c^{ij} = \{\psi^i, \psi^j\};$$

the fact that C is cosymplectic means that $\pi^\sharp(TC^\circ)$ has rank k (since TC has rank $\dim(M) - k$, so $\pi^\sharp|_{TC^\circ} : TC^\circ \to \pi^\sharp(TC^\circ)$ is an isomorphism, and $X_{\psi_i} = \pi^\sharp(d\psi_i)$, $i = 1, \ldots, k$, form a basis of $\pi^\sharp(TC^\circ)$. Hence, for $a_j \in \mathbb{R}$, the condition $c^{ij}(x)a_j = 0$ for all i means that $-d\psi^i(a_j X_{\psi^j}) = 0$ for all i, i.e, $a_j X_{\psi^j} \in TC \cap \pi^\sharp(TC^\circ) = \{0\}$, which implies that $a_j = 0$. So C being cosymplectic means that the matrix (c^{ij}) is invertible (in this case, one may also check that k is necessarily even, so C has even codimension). Let us denote by (c_{ij}) its inverse. We have the following explicit expression for the projection $\mathrm{pr}_{\pi^\sharp(TC^\circ)}$: for $Y \in TM|_C$,

$$\mathrm{pr}_{\pi^\sharp(TC^\circ)}(Y) = d\psi^i(Y)c_{ij}X_{\psi^j}.$$

To verify this formula, note that if $Y \in TC$, then the right-hand side above vanishes. On the other hand, for each X_{ψ^i} in $\pi^\sharp(TC^\circ)$,

$$\mathrm{pr}_{\pi^\sharp(TC^\circ)}(X_{\psi^l}) = d\psi^i(X_{\psi^l})c_{ij}X_{\psi^j} = c^{li}c_{ij}X_{\psi^j} = \delta^l_j X_{\psi^j} = X_{\psi^l},$$

so $\mathrm{pr}_{\pi^\sharp(TC^\circ)}$ is the identity on $\pi^\sharp(TC^\circ)$. By (1.55), we see that

$$\{f, g\}_C = (\mathrm{pr}_{TC}(X_F|_C))g = (\mathrm{pr}_{TC}(X_F|_C))G = ((\mathbb{1} - \mathrm{pr}_{\pi^\sharp(TC^\circ)})X_F|_C)G,$$

where F, G are arbitrary local extensions of f, g. So

$$\{f, g\}_C = (\{F, G\} - \{F, \psi^i\}c_{ij}\{\psi^j, G\})|_C, \tag{1.56}$$

which is Dirac's bracket formula for cosymplectic manifolds.

1.6.4 Momentum level sets

Another important class of submanifolds of Poisson manifolds is given by level sets of conserved quantities associated with symmetries, such as momentum maps.

Suppose that a Lie group G acts on a Poisson manifold M in a Hamiltonian fashion, with equivariant momentum map $J : M \to \mathfrak{g}^*$. This means that

$$u_M = X_{\langle J, u \rangle}, \tag{1.57}$$

where u_M is the infinitesimal generator of the action associated with $u \in \mathfrak{g}$, and the equivariance of J is with respect to the coadjoint G-action on $\mathfrak{g}^*$; infinitesimally, the equivariance condition implies that J is a Poisson map with respect to the Lie–Poisson structure on $\mathfrak{g}^*$ (see e.g. [37, Sec. 10] for details). We denote by $D \subseteq TM$ the distribution on M tangent to the G-orbits.

If $\mu \in \mathfrak{g}^*$ is a regular value for J, one considers the submanifold $C = J^{-1}(\mu)$, which naturally carries an action of G_μ, the isotropy group of the coadjoint action at μ. Let us suppose that this G_μ-action on $J^{-1}(\mu)$ is free and proper, so that its orbit space is a smooth manifold, and we can consider the following diagram

$$J^{-1}(\mu) \xrightarrow{\ \iota\ } M \tag{1.58}$$
$$p \downarrow$$
$$J^{-1}(\mu)/G_\mu,$$

where ι is the inclusion and p is the quotient map, which is a surjective submersion. The reduction theorem in this context (see e.g. [37, Sec. 10.4.15]) says that $J^{-1}(\mu)/G_\mu$ inherits a natural Poisson structure π_μ, with corresponding bracket $\{\cdot, \cdot\}_\mu$, characterized by the following condition:

$$\{f, g\}_\mu \circ p = \{\bar{f}, \bar{g}\} \circ \iota, \tag{1.59}$$

where $\bar{f}, \bar{g}$ are local extensions of p^*f, $p^*g \in C^\infty(J^{-1}(\mu))$ to M satisfying $d\bar{f}|_D = 0$, $d\bar{g}|_D = 0$. In this picture, both M and $J^{-1}(\mu)/G_\mu$ are Poisson manifolds, but there is no geometric interpretation of the "intermediate" manifold $C = J^{-1}(\mu)$.

When the Poisson structure on M is nondegenerate, i.e. defined by a symplectic form ω, we are in the classical setting of Marsden–Weinstein reduction. In this case $J^{-1}(\mu)$ is naturally a presymplectic manifold, with 2-form $\iota^*\omega$, and the Poisson structure π_μ on $J^{-1}(\mu)/G_\mu$ is defined by a symplectic form ω_μ; the condition relating π and π_μ is easily expressed in term of the corresponding symplectic forms as

$$p^*\omega_\mu = \iota^*\omega, \tag{1.60}$$

which determines ω_μ uniquely. As we now see, Dirac structures allow us to identify the geometric nature of $J^{-1}(\mu)$ in general and interpret (1.59) as a direct generalization of (1.60).

We start by considering the backward image of π with respect to ι. The clean-intersection condition guaranteeing the smoothness of $\mathfrak{B}_\iota(L_\pi)$ is that the distribution

$$\ker(dJ)^\circ \cap L_\pi = \mathrm{im}((dJ)^*) \cap L_\pi = \mathrm{im}((dJ)^*) \cap \ker(\pi^\sharp)$$

has constant rank over $J^{-1}(\mu)$. Since μ is a regular value of J and, for $x \in J^{-1}(\mu)$,

$$\text{im}((dJ)^*) \cap \ker(\pi^\sharp)|_x = \{\alpha \in T_x^*M \mid \alpha = (dJ)^*(u),\ \pi^\sharp(\alpha) = u_M(x) = 0\}$$
$$= (dJ)^*(\mathfrak{g}_x),$$

where $\mathfrak{g}_x$ is the isotropy Lie algebra of the G-action at x, we see that the clean-intersection condition amounts to checking that $\dim(\mathfrak{g}_x)$ is constant for $x \in J^{-1}(\mu)$. Since J is equivariant, $G_x \subseteq G_\mu$, and G_x agrees with the isotropy group of G_μ at x, which is trivial since G_μ acts on $J^{-1}(\mu)$ freely. Hence $\dim(\mathfrak{g}_x) = 0$ for $x \in J^{-1}(\mu)$. It follows that $\mathfrak{B}_\iota(L_\pi)$ makes the momentum level set $J^{-1}(\mu)$ into a Dirac manifold in such a way that ι is a b-Dirac map.

Since $\pi^\sharp(\ker(dJ)^\circ) = D$ (by (1.57)), the null distribution of $\mathfrak{B}_\iota(L_\pi)$ is (cf. (1.47))

$$T(J^{-1}(\mu)) \cap D = \ker(p),$$

where the equality follows from the equivariance of J and the fact that $\ker(p)$ agrees with the distribution tangent to the G_μ-orbits on $J^{-1}(\mu)$. By Example 1.14 and Remark 4, the forward image of $\mathfrak{B}_\iota(L_\pi)$ under p defines a Poisson structure on $J^{-1}(\mu)/G_\mu$, with bracket given by

$$\{f, g\}_\mu \circ p = \{p^*f, p^*g\}_C, \tag{1.61}$$

where $\{\cdot, \cdot\}_C$ is the Poisson bracket on admissible functions on $J^{-1}(\mu)$. By (1.50),

$$\{p^*f, p^*g\}_C = \{\widehat{p^*f}, \widehat{p^*g}\}|_C,$$

where $\widehat{p^*f}$ denotes a local extension of p^*f to M with differential vanishing on $\pi^\sharp(TC^\circ) = \pi^\sharp(\text{im}(dJ^*)) = D$. So the bracket $\{\cdot, \cdot\}_\mu$ in (1.61) is the same as the bracket defined in (1.59).

In conclusion, Dirac geometry completes the reduction diagram (1.58) in the following way: the level set $J^{-1}(\mu)$ inherits a Dirac structure $\mathfrak{B}_\iota(L_\pi)$ for which ι is a b-Dirac map, while the Poisson structure π_μ on the quotient $J^{-1}(\mu)/G_\mu$ is its forward image, so p is an f-Dirac map. As mentioned in Example 1.14, p is also a b-Dirac map, and this condition uniquely determines π_μ. We can then write the relationship between π and π_μ (cf. (1.59)) exactly as we do for symplectic forms:

$$\mathfrak{B}_\iota(L_\pi) = \mathfrak{B}_p(L_{\pi_\mu}),$$

noticing that, in the symplectic setting, $\mathfrak{B}_\iota(L_\omega) = \mathfrak{B}_p(L_{\omega_\mu})$ is nothing but (1.60).

1.7 Brief remarks on further developments

In this last section, we merely indicate some directions of more recent developments and applications of Dirac structures. Although these lecture notes have mostly focused on how Dirac structures arise in the study of constraints of classical systems on Poisson manifolds, subsequent work to Courant's paper [16] revealed connections between Dirac structures and many other subjects in mathematics and mathematical physics. Even in mechanics, (almost) Dirac structures have found applications beyond their original motivation, such as the study of implicit Hamiltonian systems (including nonholonomic systems; see [42] for a survey and original references) as well as implicit Lagrangian systems and variational principles, see e.g. [47].

Understanding the properties of the Courant bracket (1.13), used to formulate the integrability condition for Dirac structures, has led to the notion of *Courant algebroid* [33], which consists of a vector bundle $E \to M$, a symmetric nondegenerate fibrewise pairing on E, a bracket on the space $\Gamma(E)$, and a bundle map $E \to TM$ satisfying axioms that emulate those of $\mathbb{T}M \to M$ equipped with $\langle \cdot, \cdot \rangle$, $[\![\cdot, \cdot]\!]$ and $\mathrm{pr}_T : \mathbb{T}M \to TM$; see Section 1.3 (a detailed discussion can be found in [32]). Dirac structures on general Courant algebroids are used in [33] to extend the theory of Lie bialgebras, Manin triples and Poisson–Lie groups to the realm of Lie algebroids and groupoids. From another perspective, Courant algebroids admit a natural super-geometric description in terms of certain types of differential graded symplectic manifolds [38, 39]; in this context, Dirac structures may be appropriately viewed as analogs of Lagrangian submanifolds. Courant algebroids have also been studied in connection with field theories and vertex algebras (see e.g. [3, 6] and references therein).

A distinguished class of Courant algebroids, introduced by Ševera (see e.g. [40]), is defined through a modification of the Courant bracket (1.13) on $\mathbb{T}M$ by a closed 3-form $H \in \Omega^3(M)$:

$$[\![(X, \alpha), (Y, \beta)]\!]_H = [\![(X, \alpha), (Y, \beta)]\!] + i_Y i_X H. \qquad (1.62)$$

Such Courant algebroids are known as *exact*, and Dirac structures relative to $[\![\cdot, \cdot]\!]_H$ are often referred to as *H-twisted*. For example, the integrability condition for a bivector field π with respect to (1.62) becomes

$$\mathrm{Jac}_\pi(f, g, h) = H(X_f, X_g, X_h);$$

see [31]. A key example of an *H*-twisted Dirac structure is the so-called *Cartan–Dirac* structure (see e.g. [1, 8]), defined on any Lie group G whose Lie algebra is equipped with a nondegenerate Ad-invariant quadratic form; in this case, $H \in \Omega^3(G)$ is the associated Cartan 3-form.

As mentioned in Section 1.4, any Dirac structure $L \subset \mathbb{T}M$ carries a Lie algebroid structure. From a Lie-theoretic standpoint, it is natural to search for the class of Lie groupoids "integrating" Dirac structures. For Poisson structures, the corresponding global objects are the so-called *symplectic groupoids* [15] (see also [12, 19]), originally introduced as part of a quantization scheme for Poisson manifolds (see e.g. [5]). More generally, the global counterparts of Dirac structures were identified in [8], and called *presymplectic groupoids*. Passing from Dirac structures to presymplectic groupoids is key to uncovering links between Dirac structures, momentum maps and equivariant cohomology; see e.g. [8, 46].

The connection between Dirac structures and the theory of momentum maps is far-reaching. As recalled in Section 1.6, classical momentum maps in Poisson geometry are defined by Poisson maps

$$J : M \to \mathfrak{g}^*$$

from a Poisson manifold into the dual of a Lie algebra. In the last 25 years, generalized notions of momentum maps have sprung up in symplectic geometry, mostly motivated by the role of symplectic structures in different areas of mathematics and mathematical physics; see [44] for a general discussion and original references. Prominent examples include the theory of Hamiltonian actions of symplectic groupoids [36], in which arbitrary Poisson maps may be viewed as momentum maps, and the theory of G-valued moment maps, which arises in the study of moduli spaces in gauge theory [2]. These seemingly unrelated new notions of momentum map turn out to fit into a common geometric framework provided by Dirac structures and Courant algebroids. Much of the usual Hamiltonian theory in Poisson geometry carries over to the realm of Dirac structures, and the key observation is that, just as classical momentum maps are Poisson maps, many generalized momentum maps are morphisms between Dirac manifolds (in an appropriate sense, see Remark 3); see e.g. [1, 7–10].

Another viewpoint to Dirac structures that has proven fruitful is based on their description via *pure spinors* [4, 28] (see also [1]). The bundle $\mathbb{T}M$, equipped with the symmetric pairing (1.12), gives rise to a bundle $\mathrm{Cl}(\mathbb{T}M)$ of Clifford algebras, and the exterior algebra $\wedge^\bullet T^*M$ carries a natural representation of $\mathrm{Cl}(\mathbb{T}M)$, defined by the action

$$(X, \alpha) \cdot \phi = i_X \phi + \alpha \wedge \phi,$$

for $(X, \alpha) \in \mathbb{T}M$ and $\phi \in \wedge^\bullet T^*M$. In this way, $\wedge^\bullet T^*M$ is viewed as a spinor module over $\mathrm{Cl}(\mathbb{T}M)$. The subspace $L \subset \mathbb{T}M$ that annihilates a given (nonzero) ϕ is always isotropic, and ϕ is called a *pure spinor* when L is Lagrangian: $L = L^\perp$. Pure spinors corresponding to the same annihilator L must agree up to rescaling;

as a result, almost Dirac structures on M are uniquely characterized by a (generally nontrivial) smooth line bundle

$$l \subset \wedge^\bullet T^*M, \tag{1.63}$$

generated by pure spinors at all points. The integrability of L relative to the (H-twisted) Courant bracket can be naturally expressed in terms of l as well, see e.g [4, 26]. In this way, Dirac structures are represented by specific line bundles (1.63) and, by taking (local) sections, one obtains concrete descriptions of Dirac structures through differential forms. For example, the Dirac structure L_ω (1.19) is the annihilator of the differential form $e^{-\omega} \in \Omega^\bullet(M)$. This approach to Dirac structures leads to natural constructions of invariant volume forms associated with G-valued moment maps [1] and has been an essential ingredient in the study of generalized complex structures [26, 28], that we briefly recall next.

Generalized complex structures [26, 28] are particular types of complex Dirac structures $L \subset \mathbb{T}M \otimes \mathbb{C}$, with the extra property that $L \cap \overline{L} = \{0\}$; they correspond to $+i$-eigenbundles of endomorphisms $\mathcal{J} : \mathbb{T}M \to \mathbb{T}M$ satisfying $\mathcal{J}^2 = -1$ and preserving the pairing (1.12). These geometrical structures unify symplectic and complex geometries, though the most interesting examples fall in between these extreme cases (see e.g. [14]). Given a complex structure $J : TM \to TM$ and a symplectic structure $\omega \in \Omega^2(M)$, viewed as a bundle map $TM \to T^*M$, the associated generalized complex structures $\mathbb{T}M \to \mathbb{T}M$ are

$$\begin{pmatrix} J & 0 \\ 0 & -J^* \end{pmatrix}, \quad \begin{pmatrix} 0 & -\omega^{-1} \\ \omega & 0 \end{pmatrix}.$$

Besides their mathematical interest, generalized complex structures have drawn much attention due to their strong ties with theoretical physics, including e.g. topological strings and super symmetric sigma models, see e.g. [30, 48] and references therein.

Acknowledgments

I would like to thank the organizers and participants of the school in Villa de Leyva, particularly Alexander Cardona for his invitation and encouragement to have these lecture notes written up. Versions of these lectures were delivered at Porto (*Oporto Meeting on Geometry, Topology and Physics, 2009*) and Canary Islands (*Young Researchers Workshop on Geometry, Mechanics and Control, 2010*) and helped me to shape up the notes (which I hope to expand in the future); I thank the organizers of these meetings as well. I am also indebted to P. Balseiro, A. Cardona,

L. Garcia-Naranjo, M. Jotz, C. Ortiz and M. Zambon for their comments on this manuscript.

References

[1] Alekseev, A., Bursztyn, H., Meinrenken, E., Pure spinors on Lie groups. *Asterisque* **327** (2010), 129–197.

[2] Alekseev, A., Malkin, A., Meinrenken, E., Lie group valued moment maps. *J. Differential Geom.* **48** (1998), 445–495.

[3] Alekseev, A., Strobl, T., Current algebras and differential geometry. *JHEP*, **0503** (2005), 035.

[4] Alekseev, A., Xu, P., Derived brackets and Courant algebroids, unpublished manuscript.

[5] Bates, S., Weinstein, A., *Lectures on the geometry of quantization*, Berkeley Mathematics Lecture Notes 8. Providence, RI: American Mathematical Society, 1997.

[6] Bressler, P., The first Pontryagin class. *Compositio Math.* **143** (2007), 1127–1163.

[7] Bursztyn, H., Crainic, M., Dirac geometry, quasi-Poisson actions and D/G-valued moment maps. *J. Differential Geom.* **82** (2009), 501–566.

[8] Bursztyn, H., Crainic, M., Weinstein, A., Zhu, C., Integration of twisted Dirac brackets. *Duke Math. J.* **123** (2004), 549–607.

[9] Bursztyn, H., Iglesias-Ponte, D., Severa, P., Courant morphisms and moment maps. *Math. Res. Lett.* **16** (2009), 215–232.

[10] Bursztyn, H., Radko, O., Gauge equivalence of Dirac structures and symplectic groupoids. *Ann. Inst. Fourier (Grenoble)*, **53** (2003), 309–337.

[11] Cannas da Silva, A., Weinstein, A., *Geometric models for noncommutative algebras*. Berkeley Mathematics Lecture Notes 10. Providence, RI: American Mathematical Society, 1999.

[12] Cattaneo, A., Felder, G., Poisson sigma models and symplectic groupoids. In N. P. Landsman, M. Pflaum and M. Schlichenmaier (eds), *Quantization of singular symplectic quotients*, Progress in Mathematical. Basel: Birkhauser, 2001, pp. 61–93.

[13] Cattaneo, A., Zambon, M., Coisotropic embeddings in Poisson manifolds. *Trans. Am. Math. Soc.* **361** (2009), 3721–3746.

[14] Cavalcanti, G., Gualtieri, M., A surgery for generalized complex structures on 4-manifolds. *J. Differential Geom.* **76** (2007), 35–43.

[15] Coste, A., Dazord, P., Weinstein, A., *Groupoïdes symplectiques*. Publications du Département de Mathématiques. Nouvelle Série. A, Vol. 2. Lyon: Université Claude-Bernard, 1987, pp. i–ii, 1–62.

[16] Courant, T., Dirac manifolds, *Trans. Am. Math. Soc.* **319** (1990), 631–661.

[17] Courant, T., Weinstein, A., *Beyond Poisson structures*. Séminaire sudrhodanien de géométrie VIII. Travaux en Cours 27, Paris: Hermann, 1988, pp. 39–49.

[18] Crainic, M., Fernandes, R., Integrability of Lie brackets. *Ann. Math.* **157** (2003), 575–620.

[19] Crainic, M., Fernandes, R., Integrability of Poisson brackets. *J. Differential Geom.* **66** (2004), 71–137.

[20] Dirac, P., *Lectures on quantum mechanics*. New York: Belfer Graduate School of Science, Yeshiva University, 1964.

[21] Dorfman, I., *Dirac structures and integrability of evolution equations*. New York: John Wiley, 1993.

[22] Dufour, J.-P., Zung, N.-T., *Poisson structures and their normal forms*, Progress in Mathematics 242. Boston, MA: Birkhauser, 2005.

[23] Gotay, M., Coisotropic imbeddings, Dirac brackets and quantization. In M. Gotay (ed.), *Geometric quantization*, University of Calgary, 1981.

[24] Gotay, M., On coisotropic imbeddings of presymplectic manifolds. *Proc. Am. Math. Soc.* **84** (1982), 111–114.

[25] Gotay, M., Constraints, reduction and quantization. *J. Math. Phys.* **27** (1986), 2051–2066.

[26] Gualtieri, M., Generalized complex geometry. D.Phil. thesis, Oxford University, 2003. ArXiv: math.DG/0401221.

[27] Guillemin, V., Sternberg, S., Some problems in integral geometry and some related problems in microlocal analysis. *Amer. J. Math.* **101** (1979), 915–955.

[28] Hitchin, N., Generalized Calabi-Yau manifolds. *Q. J. Math.* **54** (2003), 281–308.

[29] Jotz, M., Ratiu, T., Induced Dirac structures on isotropy type manifolds. *Transform. Groups* **16** (2011), 175–191.

[30] Kapustin, A., Li,Y., Topological sigma-models with H-flux and twisted generalized complex manifolds. *Adv. Theor. Math. Phys.* **11** (2007), 269–290.

[31] Klimcik, C., Strobl, T., WZW-Poisson manifolds. *J. Geom. Phys.* **43** (2002), 341–344.

[32] Kosmann-Schwarzbach, Y., Derived brackets. *Lett. Math. Phys.* **69** (2004), 61–87.

[33] Liu, Z.-J., Weinstein, A., Xu, P., Manin triples for Lie bialgebroids. *J. Differential Geom.* **45** (1997), 547–574.

[34] Marle, C.-M., Sous-variété de rang constant d'une variété symplectique. *Asterisque* **107** (1983), 69–86.

[35] Marsden, J., Ratiu, T., *Introduction to Mechanics and Symmetry*, Text in Applied Mathematics 17. Berlin: Springer-Verlag, 1994.

[36] Mikami, K., Weinstein, A., Moments and reduction for symplectic groupoid actions. *Publ. RIMS, Kyoto Univ.* **24** (1988), 121–140.

[37] Ortega, J.-P., Ratiu, T., *Momentum maps and Hamiltonian reduction*, Progress in Mathematics 222, Boston: Birkhauser, 2004.

[38] Roytenberg, D., On the structure of graded symplectic supermanifolds and Courant algebroids, In T. Voronov (ed.), *Quantization, Poisson brackets and beyond*, Contemporary Mathematics 315. Providence, RI: American Mathematical Society, 2002.

[39] Ševera, P., Some title containing the words "homotopy" and "symplectic", e.g. this one. *Travaux mathématiques*. Fasc. XVI (2005), 121–137.

[40] Ševera, P., Weinstein, A., Poisson geometry with a 3-form background. *Prog. Theor. Phys. Suppl.* **144** (2001), 145–154.

[41] Sniatycki, J., Dirac brackets in geometric dynamics. *Ann. Inst. H. Poincaré*, A **20** (1974), 365–372.

[42] van der Schaft, A. J., Port-Hamiltonian systems: an introductory survey. *Proceedings of the International Congress of Mathematicians*, Vol. 3, Madrid, 2006, pp. 1339–1365.

[43] Weinstein, A., The local structure of Poisson manifolds. *J. Differential Geom.* **18** (1983), 523–557.

[44] Weinstein, A., The geometry of momentum, *Géometrie au XXème Siècle, Histoire et Horizons*. Paris: Hermann, 2005. ArXiv: math.SG/0208108.

[45] Weinstein, A., Symplectic categories. *Port. Math.* **67** (2010), 261–278.

[46] Xu, P., Momentum maps and Morita equivalence. *J. Differential Geom.* **67** (2004), 289–333.

[47] Yoshimura, H., Marsden, J., Dirac structures in Lagrangian mechanics I. Implicit Lagrangian systems. *J. Geom. Phys.* **57** (2006), 133–156.

[48] Zabzine, M., Lectures on generalized complex geometry and supersymmetry. *Arch. Math.* **42**: 5 (2006), 119–146.

2

Differential geometry of holomorphic vector bundles on a curve

FLORENT SCHAFFHAUSER

Abstract

These notes are based on a series of five lectures given at the 2009 Villa de Leyva Summer School on "Geometric and Topological Methods for Quantum Field Theory". The purpose of the lectures was to give an introduction to differential-geometric methods in the study of holomorphic vector bundles on a compact connected Riemann surface, as initiated in the celebrated paper of Atiyah and Bott [AB83]. We take a rather informal point of view and try to paint a global picture of the various notions that come into play in that study, taking Donaldson's theorem on stable holomorphic vector bundles as a goal for the lectures [Don83].

2.1 Holomorphic vector bundles on Riemann surfaces

2.1.1 Vector bundles

Definition

We begin by recalling the definition of a vector bundle. Standard references for the general theory of fibre bundles are the books of Steenrod [Ste51] and Husemoller [Hus93].

Definition 2.1 (Vector bundle) Let X be a topological space and let $\mathbb{K}$ be the field $\mathbb{R}$ or $\mathbb{C}$. A **topological $\mathbb{K}$-vector bundle on** X is a continuous map $p : E \longrightarrow X$ such that

Geometric and Topological Methods for Quantum Field Theory, ed. Alexander Cardona, Iván Contreras and Andrés F. Reyes-Lega. Published by Cambridge University Press. © Cambridge University Press 2013.

1. $\forall x \in X$, the fibre $E_x := p^{-1}(\{x\})$ is a finite-dimensional $\mathbb{K}$-vector space,
2. $\forall x \in X$, there exists an open neighbourhood U of x in X, an integer $r_U \geq 0$, and a homeomorphism ϕ_U such that the diagram

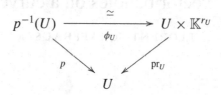

is commutative (pr_U denotes the projection onto U),
3. the induced homeomorphism $\phi_x : p^{-1}(\{x\}) \longrightarrow \{x\} \times \mathbb{K}^{r_U}$ is a $\mathbb{K}$-linear isomorphism.

Most of the time, the map p is understood, and we simply denote E a vector bundle on X. If $\mathbb{K} = \mathbb{R}$, E is called a (topological) *real* vector bundle, and if $\mathbb{K} = \mathbb{C}$, it is called a (topological) *complex* vector bundle. $p^{-1}(\{U\})$ is also denoted $E|_U$ and ϕ_U is called a *local trivialisation* of E over U. $E|_U$ is itself a vector bundle (on U). The open set $U \subset X$ is said to be trivialising for E, and the pair (U, ϕ_U) is called a (bundle) *chart*.

Example 2.2 The following maps are examples of vector bundles.

1. The *product bundle* $p : X \times \mathbb{K}^r \longrightarrow X$, where p is the projection onto X.
2. The tangent bundle $TM \longrightarrow M$ to a differentiable manifold M.
3. The Möbius bundle on S^1: let $\mathcal{M}$ be the quotient of $[0; 1] \times \mathbb{R}$ under the identifications $(0, t) \sim (1, -t)$, with projection map $p : \mathcal{M} \longrightarrow S^1$ induced by the canonical projection $[0; 1] \times \mathbb{R} \longrightarrow [0; 1]$. Observe that $\mathcal{M}$ is indeed homeomorphic to a Möbius band without its boundary circle.
4. The canonical line bundle on the n-dimensional projective space $\mathbb{R}\mathbf{P}^n$ (= the space of lines in $\mathbb{R}^{n+1}$):

$$E_{\mathrm{can}} := \{(\ell, v) \in \mathbb{R}\mathbf{P}^n \times \mathbb{R}^{n+1} \mid v \in \ell\}$$

with projection map $p(\ell, v) = l$. The fibre of p above ℓ is canonically identified with ℓ. When $n = 1$, the bundle E_{can} will be shown later to be isomorphic to the Möbius bundle (Exercise 2.1.2). The same example works with $\mathbb{C}\mathbf{P}^n$ in place of $\mathbb{R}\mathbf{P}^n$.
5. The Grassmannian of k-dimensional complex sub-spaces of $\mathbb{C}^{n+1}$, denoted $\mathrm{Gr}_k(\mathbb{C}^{n+1})$, has a complex vector bundle with k-dimensional fibres on it:

$$E_{\mathrm{can}} = \{(F, v) \in \mathrm{Gr}_k(\mathbb{C}^{n+1}) \times \mathbb{C}^{n+1} \mid v \in F\}$$

with projection map $p(F, v) = F$. The fibre of p above F is canonically identified with F. The same example works with $\mathbb{R}^{n+1}$ in place of $\mathbb{C}^{n+1}$.

It follows from the definition of a vector bundle that the $\mathbb{Z}_+$-valued map

$$x \longmapsto \text{rk}\, p^{-1}(\{x\})$$

(called the *rank function*) is a locally constant, integer-valued function on X (i.e. an element of $\check{H}^0(X; \mathbb{Z})$). In particular, if X is connected, it is a constant map, i.e. an integer.

Definition 2.3 (Rank of a vector bundle) Let X be a connected topological space. The **rank** of a $\mathbb{K}$-vector bundle $p : E \longrightarrow X$ is the dimension of the $\mathbb{K}$-vector space $p^{-1}(\{x\})$, for any $x \in X$. It is denoted rk E. A vector bundle of rank 1 is called a **line bundle**.

Definition 2.4 (Homomorphism of vector bundles) A **homomorphism**, or simply a morphism, between two $\mathbb{K}$-vector bundles $p : E \longrightarrow X$ and $p' : E' \longrightarrow X'$ is a pair (u, f) of continuous maps $u : E \longrightarrow E'$ and $f : X \longrightarrow X'$ such that

1. the diagram

$$
\begin{array}{ccc}
E & \xrightarrow{\;\;u\;\;} & E' \\
{\scriptstyle p}\downarrow & & \downarrow{\scriptstyle p'} \\
X & \xrightarrow{\;\;f\;\;} & X'
\end{array}
$$

   is commutative,
2. for all $x \in X$, the map

$$u_x : p^{-1}(\{x\}) \longrightarrow (p')^{-1}(\{f(x)\})$$

   is $\mathbb{K}$-linear.

Topological vector bundles together with their homomorphisms form a category that we denote Vect^{top}. If X is a fixed topological space, there is a category $\text{Vect}_X^{\text{top}}$ whose objects are topological vector bundles on X and whose morphisms are defined as follows.

Definition 2.5 (Homomorphisms of vector bundles on X) Let X be a fixed topological space and let $p : E \longrightarrow X$ and $p' : E' \longrightarrow X$ be two $\mathbb{K}$-vector bundles on X. A **morphism of $\mathbb{K}$-vector bundles on X** is a continous map $u : E \longrightarrow E'$ such that

1. the diagram

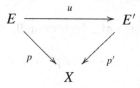

is commutative,
2. for all $x \in x$, the map

$$u_x : p^{-1}(\{x\}) \longrightarrow (p')^{-1}(\{f(x)\})$$

is $\mathbb{K}$-linear.

As usual, an isomorphism is a homomorphism which admits an inverse homomorphism. A $\mathbb{K}$-vector bundle isomorphic to a product bundle is called a *trivial bundle*.

When the base space X is a smooth manifold, one defines smooth vector bundles on X using smooth maps in place of continuous ones, and when X is a complex analytic manifold manifold, one may accordingly define holomorphic vector bundles on X (this last notion only makes sense, of course, if the field $\mathbb{K}$ in Definition 2.1 is assumed to be the field of complex numbers).

Transition maps

We shall henceforth assume that X is connected. Let E be a $\mathbb{K}$-vector bundle on X, and denote $r = \mathrm{rk}\, E$. If (U, ϕ_U) and (V, ϕ_V) are two overlapping charts in the sense that $U \cap V \neq \emptyset$, one gets a map

$$\phi_U \circ \phi_V^{-1} : \begin{array}{ccc} (U \cap V) \times \mathbb{K}^r & \longrightarrow & (U \cap V) \times \mathbb{K}^r \\ (x, v) & \longmapsto & (x, g_{UV}(x) \cdot v) \end{array}$$

where $g_{UV} : U \cap V \longrightarrow \mathrm{Aut}(\mathbb{K}^r) = \mathbf{GL}(r, \mathbb{K})$. It satisfies, for any triple of open sets (U, V, W),

$$g_{UV}\, g_{VW} = g_{UW} \tag{2.1}$$

(the product on the left-hand side being the pointwise product of $\mathbf{GL}(r, \mathbb{K})$-valued functions). Setting $U = V = W$, we obtain $(g_{UU})^2 = g_{UU}$, so

$$g_{UU} = \mathrm{I}_r$$

(the constant map equal to I_r). This in turn implies that $g_{UV}\, g_{VU} = g_{UU} = \mathrm{I}_r$, so

$$g_{VU} = g_{UV}^{-1}$$

(the map taking x to $(g_{UV}(x))^{-1} \in \mathbf{GL}(r, \mathbb{K}))$. The condition (2.1) is called a *cocycle condition*. If $(U_i)_{i \in I}$ is a covering of X by trivialising open sets, with associated local trivialisations $(\phi_i)_{i \in I}$, we get a family

$$g_{ij} : U_i \cap U_j \longrightarrow \mathbf{GL}(r, \mathbb{K})$$

of maps satisying condition (2.1): the family $(g_{ij})_{(i,j) \in I \times I}$ is called a $\mathbf{GL}(r, \mathbb{K})$-valued 1-*cocycle* subordinate to the open covering $(U_i)_{i \in I}$. It is completely determined by the transition maps $(\phi_i \circ \phi_j^{-1})_{(i,j) \in I \times I}$. Conversely, such a cocycle defines a topological $\mathbb{K}$-vector bundle of rank r

$$E := \left(\bigcup_{i \in I} \{i\} \times U_i \times \mathbb{K}^r \right) \bigg/ \sim, \tag{2.2}$$

where the equivalence relation $\sim$ identifies (i, x, v) and (j, y, w) if $y = x$ (in particular, $U_i \cap U_j \neq \emptyset$) and $w = g_{ij}(x) \cdot v$, the projection map $p : E \longrightarrow X$ being induced by the projections maps $\{i\} \times U_i \times \mathbb{K}^r \longrightarrow U_i \subset X$. In other words, a $\mathbb{K}$-vector bundle of rank r on X is a fibre bundle with typical fibre $\mathbb{K}^r$ and structure group $\mathbf{GL}(r, \mathbb{K})$, acting on $\mathbb{K}^r$ by linear transformations (see for instance [Ste51]). Two vector bundles E and E' on X, represented by two cocycles $(g_{ij})_{(i,j)}$ and $(g'_{ij})_{(i,j)}$ subordinate to a same open covering $(U_i)_{i \in I}$, are isomorphic if and only if there exists a family

$$u_i : U_i \longrightarrow \mathbf{GL}(r, \mathbb{K})$$

of maps satisfying

$$g'_{ij} = u_i g_{ij} u_j^{-1}.$$

Indeed, simply define u_i in the following way

$$U_i \times \mathbb{K}^r \xrightarrow{\phi_i^{-1}} E|_{U_i} \xrightarrow{u} E'|_{U_i} \xrightarrow{\phi'_i} U_i \times \mathbb{K}^r,$$

the map taking (x, v) to $(x, u_i(x) \cdot v)$, and check that, for all $x \in U_i \cap U_j$, $u_i(x) = g'_{ij}(x) u_j(x) g_{ij}^{-1}(x)$. This defines an equivalence relation on the set of $\mathbf{GL}(r, \mathbb{K})$-valued 1-cocycles subordinate to a given open covering $\mathcal{U} = (U_i)_{i \in I}$ of X. The set of equivalence classes for this relation is usually denoted

$$\check{H}^1_{\text{top}}(\mathcal{U}; \mathbf{GL}(r, \mathbb{K})).$$

These sets form a direct system relative to the operation of passing from an open covering of X to a finer one, and the direct limit is denoted

$$\check{H}^1_{\text{top}}(X; \mathbf{GL}(r, \mathbb{K})) := \varinjlim_{\mathcal{U}} \check{H}^1_{\text{top}}(\mathcal{U}; \mathbf{GL}(r, \mathbb{K}))$$

(see for instance [Gun66]). *This set is the set of isomorphism classes of topological* $\mathbb{K}$-*vector bundles on* X (if X is not connected, $\check{H}^1(X; \mathbf{GL}(r, \mathbb{K}))$ is the disjoint union $\sqcup_{i=1}^{k} \check{H}^1(X_i ; \mathbf{GL}(r, \mathbb{K}))$, where $\sqcup_{i=1}^{k} X_i$ is the disjoint union of connected components of X). If one considers smooth 1-cocycles instead of continuous ones, one obtains a similar description of smooth vector bundles on X. Likewise, if X is a Riemann surface, a holomorphic vector bundle of rank r on X is represented by a holomorphic 1-cocycle

$$g_{ij} : U_i \cap U_j \longrightarrow \mathbf{GL}(r, \mathbb{C})$$

in the sense that all the components of g_{ij} are holomorphic functions of one variable. An automorphism of a topological (resp. smooth, resp. holomorphic) vector bundle E on X represented by the cocycle $(g_{ij})_{(i,j)}$ may be represented by a family $(u_i : U_i \longrightarrow \mathbf{GL}(r, \mathbb{K}))_i$ of continuous (resp. smooth, resp. holomorphic) maps satisfying $u_i g_{ij} = g_{ij} u_j$ for all $(i, j) \in I \times I$ such that $U_i \cap U_j \neq \emptyset$.

Sections of a bundle

Sections of a bundle are a generalisation of mappings between two spaces X and Y in the sense that a map from X to Y is a section of the product bundle $X \times Y \longrightarrow X$.

Definition 2.6 (Sections of a vector bundle) A (global) **section** of a topological $\mathbb{K}$-vector bundle $p : E \longrightarrow X$ is a continuous map $s : X \longrightarrow E$ such that $p \circ s = \mathrm{Id}_X$. The set $\Gamma(E)$ of global sections of E is an infinite-dimensional $\mathbb{K}$-vector space and a module over the ring of $\mathbb{K}$-valued functions on X.

Local sections of E are sections $s_U : U \longrightarrow E|_U \simeq U \times \mathbb{K}^r$ of the vector bundle $E|_U$ where $U \subset X$ is a trivialising open subset. They may be seen as maps from U to $\mathbb{K}^r$. If $(g_{UV})_{(U,V)}$ is a 1-cocycle representing the vector bundle E, then a global section $s : X \longrightarrow E$ is the same as a collection $(s_U)_U$ of local sections subject to the condition

$$s_U = g_{UV} s_V$$

for any pair (U, V) of open subsets of X satisfying $U \cap V \neq \emptyset$. Smooth (resp. holomorphic) sections of a smooth (resp. holomorphic) vector bundle are defined accordingly.

Example 2.7 A section of the vector bundle $TM \longrightarrow M$ is a *vector field* on M. A section of $T^*M \longrightarrow M$ is a *differential 1-form* on M.

2.1.2 Topological classification

Evidently, if two $\mathbb{K}$-vector bundles on X are isomorphic, they have the same rank (= dimension over $\mathbb{K}$ of the typical fibre of E). If $\mathbb{K} = \mathbb{C}$ and $X = \Sigma_g$ is a compact connected oriented *surface*, isomorphism classes of *topological* or *smooth* vector bundles on X are completely classified by a pair (r, d) of integers, namely the rank and the *degree* of a complex vector bundle. We give the following definition, which makes free use of the notion of *Chern class* of a complex vector bundle (see for instance [BT82] or [Hat02]).

Definition 2.8 (Degree) Let E be a complex vector bundle on a compact connected oriented surface Σ_g. The **degree** of E is by definition the integral of the first Chern class $c_1(E) \in H^2(\Sigma_g; \mathbb{Z})$ of E:

$$\deg(E) := \int_{\Sigma_g} c_1(E) \in \mathbb{Z}.$$

The degree of a vector bundle satisfies the following relations, which are often useful in computations:

$$\deg(E^*) = -\deg(E) \quad \text{and} \quad \deg(E_1 \otimes E_2) = \deg(E_1)\,\mathrm{rk}\,(E_2) + \mathrm{rk}\,(E_1)\deg(E_2).$$

Theorem 2.9 (Topological classification of complex vector bundles on a curve) Let Σ_g be a compact connected Riemann surface, and let E, E' be two topological (resp. smooth) complex vector bundles on Σ_g. Denote $r = \mathrm{rk}\,(E)$, $r' = \mathrm{rk}\,(E')$, $d = \deg(E)$ and $d' = \deg(E')$. Then E and E' are isomorphic as topological (resp. smooth) complex vector bundles on Σ_g if and only if $r = r'$ and $d = d'$. Moreover, for any pair $(r, d) \in \mathbb{Z}_+ \times \mathbb{Z}$, there exists a complex vector bundle of rank r and degree d on Σ_g.

We refer for instance to [Tha97] for a proof of this theorem. In Section 2.3, we shall be interested in the (much more involved) classification problem for *holomorphic* vector bundles on a compact connected Riemann surface Σ_g, and the preceding result will be used to reduce it to the study of *holomorphic structures* on a given smooth complex vector bundle of *topological type* (r, d).

2.1.3 Dolbeault operators and the space of holomorphic structures

In this subsection, we only consider smooth complex vector bundles over a fixed Riemann surface Σ (although, most of the time, a similar, albeit slightly more complicated, statement holds for complex vector bundles over a higher-dimensional complex analytic manifold; see for instance [Kob87] or [Wel08] for more on this topic).

Smooth complex vector bundles and their sections

A holomorphic structure on a topological complex vector bundle is by definition a (maximal) holomorphic atlas on it (local trivialisations with holomorphic transition maps). Such a holomorphic structure defines (up to conjugation by an automorphism of the bundle) a remarkable object on the underlying *smooth* complex vector bundle: a Dolbeault operator.

In what follows, we will denote E a smooth complex vector bundle on Σ. When E is endowed with a holomorphic structure, we will designate by $\mathcal{E}$ the resulting holomorphic vector bundle. We denote $\Omega^0(\Sigma; E) = \Gamma(E)$ the complex vector space of smooth sections of E, and $\Omega^k(\Sigma; E)$ the complex vector space of E-valued, smooth, $\mathbb{R}$-linear k-forms on Σ. For any $k \geq 0$, $\Omega^k(\Sigma; E)$ is also a module over the ring $\Omega^0(\Sigma; \mathbb{C}) = C^\infty(\Sigma; \mathbb{C})$ of $\mathbb{C}$-valued smooth functions on Σ. It is important to stress that, for $k \geq 1$, $\Omega^k(\Sigma; E)$ is the space of smooth sections of the complex vector bundle $\wedge^k(T^*\Sigma) \otimes_\mathbb{R} E$. It is a complex vector space because the fibres of E are complex vector spaces, but a single element $\omega \in \Omega^k(\Sigma; E)$, when evaluated at a point $x \in \Sigma$, defines an $\mathbb{R}$-*linear* map

$$\omega_x : T_x\Sigma \wedge \cdots \wedge T_x\Sigma \longrightarrow E_x.$$

In particular, we do not restrict our attention to $\mathbb{C}$-linear such maps. Instead, we write, for instance if $k = 1$,

$$\Omega^1(\Sigma; E) = \Omega^{1,0}(\Sigma; E) \oplus \Omega^{0,1}(\Sigma; E)$$

where $\Omega^{1,0}(\Sigma; E)$ is the complex vector space of $\mathbb{C}$-linear 1-forms

$$\omega : T\Sigma \longrightarrow E$$

and $\Omega^{0,1}(\Sigma; E)$ is the complex vector space of $\mathbb{C}$-antilinear such forms. We recall that $\Omega^k(\Sigma; \mathbb{C}) = \Omega^k(\Sigma; E)$ for $E = \Sigma \times \mathbb{C}$, so the remark above is in particular valid for $\Omega^1(\Sigma; \mathbb{C})$. For $k > 1$ (and over a complex analytic manifold M of arbitrary dimension), we would have a decomposition

$$\Omega^k(M; E) = \Omega^{k,0}(M; E) \oplus \Omega^{k-1,1}(M; E) \oplus \cdots \oplus \Omega^{1,k-1}(M; E) \oplus \Omega^{0,k}(M; E)$$

where $\Omega^{p,q}(M; E)$ is the space of $\mathbb{R}$-linear $(p + q)$-forms

$$\omega : TM \wedge \cdots \wedge TM \longrightarrow E$$

which are $\mathbb{C}$-linear in p arguments and $\mathbb{C}$-antilinear in q arguments. A consequence of these decompositions is that the de Rham operator d on Σ splits into

$$d = d^{1,0} \oplus d^{0,1} : \Omega^0(\Sigma; \mathbb{C}) \longrightarrow \Omega^1(\Sigma; \mathbb{C}) = \Omega^{1,0}(\Sigma; \mathbb{C}) \oplus \Omega^{0,1}(\Sigma; \mathbb{C}).$$

That is, the exterior derivative df of a smooth function $f : \Sigma \longrightarrow \mathbb{C}$ splits into a $\mathbb{C}$-linear part $d^{1,0} f$ (also denoted ∂f) and a $\mathbb{C}$-antilinear part $d^{0,1} f$ (also denoted $\overline{\partial} f$).

Lemma 2.10 *A smooth function $f : \Sigma \longrightarrow \mathbb{C}$ is holomorphic if and only if $\overline{\partial} f = 0$.*

Proof This is a question of a purely local nature. In a holomorphic chart $z = x + iy$ of Σ, one has $df = \partial f\, dz + \overline{\partial} f\, d\overline{z}$ and $\overline{\partial} f = \frac{1}{2}(\frac{\partial f}{\partial x} + i\frac{\partial f}{\partial y})$. Write now $f = P + iQ$ with P and Q real-valued. Then $\overline{\partial} f = 0$ if and only if $\frac{\partial P}{\partial x} = \frac{\partial Q}{\partial y}$ and $\frac{\partial Q}{\partial x} = -\frac{\partial P}{\partial y}$. These are the Cauchy–Riemann equations: they mean that the Jacobian matrix of f at any given point is a similitude matrix, i.e. a complex number, which amounts to saying that f is holomorphic. $\qquad\square$

Definition 2.11 (Cauchy–Riemann operator) The operator

$$\overline{\partial} : \Omega^0(\Sigma, \mathbb{C}) \longrightarrow \Omega^{0,1}(\Sigma; \mathbb{C})$$

taking a smooth function to the $\mathbb{C}$-antilinear part of its derivative is called the **Cauchy–Riemann operator** of the Riemann surface Σ.

Observe that the Cauchy–Riemann operator satisfies the Leibniz rule

$$\overline{\partial}(fg) = (\overline{\partial} f)g + f(\overline{\partial} g).$$

We would like to have a similar characterisation for the holomorphic *sections* of an arbitrary holomorphic bundle $\mathcal{E}$ (not just $\Sigma \times \mathbb{C}$). The problem is that there is no canonically defined operator

$$D : \Omega^0(\Sigma; E) \longrightarrow \Omega^1(\Sigma; E)$$

which would play the role of the de Rham operator, so we first need to define these. Recall that the de Rham operator satisfies the Leibniz rule

$$d(fg) = (df)g + f(dg).$$

The next object, called a (linear) connection, gives a way to differentiate sections of a vector bundle E *covariantly* (in such a way that the resulting object is an E-valued 1-form, thus generalising the de Rham operator). To give a presentation of connections of a broader interest, we temporarily move back to manifolds more general than Riemann surfaces. The next definition is a bit abstract but designed to incorporate the case, for instance, of smooth complex vector bundles over real smooth manifolds.

Definition 2.12 (Linear connection) Let M be a smooth manifold over $\mathbb{K} = \mathbb{R}$ or $\mathbb{C}$. A (linear) **connection** on a smooth vector bundle $E \longrightarrow M$ is a $\mathbb{K}$-linear map

$$D : \Omega^0(M; E) \longrightarrow \Omega^1(M; E)$$

satisfying the following Leibniz rule

$$D(fs) = (df)s + f(Ds)$$

for all $f \in C^\infty(M; \mathbb{K})$ and all $s \in \Omega^0(M; E)$, where d is the de Rham operator on M.

Here $(df)s$ is the element of $\Omega^1(M; E)$ which, when evaluated at $x \in M$, is the $\mathbb{R}$-linear (*even, and this is crucial, if* $\mathbb{K} = \mathbb{C}$) map

$$\begin{array}{ccc} T_x M & \longrightarrow & E_x \\ v & \longmapsto & \underbrace{(df)_x(v)}_{\in \mathbb{K}} \cdot \underbrace{s(x)}_{\in E_x} \ . \end{array}$$

$f\, Ds$ is the element of $\Omega^1(M; E)$ which, when evaluated at $x \in M$, is the $\mathbb{R}$-linear map $f(x)(Ds)_x$. Given a section $s \in \Omega^0(M; E)$, the E-valued 1-form Ds is called the *covariant derivative* of s. We observe that the product bundle $U \times \mathbb{K}^r$ always admits a distinguished connection, called the product connection: since elements of $\Omega^1(U; U \times \mathbb{K}^r) \simeq \Omega^1(U; \mathbb{K}^r)$ are r-uplets of k-forms on U, one may define $D = d \oplus \cdots \oplus d$, the de Rham operator repeated r times. One can moreover show that any convex combination of linear connections is a linear connection. It is then a simple consequence of the existence of partitions of unity that any smooth vector bundle admits a connection (see Exercise 2.1.4). As we now show, the space of all connections is an affine space.

Proposition 2.13 The space of all linear connections on a smooth $\mathbb{K}$-vector bundle is an affine space, whose group of translations is the vector space $\Omega^1(\Sigma; \mathrm{End}(E))$.

Proof It suffices to show that the difference $D_1 - D_2$ of two linear connections defines an element of $\Omega^1(M; E)$. One has, for all $f \in C^\infty(M; \mathbb{K})$ and all $s \in \Omega^0(M; E)$,

$$(D_1 - D_2)(fs) = D_1(fs) - D_2(fs)$$

$$= (df)s + f(D_1 s) - (df)s - f(D_2 s)$$

$$= f(D_1 - D_2)s.$$

So $D_1 - D_2$ is a $C^\infty(M; \mathbb{K})$-linear map from $\Omega^0(M; E)$ to $\Omega^1(M; E)$. This is the same as an $\mathrm{End}(E)$-valued 1-form on M. $\square$

We postpone the exposition of further generalities on linear connections (curvature and the like) to Section 2.2.1, where they will be presented for a special case of linear connections called unitary connections (but seen to hold in greater generality), and we go back to generalised Cauchy–Riemann generators.

A linear connection on a smooth complex vector bundle $E \longrightarrow \Sigma$ over a Riemann surface splits into

$$D = D^{1,0} \oplus D^{0,1} \: : \: \Omega^0(\Sigma; E) \longrightarrow \Omega^1(\Sigma; E) = \Omega^{1,0}(\Sigma; E) \oplus \Omega^{0,1}(\Sigma; E).$$

Lemma 2.14 *Let D be a linear connection on $E \longrightarrow \Sigma$. The operator*

$$D^{0,1} : \Omega^0(\Sigma; E) \longrightarrow \Omega^{0,1}(\Sigma; E)$$

taking a section of E to the $\mathbb{C}$-antilinear part of its covariant derivative is $\mathbb{C}$-linear and satisfies the following Leibniz rule

$$D^{0,1}(fs) = (\overline{\partial} f)s + f(D^{0,1}s),$$

where $\overline{\partial}$ is the Cauchy–Riemann operator on Σ.

Proof $D^{0,1}$ is obviously additive. Moreover,

$$
\begin{aligned}
D(fs) &= (df)s + f(Ds) \\
&= (\partial f)s + f(D^{1,0}s) + (\overline{\partial} f)s + f(D^{0,1}s)
\end{aligned}
$$

so the $\mathbb{C}$-antilinear part of $D(fs)$ is $(\overline{\partial} f)s + f(D^{0,1}s)$. □

This motivates the following definition.

Definition 2.15 (Dolbeault operator) A **Dolbeault operator** on a smooth complex vector bundle $E \longrightarrow \Sigma$ over a Riemann surface is a $\mathbb{C}$-linear map

$$D'' : \Omega^0(\Sigma; E) \longrightarrow \Omega^{0,1}(\Sigma; E)$$

satisfying the following Leibniz rule: for all $f \in C^\infty(\Sigma; \mathbb{C})$ and all $s \in \Omega^0(\Sigma; E)$,

$$D''(fs) = (\overline{\partial} f)s + f(D''s)$$

where $\overline{\partial}$ is the Cauchy–Riemann operator of Σ.

A Dolbeault operator is also called a $(0, 1)$-connection. As in the case of connections, any smooth complex vector bundle over a complex base space admits a Dolbeault operator, and the space $\mathrm{Dol}(E)$ of all Dolbeault operators on E is an affine space, whose group of translations is the vector space $\Omega^{0,1}(\Sigma; \mathrm{End}(E))$. We now show that, given a *holomorphic* vector bundle $\mathcal{E}$ on Σ, there is a Dolbeault operator on the underlying smooth vector bundle E, whose kernel consists exactly of the holomorphic sections of $\mathcal{E}$ (much like $\ker\bar\partial$ consists of the holomorphic functions on Σ; see Lemma 2.10). We first observe that $\mathcal{G}_E$, the group[1] of all complex linear bundle automorphisms of E, acts on $\mathrm{Dol}(E)$ in the following way: if $u \in \mathcal{G}_E$ and $D'' \in \mathrm{Dol}(E)$,

$$(u \cdot D'')(s) := u\big(D''(u^{-1}s)\big) \tag{2.3}$$

is a Dolbeault operator on E (Exercise 2.1.5), and, if v is another automorphism of E,

$$(uv) \cdot D'' = u \cdot (v \cdot D'')$$

so we indeed have a group action. Moreover, D'' is a local operator in the following sense: if s_U is a *local* smooth section of E, we can define a local E-valued 1-form $D''s_U$ using bump functions on U (see the proof below). In particular, local solutions to the equation $D''s = 0$ form a sheaf on Σ.

Proposition 2.16 Let E be a smooth complex vector bundle on Σ. Given a holomorphic structure on E, denote $\mathcal{E}$ the resulting holomorphic vector bundle. Then, there exists a unique $\mathcal{G}_E$-orbit of Dolbeault operators on E such that, for any D'' in that orbit, local holomorphic sections of $\mathcal{E}$ are in bijection with local solutions to the equation $D''s = 0$.

Proof Let $(g_{ij})_{(i,j)}$ be a *holomorphic* 1-cocycle of transition maps on E. Let s be a smooth global section of E, and denote s_i the section s read in the local chart (U_i, ϕ_i). Then $s_i = g_{ij}s_j$ as maps from $U_i \cap U_j \to \mathbb{C}^r$, so

$$\bar\partial s_i = \bar\partial(g_{ij}s_j) = (\bar\partial g_{ij})s_j + g_{ij}(\bar\partial s_j) = g_{ij}(\bar\partial s_j),$$

since g_{ij} is holomorphic. This defines, for any $s \in \Omega^0(\Sigma; E)$, a global, E-valued $(0, 1)$-form $D''s$ on Σ (such that $(D''s)_i = \bar\partial s_i$). The Leibniz rule for the operator thus defined follows from the Leibniz rule for the the local operator $\bar\partial$. Let us now identify the local solutions to the equation $D''s = 0$. If σ is a smooth local section of E over an open subset U of Σ, let f be a smooth bump function whose support is contained in U. Then, by definition, there exists an open set $V \subset U$ on which f is identically 1. We may assume that V is trivialising for E. Then $\sigma|_V : V \to \mathbb{C}^r$ is a smooth local section of E over V, and it is holomorphic in V if and only if $\bar\partial(\sigma|_V) = 0$, or equivalently $(D''\sigma)_V = 0$ (observe that this last equation makes sense because we can extend σ *smoothly* to Σ using the bump function, and since $f \equiv 1$ in V, $(D''\sigma)_V$ does not depend on the extension). Using different pairs (V, f), we see that σ is holomorphic in U if and only if it is a local solution

---

[1] One may observe that there is a group bundle $\mathbf{GL}(E)$ on Σ, whose typical fibre is $\mathbf{GL}(r; \mathbb{C})$ and whose structure group is $\mathrm{Ad}\,\mathbf{GL}(r; \mathbb{C})$, such that $\mathcal{G}_E = \Gamma(\mathbf{GL}(E))$.

to the equation $D''s = 0$. Evidently, two isomorphic holomorphic structures on E determine conjugate Dolbeault operators. □

So we have an injective map

$$\{\text{holomorphic structures on } E\} \, / \, \text{isomorphism} \longrightarrow \mathrm{Dol}(E)/\mathcal{G}_E.$$

It is a remarkable fact that the image of this map can be entirely described, and that it is in fact a surjective map when the base complex analytic manifold has complex dimension 1. The problem of determining whether a Dolbeault operator comes from a holomorphic structure is a typical *integrability* question, similar to knowing whether a linear connection comes from a linear representation of the fundamental group of M at a given basepoint. The integrability conditions, too, are very similar, and we refer to [DK90] (Ch. 2, Sect. 2) for an illuminating parallel discussion of the two questions, as well as the proof of the following integrability theorem.

Theorem 2.17 (The Newlander–Nirenberg theorem in complex dimension 1) *Let* $E \longrightarrow \Sigma$ *be a smooth complex vector bundle on a Riemann surface, and let* D'' *be a Dolbeault operator on* E. *Then there exists a unique holomorphic structure on* E *such that such that the local holomorphic sections of* $\mathcal{E}$ *are in bijection with smooth local solutions to the equation* $D'' = 0$.

The proof is a question of showing that the sheaf of local solutions to the equation $D''s = 0$ is a locally free sheaf of rank $r = \mathrm{rk}\, E$ over the sheaf of holomorphic functions of Σ. It is then easy to check that two $\mathcal{G}_E$-conjugate Dolbeault operators determine isomorphic holomorphic structures, since their kernels are conjugate in $\Omega^0(\Sigma; E)$. Therefore, over a Riemann surface Σ, there is a bijection between the set of isomorphism classes of holomorphic structures on E and

$$\mathrm{Dol}(E)/\mathcal{G}_E.$$

We refer to [LPV85], Exposé 1, for an explanation of why any natural topology of this space is not Hausdorff. We conclude the present subsection by one further remark on Dolbeault operators, namely that a Dolbeault operator D'' on E induces a Dolbeault operator D''_E on $\mathrm{End}(E)$. First, note that, for all $k \geq 0$,

$$\Omega^k(\Sigma; \mathrm{End}(E)) = \Gamma(\wedge^k T^*\Sigma \otimes_{\mathbb{R}} \mathrm{End}(E)) = \mathrm{Hom}(E; \wedge^k T^*\Sigma \otimes_{\mathbb{R}} E).$$

So, given $u \in \Omega^0(\Sigma; \mathrm{End}(E))$, we need only specify $(D''_E u)(s) \in \Omega^1(\Sigma; E)$ for all $s \in \Omega^0(\Sigma; E)$ in order to completely determine $D''_E u \in \Omega^1(\Sigma; \mathrm{End}(E))$. Moreover, because $u(s)$ is locally a product between a matrix and a column vector, we want the would-be operator D_E'' on $\text{End}(E)$ to satisfy, for all $u \in \Omega^0(\Sigma; \text{End}(E))$ and all $s \in \Omega^0(\Sigma; E)$, the generalised Leibniz identity

$$D''(u(s)) = (D_E'' u)(s) + u(D''s) \tag{2.4}$$

so we *define*

$$(D_E'' u)(s) := D''(u(s)) - u(D''s).$$

Evidently, D_E'' is $\mathbb{C}$-linear in u as a map from $\Omega^0(\Sigma; \text{End}(E))$ to $\Omega^1(\Sigma; \text{End}(E))$, and, if $f \in C^\infty(\Sigma; \mathbb{C})$, one has

$$\begin{aligned}
(D_E''(fu))(s) &= D''(fu(s)) - (fu)D''s \\
&= (\bar{\partial} f) u(s) + f D''(u(s)) - f(u(D''s)) \\
&= [(\bar{\partial} f)u + f(D_E'' u)](s)
\end{aligned}$$

so D_E'' is indeed a Dolbeault operator on $\text{End}(E)$. In practice, it is simply denoted D'', which, if anything, makes (2.4) more transparent and easy to remember. As a consequence of the Leibniz identity (2.4), one can modify the way the $\mathcal{G}_E$-action on $\text{Dol}(E)$ is written:

$$\begin{aligned}
(u \cdot D'')(s) &= u(D''(u^{-1}s)) \\
&= u((D''(u^{-1}))s + u^{-1}(D''s)) \\
&= u(-u^{-1}(D''u)u^{-1}s + u^{-1}D''s) \\
&= D''s - (D''u)u^{-1}s
\end{aligned}$$

so

$$u \cdot D'' = D'' - (D''u)u^{-1}, \tag{2.5}$$

which is the way the $\mathcal{G}_E$-action on $\text{Dol}(E)$ is usually written when performing explicit computations (we refer to Exercise 2.1.6 for the computation of $D''(u^{-1})$ used in the above).

2.1.4 Exercises

Exercise 2.1.1 Show that the quotient topological space defined in (2.2) is a vector bundle of rank r on X.

Exercise 2.1.2 Show that the Möbius bundle $\mathcal{M} \longrightarrow S^1$ is isomorphic to the canonical bundle $E_{\text{can}} \longrightarrow \mathbb{R}\mathbf{P}^1$ (see Example 2.2).

Exercise 2.1.3 Show that a vector bundle of rank r, $p : E \longrightarrow X$, say, is isomorphic to the product bundle $X \times \mathbb{K}^r$ if and only if there exist r global sections

$s_1, \ldots, s_r \in \Gamma(X; E)$ such that, for all $x \in X$, $(s_1(x), \ldots, s_r(x))$ is a basis of E_x over $\mathbb{K}$.

Exercise 2.1.4 **a.** Let $(D_i)_{1 \leq i \leq n}$ be n linear connections on a vector bundle $E \longrightarrow M$, and let $(\lambda_i)_{1 \leq i \leq n}$ be n non-negative real numbers satisfying $\sum_{i=1}^{n} \lambda_i = 1$. Show that the convex combination $D = \sum_{i=1}^{n} \lambda_i D_i$ is a linear connection on E.
b. Let $(U_i)_{i \in I}$ be a covering of M by trivialising open sets for E, and let D_i be the product connection on $E|_{U_i}$. Let $(f_i)_{i \in I}$ be a partition of unity subordinate to $(U_i)_{i \in I}$. Show that $D := \sum_{i \in I} D_i$ is a well-defined map from $\Omega^0(M; E)$ to $\Omega^1(M; E)$, and that it is a linear connection on E.

Exercise 2.1.5 Let E be a smooth complex vector bundle and let D'' be a Dolbeault operator on E. Let u be an automorphism of E. Define $u \cdot D''$ by

$$(u \cdot D'')(s) = u\big(D''(u^{-1}s)\big)$$

on sections of E.
a. Show that $u \cdot D''$ is a Dolbeault operator on E.
b. Show that this defines an action of the group of automorphisms of E on the set of Dolbeault operators.

Exercise 2.1.6 Let D'' be a Dolbeault operator on a smooth complex vector bundle E, and let u be an automorphism of E. Show that

$$D''(u^{-1}) = -u^{-1}(D''u)u^{-1}.$$

2.2 Holomorphic structures and unitary connections

In this section, we study the space of holomorphic structures on a smooth complex vector bundle E over a Riemann surface Σ *in the additional presence of a Hermitian metric h on E*. This has the effect of replacing the space of Dolbeault operators by another space of differential operators: the space of unitary connections on (E, h). This new affine space turns out to have a natural structure of infinite-dimensional Kähler manifold. Moreover, the action of the group of unitary transformations of (E, h) on the space of unitary connections is a Hamiltonian action, and this geometric point of view, initiated by Atiyah and Bott in [AB83], will be key to understanding Donaldson's theorem in Section 2.3.2.

2.2.1 Hermitian metrics and unitary connections

Definition 2.18 (Hermitian metric) Let E be a smooth complex vector bundle on a smooth manifold M. A **Hermitian metric** h on E is a family $(h_x)_{x \in X}$ of maps

$$h_x : E_x \times E_x \longrightarrow \mathbb{C}$$

such that

1. $\forall (v, w_1, w_2) \in E_x \times E_x \times E_x,$

$$h(v, w_1 + w_2) = h(v, w_1) + h(v, w_2),$$

2. $\forall (v, w) \in E_x \times E_x, \forall \lambda \in \mathbb{C},$

$$h(v, \lambda w) = \lambda h(v, w),$$

3. $\forall (v, w) \in E_x \times E_x,$

$$h(w, v) = \overline{h(v, w)},$$

4. $\forall v \in E_x \setminus \{0\},$

$$h(v, v) > 0,$$

5. for any pair (s, s') of smooth sections of E, the function

$$h(s, s') : M \longrightarrow \mathbb{C}$$

is smooth.

In other words, h is a smooth family of Hermitian products on the fibres of E. A smooth complex vector bundle with a Hermitian metric is called a **smooth Hermitian vector bundle**.

Definition 2.19 A **unitary transformation** of (E, h) is an automorphism u of E satisfying, for any pair (s, s') of smooth sections of E,

$$h\big(u(s), u(s')\big) = h(s, s').$$

In other words, a unitary transformation is fibrewise an isometry. The group $\mathcal{G}_h$ of unitary transformations of (E, h) is called the (unitary) **gauge group**. There is a group bundle $\mathbf{U}(E, h)$, whose typical fibre is $\mathbf{U}(r)$ and whose structure group is $\mathrm{Ad}\mathbf{U}(r)$, such that $\mathcal{G}_h = \Gamma(\mathbf{U}(E, h))$.

A **Hermitian transformation** is an endomorphism u of E satisfying

$$h\big(u(s), s'\big) = h\big(s, u(s')\big).$$

An **anti-Hermitian transformation** is an endomorphism u of E satisfying

$$h\big(u(s), s'\big) = -h\big(s, u(s')\big).$$

The Lie algebra bundle whose sections are anti-Hermitian endomorphisms of E is denoted $\mathfrak{u}(E, h)$. Its typical fibre is the Lie algebra $\mathfrak{u}(r) = Lie(\mathbf{U}(r))$ and its structure group is $\mathrm{Ad}\mathbf{U}(r)$. As one might expect, $\Gamma(\mathfrak{u}(E, h)) = \Omega^0(M; \mathfrak{u}(E, h))$ is actually the Lie algebra of $\Gamma(\mathbf{U}(E, h)) = \mathcal{G}_h$.

Proposition 2.20 (Reduction of structure group) Let $(E \longrightarrow M)$ be a smooth complex vector bundle. Given a Hermitian metric h on E, there exists a $\mathbf{U}(r)$-valued 1-cocycle

$$g_{ij} : U_i \cap U_j \longrightarrow \mathbf{U}(r) \subset \mathbf{GL}(r, \mathbb{C})$$

representing E. Two such cocycles differ by an $\mathbf{U}(r)$-valued 0-cocycle. Conversely, an atlas of E whose transition maps are given by a unitary 1-cocycle determines a Hermitian metric on E.

More generally, if H is a subgroup of $\mathbf{GL}(r, \mathbb{C})$ and a vector bundle E can be represented by an H-valued 1-cocycle whose class modulo H-valued 0-cocycles is uniquely defined, one says that the structure group of E has been *reduced* to H. The proposition above says that a Hermitian metric is equivalent to a reduction of the structure group $\mathbf{GL}(r, \mathbb{C})$ of a complex rank r vector bundle to the maximal compact subgroup $\mathbf{U}(r)$. In the general theory of fibre bundles, the existence of such a reduction is usually deduced from the fact that the homogeneous space $\mathbf{GL}(r, \mathbb{C})/\mathbf{U}(r)$ (the space of Hermitian inner products on $\mathbb{C}^r$) is contractible (see for instance [Ste51]).

Proof of Proposition 2.20 Let h be a Hermitian metric on E. Using the Gram–Schmidt process, one can obtain an h-unitary local frame out of any given local frame of E, thereby identifying $E|_U$ with $U \times \mathbb{C}^r$ where $\mathbb{C}^r$ is endowed with its canonical Hermitian inner product. The transition functions of such an atlas have an associated 1-cocycle of transition maps which preserves the Hermitian product and is therefore $\mathbf{U}(r)$-valued. A different choice of unitary frames leads to a $\mathbf{U}(r)$-equivalent 1-cocycle. Conversely, given such a 1-cocycle, the Hermitian products obtained on the fibres of $E|_U$ and $E|_V$ respectively via the identifications with $U \times \mathbb{C}^r$ and $V \times \mathbb{C}^r$ coincide over $U \cap V$. $\square$

Since the structure group of the bundle has changed, it makes sense to ask whether there is a notion of connection which is compatible with this smaller structure group. This is usually better expressed in the language of principal bundles, but we shall not need this point of view in these notes (see for instance [KN96]).

Definition 2.21 (Unitary connection) Let (E, h) be a smooth Hermitian vector bundle on a manifold M. A linear connection

$$D : \Omega^0(M; E) \longrightarrow \Omega^1(M; E)$$

on E is called **unitary** if, for any pair (s, s') of smooth sections of E, one has

$$d\big(h(s, s')\big) = h(Ds, s) + h(s, Ds').$$

The same standard arguments as in the case of linear connections and Dolbeault operators show that a Hermitian vector bundle always admits a unitary connection (locally, the product connection satisfies the unitarity condition), and that the space $\mathcal{A}(E, h)$ of all unitary connections on (E, h) is an affine space, whose group of translations is the vector space $\Omega^1(M; \mathfrak{u}(E, h))$ of $\mathfrak{u}(E, h)$-valued 1-forms on M.

Given a k-form $\alpha \in \Omega^k(M; E)$ and a local trivialisation (U, ϕ_U) of E, let us denote $\alpha_U \in \Omega^k(U; \mathbb{C}^r)$ the k-form obtained from reading $\alpha|_U$ in the local triviali-sation $\phi_U : E|_U \xrightarrow{\sim} U \times \mathbb{C}^r$. Then, if $(g_{UV})_{(U,V)}$ is a 1-cocycle of transition maps for E, one has $\alpha_U = g_{UV}\alpha_V$. Moreover, any linear connection is locally of the form

$$(Ds)_U = d(s_U) + A_U s_U$$

where $A_U \in \Omega^1(U; \mathfrak{gl}(r, \mathbb{C}))$ is a family of matrix-valued 1-forms defined on trivi-alising open sets by $A_U s_U = (Ds)_U - d(s_U)$, d being the product de Rham operator on $\Omega^1(U; \mathbb{C}^r)$, and subject to the condition, for all $s \in \Omega^0(M; E)$,

$$\begin{aligned}
(Ds)_U &= g_{UV}(Ds)_V \\
&= g_{UV}(ds_V + A_V s_V) \\
&= g_{UV}\big(d(g_{UV}^{-1}s_U) + A_V(g_{UV}^{-1}s_U)\big) \\
&= g_{UV}\big(d(g_{UV}^{-1})s_U + g_{UV}^{-1}ds_U\big) + g_{UV}A_V g_{UV}^{-1} s_U \\
&= ds_U + \big(g_{UV}A_V g_{UV}^{-1} - (dg_{UV})g_{UV}^{-1}\big)s_U
\end{aligned}$$

so

$$A_U = g_{UV}A_V g_{UV}^{-1} - (dg_{UV})g_{UV}^{-1}. \tag{2.6}$$

The family $(A_U)_U$ subject to condition (2.6) above is sometimes called the con-nection form, even though it is *not* a global differential form on M. The connection determined by such a family is denoted d_A, or even simply A. Let us now analyse what it means for d_A to be unitary. Given a pair (s, s') of smooth sections of E, and a local chart (U, ϕ_U) of E, one has, on the one hand,

$$d\big(h(s_U, s'_U)\big) = h(ds_U, s'_U) + h(s_U, ds'_U)$$

and, on the other hand,

$$\begin{aligned}
h\big((Ds)_U, s'_U\big) &+ h\big(s_U, (Ds')_U\big) \\
&= h(ds_U, s'_U) + h(A_U s_U, s'_U) + h(s_U, ds'_U) + h(s_U, A_U s'_U)
\end{aligned}$$

so d_A is unitary if and only if

$$h(A_U s_U, s'_U) + h(s_U, A_U s'_U) = 0,$$

which means that the 1-form A_U is in fact $\mathfrak{u}(r)$-valued. Conversely, if (E, h) is represented by a unitary cocycle $(g_{UV} : U \cap V \longrightarrow \mathbf{U}(r))_{(U,V)}$ (Proposition 2.20) and $(A_U)_U$ is a family of $\mathfrak{u}(r)$-valued 1-forms satisfying condition (2.6), then there is a unique unitary connection d_A on (E, h) such that, for all $s \in \Omega^0(M; E)$, one has $(d_A s)_U = ds_U + A_U s_U$ on each U.

Just like any linear connection on a smooth complex vector bundle over a complex manifold (Lemma 2.14), a unitary connection

$$d_A : \Omega^0(M; E) \longrightarrow \Omega^1(M; E) = \Omega^{1,0}(M; E) \oplus \Omega^{0,1}(M; E)$$

splits into $d_A = d_A^{1,0} \oplus d_A^{0,1}$, where $d_A^{1,0}$ takes a section s of E to the $\mathbb{C}$-linear part of its covariant derivative, and $d_A^{0,1}$ takes s to the $\mathbb{C}$-antilinear part of $d_A s$. In particular,

$$d_A^{0,1} : \Omega^1(M; E) \longrightarrow \Omega^{0,1}(M; E)$$

is a Dolbeault operator. So, if $M = \Sigma$ is a Riemann surface, then, by the Newlander–Nirenberg theorem, $d_A^{0,1}$ determines a holomorphic structure on E. The next proposition shows what we gain by working in the presence of a Hermitian metric on E: a Dolbeault operator D'' on $E \longrightarrow M$ may be the $(0, 1)$-part of various, non-equivalent linear connections, but it is the $(0, 1)$-part of a *unique* unitary connection.

Proposition 2.22 Let (E, h) be a smooth Hermitian vector bundle on a complex manifold M, and let

$$D'' : \Omega^0(M; E) \longrightarrow \Omega^{0,1}(M; E)$$

be a Dolbeault operator on E. Then there exists a unique unitary connection

$$d_A : \Omega^0(M; E) \longrightarrow \Omega^1(M; E) = \Omega^{1,0}(M; E) \oplus \Omega^{0,1}(M; E)$$

such that $d_A^{0,1} = D''$.

Proof As for many other results in this chapter, the proof essentially boils down to linear algebra. Let $(g_{UV})_{(U,V)}$ be a unitary 1-cocycle representing (E, h). The Dolbeault operator D'' is locally of the form

$$(D''s)_U = \bar{\partial} s_U + B_U s_U$$

where $B_U \in \Omega^{0,1}(U; \mathfrak{gl}(r, \mathbb{C}))$ and $\bar{\partial}$ is the product Cauchy–Riemann operator on $\Omega^0(U; \mathbb{C}^r)$, and where the family $(B_U)_U$ satisfies

$$B_U = g_{UV} B_V g_{UV}^{-1} - (\bar{\partial} g_{UV}) g_{UV}^{-1} \tag{2.7}$$

(this does not require g_{UV} to be unitary). We then have an isomorphism of real vector spaces

$$\Omega^{0,1}(U; \mathfrak{gl}(r, \mathbb{C})) \longrightarrow \Omega^1(U; \mathfrak{u}(r))$$
$$B_U \longmapsto A_U := B_U - B_U^*$$

where $B_U^* = \overline{B_U}^t$ is the adjoint of B_U, the converse map being

$$A_U \longmapsto A_U^{0,1} = \frac{A_U(\cdot) + i A_U(i \cdot)}{2}.$$

One may observe here that

$$A_U^{1,0} = \frac{A_U(\cdot) - i A_U(i \cdot)}{2} = -B_U^*.$$

Moreover, as g_{UV} is unitary, $g_{UV}^* = g_{UV}^{-1}$ and therefore

$$A_U = B_U - B_U^*$$
$$= g_{UV} B_V g_{UV}^{-1} - (\bar{\partial} g_{UV}) g_{UV}^{-1} - g_{UV} B_V^* g_{UV}^{-1} + g_{UV}(\partial(g_{UV}^*))$$
$$= g_{UV}(B_V - B_V^*) g_{UV}^{-1} - (\partial g_{UV} + \bar{\partial} g_{UV}) g_{UV}^{-1}$$
$$= g_{UV} A_V g_{UV}^{-1} - (d g_{UV}) g_{UV}^{-1}$$

so the family $(A_U)_U$ is a unitary connection on (E, h), and $d_A^{0,1} = (B_U)_U = D''$. Conversely, if $(A_U)_U$ is a unitary connection on (E, h) such that $A_U^{0,1} = B_U$ for all U, then $A_U^{1,0} = -(A_U^{0,1})^* = -B_U^*$, so such a unitary connection is unique: $A_U = A_U^{1,0} + A_U^{0,1} = -B_U^* + B_U$. $\qquad\square$

Observe that the family $(B_U)_U$ satisfying condition (2.7) completely determines the Dolbeault operator D'', which therefore could be denoted $\bar{\partial}_B$, or even simply B.

Corollary 1 *Let Σ be a Riemann surface, and let E be a smooth complex vector bundle on Σ. Then the choice of a Hermitian metric h on E determines an isomorphism of affine spaces*

$$\mathcal{A}(E, h) \xrightarrow{\simeq} \mathrm{Dol}(E)$$
$$d_A \longmapsto d_A^{0,1}$$

between the space of unitary connections on (E, h) and the space of Dolbeault operators on E.

Recall that we denote $\mathcal{G}_h = \Gamma(\mathbf{U}(E, h))$ the group of unitary automorphisms of (E, h). It is commonly called the *unitary gauge group*. As for the group $\mathcal{G}_E = \Gamma(\mathbf{GL}(E))$ of all complex linear automorphisms of E, it is commonly called the *complex gauge group*. A good reason for this terminology is that $\mathcal{G}_E$ is actually the complexification of $\mathcal{G}_h$ (indeed $\mathbf{GL}(r, \mathbb{C})$ is the complexification of $\mathbf{U}(r)$, so the typical fibre of $\mathbf{GL}(E)$ is the complexification of the typical fibre of $\mathbf{U}(E, h)$). We saw in Section 2.1.3 that, over a Riemann surface Σ, the set of isomorphism classes of holomorphic structures on E was in bijection with the orbit space

$$\mathrm{Dol}(E)/\,\mathcal{G}_E.$$

Now if E has Hermitian metric h, we can replace, as we have just seen, $\mathrm{Dol}(E)$ with $\mathcal{A}(E, h)$, and then use the bijection between the two to transport the $\mathcal{G}_E$-action from $\mathrm{Dol}(E)$ to $\mathcal{A}(E, h)$. Computing through this procedure gives, for all $u \in \mathcal{G}_E$ and all $d_A \in \mathcal{A}(E, h)$, the relation

$$u \cdot d_A = d_A - \left[(d_A^{0,1} u) u^{-1} - \left((d_A^{0,1} u) u^{-1} \right)^* \right] \tag{2.8}$$

where $d_A^{0,1} u$ denotes the $\mathbb{C}$-antilinear part of the covariant derivative of the *endomorphism u* (this extension of a Dolbeault operator on a bundle to endomorphisms of that bundle was discussed at the end of Section 2.1.3) and α^* denotes the h-unitary adjoint of an E-valued, or an $\mathrm{End}E$-valued, k-form α (the proof of relation (2.8) is proposed as an exercise in Exercise 2.2.3). In particular, if u actually lies in $\mathcal{G}_h \subset \mathcal{G}_E$, then $u^* = u^{-1}$, and

$$(d_A^{0,1} u)^* = d_A^{1,0}(u^*) = d_A^{1,0}(u^{-1}) = -u(d_A^{1,0} u)u^{-1}$$

so

$$u \cdot d_A = d_A - (d_A^{0,1} u + d_A^{1,0} u)u^{-1}$$
$$= d_A - (d_A u)u^{-1},$$

which is nothing other than the natural action of $\mathcal{G}_h$ on $\mathcal{A}(E, h)$, defined for all $s \in \Omega^0(M; E)$ by

$$(u \cdot d_A)(s) = u\left(d_A(u^{-1}s) \right)$$

(the exact formal analogue of the $\mathcal{G}_E$-action on $\mathrm{Dol}(E)$; see equations (2.3) and (2.5)). The fact that the action of $\mathcal{G}_h$ on $\mathcal{A}(E, h)$ extends to an action of $\mathcal{G}_E = \mathcal{G}_h^{\mathbb{C}}$ is what eventually explains the relation between the symplectic picture and the geometric invariant theoretic picture for vector bundles on a curve, a relation which plays an important part in Donaldson's theorem.

To sum up, the choice of a Hermitian metric on a smooth complex vector bundle $E \longrightarrow \Sigma$ over a Riemann surface provides an identification between the set of isomorphism classes of holomorphic structures on E and the orbit space

$$\mathcal{A}(E, h)/\,\mathcal{G}_E.$$

This raises the question: what happens if we choose a different metric? If h and h' are two Hermitian metrics on E, then there exists an automorphism $u \in \mathcal{G}_E$ (in fact unique up to mutiplication by an element of $\mathcal{G}_h$) such that

$$h' = u^* h$$

(meaning that, for any pair (s_1, s_2) of smooth sections of E, one has $h'(s_1, s_2) = h(us_1, us_2)$). In particular, a linear connection D on E is h'-unitary if and only if the linear connection $u \cdot D = u(D(u^{-1} \cdot))$ is h-unitary. Indeed,

$$\begin{aligned}
d\big(h(s_1, s_2)\big) &= d\big(h'(u^{-1}s_1, u^{-1}s_2)\big) \\
&= h'\big(D(u^{-1}s_1), u^{-1}s_2\big) + h'\big(u^{-1}s_1, D(u^{-1}s_2)\big) \\
&= h\big((u \cdot D)(s_1), s_2\big) + h\big(s_1, (u \cdot D)(s_2)\big).
\end{aligned}$$

Therefore, there is a non-canonical bijection $\mathcal{A}(E, h') \simeq \mathcal{A}(E, h)$ *with the key property that it sends $\mathcal{G}_E$-orbits to $\mathcal{G}_E$-orbits.* In particular, there is a *canonical* bijection

$$\mathcal{A}(E, h')/\mathcal{G}_E \simeq \mathcal{A}(E, h)/\mathcal{G}_E.$$

This renders the choice of the metric unimportant in the whole analysis of holomorphic structures on E: the space $\mathcal{A}(E, h)$ depends on that choice, but not the space $\mathcal{A}(E, h)/\mathcal{G}_E$, which is the space of isomorphism classes of holomorphic structures on E.

2.2.2 The Atiyah–Bott symplectic form

Only from this point on does it become truly necessary to assume that the base manifold of our holomorphic bundles be a *compact, connected* Riemann surface Σ_g (g being the genus). The fact that Σ_g is of complex dimension 1 has already been used, though, for instance to show that *any* unitary connection on a smooth Hermitian vector bundle (E, h) over Σ_g defines a holomorphic structure on E. We shall now use the compactness of Σ_g to show that $\mathcal{A}(E, h)$ has a natural structure of an infinite-dimensional symplectic (in fact, Kähler) manifold. Actually, for this to be true, we would need to amend our presentation of Dolbeault operators and unitary connections to allow non-smooth such operators. Indeed, as the vector space $\Omega^1(\Sigma; \mathfrak{u}(E, h))$ on which the affine space $\mathcal{A}(E, h)$ is modelled is infinite-dimensional, we have to choose a topology on it. In order to turn the resulting topological vector space into a Banach space, we have to work with connections which are not necessarily smooth, but instead lie in a certain Sobolev completion of the space of smooth connections, and the same goes for gauge transformations. We refer to [AB83] (Section 13) and [Don83, DK90] for a discussion of this problem.

Atiyah and Bott have in particular shown that gauge orbits of such unitary connections always contain smooth connections, and that two smooth connections lying in the same gauge orbit can also be conjugated by a smooth gauge transformation. These analytic results enable us to ignore the issue of having to specify the correct connection spaces and gauge groups, and focus on the geometric side of the ideas of Atiyah–Bott and Donaldson instead.

Recall that the space $\mathcal{A}(E, h)$ of unitary connections on a smooth Hermitian vector bundle (E, h) is an affine space whose group of translations is the space $\Omega^1(\Sigma_g; \mathfrak{u}(E, h))$ of 1-forms with values in the bundle of anti-Hermitian endomorphisms of (E, h). In particular, the tangent space at A to $\mathcal{A}(E, h)$ is canonically identified with $\Omega^1(\Sigma_g; \mathfrak{u}(E, h))$. We assume throughout that the Riemann surface Σ_g comes equipped with a compatible Riemannian metric of normalised unit volume. Compatibility in the present context means that the complex structure I on each tangent plane to Σ_g is an isometry of the Riemannian metric. This defines in particular a symplectic form, also a volume form since $\dim_\mathbb{R} \Sigma_g = 2$, namely $\mathrm{vol}_{\Sigma_g} = g(I \cdot | \cdot)$. The typical fibre of $\mathfrak{u}(E, h)$ is the Lie algebra $\mathfrak{u}(r)$ of anti-Hermitian matrices of size r, so it has a canonical, positive-definite inner product

$$\kappa := -\mathrm{tr} : \begin{array}{ccc} \mathfrak{u}(r) \otimes \mathfrak{u}(r) & \longrightarrow & \mathbb{R} \\ (X; Y) & \longmapsto & -\mathrm{tr}(XY) \end{array}$$

(the restriction to $\mathfrak{u}(r)$ of $(X, Y) \mapsto -\mathrm{tr}(\overline{X}^t Y)$, the canonical Hermitian product of $\mathfrak{gl}(r, \mathbb{C})$).

Given $A \in \mathcal{A}(E, h)$ and $a, b \in T_A\mathcal{A}(E, h) \simeq \Omega^1(\Sigma_g; \mathfrak{u}(E, h))$, $a \wedge b$ is the $\mathfrak{u}(E, h) \otimes \mathfrak{u}(E, h)$-valued 2-form defined by

$$(a \wedge b)_x(v, w) = (a_x(v) \otimes b_x(w) - b_x(v) \otimes a_x(w)) \in \mathfrak{u}(r) \otimes \mathfrak{u}(r).$$

So

$$\kappa(a \wedge b)_x(v, w) := -\mathrm{tr}(a_x(v)b_x(w)) + \mathrm{tr}(b_x(v)a_x(w))$$

is an $\mathbb{R}$-valued 2-form on Σ_g. Note indeed that $\kappa(b \wedge a) = -\kappa(a \wedge b)$ because $b \wedge a = -a \wedge b$. Since Σ_g is oriented and compact, the integral

$$\omega_A(a, b) := \int_{\Sigma_g} \kappa(a \wedge b) \in \mathbb{R}$$

defines a 2-form on $\mathcal{A}(E, h)$.

Proposition 2.23 (Atiyah–Bott) The 2-form ω defined on $\mathcal{A}(E, h)$ by

$$\omega_A(a, b) := \int_{\Sigma_g} \kappa(a \wedge b)$$

is a symplectic form.

Proof ω is obviously closed, since it is constant with respect to A. To show that it is non-degenerate, we use local coordinates. The tangent vectors a and b become $\mathfrak{u}(r)$-valued 1-forms on an open subset $U \subset \Sigma_g$,

$$a = \alpha \, dx + \beta \, dy$$
$$b = \gamma \, dx + \delta \, dy$$

with $\alpha, \beta, \gamma, \delta : U \longrightarrow \mathfrak{u}(r)$ smooth functions. If $a \in \ker \omega_A$, then, for $b = *a :=$ $-\beta \, dx + \alpha \, dy$, one has

$$\kappa(a \wedge b)_{(x,y)}(v, w) = (\underbrace{\kappa(\alpha(x, y)^2 + \beta(x, y)^2)}_{\geq 0})(v_1 w_2 - v_2 w_1),$$

a positive multiple of the volume form (here we need the local coordinates (x, y) to be appropriately chosen), so

$$\int_{\Sigma_g} \kappa(a \wedge *a) \geq 0$$

and it is 0 if and only if $\alpha = \beta = 0$, i.e. $a = 0$. $\qquad \square$

Of course, there is some hidden meaning to this proof: the transformation

$$* : \alpha \, dx + \beta \, dy \longmapsto -\beta \, dx + \alpha \, dy \tag{2.9}$$

is the local expression of the *Hodge star* on $\Omega^1(\Sigma_g; \mathfrak{u}(E, h))$. It squares to minus the identity, so it is a complex structure on $\Omega^1(\Sigma_g; \mathfrak{u}(E, h))$. But in fact, the Hodge star may be defined on all non-zero homogeneous forms on Σ_g: it sends 0-forms to 2-forms and vice versa, the two transformations being inverse to one another. Locally, one has $*(f dx) = f dy$, $*(f dy) = -f dx$, $*f = f dx \wedge dy$, and $*(f dx \wedge dy) = f$. More intrisically, since Σ_g has a Riemannian metric and the fibres of $\mathfrak{u}(E, h)$ have a scalar product κ, the bundle $\bigwedge^k T^* \Sigma_g \otimes_{\mathbb{R}} \mathfrak{u}(E, h)$ has a Riemannian metric π, say. If a, are two $\mathfrak{u}(E, h)$-valued k-forms on Σ_g, i.e. two sections of $\bigwedge^k T^* \Sigma_g \otimes_{\mathbb{R}} \mathfrak{u}(E, h)$, then $\pi(a, b)$ is a smooth function on Σ_g. Now, if η is an arbitrary $\mathfrak{u}(E, h)$-valued k-form on Σ_g, $*\eta$ is defined as the unique $\mathfrak{u}(E, h)$-valued $(2 - k)$-form such that

$$\kappa(\eta \wedge *\eta) = \pi(\eta, \eta) \, \mathrm{vol}_{\Sigma_g} \tag{2.10}$$

as 2-forms on Σ_g.

Proposition 2.24 Set, for all $a, b \in T_A \mathcal{A}(E, h) \simeq \Omega^1(\Sigma_g; \mathfrak{u}(E, h))$,

$$(a \,|\, b)_{L^2} := \int_{\Sigma_g} \kappa(a \wedge *b) = \omega_A(a, *b).$$

Then $(\cdot \,|\, \cdot)_{L^2}$ is a Riemannian metric on $\mathcal{A}(E, h)$, called the L^2 metric. The Atiyah–Bott symplectic form ω, the complex structure $*$, and the metric $(\cdot \,|\, \cdot)_{L^2}$ turn $\mathcal{A}(E, h)$ into a Kähler manifold.

Proof Note that $(a \,|\, b)_{L^2} = \int_{\Sigma_g} \pi(a, b)\,\mathrm{vol}_{\Sigma_g}$. The equality with the expression in the statement of the Proposition follows from (2.10). The fact that $(\cdot \,|\, \cdot)_{L^2}$ is positive-definite has been proved in Proposition 2.23. Moreover, it is clear from either of expressions (2.9) or (2.10) that $\| * a\|_{L^2} = \|a\|_{L^2}$. The rest is the definition of a Kähler manifold (see for instance [MS98]).   □

Recall now that the gauge group $\mathcal{G}_h = \Gamma(\mathrm{U}(E, h))$ of unitary transformations of (E, h) acts on $\mathcal{A}(E, h)$ via

$$u \cdot A = A - (d_A u)u^{-1}.$$

Proposition 2.25 (Infinitesimal gauge action)   The fundamental vector field

$$\xi_A^{\#} = \frac{d}{dt}\Big|_{t=0}(\exp(t\xi)\cdot A)$$

associated to the element ξ of the Lie algebra $\Omega^0(\Sigma_g; \mathfrak{u}(E, h)) \simeq Lie(\mathcal{G}_h)$ is

$$\xi_A^{\#} = -d_A\xi \in \Omega^1(\Sigma_g; \mathfrak{u}(E, h)).$$

Proof In local coordinates, A is of the form $d + a$, where a is a $\mathfrak{u}(r)$-valued 1-form defined on an open subset $U \subset \Sigma_g$, u is a smooth map $U \longrightarrow \mathrm{U}(r)$, and $d_A u$ acts on endomorphism of $E|_U$ by $d + [a, \cdot]$ (see Exercise 2.2.6). So $u \cdot A$ is of the form

$$(d + a) - (du + [a, u])u^{-1} = d + a - (du)u^{-1} - (au - ua)u^{-1}$$

$$= d - (du)u^{-1} + uau^{-1}.$$

Setting $u = \exp(t\xi)$ and taking the derivative at $t = 0$ of $-(du)u^{-1} + uau^{-1}$, we obtain

$$-d\xi + \xi a - a\xi = -d\xi - [a, \xi],$$

which is the local expression of $-d_A\xi$.   □

Proposition 2.26 The action of $\mathcal{G}_h$ on $\mathcal{A}(E, h)$ preserves the Atiyah–Bott symplectic form and the L^2 metric on $\mathcal{A}(E, h)$.

Proof The tangent map to the action of $u \in \mathcal{G}_h$ on $\mathcal{A}(E, h)$ is the map

$$\begin{array}{ccc} \Omega^1(\Sigma_g; \mathfrak{u}(E, h)) & \longrightarrow & \Omega^1(\Sigma_g; \mathfrak{u}(E, h)) \\ a & \longmapsto & uau^{-1} \end{array}$$

so, since $\kappa = -\mathrm{tr}$ is Ad-invariant on $\mathfrak{u}(r) \otimes \mathfrak{u}(r)$,

$$\kappa\big((uau^{-1}) \wedge (ubu^{-1})\big) = \kappa\big(a \wedge b\big)$$

and therefore $u^*\omega = \omega$. Since the action is also $\mathbb{C}$-linear (see Exercise 2.2.5), it is an isometry of the L^2 metric. $\square$

Since we have a symplectic action of a Lie group $\mathcal{G}_h$ on a symplectic manifold $(\mathcal{A}(E,h), \omega)$ (albeit both infinite-dimensional), it makes sense to ask whether this action is Hamiltonian and, more importantly, to find the momentum map. To identify a possible momentum map, we need to make $(Lie(\mathcal{G}_H))$ more explicit.

Proposition 2.27 The map

$$
\begin{array}{ccc}
\Omega^2(\Sigma_g; \mathfrak{u}(E,h)) & \longrightarrow & (Lie(\mathcal{G}_h))^* \\
R & \longmapsto & (\xi \longmapsto \int_{\Sigma_g} \kappa(\xi \otimes R))
\end{array}
$$

is an isomorphism of vector spaces which is $\mathcal{G}_h$-equivariant with respect to the action $u \cdot R := \mathrm{Ad}_u \circ R$ on $\Omega^2(\Sigma_g; \mathfrak{u}(E,h))$ and the co-adjoint action on $(Lie(\mathcal{G}_h))^*$.

Proof The Lie algebra of $\mathcal{G}_h$ is $\Omega^0(\Sigma_g; \mathfrak{u}(E,h))$. It carries a Riemannian metric

$$(\lambda, \mu) \longmapsto \int_{\Sigma_g} \kappa(\lambda \wedge *\mu)$$

which canonically identifies it with its dual. Then, the Hodge star establishes an isomorphism

$$* : \Omega^2(\Sigma_g; \mathfrak{u}(E,h)) \longrightarrow \Omega^0(\Sigma_g; \mathfrak{u}(E,h)).$$

The statement on the action follows from the Ad-invariance of κ. $\square$

Now, there is a natural map from $\mathcal{A}(E,h)$ to $\Omega^2(\Sigma_g; \mathfrak{u}(E,h))$, namely the map taking a unitary connection A to its curvature F_A, which we now define.

Proposition 2.28 A unitary connection

$$d_A : \Omega^0(\Sigma_g; E) \longrightarrow \Omega^1(\Sigma_g; E)$$

on (E,h) uniquely extends to an operator

$$d_A : \Omega^k(\Sigma_g; E) \longrightarrow \Omega^{k+1}(\Sigma_g; E)$$

satisfying the generalised Leibniz rule

$$d_A(\beta \wedge \sigma) = (d\beta) \wedge \sigma + (-1)^{\deg \beta} \beta \wedge d_A \sigma$$

for all $\beta \in \Omega^j(\Sigma_g; \mathbb{C})$ and all $\sigma \in \Omega^k(\Sigma_g; E)$. The operator

$$d_A \circ d_A : \Omega^0(\Sigma_g; E) \longrightarrow \Omega^2(\Sigma_g; E)$$

is $C^\infty(\Sigma_g; \mathbb{C})$-linear, so it defines an element $F_A \in \Omega^2(\Sigma_g; \mathfrak{u}(E, h))$ called the **curvature** of A. It satisfies

$$F_{u \cdot A} = \mathrm{Ad}_u \circ F_A = u F_A u^{-1}$$

for all $u \in \mathcal{G}_h$. Moreover, if the local expression of A is $d + a$, the local expression of F_A is $da + \frac{1}{2}[a, a]$.

The following theorem is the main result of this subsection.

Theorem 2.29 (Atiyah–Bott, [AB83]) *The curvature map*

$$F : \mathcal{A}(E, h) \longrightarrow \Omega^2(\Sigma_g; \mathfrak{u}(E, h))$$

is an equivariant momentum map for the gauge action of $\mathcal{G}_h$ on $\mathcal{A}(E, h)$.

We shall need the following lemma to prove Theorem 2.29.

Lemma 2.30 *Let $A \in \mathcal{A}(E, h)$ be a unitary connection and let $b \in \Omega^1(\Sigma_g; \mathfrak{u}(E, h))$. Then $A + b$ is a unitary connection and*

$$F_{A+b} = F_A + d_A b + \frac{1}{2}[b, b].$$

Proof Since $\mathcal{A}(E, h)$ is an affine space on $\Omega^1(\Sigma_g; \mathfrak{u}(E, h))$, $A + b$ is a unitary connection. Let $d + a$ be the local expression of A, where $a \in \Omega^1(U; \mathfrak{u}(r))$. Then the local expression of F_A is $da + \frac{1}{2}[a, a]$, and the local expression of $d_A b$ is $db + [a, b]$ (as in Exercise 2.2.6). Moreover, the local expression of $A + b$ is $d + (a + b)$, so the local expression of F_{A+b} is

$$d(a + b) + \frac{1}{2}[a + b, a + b] = da + db + \frac{1}{2}[a, a] + [a, b] + \frac{1}{2}[b, b]$$

$$= \left(da + \frac{1}{2}[a, a]\right) + (db + [a, b]) + \frac{1}{2}[b, b]$$

so indeed

$$F_{A+b} = F_A + d_A b + \frac{1}{2}[b, b]. \qquad \square$$

Proof of Theorem 2.29 The equivariance of F follows from Proposition 2.28. It remains to show that F is a momentum map for the gauge action, that is, for all $\xi \in Lie(\mathcal{G}_h) = \Omega^0(\Sigma_g; \mathfrak{u}(E, h))$ and all $A \in \mathcal{A}(E, h)$,

$$\omega_A(\xi_A^\#, \cdot) = \left(d\langle F, \xi\rangle\right)_A(\cdot)$$

as linear forms on $T_A \mathcal{A}(E, h) \simeq \Omega^1(\Sigma_g; \mathfrak{u}(E, h))$. By Proposition 2.25, this is equivalent to the fact that, for all $\eta \in \Omega^1(\Sigma_g : \mathfrak{u}(E, h))$,

$$\int_{\Sigma_g} \kappa(-d_A \xi \wedge \eta) = \langle (dF)_A \cdot \eta, \xi \rangle.$$

But, by Proposition 2.30,

$$F_{A+t\eta} = F_A + td_A\eta + \frac{1}{2}t^2[\eta, \eta],$$

so

$$(dF)_A \cdot \eta = \frac{d}{dt}\bigg|_{t=0} F_{A+t\eta} = d_A\eta.$$

In other words, by Proposition 2.27, we want to show that

$$-\int_{\Sigma_g} \kappa(d_A \xi \wedge \eta) = \int_{\Sigma_g} \kappa(\xi \otimes d_A\eta). \tag{2.11}$$

But, since $\partial \Sigma_g = \emptyset$, one has

$$\int_{\Sigma_g} d\big(\kappa(\xi \otimes \eta)\big) = 0$$

on the one hand, and on the other hand,

$$d\big(\kappa(\xi \otimes \eta)\big) = \kappa(d_A \xi \wedge \eta) + \kappa(\xi \otimes d_A\eta),$$

whence relation (2.11). $\square$

2.2.3 Exercises

Exercise 2.2.1 Show that any complex vector bundle over a smooth manifold admits a Hermitian metric (as usual, use local trivialisations and a partition of unity).

Exercise 2.2.2 Let u be an endomorphism of a smooth Hermitian vector bundle (E, h). Show that there exists a unique endomorphism u^* of E such that, for all $(s, s') \in \Gamma(E) \times \Gamma(E)$,

$$h\big(u(s), s'\big) = h\big(s, u^*(s')\big),$$

where u^* is called the *adjoint* of u. A Hermitian endomorphism is self-adjoint, and an anti-Hermitian endomorphism is anti-self-adjoint.

Exercise 2.2.3 Show that, if $d_A \in \mathcal{A}(E, h)$ and $g \in \mathcal{G}_E$, then the quantity $g \cdot d_A$ defined by

$$d_A - \left[(d_A^{0,1} g)g^{-1} - \left((d_A^{0,1} g)g^{-1} \right)^* \right]$$

is a unitary connection, and this defines an action of $\mathcal{G}_E$ on $\mathcal{A}(E, h)$ making the isomorphism

$$\mathcal{A}(E, h) \simeq \mathrm{Dol}(E)$$

$\mathcal{G}_E$-equivariant.

Exercise 2.2.4 Check that relation (2.10) gives a well-defined $\mathbb{R}$-linear map

$$* : \Omega^k(\Sigma_g; \mathfrak{u}(E, h)) \mapsto \Omega^2(\Sigma_g; \mathfrak{u}(E, h))$$

satisfying $*^2 = (-1)^{k(2-k)} \mathrm{Id}$. Check that, in local coordinates, the map $*$ satisfies

$$*(\alpha \, dx + \beta \, dy) = -\beta \, dx + \alpha \, dy.$$

What about $*(\lambda \, dz + \mu \, d\bar{z})$?

Exercise 2.2.5 Show that the tangent map to the self-diffeomorphism of $\mathcal{A}(E, h)$ defined by the action of an element $u \in \mathcal{G}_h$ is $\mathbb{C}$-linear with respect to the complex structure of $\mathcal{A}(E, h)$ given on each tangent space $T_A \mathcal{A}(E, h) \simeq \Omega^1(\Sigma_g; \mathfrak{u}(E, h))$ by the Hodge star.

Exercise 2.2.6 Let A be a linear connection on a vector bundle E, and let s be a section of E. Show that if A is locally of the form

$$s \longmapsto ds + as$$

then the covariant derivative $d_A u$ of an endomorphism of E, defined at the end of Section 2.1.3 by

$$(d_A u)s = d_A(u(s)) - u(d_A s),$$

is locally of the form

$$u \longmapsto du + [a, u].$$

2.3 Moduli spaces of semi-stable vector bundles

It is sometimes important, while thinking about mathematics, to have a guiding problem to help one organise one's thoughts. For us in these notes, it is the problem of classifying holomorphic vector bundles on a smooth, irreducible complex projective curve Σ_g (= a compact connected Riemann surface of genus g). When the genus is 0 or 1, there are complete classification results for holomorphic vector bundles on Σ_g, due to Grothendieck for the case of the Riemann sphere [Gro57], and to Atiyah for the case of elliptic curves [Ati57]. There are no such classification results available for holomorphic vector bundles on a curve of genus $g > 1$. In such a situation, one generally hopes to replace the classification theorem by the construction of what is called a *moduli space*, the geometry of which can subsequently be studied. Roughly speaking, a moduli space of holomorphic vector bundles is a complex quasi-projective variety which has isomorphism classes of vector bundles over a fixed base for points, and satisfies a universal property controlling the notion of holomorphic or algebraic family of such vector bundles. We shall not get into the formal aspects of the notion of a moduli space and we refer the interested reader to [Gó1] instead. There are a few situations in which we know how to construct a moduli variety of vector bundles (i.e. give a structure of complex quasi-projective variety to *a certain set* of equivalence classes of vector bundles) and vector bundles on a smooth complex projective curve is one of those situations. The difficulty of a *moduli problem* is to understand *which set* one should try to endow with a structure of complex quasi-projective variety.

Common features of many moduli problems include:

1. Starting with a topological (or smooth) classification of the objects under study. This is typically obtained via *discrete invariants* (for vector bundles on curves: the rank and the degree) and has the virtue of dividing the moduli problem into various, more tractable moduli problems for objects of a fixed topological type.
2. Getting rid of certain objects in order to get a moduli space that admits a structure of projective algebraic variety (= a closed subspace of a projective space), or at least quasi-projective (= an open subset of a projective variety). This is where continuous invariants, called moduli, enter the picture (moduli may be thought of as some sort of local coordinates on the would-be moduli space). It is usually a difficult problem to find moduli for a class of objects, and one solution has been to use Mumford's Geometric Invariant Theory (GIT, [MFK93]) to decide which objects one should consider in order to get a nice moduli space (these objects are called *semi-stable* objects).

In fact, *stable objects* exhibit even better properties in the sense that the moduli space is then typically an orbit space (also called a geometric quotient, as opposed to a categorical quotient in the semi-stable case; see for instance [Tho06, New09]) admitting a structure of quasi-projective variety. GIT really is a way of defining quotients in algebraic geometry, and it has been applied very successfully to the study of moduli problems (Mumford's original motivation indeed). We shall not say anything else about GIT in these notes, and focus on *slope stability* for vector bundles on a curve only (it can be shown that this is in fact a GIT type of stability condition; see for instance [New09]). Nor shall we say anything about moduli functors and their coarse/fine moduli spaces (the interested reader might consult, for instance, [Muk03]).

2.3.1 Stable and semi-stable vector bundles

A basic property of holomorphic line bundles on a compact connected Riemann surface Σ_g states that they do not admit non-zero global holomorphic sections if their degree is negative (see for instance [For91], Theorem 16.5). Since a homomorphism between the holomorphic line bundles $\mathcal{L}_1$ and $\mathcal{L}_2$ is a section of the *line bundle* $\mathcal{L}_1^* \otimes \mathcal{L}_2$, a non-zero such homomorphism may only exist if $\deg(\mathcal{L}_1^* \otimes \mathcal{L}_2) \geq 0$, which is equivalent to $\deg\mathcal{L}_1 \leq \deg\mathcal{L}_2$. Semi-stable vector bundles of rank $r \geq 2$ provide a class of higher-rank vector bundles for which the statement above remains true (see Proposition 2.35). Note that, for higher-rank vector bundles, the degree of $\mathcal{E}_1^* \otimes \mathcal{E}_2$ is

$$\deg(\mathcal{E}_1^* \otimes \mathcal{E}_2) = \mathrm{rk}\,(\mathcal{E}_1)\deg(\mathcal{E}_2) - \deg(\mathcal{E}_1)\mathrm{rk}\,\mathcal{E}_2$$

so the non-negativity condition is equivalent to

$$\frac{\deg\mathcal{E}_1}{\mathrm{rk}\,\mathcal{E}_1} \leq \frac{\deg\mathcal{E}_2}{\mathrm{rk}\,\mathcal{E}_2}.$$

This motivates the following definition.

Definition 2.31 (Slope)   The slope of a non-zero complex vector bundle $E \longrightarrow \Sigma_g$ on an orientable, compact, connected surface Σ_g is the rational number

$$\mu(E) := \frac{\deg E}{\mathrm{rk}\,E} \in \mathbb{Q}.$$

We point out that no use is made of the holomorphic structures in the definition of the slope. It is a purely topological quantity that will, nonetheless, have strong holomorphic properties (another example of a topological invariant with strong holomorphic properties is the genus: on a compact, connected, orientable surface of genus g, the dimension of the space of holomorphic 1-forms is equal to g for *any* complex analytic structure on the surface).

In what follows, we call a sub-bundle $\mathcal{F} \subset \mathcal{E}$ non-trivial if it is distinct from 0 and $\mathcal{E}$. We emphasise that the definition that we give here is that of *slope* stability. However, since this is the only notion of stability that we shall consider in these notes, we will only say stable and semi-stable afterwards.

Definition 2.32 (Slope stability)   A (non-zero) holomorphic vector bundle $\mathcal{E} \longrightarrow \Sigma_g$ on a compact, connected Riemann surface Σ_g is called

1. **slope stable**, or simply stable, if for any non-trivial holomorphic sub-bundle $\mathcal{F}$, one has

$$\mu(\mathcal{F}) < \mu(\mathcal{E}),$$

2. **slope semi-stable**, or simply semi-stable, if for any non-trivial holomorphic sub-bundle $\mathcal{F}$, one has

$$\mu(\mathcal{F}) \leq \mu(\mathcal{E}).$$

A couple of remarks are in order. First, all holomorphic line bundles are stable (since they do not even have non-trivial sub-bundles), and all stable bundles are semi-stable. Second, a semi-stable vector bundle with coprime rank and degree is actually stable (this only uses the definition of slope stability and the properties of Euclidean division in $\mathbb{Z}$). Next, we have the following equivalent characterisation of stability and semi-stability, which is sometimes useful in practice.

Proposition 2.33 A holomorphic vector bundle $\mathcal{E}$ on Σ_g is stable if and only if, for any non-trivial sub-bundle $\mathcal{F} \subset \mathcal{E}$, one has $\mu(\mathcal{E}/\mathcal{F}) > \mu(\mathcal{E})$. It is semi-stable if and only if $\mu(\mathcal{E}/\mathcal{F}) \geq \mu(\mathcal{F})$ for all such $\mathcal{F}$.

Proof Denote by r, r', r'' the ranks of $\mathcal{E}, \mathcal{F}$ and $\mathcal{E}/\mathcal{F}$, respectively, and let d, d', d'' be their respective degrees. One has an exact sequence

$$0 \longrightarrow \mathcal{F} \longrightarrow \mathcal{E} \longrightarrow \mathcal{E}/\mathcal{F} \longrightarrow 0$$

so $r = r' + r''$ and $d = d' + d''$. Therefore,

$$\frac{d'}{r'} < \frac{d' + d''}{r' + r''} \Leftrightarrow \frac{d'}{r'} < \frac{d''}{r''} \Leftrightarrow \frac{d' + d''}{r' + r''} < \frac{d''}{r''}$$

and likewise with large inequalities or with equalities. This readily implies the proposition. $\square$

In a way, semi-stable holomorphic vector bundles are holomorphic vector bundles that do not admit *too many* sub-bundles, since any sub-bundle they may have is of slope no greater than their own. This turns out to have a number of interesting consequences that we now study. We recall that the category of vector bundles on a curve is a typical example of an additive category which is not Abelian: even though it admits kernels and co-kernels (hence also images and co-images), the canonical map $\mathcal{E}/\ker u \longrightarrow \operatorname{Im} u$ is in general not an isomorphism. We can, however, always compare the slopes of these two bundles.

Lemma 2.34 *If $u : \mathcal{E} \longrightarrow \mathcal{E}'$ is a non-zero homomorphism of vector bundles over Σ_g, then*

$$\mu(\mathcal{E}/\ker u) \leq \mu(\operatorname{Im} u)$$

with equality if and only if the canonical map $\mathcal{E}/\ker u \longrightarrow \operatorname{Im} u$ is an isomorphism.

One says that u is *strict* if the canonical homomorphism $\mathcal{E}/\mathrm{ker}u \longrightarrow \mathrm{Im}u$ is an isomorphism. In this case, u is injective if and only if $\mathrm{ker}u = 0$ and u is surjective if and only if $\mathrm{Im}u = \mathcal{E}'$. The proof we give below, of Lemma 2.34, requires notions on coherent modules over the sheaf O_{Σ_g}; it may be skipped upon firt reading of these notes. Recall that the category of vector bundles on Σ_g is equivalent to the category of locally free O_{Σ_g}-modules (= torsion-free coherent O_{Σ_g}-modules). Let $\mathcal{E}$ be a vector bundle on Σ_g and let $\underline{\mathcal{E}}$ be the corresponding torsion-free coherent module. Even though a coherent sub-module $\underline{\mathcal{F}}$ of $\underline{\mathcal{E}}$ is torsion-free, it only corresponds to a sub-bundle $\mathcal{F}$ of $\mathcal{E}$ if the coherent module $\underline{\mathcal{E}}/\underline{\mathcal{F}}$ is also torsion-free (and the latter then corresponds to the vector bundle $\mathcal{E}/\mathcal{F}$). This is equivalent to saying that $\underline{\mathcal{F}}$ is locally a direct summand of $\underline{\mathcal{E}}$. Given a coherent sub-module $\underline{\mathcal{F}}$ of $\underline{\mathcal{E}}$, there exists a smallest coherent sub-module $\underline{\widetilde{\mathcal{F}}}$ containing $\underline{\mathcal{F}}$ and such that $\underline{\mathcal{E}}/\underline{\widetilde{\mathcal{F}}}$ is torsion-free, namely the pre-image of the torsion sub-module of $\underline{\mathcal{E}}/\underline{\mathcal{F}}$. Then $\underline{\widetilde{\mathcal{F}}}/\underline{\mathcal{F}}$ has finite support and $\underline{\mathcal{F}}$ and $\underline{\widetilde{\mathcal{F}}}$ have the same rank. Moreover, $\mathrm{deg}\underline{\mathcal{F}}$ (which is well-defined since $\mathcal{F}$ is locally free) satisfies $\mathrm{deg}\underline{\mathcal{F}} \leq \mathrm{deg}\underline{\widetilde{\mathcal{F}}}$, with equality if and only if $\underline{\widetilde{\mathcal{F}}} = \underline{\mathcal{F}}$. It is convenient to call the *sub-bundle* $\widetilde{\mathcal{F}}$ corresponding to $\underline{\widetilde{\mathcal{F}}}$ the sub-bundle of $\mathcal{E}$ *generated* by $\underline{\mathcal{F}} \subset \underline{\mathcal{E}}$.

Proof of Lemma 2.34 Recall that the category of coherent O_{Σ_g}-modules is Abelian. In particular, the isomorphism theorem between co-images and images holds in that category. Let $\underline{u} : \underline{\mathcal{E}} \longrightarrow \underline{\mathcal{E}}'$ be the homomorphism of coherent O_{Σ_g}-modules associated to $u : \mathcal{E} \longrightarrow \mathcal{E}'$. Then, on the one hand, the locally free O_{Σ_g}-module associated to $\mathrm{ker}u$ is $\mathrm{ker}\underline{u}$ and the locally free O_{Σ_g}-module associated to $\mathcal{E}/\mathrm{ker}u$ is $\underline{\mathcal{E}}/\mathrm{ker}\underline{u}$. On the other hand, $\mathrm{Im}u$ is the vector bundle *generated* by $\mathrm{Im}\underline{u} \simeq \underline{\mathcal{E}}/\mathrm{ker}\underline{u}$. So $\mathrm{rk}\,(\mathrm{Im}u) = \mathrm{rk}\,(\mathrm{Im}\underline{u})$ and $\mathrm{deg}(\mathrm{Im}u) \geq \mathrm{deg}(\mathrm{Im}\underline{u})$, with equality if and only if u is strict. Therefore $\mu(\mathrm{Im}u) \geq \mu(\mathrm{Im}\underline{u})$, with equality if and only if u is strict. So

$$\mu(\mathcal{E}/\mathrm{ker}u) = \mu(\underline{\mathcal{E}}/\mathrm{ker}\underline{u}) = \mu(\mathrm{Im}\underline{u}) \leq \mu(\mathrm{Im}u)$$

with equality if and only if $\mathcal{E}/\mathrm{ker}u \simeq \mathrm{Im}u$. □

This immediately implies the result alluded to in the introduction to the present subsection.

Proposition 2.35 Let $\mathcal{E}$ and $\mathcal{E}'$ be two semi-stable vector bundles such that $\mu(\mathcal{E}) > \mu(\mathcal{E}')$. Then any homomorphism $u : \mathcal{E} \longrightarrow \mathcal{E}'$ is zero.

Proof If u is non-zero, then, since $\mathcal{E}$ is semi-stable, Proposition 2.33 and Lemma 2.34 imply that

$$\mu(\mathrm{Im}u) \geq \mu(\mathcal{E}/\mathrm{ker}u) \geq \mu(\mathcal{E}) > \mu(\mathcal{E}'),$$

which contradicts the semi-stability of $\mathcal{E}'$. □

We now focus on the category of semi-stable vector bundles of fixed slope $\mu \in \mathbb{Q}$. Unlike the category of all vector bundles on Σ_g, this is an Abelian category: it is additive, and we prove below that it admits kernels and co-kernels and that the isomorphism theorem holds.

Proposition 2.36 Let $u : \mathcal{E} \longrightarrow \mathcal{E}'$ be a non-zero homomorphism of semi-stable vector bundles of slope μ. Then $\ker u$ and $\operatorname{Im} u$ are semi-stable vector bundles of slope μ, and the natural map $\mathcal{E}/\ker u \longrightarrow \operatorname{Im} u$ is an isomorphism. In particular, the category of semi-stable vector bundles of slope μ is Abelian.

Proof Since $u \neq 0$, $\operatorname{Im} u$ is a non-zero sub-bundle of $\mathcal{E}'$, so $\mu(\operatorname{Im} u) \leq \mu(\mathcal{E}') = \mu$. But, by Lemma 2.34,

$$\mu(\operatorname{Im} u) \geq \mu(\mathcal{E}/\ker u) \geq \mu(\mathcal{E}) = \mu.$$

So $\mu(\operatorname{Im} u) = \mu$ and $\mu(\mathcal{E}/\ker u) = \mu$. In particular, by Lemma 2.34, $\mathcal{E}/\ker u \simeq \operatorname{Im} u$. Consider now the exact sequence

$$0 \longrightarrow \ker u \longrightarrow \mathcal{E} \longrightarrow \mathcal{E}/\ker u \longrightarrow 0.$$

Since $\mu(\mathcal{E}) = \mu(\mathcal{E}/\ker u) = \mu$, one also has $\mu(\ker u) = \mu$. Finally, since a sub-bundle of $\ker u$ (resp. $\operatorname{Im} u$) is also a sub-bundle of $\mathcal{E}$ (resp. $\mathcal{E}'$), its slope is no greater than $\mu(\mathcal{E}) = \mu = \mu(\ker u)$ (resp. $\mu(\mathcal{E}') = \mu = \mu(\operatorname{Im} u)$), so $\ker u$ (resp. $\operatorname{Im} u$) is semi-stable. $\square$

As an easy consequence of the above, the following result shows that, by considering only stable bundles of the same slope, we can better control the homomorphisms between them.

Proposition 2.37 Let $\mathcal{E}$ and $\mathcal{E}'$ be two stable vector bundles on Σ_g such that $\mu(\mathcal{E}) = \mu(\mathcal{E}')$, and let $u : \mathcal{E} \longrightarrow \mathcal{E}'$ be a non-zero homomorphism. Then u is an isomorphism.

Proof Recall that $\ker u \neq \mathcal{E}$ by assumption. Since $u : \mathcal{E} \longrightarrow \mathcal{E}'$ is a non-zero homomorphism between semi-stable bundles of the same slope, Proposition 2.36 implies that u is strict and that $\ker u$ is either 0 or has slope equal to $\mu(\mathcal{E})$. Since $\mathcal{E}$ is actually stable, $\ker u$ must be 0. Since u is strict, this implies that u is injective. Likewise, $\operatorname{Im} u \neq 0$ by assumption, and has slope equal to $\mu(\mathcal{E}')$ by Proposition 2.36. Since $\mathcal{E}'$ is actually stable, this forces $\operatorname{Im} u$ to be equal to $\mathcal{E}'$. Then, again since u is strict, $\operatorname{Im} u = \mathcal{E}'$ implies that u is surjective. Therefore, u is an isomorphism. $\square$

Note that a vector bundle always has non-trivial automorphisms (multiplication by a non-zero scalar on the fibres). When these are all the automorphisms of a given bundle, it is called a *simple* bundle. We now show that stable implies simple.

Proposition 2.38 If $\mathcal{E}$ is a stable vector bundle on Σ_g, then $\operatorname{End}\mathcal{E}$ is a field, isomorphic to $\mathbb{C}$. In particular, $\operatorname{Aut}\mathcal{E} \simeq \mathbb{C}^*$.

Proof Let u be a non-zero endomorphism of $\mathcal{E}$. By Proposition 2.37, u is an automorphism of $\mathcal{E}$, so $\operatorname{End}\mathcal{E}$ is a field, which contains $\mathbb{C}$ as its sub-field of scalar endomorphisms. Then, for any $u \in \operatorname{End}\mathcal{E}$, the sub-field $\mathbb{C}(u) \subset \operatorname{End}\mathcal{E}$ is a commutative field, and the Cayley–Hamilton theorem shows that u is algebraic over $\mathbb{C}$. Since $\mathbb{C}$ is algebraically closed, this shows that $u \in \mathbb{C}$. So $\operatorname{End}\mathcal{E} \simeq \mathbb{C}$ (in particular, the field $\operatorname{End}\mathcal{E}$ is commutative) and therefore $\operatorname{Aut}\mathcal{E} \simeq \mathbb{C}^*$. $\square$

Corollary 2 *A stable vector bundle is indecomposable: it is not isomorphic to a direct sum of non-trivial sub-bundles.*

Proof The automorphism group of a direct sum $\mathcal{E} = \mathcal{E}_1 \oplus \mathcal{E}_2$ contains $\mathbb{C}^* \times \mathbb{C}^*$, so $\mathcal{E}$ cannot be simple. Then, by Proposition 2.38, it cannot be stable. $\square$

The following result is key to understanding semi-stable bundles: these are extensions of stable bundles of the same slope.

Theorem 2.39 (Seshadri, [Ses67]) *The simple objects in the category of semi-stable bundles of slope μ are the stable bundles of slope μ. Any semi-stable holomorphic vector bundle of slope μ on Σ_g admits a filtration*

$$0 = \mathcal{E}_0 \subset \mathcal{E}_1 \subset \cdots \subset \mathcal{E}_k = \mathcal{E}$$

by holomorphic sub-bundles such that, for all $i \in \{1, \ldots, k\}$,

1. $\mathcal{E}_i/\mathcal{E}_{i-1}$ is stable,
2. $\mu(\mathcal{E}_i/\mathcal{E}_{i-1}) = \mu(\mathcal{E})$.

*Such a filtration is called a **Jordan–Hölder filtration of length k of $\mathcal{E}$.***

Proof Recall that a simple object in an Abelian category is an object with no non-trivial sub-object. In particular, a stable bundle $\mathcal{E}$ is simple in that sense (it contains no non-trivial sub-bundle *of slope equal to* $\mu(\mathcal{E})$). Conversely, if a semi-stable bundle $\mathcal{E}$ is simple in that sense, then any non-trivial sub-bundle $\mathcal{F} \subset \mathcal{E}$ satisfies $\mu(\mathcal{F}) \leq \mu(\mathcal{E})$ because $\mathcal{E}$ is semi-stable, and $\mu(\mathcal{F}) \neq \mu(\mathcal{E})$ because $\mathcal{E}$ has no non-trivial sub-objects in the category of semi-stable bundles with slope μ.

To prove the existence of a Jordan–Hölder filtration for a semi-stable bundle $\mathcal{E}$, observe that increasing and decreasing sequences of sub-bundles of $\mathcal{E}$ are stationary because of the bounds on the rank. If $\mathcal{E}$ is not a simple object, there exists a non-trivial sub-bundle $\mathcal{E}'$ of $\mathcal{E}$ which is semi-stable and of slope μ. If $\mathcal{E}'$ is not a simple object, we can go on and find a decreasing sequence of non-trivial (semi-stable) sub-bundles (of slope μ) in $\mathcal{E}$. Such a sequence is stationary, and we call $\mathcal{E}_1$ the final term: it is a simple sub-object of $\mathcal{E}$, so it is a stable bundle of slope μ. In particular, $\mathcal{E}/\mathcal{E}_1$ is semi-stable and also has slope μ (see Exercise 2.3.3). So there is a sub-bundle $\mathcal{E}_2/\mathcal{E}_1$ which is stable and of slope μ. This gives an increasing sequence

$$0 = \mathcal{E}_0 \subset \mathcal{E}_1 \subset \mathcal{E}_2 \subset \cdots$$

of (semi-stable) sub-bundles of $\mathcal{E}$ (of slope μ) whose successive quotients are stable bundles of slope μ. Such a sequence is stationary, so there is a k such that $\mathcal{E}_k = \mathcal{E}$, and the resulting filtration of $\mathcal{E}$ is a Jordan–Hölder filtration. $\qquad\square$

One may observe that, to show the existence of a filtration whose successive quotients are simple objects in the category of semi-stable bundles of slope μ, the proof only used that decreasing and increasing sequences *of such bundles* were stationary. An Abelian category satisfying these properties is called *Artinian* (decreasing sequences of sub-objects are stationary) and *Noetherian* (increasing sequences of sub-objects are stationary).

Observe that if a bundle is stable, it admits a Jordan–Hölder filtration of length 1, namely $0 = \mathcal{E}_0 \subset \mathcal{E}_1 = \mathcal{E}$. In general, there is no unicity of the Jordan–Hölder filtration, but the isomorphism class of the graded object associated to a filtration is unique, as shown by the next result. In particular, the lengths of any two Jordan–Hölder filtrations of $\mathcal{E}$ are equal and a semi-stable bundle is stable if and only if its Jordan–Hölder filtrations have length 1.

Proposition 2.40 (Seshadri, [Ses67])   Any two Jordan–Hölder filtrations

$$(S) : 0 = \mathcal{E}_0 \subset \mathcal{E}_1 \subset \cdots \subset \mathcal{E}_k = \mathcal{E}$$

and

$$(S') : 0 = \mathcal{E}_0' \subset \mathcal{E}_1' \subset \cdots \subset \mathcal{E}_l' = \mathcal{E}$$

of a semi-stable vector bundle $\mathcal{E}$ have same length $k = l$, and the associated graded objects

$$\mathrm{gr}(S) := \mathcal{E}_1/\mathcal{E}_0 \oplus \cdots \oplus \mathcal{E}_k/\mathcal{E}_{k-1}$$

and

$$\mathrm{gr}(S') := \mathcal{E}_1'/\mathcal{E}_0' \oplus \cdots \oplus \mathcal{E}_k'/\mathcal{E}_{k-1}'$$

satisfy

$$\mathcal{E}_i/\mathcal{E}_{i-1} \simeq \mathcal{E}_i'/\mathcal{E}_{i-1}'$$

for all $i \in \{1, \ldots, k\}$.

Proof Assume for instance that $l < k$. Then there exists an $i \in \{1, \ldots, k\}$ such that $\mathcal{E}'_1 \subset \mathcal{E}_i$ and $\mathcal{E}'_1 \not\subset \mathcal{E}_{i-1}$. So the map $\mathcal{E}_1 \hookrightarrow \mathcal{E}_i \longrightarrow \mathcal{E}_i / \mathcal{E}_{i-1}$ is a non-zero morphism between stable bundles of slope μ. By Proposition 2.37, it is an isomorphism. So $\mathcal{E}'_1 \cap \mathcal{E}_{i-1} = 0$ and $\mathcal{E}_i = \mathcal{E}_{i-1} \oplus \mathcal{E}'_1$. Then,

$$(S_1): 0 \subset \mathcal{E}'_1 \subset \mathcal{E}'_1 \oplus \mathcal{E}_1 \subset \cdots \subset \mathcal{E}'_1 \oplus \mathcal{E}_{i-1} \subset \mathcal{E}_{i+1} \subset \cdots \subset \mathcal{E}_k = \mathcal{E}$$

is a Jordan–Hölder filtration of length k of $\mathcal{E}$. Since (S') and (S_1) have the same first term, they induce Jordan–Hölder filtrations of $\mathcal{E}/\mathcal{E}'_1$, of lengths $l - 1$ and $k - 1$, respectively, with $l - 1 < k - 1$. Repeating this process $l - 1$ more times, we eventually reach $\mathcal{E}/\mathcal{E}'_{l-1}$ with a Jordan–Hölder filtration of length $k - l > 0$. In particular, if the inclusions $\mathcal{E}'_{l-1} \subset \mathcal{E}_{k-1} \subset \mathcal{E}_k = \mathcal{E}$ are strict, there is a sub-bundle of $\mathcal{E}_{k-1}/\mathcal{E}'_{l-1}$ contradicting the stability of $\mathcal{E}/\mathcal{E}'_{l-1}$. So $l = k$.

Then we prove the second assertion by induction on the length k of Jordan–Hölder filtrations of $\mathcal{E}$. If $k = 1$, it is obvious. If $k > 1$, consider again the filtration (S_1). It satisfies $\mathrm{gr}(S_1) \simeq \mathrm{gr}(S)$. Moreover, (S_1) and (S') have the same first term, so they induce Jordan–Hölder filtrations $(\overline{S_1})$ and $(\overline{S'})$ of length $k - 1$ of $\mathcal{E}/\mathcal{E}'_1$. By the induction hypothesis $\mathrm{gr}(\overline{S_1}) \simeq \mathrm{gr}(\overline{S'})$. So

$$\mathrm{gr}(S) \simeq \mathrm{gr}(S_1) \simeq \mathrm{gr}(\overline{S_1}) \oplus \mathcal{E}'_1 \simeq \mathrm{gr}(\overline{S'}) \oplus \mathcal{E}'_1 \simeq \mathrm{gr}(S').$$

This motivates the following definition. □

Definition 2.41 (Poly-stable bundles) A holomorphic vector bundle $\mathcal{E}$ on Σ_g is called poly-stable if it is isomorphic to a direct sum

$$\mathcal{F}_1 \oplus \cdots \oplus \mathcal{F}_k$$

of stable vector bundles of the same slope.

Evidently, a stable bundle is poly-stable. We point out that a poly-stable vector bundle of rank r admits a reduction of its structure group $\mathbf{GL}(r, \mathbb{C})$ to a subgroup of the form $\mathbf{GL}(r_1, \mathbb{C}) \times \cdots \times \mathbf{GL}(r_k, \mathbb{C})$, with $r_1 + \cdots + r_k = r$.

Proposition 2.42 Let $\mathcal{F}_1, \ldots, \mathcal{F}_k$ be stable vector bundles of slope μ. Then $\mathcal{E} := \mathcal{F}_1 \oplus \cdots \oplus \mathcal{F}_k$ is a semi-stable vector bundle of slope μ.

Proof $\mathcal{E}$ is a (trivial) extension of semi-stable vector bundles of slope μ, so it is semi-stable of slope μ (see Exercise 2.3.3). □

The graded object associated to any Jordan–Hölder filtration of a semi-stable vector bundle $\mathcal{E}$ is a poly-stable vector bundle (since it is a direct sum of simple objects in the category of semi-stable bundles of slope $\mu(\mathcal{E})$, it is a semi-simple object in that category). By Proposition 2.40, its isomorphism class is uniquely defined; it is usually denoted $\mathrm{gr}(\mathcal{E})$ and it is a graded isomorphism class of poly-stable vector bundles. The following notion is due to Seshadri (he used it to compactify the quasi-projective moduli variety of stable bundles constructed by Mumford).

Definition 2.43 (*S*-equivalence class, [Ses67]) The graded isomorphism class $\mathrm{gr}(\mathcal{E})$ associated to a semi-stable vector bundle $\mathcal{E}$ is called the **S-equivalence class** of $\mathcal{E}$. If $\mathrm{gr}(\mathcal{E}) \simeq \mathrm{gr}(\mathcal{E}')$, we say that $\mathcal{E}$ and $\mathcal{E}'$ are *S*-equivalent, and we write $\mathcal{E} \sim_S \mathcal{E}'$.

This defines an equivalence relation between semi-stable bundles of a given fixed slope. If two bundles of slope μ are *S*-equivalent, they have the same rank and the same degree (because the rank and degree of $\mathcal{F}_1 \oplus \cdots \oplus \mathcal{F}_k$ are equal to those of $\mathcal{E}$; see Exercise 2.3.4). The important point is that *two non-isomorphic semi-stable vector bundles may be S-equivalent.* Two *S*-equivalent stable bundles, however, are isomorphic, by definition of the *S*-equivalence class.

Definition 2.44 (Moduli set of semi-stable vector bundles) The set $\mathfrak{M}(r, d)$ of *S*-equivalence classes of semi-stable holomorphic vector bundles of rank r and degree d on Σ_g is called the **moduli set** of semi-stable vector bundles of rank r and degree d. It contains the set $\mathcal{N}(r, d)$ of isomorphism classes of stable vector bundles of rank r and degree d. When r and d are coprime, every semi-stable bundle is in fact stable and these two sets coincide.

Equivalently, $\mathfrak{M}(r, d)$ is the set of isomorphism classes of poly-stable holomorphic vector bundles of rank r and degree d. This will be important in Section 2.3.2, where Donaldson's theorem will be presented. The following theorem is the main result of the basic theory of vector bundles on a curve. It is due to Mumford for the first part [Mum63] and Seshadri for the second part [Ses67].

Theorem 2.45 (Mumford-Seshadri, [Mum63, Ses67]) *Let $g \geq 2, r \geq 1$ and $d \in \mathbb{Z}$.*

1. *The set $\mathcal{N}(r, d)$ of isomorphism classes of stable holomorphic vector bundles of rank r and degree d admits a structure of smooth, complex quasi-projective variety of dimension $r^2(g - 1) + 1$.*
2. *The set $\mathfrak{M}(r, d)$ of S-equivalence classes of semi-stable holomorphic vector bundles of rank r and degree d admits a structure of complex projective variety of dimension $r^2(g - 1) + 1$. $\mathcal{N}(r, d)$ is an open dense sub-variety of $\mathfrak{M}(r, d)$.*

In particular, when $r \wedge d = 1$, $\mathfrak{M}(r, d) = \mathcal{N}(r, d)$ is a smooth complex projective variety.

For general r and d, it can in fact be shown that the set of *isomorphism* classes of semi-stable vector bundles of rank r and degree d does not admit such an algebraic structure [Ses82]. In other words, to obtain a moduli variety, we *have to* identify *S*-equivalent, possibly non-isomorphic, objects.

2.3.2 Donaldson's theorem

In [Don83], Donaldson proposed a differential-geometric proof of the celebrated Narasimhan–Seshadri theorem [NS65] which magnificently complemented the symplectic approach to holomorphic vector bundles on a curve of Atiyah and Bott. Donaldson's theorem echoes, in an infinite-dimensional setting, a result by Kempf and Ness, relating semi-stable closed orbits of the action of a complex reductive group to the action of a maximal compact subgroup of that group. Thanks to a differential-geometric characterisation of stability, Donaldson's theorem establishes a homeomorphism between the moduli space $\mathfrak{M}(r, d)$ and the symplectic quotient $F^{-1}(\{*i2\pi \frac{d}{r\mathrm{Id}_E}\})/\mathcal{G}_h$.

Theorem 2.46 (Donaldson, [Don83]) *Fix a smooth Hermitian vector bundle (E, h) of rank r and degree d. Let $\mathcal{E}$ be a holomorphic vector bundle of rank r and degree d, and let $O(\mathcal{E})$ be the corresponding orbit of unitary connections on (E, h). Then $\mathcal{E}$ is stable if and only if $O(\mathcal{E})$ contains a unitary connection A satisfying:*

1. $\mathrm{Stab}_{\mathcal{G}_E}(A) \simeq \mathbb{C}^*$.
2. $F_A = *i2\pi \frac{d}{r}\mathrm{Id}_E$.

Moreover, such a connection, if it exists, is unique up to an element of the unitary gauge group $\mathcal{G}_h$.

Indeed, since we know that isomorphism classes of holomorphic vector bundles of rank r and degree d are in one-to-one correspondence with *complex* gauge group orbits of unitary connections on (E, h), it seems natural to look for which unitary connections or, more accurately, which *orbits* of unitary connections correspond to isomorphism classes of *stable* holomorphic vector bundles of rank r and degree d. Donaldson's theorem states that these orbits are precisely the complex gauge orbits of unitary connections which are both *irreducible* (condition (1): $\mathcal{E} = (E, A)$ is an indecomposable holomorphic vector bundle) and *minimal Yang–Mills connections* (condition (2): A is an absolute minimum of the Yang–Mills functional $A \longmapsto \int_{\Sigma_g} \|F_A\|^2 \mathrm{vol}_{\Sigma_g}$; see [AB83, Don83]). Moreover, any such complex gauge orbit contains a unique *unitary* gauge orbit.

Corollary 3 (The Narasimhan–Seshadri theorem, [NS65]) *Graded isomorphism classes of poly-stable vector bundles of rank r and degree d are in one-to-one* correspondence with unitary gauge orbits of minimal Yang–Mills connections:

$$\mathcal{M}_{\Sigma_g}(r, d) \simeq F^{-1}\left(\left\{*i2\pi\frac{d}{r}\mathrm{Id}_E\right\}\right)/\mathcal{G}_h.$$

2.3.3 Exercises

Exercise 2.3.1 Show that a semi-stable holomorphic which has coprime rank and degree is in fact stable.

Exercise 2.3.2 Show that $\mu(\mathcal{E}^*) = -\mu(\mathcal{E})$ and $\mu(\mathcal{E} \otimes \mathcal{E}') = \mu(\mathcal{E}) + \mu(\mathcal{E}')$. Compute $\mu(\mathrm{Hom}(\mathcal{E}, \mathcal{E}'))$.

Exercise 2.3.3 Consider the extension (short exact sequence)

$$0 \longrightarrow \mathcal{E}' \longrightarrow \mathcal{E} \longrightarrow \mathcal{E}'' \longrightarrow 0$$

of $\mathcal{E}''$ by $\mathcal{E}'$.
a. Assume that $\mathcal{E}'$ and $\mathcal{E}''$ are semi-stable and both have slope μ. Show that $\mu(\mathcal{E}) = \mu$ and that $\mathcal{E}$ is semi-stable.
b. Show that if $\mathcal{E}'$ and $\mathcal{E}''$ are stable and have the same slope, $\mathcal{E}$ is not stable. *Hint*: By **a**, $\mu(\mathcal{E}) = \mu(\mathcal{E}')$ and $\mathcal{E}'$ is a sub-bundle of $\mathcal{E}$.
c. Let μ, μ', μ'' be the respective slopes of the bundles $\mathcal{E}$, $\mathcal{E}'$, $\mathcal{E}''$. Show that

$$\mu' < \mu \Leftrightarrow \mu' < \mu'' \Leftrightarrow \mu < \mu'',$$

$$\mu' = \mu \Leftrightarrow \mu' = \mu'' \Leftrightarrow \mu = \mu'',$$

$$\mu' > \mu \Leftrightarrow \mu' > \mu'' \Leftrightarrow \mu > \mu''.$$

c. Suppose that the three bundles $\mathcal{E}$, $\mathcal{E}'$ and $\mathcal{E}''$ have the same slope. Show that $\mathcal{E}$ is semi-stable if and only if $\mathcal{E}'$ and $\mathcal{E}''$ are semi-stable.

Exercise 2.3.4 Let $\mathcal{E}$ and $\mathcal{E}'$ be two semi-stable bundles of slope μ and assume that $\mathcal{E}$ and $\mathcal{E}'$ are S-equivalent. Show that $\mathrm{rk}\,\mathcal{E} = \mathrm{rk}\,\mathcal{E}'$ and $\deg\mathcal{E} = \deg\mathcal{E}'$. *Hint*: Consider the poly-stable object $\mathcal{F}_1 \oplus \cdots \oplus \mathcal{F}_k$ associated to an arbitrary Jordan–Hölder filtration of $\mathcal{E}$, and show that $\deg(\mathcal{F}_1) + \cdots + \deg(\mathcal{F}_k) = \deg(\mathcal{E})$. Beware that the direct sum $\mathcal{F}_1 \oplus \cdots \oplus \mathcal{F}_k$ is not isomorphic to $\mathcal{E}$ in general.

Exercise 2.3.5 Let $\mathcal{E}$ be a holomorphic vector bundle and let $\mathcal{L}$ be a holomorphic line bundle.
a. Show that $\mu(\mathcal{E} \otimes \mathcal{L}) = \mu(\mathcal{E}) + \mu(\mathcal{L})$.
b. Show that $\mathcal{E}$ is stable (resp. semi-stable) if and only if $\mathcal{E} \otimes \mathcal{L}$ is stable (resp. semi-stable). *Hint*: Sub-bundles of $\mathcal{E} \otimes \mathcal{L}$ are of the form $\mathcal{F} \otimes \mathcal{L}$, where $\mathcal{F}$ is a sub-bundle of $\mathcal{E}$.

References

[AB83]   M. F. Atiyah and R. Bott. The Yang–Mills equations over Riemann surfaces. *Phil. Trans. R. Soc. Lond. A* **308**(1505): 523–615, 1983.

[Ati57]   M. F. Atiyah. Vector bundles over an elliptic curve. *Proc. Lond. Math. Soc.* **7**(3): 414–452, 1957.

[BT82]   Raoul Bott and Loring W. Tu. *Differential forms in algebraic topology*, Graduate Texts in Mathematics 82. New York: Springer-Verlag, 1982.

[DK90]   S. K. Donaldson and P. B. Kronheimer. *The geometry of four-manifolds*. Oxford Mathematical Monographs. Oxford: Clarendon Press; New York: Oxford University Press, 1990. Oxford Science Publications.

[Don83]   S. K. Donaldson. A new proof of a theorem of Narasimhan and Seshadri. *J. Diff. Geom.* **18**(2): 269–277, 1983.

[For91]   Otto Forster. *Lectures on Riemann surfaces*, Graduate Texts in Mathematics 81. New York: Springer-Verlag, 1991. Translated from the 1977 German original by Bruce Gilligan. Reprint of the 1981 English translation.

[Gó1]   Tomás L. Gómez. Algebraic stacks. *Proc. Indian Acad. Sci., Math. Sci.*, **111**(1): 1–31, 2001.

[Gro57]   A. Grothendieck. Sur la classification des fibres holomorphes sur la sphère de Riemann. *Am. J. Math.* **79**: 121–138, 1957.

[Gun66]   R. C. Gunning. *Lectures on Riemann surfaces*. Princeton, NJ: Princeton University Press, 1966.

[Hat02]   Allen Hatcher. *Algebraic topology*. Cambridge: Cambridge University Press, 2002.

[Hus93]   Dale H. Husemoller. *Fibre bundles*, 3rd edn. Berlin: Springer-Verlag, 1993.

[KN96]   Shoshichi Kobayashi and Katsumi Nomizu. *Foundations of differential geometry*, Vol. I. Wiley Classics Library. New York: John Wiley & Sons Inc., 1996. Reprint of the 1963 original.

[Kob87]   Shoshichi Kobayashi. *Differential geometry of complex vector bundles*. Princeton, NJ: Princeton University Press; Tokyo: Iwanami Shoten Publishers, 1987.

[LPV85]   J. Le Potier and J. L. Verdier. Variété de modules de fibres stables sur une surface de Riemann: résultats d'Atiyah et Bott. In *Moduli of stable bundles over algebraic curves (Paris, 1983)*, Progress in Mathematics 5. Boston, MA: Birkhäuser Boston, 1985, pp. 5–28.

[MFK93]   D. Mumford, J. Fogarty, and F. Kirwan. *Geometric invariant theory*, 3rd *enl. ed.* Berlin: Springer-Verlag, 1993.

[MS98]   Dusa McDuff and Dietmar Salamon. *Introduction to symplectic topology*, 2nd edn. Oxford Mathematical Monographs. Oxford: Clarendon Press; New York: Oxford University Press, 1998.

[Muk03]   Shigeru Mukai. *An introduction to invariants and moduli*, tran., W. M. Oxbury. Cambridge: Cambridge University Press, 2003.

[Mum63]   David Mumford. Projective invariants of projective structures and applications. In *Proceedings of the International Congress of Mathematicians (Stockholm, 1962)*, Djursholm: Institut Mittag-Leffler, 1963, pp. 526–530.

[New09]   P. E. Newstead. Geometric invariant theory. In *Moduli spaces and vector bundles*, London Mathematical Society Lecture Note Series 359. Cambridge: Cambridge University Press, 2009, pp. 99–127.

[NS65]   M. S. Narasimhan and C. S. Seshadri. Stable and unitary vector bundles on a compact Riemann surface. *Ann. Math.* (2)**82**: 540–567, 1965.

[Ses67] C. S. Seshadri. Space of unitary vector bundles on a compact Riemann surface. *Ann. Math.* (2)**85**: 303–336, 1967.

[Ses82] C. S. Seshadri. *Fibres vectoriels sur les courbes algébriques*, volume 96 of *Astérisque*. Paris: Société Mathématique de France, 1982. Notes written by J.-M. Drezet from a course at the École Normale Supérieure, June 1980.

[Ste51] Norman Steenrod. *The topology of fibre bundles*. Princeton, NJ: Princeton University Press, 1951.

[Tha97] Michael Thaddeus. An introduction to the topology of the moduli space of stable bundles on a Riemann surface. Andersen, Jørgen Ellegaard *et al.* (eds), *Geometry and physics: Proceedings of the conference at Aarhus University, Aarhus, Denmark, 1995*. Lecture Notes in Pure and Applied Mathematics 184. New York: Marcel Dekker, 1997, pp. 71–99.

[Tho06] R. P. Thomas. Notes on GIT and symplectic reduction for bundles and varieties. In *Surveys in differential geometry*, Vol. 10, Somerville, MA: International Press, 2006, pp. 221–273.

[Wel08] Raymond O. Wells, Jr. *Differential analysis on complex manifolds. With a new appendix by Oscar Garcia-Prada*, 3rd edn. New York: Springer, 2008.

3

Paths towards an extension of Chern–Weil calculus to a class of infinite dimensional vector bundles

SYLVIE PAYCHA

Abstract

We discuss possible extensions of the classical Chern–Weil formalism to an infinite dimensional setup using regularised traces on classical pseudodifferential operators. This is largely based on joint work with Simon Scott, Jouko Mickelsson and Steven Rosenberg.

Introduction

Classical Chern–Weil formalism relates geometry to topology, assigning to the curvature of a connection, de Rham cohomology classes of the underlying manifold. This theory, which was developed in the 1940s by Shiing-Shen Chern [C2] and André Weil[1] and can be seen as a generalisation of the Chern–Gauss–Bonnet theorem [C1], was an important step in the theory of characteristic classes.

Recall that, given a topological group G, a characteristic class for a principal G-bundle $P \to X$ with coefficients in a commutative ring Λ with unit is given by an element f^*c in $H^*(X, \Lambda)$, where c is an element of the cohomology algebra $H^*(BG, \Lambda)$, BG stands for the classifying space of G and the classifying map f is determined up to homotopy by the equivalence class of the principal bundle so that $[f] \in [X, BG]$ (see e.g. [BT], [MS]).

When G is a Lie group with Lie algebra $\mathfrak{g}$ and the coefficient ring Λ is $\mathbb{R}$ or $\mathbb{C}$, the Chern–Weil approach provides a geometric construction of characteristic classes by assigning to an $\mathrm{Ad}(G)$-invariant polynomial I on $\mathfrak{g}$ a closed form $I(\Omega, \ldots, \Omega)$, where Ω is the curvature of a $\mathfrak{g}$-valued connection one-form on P descending to

---

[1] In an unpublished paper.

Geometric and Topological Methods for Quantum Field Theory, ed. Alexander Cardona, Iván Contreras and Andrés F. Reyes-Lega. Published by Cambridge University Press. © Cambridge University Press 2013.

X (see e.g. [BT], [MS]). When G is compact,[2] the Chern–Weil homomorphism $\Phi : \text{Inv}_G (S\,\mathfrak{g}^*) \to H^*(BG, \Lambda)$ from the invariants of the symmetric algebra $S\,\mathfrak{g}^*$ of the dual $\mathfrak{g}^*$ of $\mathfrak{g}$ seen as a G-module for the adjoint action to the cohomology group $H^*(BG, \Lambda)$ built this way is actually an isomorphism [Bott].

When G is a matrix group, $\text{Ad}(G)$-invariant polynomials on $\mathfrak{g}$ can be built from the trace on matrices in view of the $\text{Ad}(G)$-invariance of the trace. When G is the linear group of invertible n by n matrices with real or complex coefficients, we obtain all invariant polynomials in this way, which turn out to be generated by the monomials $X \mapsto \text{tr}(X^j)$ (see e.g. Appendix A of [BGJ]).

If G is the compact matrix Lie group $U_n(\mathbb{C})$ of unitary n by n matrices, then the cohomology group $H^*(BG, \mathbb{C})$ is generated by Chern classes given by the de Rham classes of closed forms $\text{tr}(\Omega^j)$. When G is the group $O_n(\mathbb{R})$ of orthogonal n by n matrices, a set of generators of $H^*(BG, \mathbb{R})$ is given by Pontrjagin classes which correspond to the de Rham classes of closed forms $\text{tr}(\Omega^{\wedge 2j})$, where the exponent $\wedge 2j$ stands for the antisymmetric product of antisymmetric matrices (see e.g. [BT]).

This survey, based on work on the subject by various authors ([LRST], [PR1], [PR2], [PS], [MP]) suggests a few steps towards a generalisation of the Chern–Weil approach to a class of infinite rank vector bundles modelled on the space $C^\infty(M, E)$ of smooth sections of a vector bundle $E \to M$ over a closed manifold M. The first obvious infinite dimensional counterpart for the invertible matrix group (which hosts structure groups in the finite dimensional case) is the gauge group $\mathcal{G}(E)$ of the vector bundle E. When M reduces to a point, the gauge group yields back a matrix group. "Inflating" a point to a manifold M gives rise to new degrees of freedom, diffeomorphisms of the manifold M. We therefore consider the diffeomorphism group $\text{Diff}(M, E)$ of the bundle $E \to M$, which encompasses both the gauge group of E and the diffeomorphisms of M. But, unlike in the finite dimensional case where connection one-forms take values in the Lie algebra of the structure group, here they take values in yet another algebra, the algebra $\text{Cl}(M, E)$ of classical pseudodifferential operators acting on smooth sections of E. In some cases, it is enough to consider the Lie algebra $\text{Cl}^0(M, E)$ of zero-order classical pseudodifferential operators acting on smooth sections of E. The group $\text{Diff}(M, E)$ acts smoothly on both these Lie algebras.

In Part 1, we describe three different infinite dimensional Lie groups and their corresponding Lie algebras, the gauge group $\mathcal{G}(E)$ of the bundle E over M, its diffeomorphism group $\text{Diff}(M, E)$ and the Lie group $\text{Cl}^{0,\times}(M, E)$ of zero-order classical pseudodifferential operators acting on smooth sections of E. The

---

[2] When G is compact, in the continuous (or equivalently smooth) cohomology of G with values in the symmetric algebra $S\,\mathfrak{g}^*$, $H_c^0(G, S\,\mathfrak{g}^*)$ coincides with the invariants $\text{Inv}_G(S\,\mathfrak{g}^*)$ whereas $H_c^i(G, S\,\mathfrak{g}^*)$ vanishes for positive i.

infinite dimensional vector bundles with structure group Diff(M, E) are described in Section 3.10.

As in the finite dimensional Chern–Weil formalism recalled in Section 3.6, we first build Diff(M, E) Ad-invariant polynomials on Cl(M, E) (similarly on Cl$^0(M, E)$), which yield closed forms with de Rham class independent of the connection (Theorem 3.33 in Section 3.8).

For this purpose we recall the classification of Ad-invariant linear forms on Cl(M, E), resp. Cl$^0(M, E)$ (see Theorem 3.9 of Section 3.3): they are generated by the noncommutative residue introduced by Adler [Ad] and Manin [Man], later generalised by Guillemin [Gu] and Wodzicki [W1] (see [Kas] for a survey), resp. two types of traces, the noncommutative residue and leading symbol traces used in [PR1, PR2]. Mimicking the finite dimensional setup, we apply these traces to powers of the curvature tensor to build Ad-invariant polynomials.

Interesting related issues arise along the way such as the topology of the Fréchet–Lie group Cl$^{0,\times}(M, E)$ of invertible zero-order pseudodifferential operators; we report on results by Rochon, who describes its stable homotopy groups (Theorem 3.8 in Section 3.4). Homotopy groups of Cl$^{0,\times}(M, E)$ are related to those of the gauge group of the pull-back bundle p^*E over the cotangent sphere bundle of M via the canonical projection $p : T^*M \to M$ onto the base manifold. The first and second homotopy groups come into play when attempting to lift central extensions of the Lie algebra Cl$^0(M, E)$ to the Lie group Cl$^{0,\times}(M, E)$ [N1].

To our knowledge, central extensions of Cl$^0(M, E)$ and Cl$^{0,\times}(M, E)$ have not yet been classified; however, one can build explicit ones from the noncommutative residue and the leading symbol trace using an outer derivation on Cl$^0(M, E)$, the resulting extension being independent of the specific outer derivation one chooses among a given class (Proposition 7 in Section 3.6). The 2-cocycle built from the noncommutative residue called the Radul cocycle turns out to be trivial (Theorem 3.17 in Section 3.3) as a result of the existence of some linear extensions to the algebra Cl$^0(M, E)$ of the L^2-trace on smoothing operators which we call regularised traces (Section 3.7). In contrast with the noncommutative residue and leading symbol traces which vanish on smoothing pseudodifferential operators (we call them *singular traces*), regularised traces coincide with the usual trace on smoothing operators.

One can build Chern–Weil classes from the noncommutative residue [MRT] or leading symbol traces [LRST] following the same construction as in the finite dimensional situation. We shall call these two types of Chern–Weil classes singular Chern–Weil classes since they are built from singular traces. The leading order Chern forms built from leading order traces detect cohomology on mapping spaces; see Theorem 4.7 in [LRST] where the resulting leading order Chern classes are expressed in terms of Chern classes on the target space.[3] To our knowledge, there is not yet any evidence that there exists a nonzero Wodzicki–Chern class; this is still an open question. We instead turn to regularised traces as a substitute for the matrix trace used to build Ad-invariants in the finite dimensional Chern–Weil formalism. The price to pay for choosing regularised traces on $Cl^0(M, E)$ as a substitute for genuine traces such as the noncommutative residue and the leading symbol traces is that analogues of Chern–Weil invariant polynomials do not give rise to closed forms. Implementing techniques borrowed from the theory of classical pseudodifferential calculus, one measures the obstructions to the closedness in terms of noncommutative residues (Theorem 3.39 in Section 3.13.5); these are interesting in their own right since they are related to anomalies in quantum field theory (see e.g. [CDP] and references therein). In specific situations such as in Hamiltonian gauge theory described in Section 3.14 where we need to build Chern classes on pseudodifferential Grassmannians, the very locality of the noncommutative residue can provide a way to build counterterms, and thereby to renormalise the original non-closed forms in order to turn them into closed ones (see Theorem 3.41). Introducing counterterms in this geometric setup is reminiscent of renormalisation methods used by physicists to cure anomalies by introducing appropriate counterterms.

Loop groups [F] also provide an interesting geometric setup since obstructions to the closedness can vanish, thus leading to closed forms (Proposition 16 in Section 3.15). On infinite rank vector bundles associated with a family of Dirac operators on even dimensional closed spin manifolds, these obstructions can be circumvented by an appropriate choice of regularised trace involving the very superconnection which gives rise to the curvature. Chern–Weil forms associated with a superconnection yield back even degree homogeneous components of the Chern character associated with the family of Dirac operators that arises in the family index theorem (Theorem 3.45 in Section 3.15).

These are only a few steps towards a better understanding of the geometry and topology of a class of infinite dimensional manifolds and vector bundles; much is yet left to be done.

The chapter is organised in four parts: a first one which describes some useful infinite dimensional groups to serve as potential structure groups for certain infinite dimensional vector bundles arising in the sequel, a second one which investigates traces on classes of pseudodifferential operators and the obstructions that prevent certain linear forms from defining traces, a third one in which singular Chern–Weil forms are built on a class of infinite dimensional vector bundles from singular

---

[3] Theorem 4.7 in [LRST] actually holds on the component of the constant maps.

traces, and a last part which explores ways to circumvent anomalies arising from building regularised traces, analogues of Chern–Weil forms in infinite dimensions.

Part 1: Some useful infinite dimensional Lie groups

The first question that arises in an attempt to generalise classical Chern–Weil theory to infinite dimensions is what to substitute for the matrix structure group $G = \mathrm{CL}_k(\mathbb{C})$. Viewing it as $C^\infty(\{*\}, G)$, where $\{*\}$ is a point, we want to "inflate" the point to a smooth manifold M. This very non-canonical step leads us to the mapping group $C^\infty(M, G)$ as a first possible generalisation. But this is only an instance of the more general gauge group $\mathcal{G}(E)$ of a vector bundle E over M with structure group G. We shall then describe two other groups which contain the gauge group, the diffeomorphism group $\mathrm{Diff}(M, E)$ of E, and the group $\mathrm{Cl}(M, E)^{0,\times}$ of zero-order invertible pseudodifferential operators acting on sections of E. $\mathrm{Diff}(M, E)$ will later play the role of a structure group, while the Lie algebra $\mathrm{Cl}(M, E)^0$ of $\mathrm{Cl}(M, E)^{0,\times}$ will act as a receptacle for connection one-forms.

3.1 The gauge group of a bundle

We recall the concept of gauge group (see [AB], [AS], [CM], [Mich], [Woc]), which falls in the more general class of infinite dimensional Fréchet–Lie groups, a class widely studied in the literature; see e.g. [OMYK], [KM].

Let G be a smooth finite dimensional Lie group, and let M be a closed smooth manifold. The set $C^\infty(M, G)$ of smooth maps from M to G is an infinite dimensional Fréchet–Lie group with respect to pointwise multiplication, with Lie algebra

$$\mathrm{Lie}\left(C^\infty(M, G)\right) := C^\infty(M, \mathfrak{g}),$$

where $\mathfrak{g}$ is the Lie algebra of G. The bracket on $\mathrm{Lie}\left(C^\infty(M, G)\right)$ is defined pointwise by the bracket on $\mathfrak{g}$.

The group $C^\infty(M, G)$ is a particular instance of the more general notion of gauge group. Let $\pi : P \to M$ be a principal G-bundle over a closed manifold M and let us introduce the following bundle of groups (it is not a principal bundle)

$$\mathrm{Ad}P := P \times_G G,$$

where G acts on G by the adjoint group action $\mathrm{Ad}_g(\gamma) = g^{-1}\gamma\, g$. The set of G-invariant G-valued functions on P

$$\mathcal{G}(P) := C^\infty(P, \mathrm{Ad}P)$$

$$:= \{f \in C^\infty(P, G),\, f(p \cdot g) = g^{-1} \cdot f(p) \cdot g\ \forall p \in P, \forall g \in G\},$$

is a Fréchet–Lie group when equipped with the pointwise multiplication. The adjoint action $\mathrm{Ad} : G \to \mathrm{Aut}(G)$ induces an adjoint action $\mathrm{ad} : G \mapsto \mathrm{End}(\mathfrak{g})$ and we set

$$\mathrm{ad} P := P \times_G \mathfrak{g}.$$

The Lie algebra of the gauge group $\mathcal{G}(P)$ is the gauge algebra given by

$$\mathrm{Lie}\,(\mathcal{G}(P)) := C^\infty(P, \mathrm{ad}P)$$

$$:= \{\sigma \in C^\infty(P, \mathfrak{g}), \sigma(p \cdot g) = \mathrm{Ad}(g^{-1})\sigma(p) \forall p \in P, g \in G\},$$

equipped with the pointwise Lie bracket. Here $\mathfrak{g}$ is the Lie algebra of G.
The gauge group can be seen as the group of smooth vertical bundle automorphisms of P.

Proposition 1

$$\mathcal{G}(P) \simeq C^\infty(M, \mathrm{Aut}(P)) =: \mathrm{Aut}(M, P),$$

where

$$\mathrm{Aut}(P) := \{\alpha : P \to P \text{ isomorphism}, \alpha(p \cdot g) = \alpha(p) \cdot g \,\forall g \in G\}.$$

Proof An element $f \in C^\infty(P, G)$ acts pointwise on P by an automorphism $\alpha_f :$ $p \mapsto p \cdot f(p)$, with inverse $\alpha_{f^{-1}} : p \mapsto p \cdot (f(p))^{-1}$. It is G-equivariant since $\alpha_f(p \cdot g) = \alpha_{f \cdot g}$ for any $g \in G$. Conversely, a G-equivariant automorphism $\alpha :$ $P \to P$ covering the identity induces a map $f_\alpha : P \to G$ uniquely defined by $\alpha(p) = p \cdot f_\alpha(p)$. $\square$

Example 3.1 When $P = M \times G$ is a trivial principal bundle over M, then $\mathcal{G}(P) = C^\infty(M, G)$ and $\mathrm{Lie}\,(\mathcal{G}(P)) = C^\infty(M, \mathfrak{g})$.

To a vector bundle $E \to M$ over M of rank k over C, we associate a gauge group via its frame bundle $\mathrm{GL}(E)$, which is the $GL_k(\mathbb{C})$-principal bundle over M with fibre above a point x, given by:

$$\mathrm{GL}_x(E) := \{\alpha_x : \mathbb{C}^k \to E_x, \alpha_x \text{ isomorphism}\}.$$

The gauge group of the vector bundle E corresponds to the gauge group of the principal bundle $\mathrm{GL}(E)$:

$$\mathcal{G}(E) := \mathcal{G}(\mathrm{GL}(E))$$

$$= \{\sigma \in C^\infty(\mathrm{GL}(E), \mathrm{GL}_k(\mathbb{C})), \sigma(p \cdot g) = g^{-1} \cdot \sigma(p) \cdot g \,\forall p \in \mathrm{GL}(E),$$

$$\forall g \in \mathrm{Gl}_k(\mathbb{C})\},$$

Proposition 2

$$\mathcal{G}(E) \simeq C^\infty(M, \mathrm{GL}(E)) \simeq C^\infty(M, \mathrm{Aut}(E)) =: \mathrm{Aut}(M, E),$$

where the automorphism bundle $\mathrm{Aut}(E)$ is a bundle of groups over M given by

$$\mathrm{Aut}(E) := \{\alpha \in E^* \times E, \alpha_x : E_x \to E_x \text{ isomorphism } \forall x \in M\}.$$

Proof By the previous proposition applied to $P = \mathrm{GL}(E)$ we have $\mathcal{G}(\mathrm{GL}(E))$ $\simeq C^\infty(M, \mathrm{Aut}(\mathrm{GL}(E)))$. Since $\mathrm{Aut}(\mathrm{GL}(E)) \simeq \mathrm{Aut}(E)$ as bundles, it follows that $\mathcal{G}(\mathrm{GL}(E)) \simeq C^\infty(M, \mathrm{Aut}(E))$.

Let us now prove that $\mathcal{G}(\mathrm{GL}(E)) \simeq C^\infty(M, \mathrm{GL}(E))$. There is a natural isomorphism between $\mathcal{G}(\mathrm{GL}(E))$ and $C^\infty(M, \mathrm{GL}(E))$, which sends $s \in \mathcal{G}(\mathrm{GL}(E))$ to $\tilde{s}$ defined by

$$\tilde{s}(x) = [p, s(p)],$$

where p is any element of $\pi^{-1}(x)$ and $[p, s(p)]$ is an element of $\mathrm{GL}(E) \times_G G$ corresponding to $(p, s(p)) \in \mathrm{GL}(E) \times G$. Modifying the lifting $p \cdot g$ in $\pi^{-1}(x)$ does not affect $[p, s(p)]$ since

$$[p \cdot g, s(p \cdot g)] = [p \cdot g, g^{-1} s(p) g] = [p, s(p)].$$

The map is clearly onto; to $\tilde{s}(x)$ we associate the map $s : \mathrm{GL}(E) \to \mathrm{GL}_k(\mathbb{C})$ defined by the condition

$$\tilde{s}(x) = [p, s(p)], \quad \forall p \in \pi^{-1}(x).$$

It is injective for

$$\tilde{s}_1 = \tilde{s}_2 \iff ([p, s_1(p)] = [p, s_2(p)] \ \forall p \in \mathrm{GL}(E)) \iff s_1 = s_2. \qquad \square$$

3.2 The diffeomorphism group of a bundle

We now enlarge the group $\mathrm{Aut}(M, E)$ to the group $\mathrm{Diff}(M, E)$ of smooth diffeomorphisms of E

$$\mathrm{Diff}(M, E) = \{\Phi = (\alpha, f), f \in \mathrm{Diff}(M), \alpha \in C^\infty(M, \mathrm{Isom}(f^*E, E))\}. \quad (3.1)$$

Here,

$$\mathrm{Diff}(M) := \{f \in C^\infty(M, M), \exists f^{-1} \in C^\infty(M, M)\}$$

is the group of smooth diffeomorphisms of M, which by [Le], [Om] (see [N2] for a review) is a Fréchet–Lie group with Lie algebra $C^\infty(M, TM)$, where TM is the tangent bundle to M. The group $\mathrm{Aut}(M, E)$ embeds in $\mathrm{Diff}(M, E)$

by $\iota : \alpha \mapsto (\alpha, Id)$ and corresponds to the kernel of the natural projection map $p : \mathrm{Diff}(M, E) \to \mathrm{Diff}(M)$, so that we have an exact sequence:

$$\{Id\} \longrightarrow \mathrm{Aut}(M, E) \longrightarrow \mathrm{Diff}(M, E) \longrightarrow \mathrm{Diff}(M) \to \{Id\}.$$

Lemma 3.2 *The group* $\mathrm{Diff}(M, E)$ *is a Fréchet–Lie group with Lie algebra* $C^\infty(M, TM \times \mathrm{End}(E))$ *equipped with the product bracket.*

Proof We follow [AS] to equip $\mathrm{Diff}(M, E)$ with a Fréchet–Lie group structure inherited from the Fréchet–Lie group structures of the diffeomorphism group $\mathrm{Diff}(M)$ of M and the automorphism group $\mathrm{Aut}(M, E)$ of E.

Using a connection on E and with the notations of (3.1) we can identify the fibres $E_{f(x)}$ and E_x for f in a small neighbourhood U of the identity in $\mathrm{Diff}(M)$ by an isomorphism i_U. A neighbourhood of the identity in $\mathrm{Diff}(M, E)$ is therefore given by

$$f^{-1}(U) \cap \alpha^{-1} \circ i_U(V)$$

for some open neighbourhoods U, resp. V, of the identity in $\mathrm{Diff(M)}$, resp. in $\mathrm{Aut}(M, E)$. Consequently, locally around the identity, $\mathrm{Diff}(M, E)$ is isomorphic to $C^\infty(M, TM \times \mathrm{End}(E))$, which is a Fréchet linear space. The local charts described above are smooth.

The composition $(\alpha, f) \circ (\beta, g) = (\alpha \circ \beta; f \circ g)$, where $\beta : (f \circ g)^* E \to f E$ and $\alpha : f^* E \to E$ are isomorphisms of vector bundles, and the inversion operation $(\alpha, f)^{-1} = (\alpha^{-1}, f^{-1})$ with $\alpha^{-1} : E \to f^* E$, the inverse of α, are smooth on $\mathrm{Diff}(M, E)$. Thus, $\mathrm{Diff}(M, E)$ is a Fréchet–Lie group with Fréchet–Lie algebra $C^\infty(M, TM \times \mathrm{End}(E))$.   □

Remark 1 Using a connection ∇ on $E \to M$, elements in $C^\infty(M, TM \times \mathrm{End}(E))$ can be seen as differential operators via the map

$$(U, \alpha) \mapsto [\sigma \mapsto (\nabla_U(\sigma) + \alpha(\sigma))],$$

where ∇_U is viewed as a differential operator of order 1, and the automorphism α as a multiplication operator.

3.3 The algebra of zero-order classical pseudodifferential operators

The algebra of matrices $\mathrm{gl}_d(\mathbb{C})$, which plays a fundamental role for ordinary Chern–Weil calculus, is replaced by the algebra of zero-order classical pseudodifferential operators on a closed manifold M acting on $\mathbb{C}^d$-valued smooth functions on M. This algebra contains the algebra $\mathrm{Map}(M, \mathrm{gl}_d(\mathbb{C}))$ of smooth maps on M with values in $\mathrm{gl}_d(\mathbb{C})$. One can think of $\mathrm{gl}_d(\mathbb{C})$ as what remains of the infinite dimensional algebra of zero-order classical pseudodifferential operators on M when M is reduced to a point $\{*\}$.

We briefly recall the definition of *classical pseudodifferential operators* (ψdos) on closed manifolds, referring the reader to [H], [Sh], [T], [Tr] for further details. These generalise differential operators P with smooth coefficients which are linear operators acting on smooth functions on an open subset U of $\mathbb{R}^n$ defined by

$$Pu(x) = \sum_{|\alpha| \le a} c_\alpha(x) \, D_x^\alpha u(x), \quad u \in C^\infty(U), c_\alpha \in C^\infty(U),$$

where $a \in \mathbb{N}_0$ stands for the order of a and $D_x^\alpha := (-i)^{|\alpha|} \partial_{x_1}^{\alpha_1} \cdots \partial_{x_n}^{\alpha_n}$ where $\alpha = (\alpha_1, \ldots, \alpha_n)$ is a multi-index. Unlike a differential operator, which is local in as far as the support of Pu is contained in that of u, a pseudodifferential operator P is only expected to be pseudolocal, i.e. the singular support of Pu is contained in the singular support of u.

To the differential operator P one associates the polynomial $\sigma(P)(x, \xi) = \sum_{|\alpha| \le a} c_\alpha(x) \xi^\alpha$ called the symbol of P, which relates to P via a Fourier transform $Pu(x) = \int e^{i \langle x \cdot \xi \rangle} \sigma(P)(x, \xi) \hat{u}(\xi) \, d\xi$ for all x in U. Similarly, we define a pseudo-differential operator via its symbol and first describe the class of symbols under consideration.

Given a complex number a, we consider the space of symbols $S^a(U)$, which consists of smooth functions $\sigma(x, \xi)$ on $U \times \mathbb{R}^n$ such that for any compact subset K of U and any two multi-indices $\alpha = (\alpha_1, \ldots, \alpha_n), \beta = (\beta_1, \ldots, \beta_n)$ in $\mathbb{N}^n$, there exists a constant $C_{K\alpha\beta}$ satisfying

$$|\partial_x^\alpha \partial_\xi^\beta \sigma(x, \xi)| \le C_{K\alpha\beta}(1 + |\xi|)^{\text{Re}(a) - |\beta|}$$

for all (x, ξ) in $K \times \mathbb{R}^n$, where $|\beta| = \beta_1 + \cdots + \beta_n$.

If $\text{Re}(a_1) < \text{Re}(a_2)$, then $S^{a_1}(U) \subset S^{a_2}(U)$.

Remark 2 If a is real then a corresponds to the order of $\sigma \in S^a(U)$. The notion of order extends to complex values for classical pseudodifferential symbols (see below).

The *symbol product* on symbols is defined as follows: if σ_1 lies in $S^{a_1}(U)$ and σ_2 lies in $S^{a_2}(U)$, then

$$\sigma_1 \star \sigma_2(x, \xi) \sim \sum_{\alpha \in \mathbb{N}^n} \frac{(-i)^{|\alpha|}}{\alpha!} \partial_\xi^\alpha \sigma_1(x, \xi) \partial_x^\alpha \sigma_2(x, \xi), \tag{3.2}$$

where $\sim$ means that for any integer $N \ge 1$ we have

$$\sigma_1 \star \sigma_2(x, \xi) - \sum_{|\alpha| < N} \frac{(-i)^{|\alpha|}}{\alpha!} \partial_\xi^\alpha \sigma_1(x, \xi) \partial_x^\alpha \sigma_2(x, \xi) \in S^{a_1 + a_2 - N}(U).$$

In particular, $\sigma_1 \star \sigma_2$ lies in $S^{a_1 + a_2}(U)$.

We denote by $S^{-\infty}(U) := \bigcap_{a \in \mathbb{C}} S^a(U)$ the algebra of smoothing symbols on U and let $S(U)$ be the algebra $\bigcup_{a \in \mathbb{C}} S^a(U)$.

A *classical symbol* of complex order a is a symbol $\sigma \in S^a(U)$ such that

$$\forall N \in \mathbb{N}, \quad \sigma - \sum_{j < N} \psi(\xi)\sigma_{a-j}(x, \xi) \in S^{a-N}(U),$$

where $\sigma_{a-j}(x, \xi)$ is a positively homogeneous function on $U \times \mathbb{R}^n$ of degree $a - j$, i.e. $\sigma_{a-j}(x, t\xi) = t^{a-j}\sigma_{a-j}(x, \xi)$ for all positive numbers t. Here ψ in $C^\infty(\mathbb{R}^n)$ is an excision function which vanishes for $|\xi| \leq \frac{1}{2}$ and such that $\psi(\xi) = 1$ for $|\xi| \geq 1$.

We write for short

$$\sigma(x, \xi) \sim \sum_{j=0}^{\infty} \psi(\xi)\,\sigma_{a-j}(x, \xi). \tag{3.3}$$

We call a the order of the classical symbol σ and denote by $CS^a(U)$ the subset of classical symbols of order a. The positively homogeneous component $\sigma_a(x, \xi)$ of degree a corresponds to the *leading symbol* of σ.

Example 3.3 A smooth function h in $C^\infty(U)$ can be viewed as a multiplication operator $f \mapsto h\,f$ on smooth functions f in $C^\infty(U)$ and hence as a zero-order classical symbol.

A *log-polyhomogeneous symbol* of complex order a and logarithmic type $k \in \mathbb{Z}_{\geq 0}$ is a symbol $\sigma \in S^a(U)$ such that

$$\sigma(x, \xi) \sim \sum_{l=0}^{k} \sigma_l(x, \xi) \log |\xi|^l, \quad \forall (x, \xi) \in T^*U, \tag{3.4}$$

with $\sigma_l \in CS^a(U)$. We call a the order and k the logarithmic type of the symbol σ and denote by $CS^{a,k}(U)$ the subset of log-polyhomogneous symbols of order a and type k.

The symbol product of two classical (log-polyhomogeneous) symbols is a classical (log-polyhomogeneous) symbol and we denote by

$$CS(U) = \left\langle \bigcup_{a \in \mathbb{C}} CS^a(U) \right\rangle; \quad CS^{*,k}(U) = \left\langle \bigcup_{a \in \mathbb{C}} CS^{a,k}(U) \right\rangle$$

the algebra generated by all classical (log-polyhomogneous) symbols on U.

Remark 3 Note that the union itself is not an algebra since the sum of two classical pseudodifferential operators is a classical pseudodifferential operator only if their orders differ by an integer.

We can associate to a symbol σ in $S(U)$ the continuous operator $\mathrm{Op}(\sigma)$: $C_c^\infty(U) \to C^\infty(U)$ defined for u in $C_c^\infty(U)$ – the space of smooth compactly supported functions on U – by

$$(\mathrm{Op}(\sigma)u)(x) = \int e^{ix\cdot\xi}\sigma(x,\xi)\widehat{u}(\xi)d\xi,$$

where $d\xi := \frac{1}{(2\pi)^n}d\xi$ with $d\xi$ the ordinary Lebesgue measure on $T_x^*U \simeq \mathbb{R}^n$ and where $\widehat{u}(\xi)$ is the Fourier transform of u. Since

$$(\mathrm{Op}(\sigma)u)(x) = \int\int e^{i(x-y)\cdot\xi}\sigma(x,\xi)u(y)d\xi dy,$$

$\mathrm{Op}(\sigma)$ is an operator with Schwartz kernel given by

$$k(x,y) = \int e^{i(x-y)\cdot\xi}\sigma(x,\xi)d\xi,$$

which is smooth off the diagonal.

A pseudodifferential operator A on U is an operator which can be written in the form $A = \mathrm{Op}(\sigma) + R$, where σ lies in $S(U)$ and R is a smoothing operator, i.e. R has a smooth kernel. If σ is a classical symbol of order a, then A is called a classical pseudodifferential operator (ψdo) of order a. The product on symbols induces a composition $\mathrm{Op}(\sigma_1 \star \sigma_2) = \mathrm{Op}(\sigma_1)\,\mathrm{Op}(\sigma_2)$. This in turn induces a composition on properly supported operators. A ψdo A on U is called properly supported if, for any compact $C \subset U$, the set $\{(x,y) \in \mathrm{Supp}(K_A),\, x \in C \text{ or } y \in C\}$ is compact, where $\mathrm{Supp}(K_A)$ denotes the support of the Schwartz kernel of A, i.e. a distribution on $U \times U$ such that, for $u \in C_c^\infty(U)$, $Au(x) = \int K_A(x,y)u(y)dy$. A properly supported ψdo maps $C^\infty(U)$ into itself and admits a symbol given by $\sigma(A)(x,\xi) = e^{-ix\cdot\xi}Ae^{ix\cdot\xi}$. The composition AB of two properly supported ψdos is a properly supported ψdo and $\sigma(AB) \sim \sigma(A) \star \sigma(B)$.

More generally, let M be a smooth closed manifold of dimension n and let π: $E \to M$ be a smooth vector bundle of rank d over M; an operator $A : C^\infty(M,E) \to C^\infty(M,E)$ is a (classical) pseudodifferential operator of order a if given a local trivialising chart (U,ϕ) on M, for any localisation $A_v = \chi_v^2 A\chi_v^1 : C_c^\infty(U,\mathbb{C}^d) \to C_c^\infty(U,\mathbb{C}^d)$ of A where $\chi_v^i \in C_c^\infty(U)$, the operator $\phi_*(A_v) := \phi A_v \phi^{-1}$ from the space $C_c^\infty(\phi(U),\mathbb{C}^d)$ into $C^\infty(\phi(U),\mathbb{C}^d)$ is a (classical) pseudodifferential operator of order a.

Whereas the symbol $\sigma(A)$ of a classical pseudodifferential operator A of order a is only locally defined, the *leading symbol* $\sigma_L(A) = \sigma_a(A)$ is independent of the choice of local chart and hence globally defined. The operator A is said to be *elliptic* if $\sigma_L(A)(x,\xi)$ is invertible as an endomorphism of the fibre E_x of E over x for any (x,ξ) in $T^*M \setminus \{0\}$.

Example 3.4 A smooth section f in $C^\infty(M, \text{End}(E))$ can be viewed as a multiplication operator $u \mapsto f u$ on smooth sections u of E and hence as a zero-order classical ψdo.

Let $\text{Cl}^a(M, E)$ denote the set of classical pseudodifferential operators of order a acting on smooth sections of E.

Remark 4 It is useful to keep in mind that an operator in $\text{Cl}^a(M, E)$ with real order a induces a bounded operator from a Sobolev closure $H^s(M, E)$ of $C^\infty(M, E)$ for any real number s, to $H^{s-a}(M, E)$ so that $\text{Cl}^a(M, E)$ lies in $\mathcal{L}\left(H^s(M, E), H^{s-a}(M, E)\right)$ (see e.g. Lemma 1.3.5 in [Gi]). Here $\mathcal{L}(B_1, B_2)$ stands for the space of bounded linear maps from the Banach space B_1 to the Hilbert space B_2. An elliptic operator of order a furthermore induces a Fredholm operator from $H^s(M, E)$ to $H^{s-a}(M, E)$ for any $s \in \mathbb{R}$ (see e.g. Lemma 1.4.5 in [Gi]). An elliptic operator A in $\text{Cl}^a(M, E)$ therefore has finite dimensional kernel $\text{Ker}(A)$ and cokernel $\text{Ker}(A^*)$, where A^* stands for the formal adjoint of A and these operators have closed range $R(A)$ and $R(A^*)$ (see e.g. Lemma 1.4.2 in [Gi]). Moreover, $H^s(M, E) = \text{Ker}(A) \oplus R(A^*)$ and $H^{s-a}(M, E) = \text{Ker}(A) \oplus R(A^*)$ for any s in $\mathbb{R}$, where the direct sums are actually orthogonal sums. It follows that if A is an essentially self-adjoint elliptic operator, then the operator $A + \pi_A$ is invertible, where π_A stands for the L^2-orthogonal projection onto $\text{Ker}(A)$.

If A_1 lies in $\text{Cl}^{a_1}(M, E)$ and A_2 lies in $\text{Cl}^{a_2}(M, E)$, then the product $A_1 A_2$ lies in $\text{Cl}^{a_1+a_2}(M, E)$ and we denote by

$$\text{Cl}(M, E) := \left\langle \bigcup_{a \in \mathbb{C}} \text{Cl}^a(M, E) \right\rangle, \tag{3.5}$$

the algebra generated by all classical pseudodifferential operators acting on smooth sections of E, and by $\text{Cl}^{-\infty}(M, E) := \bigcup_{a \in \mathbb{C}} \text{Cl}^a(M, E)$ the algebra of smoothing operators.

It follows from the above discussion that

$$C^\infty(M, \text{End}(E)) \subset \text{Cl}^0(M, E) \subset \text{Cl}(M, E).$$

Remark 5 When E is the trivial bundle $M \times \mathbb{R}$, we drop E in the notation, writing $\text{Cl}^a(M)$, $\text{Cl}^{-\infty}(M)$ and $\text{Cl}(M)$ instead of $\text{Cl}^a(M, E)$, $\text{Cl}^{-\infty}(M, E)$ and $\text{Cl}(M, E)$.

Remark 6 When M reduces to a point, then E is a vector space and we have

$$C^\infty(M, \text{End}(E)) = \text{Cl}(M, E) = \text{Cl}^0(M, E) = \text{Cl}^{-\infty}(M, E) = \text{End}(E).$$

For any complex number a, the linear space $\text{Cl}^a(M, E)$ of classical pseudodifferential operators of order a can be equipped with a Fréchet topology. For this, one equips the set $CS^a(U, V) = CS^a(U) \otimes \mathrm{End}(V)$ of classical symbols of order a on an open subset U of $\mathbb{R}^n$ with values in a euclidean space V (with norm $\| \cdot \|$) with a Fréchet structure. The following semi-norms labelled by multi-indices α, β and integers $j \geq 0$, N give rise to a Fréchet topology on $CS^a(U, V)$ (see [H]):

$$\sup_{x \in K, \xi \in \mathbb{R}^n} (1 + |\xi|)^{-\mathrm{Re}(a)+|\beta|} \| \partial_x^\alpha \partial_\xi^\beta \sigma(x, \xi) \|;$$

$$\sup_{x \in K, \xi \in \mathbb{R}^n} (1 + |\xi|)^{-\mathrm{Re}(a)+N+|\beta|} \left\| \partial_x^\alpha \partial_\xi^\beta \left(\sigma - \sum_{j=0}^{N-1} \psi(\xi) \sigma_{a-j} \right)(x, \xi) \right\|;$$

$$\sup_{x \in K, |\xi|=1} \| \partial_x^\alpha \partial_\xi^\beta \sigma_{m-j}(x, \xi) \|, \tag{3.6}$$

where K is any compact set in U.

This Fréchet structure on $CS^a(U, V)$ induces one on $\mathrm{Cl}^a(M, E)$. Indeed, given an atlas $(U_i, \phi_i)_{i \in I}$ on M and local trivialisations $E_{|U_i} \simeq U_i \times V, i \in I$ on the charts where V is the model fibre of E, an operator A in $\mathrm{Cl}^a(M, E)$ can be written $A = \sum_{i \in I} (A_i + R_i)$, where R_i is a smoothing operator with smooth kernel K_i with compact support in $U_i \times U_i$ and the operators A_i are classical pseudodifferential operators properly supported[4] in U_i which can therefore be written $\mathrm{Op}(\sigma_i)$ for some classical symbols σ_i in $CS^a(U_i, V)$, i varying in I. The semi-norms given by the supremum norm of each K_i and its partial derivatives combined with the semi-norms on $\mathrm{Op}(\sigma_i)$ induced by the ones described above on the symbols σ_i, provide a Fréchet topology on $\mathrm{Cl}^a(M, E)$.

Using a partition of the unity combined with the Fourier representation

$$Au(x) = \int_{\mathbb{R}^n} e^{i x \cdot \xi} \sigma(A)(x, \xi) \, \hat{u}(\xi) \, d\xi$$

of a classical pseudodifferential operator $A \in \mathrm{Cl}^a(U, E)$, the symbol $\sigma(A)(x, \xi)$ of which has compact support in the first variable contained in a trivialising open set $U \subset M$, one can show that the inclusion (see Remark 4)

$$\mathrm{Cl}^a(M, E) \subset \mathcal{L}\left(H^s(M, E); H^{s-a}(M, E) \right) \tag{3.7}$$

is continuous for any $(a, s) \in \mathbb{R}^2$ with respect to the operator norm on $\mathcal{L}(H^s(M, E); H^{s-a}(M, E))$ (see e.g. [Gi]). In particular, there is a continuous inclusion map:

$$i : \mathrm{Cl}^0(M, E) \to \mathcal{L}(L^2(M, E)), \tag{3.8}$$

where we have set $\mathcal{L}(H) := \mathcal{L}(H; H)$ and $L^2(M, E) = H^0(M, E)$.

---

[4] An operator A in $\mathrm{Cl}(M, E)$ is properly supported in U if whenever a smooth section u in $C^\infty(M, E)$ has support in some compact subset of U then its image Au also has support in some compact subset of U and the same property holds for the formal adjoint of A.

Proposition 3 $\mathrm{Cl}^0(M, E)$ is a Fréchet–Lie algebra.

Proof Since $\sigma(AB) \sim \sigma(A) \star \sigma(B)$ for two operators A, B with symbols $\sigma(A)$, $\sigma(B)$, the product map on $\mathrm{Cl}^0(M, E)$ is smooth as a consequence of the smoothness of the symbol product induced on $CS(U) \otimes \mathrm{End}(V)$ for any vector space V, by the symbol product (3.2) on $CS(U)$ and the composition on $\mathrm{End}(V)$. It follows that the bracket is a continuous bilinear map on $\mathrm{Cl}^0(M, E)$. $\qquad\square$

We shall also need a slight extension of the algebra of classical pseudodifferential operators, namely the algebra of log-polyhomogeneous operators

$$\mathrm{Cl}^{*,*}(M, E) = \bigcup_{a \in \mathbb{C}} \bigcup_{k \in \mathbb{Z}_{\geq 0}} \mathrm{Cl}^{a,k}(M, E), \tag{3.9}$$

where for $(a, k) \in \mathbb{C} \times \mathbb{Z}_{\geq 0}$, the set $\mathrm{Cl}^{a,k}(M, E)$ consists of pseudodifferential operators acting on $C^\infty(M, E)$, whose symbol is log-polyhomogeneous (see (3.4)) of order a and log-type k. Since we shall only make a marginal use of this algebra, we refer the reader to [L] for its detailed description.

Example 3.5 The logarithm $\log Q$ of an admissible classical pseudodifferential operator Q lies in $\mathrm{Cl}^{0,1}(M, E)$.

3.4 The group of invertible zero-order ψdos

The Lie algebra $\mathrm{Cl}^0(M, E)$ offers a natural generalisation of the algebra $\mathrm{End}(V)$ with V the typical fibre of E. The corresponding Lie group of invertible zero-order ψdos offers a natural generalisation of the group $\mathrm{GL}(V)$ of invertible linear transformations of the vector space V. It is contained in the larger group of invertible Fourier integral operators of order zero which in the case of a trivial vector bundle $E \simeq M \times \mathbb{C}$ has Lie algebra given by the algebra of first-order pseudodifferential operators [Schm]. This relies on foundational results by Adams, Ratiu and Schmid in [ARS] on the topological structure of the whole group of invertible Fourier integral operators. Here, we focus on invertible zero-order classical pseudodifferential operators.

Let

$$\mathrm{Cl}^{0,\times}(M, E) := \{A \in \mathrm{Cl}^0(M, E), \exists A^{-1} \in \mathrm{Cl}^0(M, E)\}$$

be the *group of invertible zero-order classical pseudodifferential operators* which is strictly contained in the intersection $\mathrm{Cl}^0(M, E) \cap \mathrm{Cl}^\times(M, E)$ where

$$\mathrm{Cl}^\times(M, E) = \{A \in \mathrm{Cl}(M, E), \exists A^{-1} \in \mathrm{Cl}(M, E)\}$$

is the group of invertible classical pseudodifferential operators.

Remark 7 It is useful to note that $\mathrm{Cl}^{\times}(M, E)$ acts on $\mathrm{Cl}^{a}(M, E)$ for any complex number a by the adjoint action defined for an operator P in $\mathrm{Cl}^{\times}(M, E)$ by the left action

$$\mathrm{Cl}^{a}(M, E) \to \mathrm{Cl}^{a}(M, E)$$

$$A \mapsto \mathrm{Ad}_{P} A := P A P^{-1} \tag{3.10}$$

and specifically on the algebra $\mathrm{Cl}^{0}(M, E)$.

Following [KM] we say that a Lie group $\mathcal{G}$ admits an *exponential map* if there exists a smooth map

$$\mathrm{Exp} : \mathrm{Lie}\,(\mathcal{G}) \to \mathcal{G}$$

such that $t \mapsto \mathrm{Exp}(t\,X)$ is a one-parameter subgroup with tangent vector X. Then $\mathrm{Exp}(0) = e_{\mathcal{G}}$ and Exp induces the identity map $D_{e}\mathrm{Exp} = \mathrm{Id}_{\mathrm{Lie}(\mathcal{G})}$ on the corresponding Lie algebra.

Regularity (in Milnor's sense [Mil]; see also [KM]) of a Lie group ensures the existence of a smooth exponential map (see [Mil], [GN], [N2]). A Lie group modelled on a locally convex space is a *regular Lie group* if, for each smooth curve $u : [0, 1] \to \mathfrak{g}$, there exists a smooth curve $\gamma_{u} : [0, 1] \to G$ (which is unique; see e.g. Lemma 38.3 in [KM]) which solves the initial value problem $\dot{\gamma} = \gamma \cdot u; \gamma(0) = 1_{G}$ with 1_{G} the identity element of G with smooth "evolution" map

$$C^{\infty}([0, 1], \mathfrak{g}) \to G$$

$$u \mapsto \gamma_{u}(1).$$

Example 3.6 Any Banach–Lie group is regular (see e.g. [KM]).

Example 3.7 Gauge groups $\mathcal{G}(P)$, introduced in Section 3.1, provide examples of Fréchet–Lie groups with exponential map.

Let us quote the following result which belongs to folklore knowledge.

Proposition 4 $\mathrm{Cl}^{0,\times}(M, E)$ is a regular Fréchet–Lie group with Lie algebra $\mathrm{Cl}^{0}(M, E)$ and admits an exponential map.

Proof Let us first show that the group $\mathcal{A}^{\times} = \mathrm{Cl}^{0,\times}(M, E)$ is an open subset in the Fréchet–Lie algebra $\mathcal{A} = \mathrm{Cl}^{0}(M, E)$, from which it will follow that it is a Fréchet manifold modelled on $\mathrm{Cl}^{0}(M, E)$.

Zero-order classical pseudodifferential operators, as well as being bounded (see inclusion (3.8)), turn out by (3.7) to correspond to all bounded classical pseudo-differential operators so that $\mathcal{A} = \mathrm{Cl}(M, E) \cap \mathcal{B}$ and

$$\mathcal{A}^{\times} = \{A \in \mathrm{Cl}(M, E), A^{-1} \in \mathrm{Cl}(M, E)\} \cap \mathcal{B}^{\times},$$

where we have set $\mathcal{B} := \mathcal{L}(L^2(M, E))$ and $\mathcal{B}^\times := \{B \in \mathcal{B}, \exists B^{-1} \in \mathcal{B}\}$. The algebra $\mathcal{B}$ being a Banach algebra, $\mathcal{B}^\times := \{B \in \mathcal{B}, \exists B^{-1} \in \mathcal{B}\}$ is open in $\mathcal{B}$ by the local inverse theorem.

We want to build an open neighbourhood in $\mathcal{A}^\times$ of any operator A in $\mathcal{A}^\times$. Since the operator A lies in $\mathcal{B}^\times$, there is a small neighbourhood U of A contained in $\mathcal{B}^\times$. The inclusion map $i : \mathcal{A} \to \mathcal{B}$ being continuous, $V := i^{-1}(U)$ yields an open neighbourhood in $\mathcal{A}$ of operators invertible as operators in $\mathcal{B}$ with inverse in $\mathcal{B}$. But if A lies in $\mathcal{A}$ then its inverse $B = A^{-1}$ in $\mathcal{B}$ also lies in $\mathcal{A}$. More precisely, if an operator A of order a in $\mathrm{Cl}^{0,\times}(M, E)$ has symbol $\sigma(A)$, its inverse A^{-1} has symbol $\tau(A)$ of order $-a$ with positively homogeneous components given by

$$\tau_{-a}(A) = (\sigma_a(A))^{-1},$$

$$\tau_{-a-j}(A) = -\tau_{-a}(A) \sum_{k+l+|\alpha|=j, l, j} \frac{(-i)^{|\alpha|}}{\alpha!} \partial_\xi^\alpha \sigma_{a-k}(A) \partial_x^\alpha \tau_{-a-l}(A).$$

We now need to check the smoothness of the product and inversion maps. By Proposition 3 we already know that the product map is smooth. The smoothness of the inversion $A \mapsto A^{-1}$ on $\mathrm{Cl}^{0,\times}(M, E)$ follows from its continuity by a result of Glöckner [Gl]. The continuity property can be seen on direct inspection from the expression of the symbol of the inverse in terms of the homogeneous components of the symbol of the original operator as described above.

Let us now check that the group $\mathcal{A}^\times = \mathrm{Cl}^{0,\times}(M, E)$ is regular using the regularity of the Banach–Lie group $\mathcal{B}^\times$ defined above. Given a smooth curve $U : [0, 1] \to \mathrm{Cl}^0(M, E)$, the first-order differential equation:

$$\Gamma_U^{-1} \dot{\Gamma}_U = U, \quad \Gamma(0) = I,$$

which can be viewed as a differential equation in $\mathcal{B}$ via the inclusion $\mathcal{A} \subset \mathcal{B}$, has a unique solution $\Gamma_U(t)$ in $\mathcal{B}^\times$ by the regularity of $\mathcal{B}^\times$. A similar argument to the one above shows that this solution actually lies in $\mathcal{A}^\times$ with symbol $\gamma_u(t)$ a solution of $\dot{\gamma} \sim \gamma \star u$, where $u(t)$ stands for the symbol of U. The homogeneous components $(\gamma_u(t))_{-j}$, $j \in \mathbb{N}$, of the symbol $\gamma_u(t)$ are defined inductively on j by a countable set of equations

$$\dot{\gamma}_0 = \gamma_0 u_0$$

$$\dot{\gamma}_{-1} = \gamma_{-1} u_0 + \gamma_0 u_{-1} - i \partial_\xi \gamma_0 \partial_x u_0$$

$$\dot{\gamma}_{-2} = \gamma_{-2} u_0 + \gamma_0 u_{-2} + \gamma_{-1} u_{-1} - i \partial_\xi \gamma_{-1} \partial_x u_0 - i \partial_\xi \gamma_0 \partial_x u_{-1}$$

$$+ \frac{1}{2} \sum_{|\alpha|=2} \partial_\xi^\alpha \gamma_0 \partial_x^\alpha u_0$$

$$\vdots$$

The "evolution" map $U \mapsto \Gamma_U(1)$ is smooth as can be seen from direct inspection on the symbol level. □

Remark 8 Since $\mathrm{Cl}^0(M, E)$ is a topological algebra $\mathcal{A}$ such that the group $\mathcal{A}^\times$ of invertible elements is open and the inverse map continuous, it is a *good topological algebra* in the sense of Bost [Bo].

The topology of $\mathrm{Cl}^{0,\times}(M, E)$ has been investigated in various contexts. Recall (see e.g. [Ka]) that the fundamental group $\pi_1(\mathrm{GL}_d(\mathbb{C}))$ is generated by the homotopy classes $[l]$ of the loops

$$l(t) = e^{2i\pi t\, p}$$

where $p : \mathbb{C}^d \to \mathbb{C}^d$ is a projector.

A similar statement holds for the fundamental group of $\mathrm{Cl}^{0,\times}(M, E)$ with the projectors p replaced by pseudodifferential operators introduced by Burak [Bu], later used by Wodzicki [W1] and further investigated by Ponge [Po1] which encode the *spectral asymmetry* of elliptic classical pseudodifferential operators:

$$P_{\theta,\theta'}(Q) := \frac{1}{2i\pi} \int_{C_{\theta,\theta'}} \lambda^{-1}\, Q\, (Q - \lambda)^{-1}\, d\lambda,$$

where

$$C_{\theta,\theta'} := \{\rho\, e^{i\theta}, \infty > \rho \geq r\} \cup \{r\, e^{it}, \theta \leq t \leq \theta'\} \cup \{\rho\, e^{i\theta'}, r \leq \rho < \infty\},$$

with Q an elliptic operator in $\mathrm{Cl}(M, E)$ of positive order and whereby r is chosen small enough so that no nonzero eigenvalue of Q lies in the disc $|\lambda| \leq r$. It turns out that $P_{\theta,\theta'}(Q)$ is a bounded ψdo projection on $L^2(M, E)$ (see [BL] and [Po1]) and either a zero-order pseudodifferential operator or a smoothing operator. For Q of order q with leading symbol $\sigma_L(Q)$, the leading symbol of $P_{\theta,\theta'}(Q)$ reads:

$$p_{\theta,\theta'}(\sigma_L(Q)) := \frac{1}{2i\pi} \int_{C_{\theta,\theta'}} \lambda^{-1}\, \sigma_L(Q)\, (\sigma_L(Q) - \lambda)^{-1}\, d\lambda.$$

The following proposition (see [KV1]; see also [LP]) shows that these pseudodifferential projectors generate the fundamental group $\pi_1\left(\mathrm{Cl}^{0,\times}(M, E)\right)$. Let $\mathrm{GL}_\infty(\mathcal{A})$ be the direct limit[5] of linear groups $\mathrm{GL}_n(\mathcal{A})$.

Proposition 5 $\pi_1\left(\mathrm{GL}_\infty\left(\mathrm{Cl}^0(M, E)\right)\right)$ is generated by the homotopy class of loops

$$L_{\theta,\theta'}^Q(t) := e^{2i\pi\, t\, P_{\theta,\theta'}(Q)}, \tag{3.11}$$

where $Q \in \mathrm{Cl}(M, E)$ is any elliptic operator with positive order.

---

[5] A natural embedding $\mathrm{GL}_n(\mathcal{A}) \to \mathrm{GL}_{n+1}(\mathcal{A})$ of an $n \times n$ matrix g belonging to $\mathrm{GL}_n(\mathcal{A})$ inside $\mathrm{GL}_{n+1}(\mathcal{A})$ is obtained by inserting g in the upper left corner, 1 in the lower right corner and filling the other slots in the last line and column with zeroes.

Proof We take the proof of Proposition 5 from [LP].[6] For any algebra $\mathcal{A}$, let $K_0(\mathcal{A})$ denote the group of formal differences of homotopy classes of idempotents in the direct limit $\mathrm{gl}_\infty(\mathcal{A})$ of matrix algebras $\mathrm{gl}_n(\mathcal{A})$.[7] When $\mathcal{A}$ is a *good topological* algebra [Bo], the Bott periodicity isomorphism:

$$K_0(\mathcal{A}) \longrightarrow \pi_1\,(GL_\infty(\mathcal{A}))$$
$$[P] \longmapsto \qquad e^{2i\pi t P} \qquad\qquad (3.12)$$

holds. Since for any vector bundle E over M, the algebra $\mathrm{Cl}^0(M, E)$ is a good topological algebra, applying (3.12) to $\mathcal{A} = \mathrm{Cl}^0(M, E)$ reduces the proof down to checking that $K_0(\mathrm{Cl}^0(M, E))$ is generated by idempotents $P_{\theta,\theta'}(Q)$.

The exact sequence

$$0 \longrightarrow \mathrm{Cl}^{-1}(M, E) \longrightarrow \mathrm{Cl}^0(M, E) \xrightarrow{\sigma_0} C^\infty(S^*M, p^*(\mathrm{End}E)) \longrightarrow 0, \quad (3.13)$$

where $p : S^*M \to M$ is the canonical projection of the cotangent sphere to the base manifold M and σ_0 is the leading symbol map, gives rise to a six-term exact sequence in K-theory:

$$K_0(\mathrm{Cl}^{-1}(M, E)) \quad \to K_0(\mathrm{Cl}^0(M, E)) \xrightarrow{\sigma_0} K_0(C^\infty(S^*M, p^*(\mathrm{End}E)))$$
$$\uparrow \mathrm{Ind} \qquad\qquad\qquad\qquad\qquad\qquad\qquad\qquad \downarrow 0$$
$$K_1(C^\infty(S^*M, p^*(\mathrm{End}E))) \xleftarrow{\sigma_0} K_1(\mathrm{Cl}^0(M, E)) \leftarrow K_1(\mathrm{Cl}^{-1}(M, E)) = 0.$$
$$(3.14)$$

On the other hand, on the grounds of results by Kontsevich and Vishik (see the proof of Lemma 4.2 in [KV1]), $K_0(C^\infty(S^*M, p^*(\mathrm{End}E)))$ is generated by the classes $p_{\theta,\theta'}(\sigma_L(Q))$, where as before $\sigma_L(Q)$ is the leading symbol of an elliptic operator $Q \in \mathrm{Cl}(M, E)$; this combined with the surjectivity of the map σ_0 in the diagram (3.14) yields the result. $\qquad\square$

Remark 9 When M reduces to a point $\{*\}$ then $\sigma_L(Q)$ reduces to a $d \times d$ matrix q with d the rank of E, and $\Pi_{\theta,\theta'}$ reduces to

$$\pi_{\theta,\theta'} = \frac{1}{2i\pi} \int_{C_{\theta,\theta',r}} \lambda^{-1}\, q\, (q - \lambda)^{-1}\, d\lambda,$$

which is a finite dimensional projector. Hence, the generators $[L^Q_{\theta,\theta'}(t)]$ reduce to generators $[l^q_{\theta,\theta'}(t)] = [e^{2i\pi\, t\, p_{\theta,\theta'}(q)}]$ built from projectors $\pi_{\theta,\theta'}(q)$.

---

[6] As pointed out to us by R. Ponge, the proof can probably be shortened using results of [BL] to show directly that $K_0(\mathrm{Cl}^0(M, E))$ is generated by idempotents $P_{\theta,\theta'}(Q)$.

[7] A natural embedding $\mathrm{gl}_n(\mathcal{A}) \to \mathrm{gl}_{n+1}(\mathcal{A})$ of an $n \times n$ matrix $1 \in \mathrm{gl}_n(\mathcal{A})$ in $\mathrm{gl}_{n+1}(\mathcal{A})$ is obtained inserting A in the upper left corner and filling the last line and column with zeroes.

The exact sequence (3.13) on the Lie algebra level lifts to the following exact sequence on the group level:

$$1 \to \mathcal{A}_{\mathrm{Id}}^{\times} \to \mathcal{A}^{\times} \to C^{\infty}\left(S^*M, \mathrm{Aut}(p^*E)\right) \to 1 \qquad (3.15)$$

where we have set

$$\mathcal{A}_{\mathrm{Id}}^{\times} := \{A \in \mathrm{Cl}^{0,\times}(M, E), \sigma_L(A) = \mathrm{Id}_{p^*E}\}$$
$$= \{\mathrm{Id}_{p^*E} + B \text{ invertible}, B \in \mathrm{Cl}^{-1}(M, E)\}.$$

The group $\mathcal{A}_{\mathrm{Id}}^{\times}$ turns out to be a classifying space for odd K-theory; its stable homotopy groups $\pi^k(\mathcal{A}_{\mathrm{Id}}^{\times})$ are exactly those given by Bott periodicity for the group $\mathrm{GL}^{\infty}(L^2(M, E))$ of invertible operators acting on $L^2(M, E)$ which differ from the identity by a finite rank operator or equivalently for the group $\mathrm{GL}_c(L^2(M, E))$ of invertible operators acting on $L^2(M, E)$ which differ from the identity by a compact operator, namely 0 for even k and $\mathbb{Z}$ for odd k (see Proposition 15.4 in [BW]). Higher stable homotopy groups of $\mathcal{A}^{\times}$ were derived in [Ro] from those of $\mathcal{A}_{\mathrm{Id}}^{\times}$ using an appropriate long exact sequence.

Theorem 3.8 ([Ro] Theorem 1) *The homotopy groups*

$$\pi^k\left(\lim_{n\to\infty} \mathrm{Cl}^{0,\times}(M, \underline{\mathbb{C}}^n)\right),$$

where the telescopic limit is defined as in [BT] via the inclusion of $\mathrm{Cl}^{0,\times}(M, \underline{\mathbb{C}}^n)$ *in* $\mathrm{Cl}^{0,\times}(M, \underline{\mathbb{C}}^{n+1})$, *are given by*

1. *for odd k by $K_0(C^{\infty}(S^*M))$,*
2. *for even k by $\ker\left[\mathrm{ind}_t \circ \delta : K^1(S^*M) \to \mathbb{Z}\right]$, where*

$$\delta : K^1(S^*M) \to K_c^0(T^*M) \simeq K^0\left(\overline{T^*M}, S^*M\right)$$

is the boundary morphism associated with the six-term exact sequence corresponding to the embedding of S^*M in the radial compactification $\overline{T^*M}$ of the cotangent vector bundle and where $\mathrm{ind}_t : K_c^0(T^*M) \to \mathbb{Z}$ is the Atiyah–Singer topological index.

Remark 10 The short exact sequence (3.13) leads to a long exact sequence for homotopy groups:

$$\cdots \pi_k\left(\mathcal{A}_{\mathrm{Id}}^{\times}\right) \to \pi_k\left(\mathcal{A}^{\times}\right) \xrightarrow{\sigma_0} \pi_k(\mathcal{B}^{\times}) \xrightarrow{\delta} \pi_{k-1}\left(\mathcal{A}_{\mathrm{Id}}^{\times}\right) \to \cdots$$
$$\cdots \to \pi_1\left(\mathcal{B}^{\times}\right) \xrightarrow{\delta} \pi_0\left(\mathcal{A}_{\mathrm{Id}}^{\times}\right) \to \pi_0(\mathcal{A}^{\times}) \xrightarrow{\sigma_0} \pi_0\left(\mathcal{B}^{\times}\right) \qquad (3.16)$$

which shows that the topology of $\mathcal{A}^{\times}$ is closely related to that of the gauge group

$$C^{\infty}\left(S^*M, p^*(\mathrm{Aut}(E))\right) = \{\sigma \in \mathcal{B} := C^{\infty}\left(S^*M, p^*(\mathrm{End}(E))\right) : \exists \sigma^{-1} \in \mathcal{B}\}.$$

Gauge groups of principal bundles play an important role in Yang–Mills theory and their homotopy groups are directly related to the homotopy groups of the moduli space of connections on the principal bundle. Homotopy groups of gauge groups for principal bundles over a closed orientable surface or a sphere were investigated by Wockel in [Woc] using a long exact homotopy sequence induced by evaluation maps which send an element of the gauge group to an element of the structure group of the principal bundle in which the connecting homomorphism is given by the Samelson product.

Part 2: Traces and central extensions

Having chosen $Cl^0(M, E)$ as a potential infinite dimensional substitute for the Lie algebra $gl_d(\mathbb{C})$, it remains to find linear forms on $Cl^0(M, E)$ as a substitute for the trace on matrices.

3.5 Traces on zero-order classical ψdos

The ordinary trace on matrices extends to a trace on smoothing operators:

$$tr_{L^2} : Cl^{-\infty}(M, E) \to \mathbb{C}$$

$$A \mapsto \int_M tr_x(k_A(x, x)) \, dx = \int_{T^*M} tr_x(\sigma(A)(x, \xi)) \, dx \, d\xi,$$

where k_A stands for the Schwartz kernel of A, $\sigma(A)$ for the symbol of A and tr_x for the fibrewise trace defined previously using the ordinary trace on matrices.

But it does not further extend to a trace on $Cl^0(M, E)$, i.e. to a linear form $\lambda : Cl^0(M, E) \to \mathbb{C}$ which vanishes on brackets

$$\partial\lambda(A, B) := \lambda([A, B]) = 0 \quad \forall A, B \in Cl^0(M, E).$$

If $E = M \times \mathbb{C}$ then, as was pointed out by M. Lesch [L], choosing an elliptic operator $A \in Cl^0(M)$ with nonzero Fredholm index and B a parametrix of A, the operators $I - BA$ and $I - AB$ are smoothing, which leads to the following contradiction

$$\lambda([A, B]) = tr_{L^2}(I - BA) - tr_{L^2}(I - AB) = ind(A) \neq 0.$$

A well known result by Wodzicki [W1] (see also [W2] and [Kas] for a review) and proved independently by Guillemin [Gu] gives the uniqueness (up to a multiplicative factor) of a trace on the whole algebra $Cl(M, E)$ of classical pseudodifferential operators.[8]

---

[8] Since then other proofs, in particular a homological proof on symbols in [BG] (see also [P2] for another alternative proof) and various extensions of this uniqueness result, were derived; see [FGLS] for a generalisation

Indeed, Wodzicki showed that, on a connected closed manifold of dimension $n > 1$, any trace on $\mathrm{Cl}(M, E)$ is proportional to the *noncommutative residue* defined as follows. The residue density at point $x \in M$

$$\omega_{\mathrm{res}}(A)(x) := \left(\int_{S_x^* M} \mathrm{tr}_x \left(\sigma_{-n}(A)(x, \xi) \right) \, d_S \xi \right) dx,$$

where $S_x^* M \subset T_x^* M$ is the cotangent unit sphere,[9] is globally defined so that the noncommutative residue

$$\mathrm{res}(A) := \int_M \omega_{\mathrm{res}}(A)(x) := \int_M \mathrm{res}_x(A) \, dx \qquad (3.17)$$

is well defined on $\mathrm{Cl}(M, E)$. In dimension 1, i.e. when $M = S^1$, there are two noncommutative residues according to which circle in the cosphere bundle $S^* S^1$ one chooses. Wodzicki's noncommutative residue actually generalises to higher dimensions the notion of residue previously introduced by Adler and Manin in the one dimensional case (see [Ad], [Man]).

Restricting to zero-order classical pseudodifferential operators allows for another type of trace, *leading symbol traces*, associated with any linear form τ on $C^\infty(S^* M)$ introduced in [PR1, PR2] in relation to Chern–Weil forms:

$$\mathrm{Tr}_0^\tau(A) := \tau(\mathrm{tr}_x(\sigma_0(A)(x, \xi))).$$

Whenever $\tau(1) \neq 0$ we set:

$$\mathrm{tr}_0^\tau(A) := \frac{\mathrm{Tr}_0^\tau(A)}{\tau(1)} = \frac{\tau\left(\mathrm{tr}_x\left(\sigma_0(A)(x, \xi)\right)\right)}{\tau(1)}. \qquad (3.18)$$

Theorem 3.9 [LP] *Let M have dimension larger than 1. Traces on the algebra $\mathrm{Cl}^0(M, E)$ are linear combinations of the noncommutative residue and leading symbol traces.*

We refer the reader to [LP] for a proof using a spectral sequence argument. It has since then been pointed out to us by R. Ponge that Wodzicki had already classified traces on zero-order operators in [W3] using a spectral sequence argument. Other proofs are since then available, using commutators and pseudodifferential operator

---

to manifolds with boundary, see [Schr] for a generalisation to manifolds with conical singularities (both of which prove uniqueness up to smoothing operators), see [L] for an extension to log-polyhomogeneous operators as well as for an argument due to Wodzicki to get uniqueness on the whole algebra of classical operators, see [Po2] for an extension to Heisenberg manifolds.

[9] Here dx stands for the volume form $dx_1 \wedge \cdots \wedge dx_n$ on the n-dimensional manifold M and $d_S\xi := \frac{d_S\xi}{(2\pi)^n}$ stands for the (normalised) volume measure on the cotangent unit sphere $S^* M$ induced by the canonical volume measure on the cotangent bundle $T^* M$ and $(\cdot)_{-n}$ denotes the degree $-n$ positively homogeneous component of the symbol.

kernels [Po3], Poisson brackets [LN] and references therein or Stokes' property in the spirit of [P2] (see also [P3]).

Remark 11 When M reduces to a point so that $n = 0$ and A is a matrix, then both res(A) and tr$_0^\tau(A)$ are proportional to the ordinary matrix trace.

3.6 Logarithms and central extensions

Logarithms of classical pseudodifferential operators are not classical unless the operator has order zero; they are nevertheless useful to build central extensions of the algebra of zero-order classical operators. Not every classical operator has a logarithm, hence the need to introduce a class of admissible operators.

Let us first recall that if the leading symbol $\sigma_L(A)$ of an invertible elliptic operator $A \in \text{Cl}(M, E)$ has no eigenvalue in a conical neighbourhood Λ_θ of a ray $L_\theta := \{z \in \mathbb{C}, \text{Arg}z = \theta\}$ with vertex 0, then there are at most a finite number of eigenvalues of the operator A in Λ_θ.[10]

Definition 3.10 We call an operator A in $\text{Cl}(M, E)$ *admissible* with *spectral cut* θ if

1. for every (x, ξ) in $T^*M \setminus \{0\}$, the leading symbol $\sigma_L(A)(x, \xi)$ has no eigenvalue in a conical neighbourhood Λ_θ of the ray L_θ,
2. A has no eigenvalue on the ray L_θ.

In particular such an operator is invertible and elliptic.

The *logarithm of an admissible operator* A in $\text{Cl}(M, E)$ with spectral cut θ is defined in terms of the derivative at $z = 0$ of its complex power [Se]:

$$\log_\theta A = \partial_z A_\theta^z |_{z=0},$$

where A_θ^z is the complex power of A defined using a Cauchy integral on a contour Γ_θ around the spectrum of A.

The logarithm $\log_\theta A$ is not far from being a zero-order classical pseudodifferential operator since its symbol is of the type $a \log |\xi| + \sigma^0$, where a is the order of A and σ^0 is a zero-order classical symbol with leading part given by the logarithm $\log_\theta(\sigma_L(A))$ of the leading symbol of A. It follows that $[\log_\theta A, B]$ is classical of order zero if B is, so that the derivation $\text{ad}_{\log_\theta A}$ induces an endomorphism of $\text{Cl}^0(M, E)$.

We now want to build central extensions of the Lie algebra $\text{Cl}^0(M, E)$, before which we first recall some basic definitions.

---

[10] A proof can be found in [Sh] which holds for differential operators; the result can nevertheless easily be extended to classical pseudodifferential operators (see e.g. footnote 6 in [KV1] and Proposition 1.4.1 in [D]).

Definition 3.11 A continuous $\mathbb{C}$-valued 2-*cocycle* on a complex topological Lie algebra L is a continuous skew-symmetric form $c : L \times L \to \mathbb{C}$ with the property:

$$c([u, v], w) + c([v, w], u) + c([w, u], v) = 0.$$

It is a *coboundary* whenever there is a continuous linear form $\lambda : L \to \mathbb{C}$ such that $c(u, v) = \lambda([u, v])$.

The *Lie algebra central extension* of L by c, denoted by $\tilde{L} := L \oplus_c \mathbb{C}$, stands for the topological Lie algebra with underlying topological vector space $L \times \mathbb{C}$ and bracket given by:

$$[(u, a), (v, b)] = ([u, v], c(u, v)) \quad \forall(u, v) \in L^2, \forall(a, b) \in \mathbb{C}^2.$$

Two central extensions defined by 2-cocycles c_1 and c_2 are equivalent whenever the cocycle $c_1 - c_2$ is a coboundary. A central extension with cocycle c is said to be trivial whenever c is a coboundary. The second cohomology group $H^2(L)$ of a topological Lie algebra L, i.e. the set of classes $[c]$ of 2-cocycles modulo equivalence corresponds to the set of nonequivalent central extensions.

A 2-cocycle $c(A, B) = T(A \, \delta B)$ can be built from a trace functional T and a derivation δ on the Lie algebra (see e.g. [Rog], [GR]).

Proposition 6 Let L be a complex topological associative algebra. Let $T : L \to \mathbb{C}$ be a continuous trace on L,[11] and let $\partial : L \to L$ be a continuous derivation on L.[12]

If the linear form T fulfils a Stokes' type property, i.e. provided $T \circ \partial = 0$ then

$$(u, v) \mapsto c(u, v) := T(u \, \partial v)$$

defines a 2-cocycle on L.

Proof The bilinear form c is skew-linear since

$$c(u, v) = T(u \, \partial v) = T(\partial(uv)) - T(\partial u \, v) = -T(v \, \partial u) = -c(u, v),$$

for any (u, v) in L^2.

Moreover, for any (u, v, w) in L^3 we have:

$$c([u, v], w) + c([v, w], u) + c([w, u], v)$$

$$= T([u, v] \, \partial w) + T([v, w] \, \partial u) + T([w, u] \, \partial v)$$

$$= -T(\partial[u, v] \, w) + T([v, w] \, \partial u) + T([w, u] \, \partial v)$$

$$= -T([\partial u, v] \, w) - T([u, \partial v] \, w) + T([v, w] \, \partial u) + T([w, u] \, \partial v),$$

---

[11] That is, a continuous linear form on L which vanishes on Lie brackets $T([u, v]) = 0$ for $u, v \in L$.
[12] That is, a continuous linear map such that $\partial(u \, v) = \partial u \, v + u \, \partial v$.

which vanishes. Here we used the cyclicity of the map T combined with "Stokes' property".   □

Remark 12 In particular, outer derivations on the Lie algebra can be useful to build central extensions.

We know from the previous section that the noncommutative residue res and leading symbol traces tr_τ^0 define traces on $\mathrm{Cl}^0(M, E)$; it is easy to see from their expression that they are both continuous with respect to (w.r.t.) the Fréchet topology.

On the other hand, for any admissible operator Q in $\mathrm{Cl}(M, E)$ of positive order with spectral cut θ, the linear map

$$\mathrm{ad}_{\log_\theta(Q)} : \mathrm{Cl}^0(M, E) \to \mathrm{Cl}^0(M, E)$$

$$A \mapsto [\log_\theta Q, A]$$

yields a continuous outer derivation on $\mathrm{Cl}^0(M, E)$. Since both the leading symbol trace (as can be seen from computing the leading symbol) and the Wodzicki residue (see e.g. [Ok]) vanish on brackets $[\log_\theta Q, A]$, we can build two types of Lie cocycles on $\mathrm{Cl}^0(M, E)$:

$$(A, B) \mapsto c_{\mathrm{res}}^{Q,\theta}(A, B) := \mathrm{res}\left(A\,\mathrm{ad}_{\log_\theta Q} B\right)$$

$$(A, B) \mapsto c_{0,\tau}^{Q,\theta}(A, B) := \mathrm{tr}_\tau^0\left(A\,\mathrm{ad}_{\log_\theta Q} B\right),$$

each of which gives rise to a central extension, namely a Lie algebra $\mathrm{Cl}(M, E) \times \mathbb{C}$ with two types of Lie brackets:

$$[(A, a), (B, b)]_{\mathrm{res}}^{Q,\theta} := \left([A, B],\ c_{\mathrm{res}}^{Q,\theta}(A, B)\right),$$

$$[(A, a), (B, b)]_{0,\tau}^{Q,\theta} := \left([A, B],\ c_{0,\tau}^{Q,\theta}(A, B)\right).$$

The first type of central extension has been the object of many investigations in mathematics and physics, starting from work of Kravchenko and Khesin [KK], who determined all outer derivations on the algebra $\mathcal{F}CS(S^1) = C^\infty(S^1) \otimes \mathbb{C}[\xi, \xi^{-1}]$ of formal classical pseudodifferential symbols on the circle S^1 equipped with the product

$$(F \circ G)(x, \xi) := \sum_{k=0}^{\infty} \frac{: \partial_\xi^k F(x, \xi)\, \partial_x^k G(x, \xi) :}{k!},$$

with the ordered product $:\ :$ defined by $: f(x)\xi^k g(x)\xi^l := f(x) g(x) \xi^{k+l}$. These outer derivations turn out to be of two types ad_x and $\mathrm{ad}_{\log \xi}$ and therefore lead to two types of cocycles $c_1(F, G) = \mathrm{Tr}(F [\log \xi, G])$ and $c_2(A, B) = \mathrm{Tr}(F [x, G])$ built from the Adler trace defined by $\mathrm{Tr}(F) := \int_{S^1} f_{-1}(x)\, dx$

(see [Ad]) for an element $F(x, \xi) = \sum_{k \in \mathbb{Z}} f_k(x) \xi^k$ of $\mathcal{F}CS(S^1)$. The restriction to the algebra $\text{Vect}(S^1)$ of smooth vector fields on S^1 of the non-trivial central extension of $\mathcal{F}CS(S^1)$ by c_1 via the natural embedding $\text{Vect}(S^1) \to \mathcal{F}CS(S^1)$, yields back the celebrated Virasoro algebra (see e.g. [Ro] for a historical review). Similar constructions were later carried out in higher dimensions by Radul [Ra], who gave his name to cocycles of the type $\text{res}(A [\log_\theta(Q), B])$ called *Radul cocycles*, the noncommutative residue playing in higher dimensions the role of the Adler trace in the one dimensional case; as we shall see in the next section, these central extensions turn out to be trivial. As we previously pointed out, in dimension 1, $M = S^1$, the cotangent sphere reduces $S^*M = S^*S^1$ to two circles at the north and south pole, thus leading to two types of residues res_+ and res_- according to which of these poles the evaluation of the degree -1 homogeneous component of the symbol is taken at; letting $D = -i\partial_x$ be the Dirac operator on the circle, we have $\text{res}_- = \text{res}_+ \circ F$ where we have set $F = (D + \pi_D)(|D + \pi_D|)^{-1}$ with π_D the orthogonal projection onto the kernel of D w.r.t. the natural L^2-inner product on $L^2(S^1)$.[13] Different linear combinations $\text{res}_{\alpha,\beta} := \alpha \, \text{res}_+ + \beta \, \text{res}_-$ of these residues can lead to non-equivalent cocycles of the type $c^Q_{\text{res}_{\alpha,\beta}} := \text{res}_{\alpha,\beta}(A [\log_\theta Q, B])$; whereas $c^Q_{\text{res}_{1,-1}}$ is a trivial cocycle, in contrast $c^Q_{\text{res}_{1,1}}$ with $Q = |D| + \pi_{|D|}$ induces on the loop algebra $C^\infty(S^1, \text{End}(E)) \subset \text{Cl}(S^1, E)$ the non-trivial central term of a Kac–Moody algebra.

A natural generalisation to higher odd dimensions is Mickelsson's *twisted Radul cocycle* $(A, B) \mapsto \text{res}(FA [\log_\theta Q, B])$, where F is defined as before by $F = (D + \pi_D)(|D + \pi_D|)^{-1}$, $Q = |D|$ with D a Dirac type operator; it leads to non-trivial central extensions and plays a role in gauge anomalies [Mick1, Mick2, CFNW, LMR]. Related cocycles are used in noncommutative geometry where the noncommutative residue is replaced by the Dixmier trace.

The second type of cocycle, which reads:

$$c^{Q,\theta}_{0,\tau}(A, B) = \text{tr}^0_\tau \left(A \, \text{ad}_{\log_\theta Q} B \right) = \tau \left(\text{tr} \left(\sigma_0(A \, \text{ad}_{\log_\theta Q} B) \right) \right)$$
$$= \tau \left(\text{tr} \left(\sigma_0(A) \, \text{ad}_{\sigma_0(\log_\theta Q)} \sigma_0(B) \right) \right) = \tau \left(c^{\sigma_L(Q),\theta}_\tau (\sigma_0(A), \sigma_0(B)) \right),$$

corresponds to the pull-back by the leading symbol map

$$\text{Cl}^0(M, E) \to C^\infty(S^*M, \text{End}(p^*E))$$

$$A \mapsto \sigma_0(A),$$

---

[13] Here we have used the fact recalled in Remark 4, that D being an essentially self-adjoint operator on a compact manifold has finite dimensional kernel orthogonal to its range.

of the cocycle

$$(\rho_1, \rho_2) \mapsto c_\tau^{\sigma_L(Q),\theta}(\rho_1, \rho_2) := \tau\left(\mathrm{tr}\left(\rho_1\left[\log_\theta \sigma_L(Q), \rho_2\right]\right)\right)$$

on $C^\infty(S^*M, \mathrm{End}(p^*E))$. As before, p^*E stands for the pull-back of E by the canonical projection $p : S^*M \to M$ and we have used the fact that the zero-order part of the symbol of $\log_\theta Q$ is $\log_\theta \sigma_L(Q)$.

Proposition 7 Given a continuous trace T on $\mathrm{Cl}^0(M, E)$, two invertible admissible operators Q_1 and Q_2 in $\mathrm{Cl}^0(M, E)$ with spectral cuts θ_1 and θ_2 give rise to equivalent central extensions. In particular, the cohomology classes corresponding to $c_{\mathrm{res}}^{Q,\theta}$ and $c_{0,\tau}^{Q,\theta}$ are independent of the choice of Q and of the choice of spectral cut θ.

Proof A modification of the spectral cut introduces an extra pseudodifferential projection which is a zero-order classical pseudodifferential operator (see e.g. [OP] and references therein), i.e. $\log_{\theta_1} Q - \log_{\theta_2} Q$ lies in $\mathrm{Cl}^0(M, E)$ for two different spectral cuts θ_1 and θ_2; $\frac{\delta_{\log_{\theta_1}} Q}{q} - \frac{\delta_{\log_{\theta_2}} Q}{q}$ and therefore becomes an inner derivation on $\mathrm{Cl}^0(M, E)$. Consequently, given a trace T on $\mathrm{Cl}^0(M, E)$, and setting $\Delta := \log_{\theta_1} Q - \log_{\theta_2} Q$, we have

$$T(A\,\delta_{\log_{\theta_1} Q} B) - T(A\,\delta_{\log_{\theta_2} Q} B) = T\left(A\left(\delta_{\log_{\theta_1} Q} - \delta_{\log_{\theta_2} Q}\right)(B)\right)$$
$$= T([A, \Delta B]) + T([\Delta, BA]) - T([A, B]\Delta)$$
$$= -\delta T(\Delta \cdot)(A, B).$$

Similarly, $\frac{\log Q_1}{q_1} - \frac{\log Q_2}{q_2}$ lies in $\mathrm{Cl}^0(M, E)$, where q_1 and q_2 are the orders of two different weights Q_1 and Q_2 respectively,[14] since the symbol of $\frac{\log Q_i}{q_i}$ is of the type $\log |\xi| + \frac{\sigma_{Q_i}^0}{q_i}$ for some zero-order symbol $\sigma_{Q_i}^0$. As a result, $\frac{\delta_{\log Q_1}}{q_1} - \frac{\delta_{\log Q_2}}{q_2}$ becomes an inner derivation on $\mathrm{Cl}^0(M, E)$. Consequently, given a trace T on $\mathrm{Cl}^0(M, E)$,

$$\frac{T(A\,\delta_{\log Q_1} B)}{q_1} - \frac{T(A\,\delta_{\log Q_2} B)}{q_2} = T\left(A\left(\frac{\delta_{\log Q_1}}{q_1} - \frac{\delta_{\log Q_2}}{q_2}\right)(B)\right)$$
$$= T([A, \Delta B]) + T([\Delta, BA]) - T([A, B]\Delta)$$
$$= -\delta T(\Delta \cdot)(A, B),$$

where we have now set $\Delta := \frac{\log Q_1}{q_1} - \frac{\log Q_2}{q_2}$.

In both cases Δ lies in $\mathrm{Cl}^0(M, E)$ and the map $A \mapsto T(\Delta A)$ defines a continuous linear form on $\mathrm{Cl}^0(M, E)$ from which we infer that the central extensions given by the two cocycles $c_1(A, B) = T(A\,\delta_{\log_{\theta_1} Q} B)$ and $c_2(A, B) = T(A\,\delta_{\log_{\theta_2} Q} B)$ on the one hand and $c_1(A, B) = T(A\,\delta_{\log Q_1} B)$ and $c_2(A, B) = T(A\,\delta_{\log Q_2} B)$ on the

---

[14] Which we assume to have same spectral cut, so we drop the explicit mention of the spectral cut.

other hand are equivalent. This holds in particular for $T = \text{res}$, the noncommutative residue, and $T = \text{tr}_0^\tau$, the leading symbol trace, which ends the proof. ☐

It would be interesting to classify all central extensions of the Lie algebra of zero-order classical pseudodifferential operators which, unlike central extensions of the gauge Lie algebra $C^\infty(M, \text{End}(E))$, have not to my knowledge been studied in a systematic way. Whereas central extensions of $C^\infty(M, \mathbf{g})$ for simple Lie algebras $\mathbf{g}$ are known to be classified by one-dimensional closed currents C on M giving rise to cocycles $\int_C \langle u, dv \rangle$ where $\langle \cdot, \cdot \rangle$ stands for the universal symmetric bilinear form on $\mathbf{g}$ (see Proposition 4.2.8 in [PS]), to my knowledge, the classification of central extensions of $\text{Cl}(M) \otimes \mathbf{g}$ has not yet been carried out.

A related issue is to determine all central extensions of the group $\text{Cl}^{0,*}(M, E)$ of invertible zero-order classical pseudodifferential operators. As well as the knowledge of the second Lie algebra cohomology of $\text{Cl}^0(M, E)$, integrating a Lie algebra extension to a group extension requires a good knowledge of the topology of the group $\text{Cl}^{0,*}(M, E)$ and in particular of its first and second homotopy groups. In the same way that the homomorphism (see e.g. Proposition 2 in [LP])

$$\text{per}_\lambda : \pi_1\left(\text{Cl}^{0,\times}(M, E)\right) \to \mathbb{C}, \quad \gamma \mapsto \int_\gamma \theta_\lambda,$$

where $\theta_\lambda : u \mapsto \lambda(\gamma^{-1} d\gamma(u))$ is the left-invariant 1-form on $\text{Cl}^{0,\times}(M, E)$ associated to λ, can be an obstruction to lifting a continuous 1-cocycle λ (i.e. a trace) on $\text{Cl}^0(M, E)$ to a continuous 1-cocycle (i.e. a multiplicative determinant) on $\text{Cl}^{0,*}(M, E)$, the period homomorphism

$$\text{per}_c : \pi_2\left(\text{Cl}^{0,\times}(M, E)\right) \to \mathbb{C}, \quad \Sigma \mapsto \int_\Sigma \Omega_c,$$

where Ω_c is the left invariant two-form on $\text{Cl}^{0,\times}(M, E)$ associated to a continuous 2-cocycle c, can be an obstruction to lifting c to a continuous cocycle on $\text{Cl}^{0,\times}(M, E)$ (see Theorem 7.12 in [N1]).

3.7 Linear extensions of the L^2-trace

Given an admissible operator Q in $\text{Cl}(M, E)$, let us build a linear form tr_θ^Q on $\text{Cl}^0(M, E)$ with coboundary given by $c_{\text{res}}^{Q,\theta}$ thus showing that the central extension given by a Radul cocycle is trivial. Such a linear form is a linear extension of the L^2-trace on smoothing operators.

Both the noncommutative residue and the leading symbol traces clearly vanish on smoothing operators so that neither of them extends the ordinary trace on smoothing operators. If we insist on building linear forms on $\text{Cl}^0(M, E)$ that extend the ordinary trace on smoothing operators, we need to drop the requirement that it vanishes on brackets. The linear forms we are about to describe are actually defined on the whole algebra $Cl(M, E)$.

We use the unique extension [MSS] (see also [P2] where the uniqueness of the noncommutative residue and the canonical trace are handled simultaneously), called the canonical trace, of the trace on smoothing operators to the set

$$Cl^{\notin \mathbb{Z}}(M, E) := \bigcup_{a \notin \mathbb{Z}} Cl^a(M, E) \qquad (3.19)$$

of *non-integer order operators* in $Cl(M, E)$. It was popularised by Kontsevich and Vishik in [KV1, KV2] even though it was known long before by Wodzicki and Guillemin and is defined as follows.

For any A in $Cl(M, E)$, for any x in M, one can infer from (3.3) (see e.g. [L]) that the integral $\int_{B_x(0,R)} tr_x \sigma(A)(x, \xi) d\xi$ of the fibrewise trace $tr_x \sigma(A)$ of its symbol $\sigma(A)$ over the ball $B_x(0, R)$ of radius R and centred at 0 in the cotangent bundle $T_x^* M$, has an asymptotic expansion in decreasing powers of R which is polynomial in $\log R$. Picking the constant term yields the cut-off regularised integral

$$\fint_{T_x^* M} tr_x \left(\sigma(A)(x, \xi) \right) d\xi := \mathrm{fp}_{R \to \infty} \int_{B_x(0,R)} tr_x \left(\sigma(A)(x, \xi) \right) d\xi$$

which clearly coincides with the ordinary integral on smoothing symbols.

Proposition 8 [KV1, KV2]   Whenever the operator A in $Cl(M, E)$ has non-integer order or has order $< -n$ then

$$\omega_{KV}(A)(x) := \left(\fint_{T_x^* M} tr_x \left(\sigma(A)(x, \xi) \right) d\xi \right) dx,$$

where as before dx stands for $dx_1 \wedge \cdots \wedge dx_n$, defines a global density on the n-dimensional manifold M so that the *canonical trace* [KV1, KV2] (see also [L] for an extension to log-polyhomogeneous operators):

$$TR(A) := \int_M \omega_{KV}(A)(x) := \int_M TR_x(A) \, dx$$

makes sense.[15] The canonical trace vanishes on brackets of non-integer order or of order $< -n$ [KV1, KV2] (see also [L]), i.e.

$$TR([A, B]) = 0 \; \forall A, B \in Cl(M, E) \text{ s.t. } [A, B] \in Cl^{\notin \mathbb{Z}}(M, E) \cup Cl^{< -n}(M, E).$$

---

[15] However, in general $\omega_{KV}(A)(x)$ is only locally defined and does not integrate over M to a well-defined linear form.

Remark 13 For any smoothing operator

$$\text{TR}(A) = \int_{T_x^* M} \text{tr}_x \left(\sigma(A)(x, \xi)\right) d\xi \, dx = \text{tr}(A)$$

so that the canonical trace extends the ordinary trace on smoothing operators.

Unfortunately, the operators one comes across in infinite dimensional geometry as well as in quantum field theory are typically integer order operators such as the Laplace operator, the Dirac operator, the Green operator, etc., so that we cannot implement the canonical trace on such operators.

In order to match the canonical trace with our needs in spite of this apparent discrepancy, we perturb the operators holomorphically $A \mapsto A(z)$ thereby perturbing their order $a \mapsto a(z)$ and we define regularised traces of such operators as finite parts at $z = 0$ of $\text{TR}(A(z))$.

To carry out this construction we need the notion of *holomorphic family of symbols* which we now recall.

Definition 3.12 Let Ω be a domain of $\mathbb{C}$. A family $(\sigma(z))_{z \in \Omega} \subset CS(U)$ is holomorphic when

1. the order $a(z)$ of $\sigma(z)$ is holomorphic on Ω;
2. for $(x, \xi) \in U \times \mathbb{R}^n$, the function $z \to \sigma(z)(x, \xi)$ is holomorphic on Ω and $\forall k \geq 0$, $\partial_z^k \sigma(z) \in S^{a(z)+\epsilon}(U)$ for any $\epsilon > 0$, with a uniform symbol estimate in z on compact subsets of Ω;
3. for any integer $j \geq 0$, the homogeneous symbol $\sigma(z)_{a(z)-j}(x, \xi)$ is holomorphic on Ω.

It leads to the following notion of *holomorphic family of classical pseudodifferential operators*.

Definition 3.13 A family $z \mapsto A(z)$ in $\text{Cl}^{a(z)}(M, E)$ of log-classical ψdos parametrised by a domain Ω of $\mathbb{C}$ is holomorphic if in each local trivialisation of E one has

$$A(z) = \text{Op}(\sigma(z)) + R(z)$$

with $\sigma(z)$ a holomorphic family of classical symbols of order $a(z)$[16] and $R(z)$ a smoothing operator with Schwartz kernel $R(z, x, y) \in C^\infty(\Omega \times M \times M, \text{End}(V))$ holomorphic in z where V is the model space of E.

---

[16] In applications the order is affine in z.

A holomorphic family of classical operators of holomorphic order $a(z)$ parametrised by Ω has integer order no larger than $-n$ on the set $\Omega \cap a^{-1}(\mathbb{Z} \cap [-n, \infty[)$. Outside that set, the canonical trace TR $(A(z))$ is therefore well defined.

Let us recall a result of Wodzicki, Guillemin and popularised by Kontsevich and Vishik in [KV1, KV2] which relates the complex residue of the canonical trace of a holomorphic family at a pole with the noncommutative residue of the family at this pole.

Theorem 3.14 [KV1, KV2], [L] *Let $z \mapsto A(z)$ be a holomorphic family of operators in $Cl^{a(z)}(M, E)$ parametrised by a domain Ω of $\mathbb{C}$. The map*

$$z \mapsto TR(A(z))$$

is meromorphic with poles of order 1 at points $z_j \in \Omega \cap a^{-1}([-n, +\infty[\cap \mathbb{Z})$ such that $a'(z_j) \neq 0$. Moreover, the pole at such a point z_j is proportional to the Wodzicki residue of $A(z_j)$:

$$\text{Res}_{z=z_j} TR(A(z)) = -\frac{1}{a'(z_j)} \text{res}\left(A(z_j)\right). \tag{3.20}$$

Definition 3.15 A holomorphic *regularisation scheme* on $Cl(M, E)$ is a linear map $\mathcal{R}$ which sends an operator A in $Cl(M, E)$ to a holomorphic family $\mathcal{R}(A)(z) = A(z)$ of operators in $Cl(M, E)$ such that $A(0) = A$ and $A(z)$ has holomorphic order $z \mapsto a(z)$, $a'(0)$ nonzero.

ζ- regularisation

$$\mathcal{R} : A \mapsto A(z) := A \, Q_\theta^{-z}, \tag{3.21}$$

with Q an admissible operator in $Cl(M, E)$ with positive order q and spectral cut θ, yields typical and very useful examples of holomorphic regularisations. Moreover, applying equation (3.20) to $A(z) = A \, Q_\theta^{-z}$ yields

$$\text{Res}_{z=0} TR(A \, Q_\theta^{-z}) = -\frac{1}{q} \text{res}(A).$$

On the basis of the results of the previous section, given a holomorphic regularisation $\mathcal{R} : A \mapsto A(z)$, we can pick the finite part in the Laurent expansion TR $(A(z))$ and set the following definition.

Definition 3.16 A holomorphic regularisation scheme $\mathcal{R} : A \mapsto A(z)$ on $Cl(M, E)$ induces a linear form:

$$tr^{\mathcal{R}} : Cl(M, E) \to \mathbb{C}$$

$$A \mapsto tr^{\mathcal{R}}(A) := \text{fp}_{z=0} TR(A(z))$$

called $\mathcal{R}$-*regularised trace*.[17] When $\mathcal{R}$ is a ζ-*regularisation* (3.21) determined by an admissible operator Q with spectral cut θ, the linear form $\mathrm{tr}^{\mathcal{R}}$ is called Q-*weighted trace* and denoted by tr_θ^Q.

The following theorem says that Radul cocycles are coboundaries of weighted traces.

Theorem 3.17 [MN], [CDMP] *Let $Q \in \mathrm{Cl}(M, E)$ be an admissible operator of positive order q with spectral cut θ. The coboundary of the Q-weighted trace reads:*

$$\partial \mathrm{tr}_\theta^Q (A, B) = \frac{1}{q} \mathrm{res}\left(A \left[\log_\theta Q, B\right] \right) = \frac{1}{q} c_{\mathrm{res}}^{Q,\theta}(A, B). \qquad (3.22)$$

for any two operators A and B in $\mathrm{Cl}(M, E)$.

Proof Using the vanishing of the canonical trace on non-integer order brackets we can write

$$\mathrm{TR}\left([A, B] Q_\theta^{-z}\right) = \mathrm{TR}\left(A B Q_\theta^{-z} - B A Q_\theta^{-z}\right)$$
$$= \mathrm{TR}\left(A B Q_\theta^{-z} - A Q_\theta^{-z} B\right)$$
$$= \mathrm{TR}(A [B, Q_\theta^{-z}]).$$

The family $C(z) := \frac{A[B, Q_\theta^{-z}]}{z} \in \mathrm{Cl}(M, E)$ is a holomorphic family of order $a - b - q z$ and $C(0) = -A [B, \log_\theta Q]$. By (3.20) applied to the holomorphic family $C(z)$ with $z_j = 0$ we get:

$$\mathrm{tr}_\theta^Q ([A, B]) = \mathrm{fp}_{z=0} \mathrm{TR}\left(A B Q_\theta^{-z} - B A Q_\theta^{-z}\right)$$
$$= \mathrm{fp}_{z=0} \mathrm{TR}\left(A [B, Q_\theta^{-z}]\right)$$
$$= \mathrm{Res}_{z=0} \mathrm{TR} \frac{\left(A [B, Q_\theta^{-z}]\right)}{z}$$
$$= -\frac{1}{q} \mathrm{res}(A [B, \log_\theta Q]). \qquad \square$$

The following result measures the difference between the regularised trace and the (generally non-existing) canonical trace.

Theorem 3.18 *Let $\mathcal{R} : A \mapsto A(z)$ be a holomorphic regularisation with non-constant order $a(z)$ affine in z.*

The linear form $\mathrm{tr}^{\mathcal{R}}$ extends the usual trace defined on operators of order smaller than $-n$ as well as the canonical trace TR defined on non-integer order operators to ψdos of all orders.

---

[17] It carries this name because it extends the ordinary trace on smoothing operators and in spite of the fact that it does not vanish on brackets, as we shall soon see.

Moreover [PS1],

$$\mathrm{tr}^{\mathcal{R}}(A) = \int_M dx \left(\mathrm{TR}_x(A) - \frac{1}{a'(0)} \mathrm{res}_x \left(A'(0) \right) \right) \tag{3.23}$$

where, in spite of the fact that $A'(0)$ is no longer expected to be classical,[18] its residue density is defined in a similar manner:

$$\mathrm{res}_x(A'(0)) := \int_{S_x^* M} \mathrm{tr}_x(\sigma_{-n}(A'(0)))(x, \xi)\, d\xi.$$

If $\mathcal{R}$ is ζ-regularisation with weight Q in $\mathrm{Cl}(M, E)$ of positive order q and spectral cut θ, then $A'(0) = -A \log_\theta Q$. Formula (3.23) reads:

$$\mathrm{tr}_\theta^Q(A) = \int_M dx \left(\mathrm{TR}_x(A) - \frac{1}{q} \mathrm{res}_x \left(A \log_\theta Q \right) \right). \tag{3.24}$$

Corollary 1 *With the same notation as in Theorem 3.18, if the residue density $\mathrm{res}_x \left(A'(0) \right)$ vanishes, then $\mathrm{TR}_x(A)\, dx$ defines a global density and*

$$\mathrm{tr}^{\mathcal{R}}(A) = \mathrm{TR}(A)$$

independently of the choice of regularisation scheme.

Let $\mathcal{R}$ be ζ-regularisation with weight Q in $\mathrm{Cl}(M, E)$ of positive order q and spectral cut θ. If the residue density $\mathrm{res}_x \left(A \log Q \right)$ vanishes, then $\mathrm{TR}_x(A)\, dx$ defines a global density and

$$\mathrm{tr}^Q(A) = \mathrm{TR}(A)$$

independently of the choice of weight Q.

The case of operators of order $-n$ is particularly interesting.

Proposition 9 Let M be a closed Riemannian manifold and $E \to M$ a vector bundle over M. Provided[19] there is an admissible operator Q in $\mathrm{Cl}(M, E)$ with leading symbol $\sigma_L(Q)(x, \xi) = |\xi|^q$, where $q > 0$ is the order of Q, the canonical trace TR is well defined on $\mathrm{Cl}^{-n}(M, E)$ and

$$\mathrm{TR}(A) = \mathrm{tr}^Q(A) \quad \forall A \in \mathrm{Cl}^{-n}(M, E) \tag{3.25}$$

uniquely extends the ordinary trace on the algebra $Cl^{<-n}(M, E)$ of classical pseudodifferential operators of order $< -n$, where n is the dimension of M.

---

[18] It is log-polyhomogeneous of log type 1 [PS1], meaning by this that the asymptotic expansion (3.3) might present a logarithmic divergence $\log |\xi|$ in $|\xi|$ as $|\xi| \to \infty$.

[19] As will become clear from the examples below, this condition is fulfilled whenever E is Hermitian, in which case Q corresponds to a generalised Laplacian built from the Levi-Civita and a unitary connection on the bundle.

Proof An operator A in $\mathrm{Cl}(M, E)$ can be embedded in the holomorphic family $A(z) := A \, Q^{-z}$ in $\mathrm{Cl}(M, E)$ with non-constant order $-qz$ where q is the order of Q. In particular, $A(0) = A$. On the one hand, the symbol of $\log Q$ differs from $q \log |\xi| + \log \sigma_L(Q)(\xi/|\xi|)$ by a symbol of negative order. On the other hand, $\sigma_L(Q)(x, \xi) = |\xi|^q$ implies that $\sigma_L(Q)(x, \xi/|\xi|) = 1$. Thus, for A of order $-n$ we have

$$\mathrm{res}_x\left(A'(0)\right) = -\int_{S_x^*M} \mathrm{tr}_x\left(\sigma_{-n}\left(A \log Q\right)\right)(x, \xi) \, d\xi$$

$$= -\int_{S_x^*M} \mathrm{tr}_x\left(\sigma_{-n}(A \log Q)(x, \xi)\right) d\xi$$

$$= -\int_{S_x^*M} \mathrm{tr}_x\left(\sigma_{-n}(A) \, \sigma_0(\log Q)(x, \xi)\right) d\xi$$

$$= -\int_{S_x^*M} \mathrm{tr}_x\left(\sigma_{-n}(A)(x, \xi) \, \log \sigma_L(Q)(x, \xi)\right) d\xi$$

$$= 0.$$

Hence, by Corollary 1, the canonical trace $\mathrm{TR}(A)$ of an operator A in $\mathrm{Cl}(M, E)$ of order $-n$ is well defined and $\mathrm{TR}(A) = \mathrm{tr}^Q(A)$. This extends TR to a linear form on $\mathrm{Cl}^{-n}(M, E)$. This linear extension defines a trace since the bracket of two operators A and B of order $-n$ is an operator of order $-2n$ and

$$\mathrm{TR}([A, B]) = \mathrm{tr}^Q([A, B]) = -\frac{1}{q}\mathrm{res}(A \, [B, \log Q]) = 0,$$

for $A \, [B, \log Q]$ has order no larger than $-2n < -n$. $\qquad\square$

Let us describe a first rather trivial geometric instance of such a situation.

Example 3.19 Let M be a Riemannian manifold, $E = M \times \mathbb{C}$ the trivial bundle, Δ the Laplace–Beltrami operator and $Q = \Delta + 1$, then for any smooth function f on M, the canonical trace

$$\mathrm{TR}(\mathcal{M}_f (\Delta + 1)^{-\frac{n}{2}}) = \mathrm{fp}_{z=0}\mathrm{TR}\left(\mathcal{M}_f (\Delta + 1)^{-\frac{n+z}{2}}\right) \qquad (3.26)$$

is well defined, where $\mathcal{M}_f$ denotes the multiplication operator associated with f.

This easily generalises to the following geometric setup.

Example 3.20 Let M be a closed Riemannian smooth manifold and E a smooth Hermitian vector bundle over M. A connection ∇^E on E combined with the Levi-Civita connection ∇^M on M gives rise to a connection $\nabla^{T^*M \otimes E}$ on the tensor bundle

$T^*M \otimes E$ from which we build a generalised Laplacian:

$$\Delta^E := -\mathrm{tr}\left(\nabla^{T^*M \otimes E}\nabla^E\right).$$

It is a second-order elliptic differential operator. If the connection is Hermitian, it is moreover a non-negative differential operator for the Hermitian product on sections of E induced by the metric on M and the Hermitian structure on E. The underlying manifold M being closed, the operator Δ^E has finite dimensional kernel $\mathrm{Ker}(\Delta^E)$; let π_E denote the orthogonal projection onto this kernel. We can therefore build an invertible operator $Q = \Delta^E + \pi_E$ which yields an admissible operator of order 2.

By Proposition 9, TR defines a trace on $\mathrm{Cl}(M, E)$ and

$$\mathrm{TR}(A) = \mathrm{tr}^{\Delta^E + \pi_E}(A) \quad \forall A \in \mathrm{Cl}^{-n}(M, E).$$

A second geometric instance arises in the context of loop groups.

Example 3.21 Let G be a semi-simple compact Lie group with a negative definite Killing form and such that the adjoint representation ad on the Lie algebra $\mathrm{Lie}(G)$ is antisymmetric for this bilinear form. Let $\mathcal{G}$ be an infinite dimensional Hilbert–Lie group with Lie algebra given by the H^s-Sobolev completion $H^s(S^1, \mathrm{Lie}(G))$ of the space $C^\infty(M, \mathrm{Lie}(G))$ of smooth maps on M with values in the Lie algebra. It was shown by D. Freed [F] that the curvature Ω^s of the Levi-Civita connection on $\mathcal{G}$ equipped with the H^s-left invariant metric has Ricci tensor $R^s(X, Y) : Z \mapsto \Omega^s(Z, X)Y$ which lies in $\mathrm{Cl}(S^1, \mathrm{Lie}(G))$ and has order max $(-1, -2s)$ so that it is of order -1 for $s = \frac{1}{2}$. It was later shown in [La] (Theorem 1) that the curvature takes values in classical pseudodifferential operators of order at most -2 for any Sobolev parameter $s > 1$ and that this bound is exact in a specific example. D. Freed further showed that $\mathrm{tr}_{\mathrm{Lie}(G)}(R^s(X, Y))$ has order $-2q = -2$ with $q = \min(1, 2s) = 1$. When $s = \frac{1}{2}$, $\mathrm{tr}_{\mathrm{Lie}(G)}(R^s(X, Y))$ therefore defines a trace-class operator whose trace gives rise to Freed's conditioned trace:

$$\mathrm{Tr}_{\mathrm{cond}}(R^{\frac{1}{2}}) := \mathrm{Tr}\left(\mathrm{tr}_{\mathrm{Lie}(G)}(R^{\frac{1}{2}}(X, Y))\right). \tag{3.27}$$

In [CDMP], Freed's conditioned trace is viewed as a weighted trace using a scalar weight $Q = (\Delta + \pi_\Delta) \otimes 1_{\mathrm{Lie}(G)}$, where Δ is the Laplace–Beltrami operator on S^1 and π_Δ is the orthogonal projection onto its kernel.

We infer from Proposition 9, that the canonical trace $\mathrm{TR}(R^{\frac{1}{2}})$ is well defined and

$$\mathrm{TR}(R^{\frac{1}{2}}) = \mathrm{tr}^Q(R^{\frac{1}{2}}) = \mathrm{Tr}_{\mathrm{cond}}(R^{\frac{1}{2}}). \tag{3.28}$$

Part 3: Singular Chern–Weil classes

We construct Chern–Weil classes in infinite dimensions following the usual finite dimensional scheme up to the fact that we use the only traces on $\mathrm{Cl}^0(M, E)$, namely linear combinations of the Wodzicki residue and leading symbol traces, as a substitute for the usual trace on matrices. Let us first recall the finite dimensional setup.

3.8 Chern–Weil calculus in finite dimensions

Let $E \to X$ be a vector bundle over a d-dimensional manifold X with structure group G a subgroup of the linear group $\mathrm{Gl}_d(\mathbb{C})$ and let $\mathcal{A} = \mathrm{End}(E)$ be the bundle of endomorphisms of E over X. Let $\Omega(X, \mathcal{A})$ denote the algebra of exterior forms on X with values in $\mathcal{A}$ equipped with the product induced from the wedge product on forms and the product in $\mathcal{A}$. If σ is a section of E over X and α lies in $\Omega^k(X, \mathcal{A})$ then $\alpha(\sigma)$ lies in $\Omega^k(X, E)$.

If ∇ is a connection on E then its *curvature* $\Omega = \nabla^2$ lies in $\Omega^2(X, \mathcal{A})$. More generally, if $\mathcal{C}(E)$ is the space of connections on E, to an analytic map $f(z) = \sum_{i=0}^{\infty} \frac{f^{(i)}(0)}{i!} z^i$ on $\mathbb{C}$ we assign a map

$$f : \mathcal{C}(E) \to \Omega(X, \mathcal{A})$$

$$\nabla \mapsto f(\nabla^2) = \sum_{i=0}^{\infty} \frac{f^{(i)}(0)}{i!} \nabla^{2i}.$$

Remark 14 This sum is actually finite since $\nabla^{2i} = 0$ for any i larger than $\frac{d}{2}$.

The connection ∇ extends to a map

$$C^{\infty}(X, TX) \times \Omega(X, \mathcal{A}) \to \Omega(X, \mathcal{A})$$

$$(U, \alpha) \mapsto \left(\sigma \mapsto [\nabla_U, \alpha](\sigma) := \nabla_U(\alpha(\sigma)) + (-1)^{|\alpha|+1}\alpha(\nabla_U\sigma)\right).$$

Here σ stands for a section of E over X and $|\alpha|$ for the degree of the form.

The trace $\mathrm{tr} : \mathrm{gl}_d(\mathbb{C}) \to \mathbb{C}$ on the algebra $\mathrm{gl}_d(\mathbb{C})$ of $d \times d$ matrices with complex coefficients extends to a trace on $\mathrm{End}(E)$ by

$$\mathrm{tr} : \mathrm{End}(E) \to X \times \mathbb{C}$$

$$(x, A) \mapsto (x, \mathrm{tr}(A))$$

where (x, A) is an element of $\mathrm{End}(E)_{|_U} \simeq U \times \mathrm{gl}_d(\mathbb{C})$ in a local trivialisation of E over a local chart U containing x and where tr on the right-hand side is the ordinary trace on matrices. This is a well-defined bundle morphism since

$$\mathrm{tr}(C^{-1} A C) = \mathrm{tr}(A) \quad \forall C \in \mathrm{Gl}_d(\mathbb{C}), \forall A \in \mathrm{gl}_d(\mathbb{C}). \tag{3.29}$$

Similarly, to a form $\alpha \in \Omega(X, \mathcal{A})$ locally written $\alpha(x) = A(x)\,dx_1 \wedge \cdots \wedge dx_n$ corresponds a form $\mathrm{tr}(\alpha)(x) := \mathrm{tr}(A(x))\,dx_1 \wedge \cdots \wedge dx_n$ in $\Omega(X)$.

From the fact that the trace tr obeys the following properties

$$[d, \text{tr}](\alpha) := d\,\text{tr}(\alpha) - \text{tr}([\nabla, \alpha]) = 0 \quad \forall \alpha \in \Omega(X, \mathcal{A}) \qquad (3.30)$$

and

$$\partial\text{tr}(\alpha, \beta) := \text{tr}\left(\alpha \wedge \beta + (-1)^{|\alpha||\beta|}\beta \wedge \alpha\right) = 0 \quad \forall \alpha, \beta \in \Omega(X, \mathcal{A}), \qquad (3.31)$$

we infer the subsequent useful lemma.

Lemma 3.22 *For any α in $\Omega(X, \mathcal{A})$*

$$[\nabla, \text{tr}](\alpha) := d\,\text{tr}(\alpha) - \text{tr}([\nabla, \alpha]) = 0. \qquad (3.32)$$

Proof In a local chart above an open subset U of X,

$$[\nabla, \alpha] = d\alpha + \theta \wedge \alpha + (-1)^{|\alpha|+1}\alpha \wedge \theta$$

for some 1-form $\theta \in \Omega^1(U, \mathcal{A})$ so that we can write

$$\begin{aligned}
[\nabla, \text{tr}](\alpha) &= d\,\text{tr}(\alpha) - \text{tr}([\nabla, \alpha]) \\
&= d\,\text{tr}(\alpha) - \text{tr}\left(d\alpha + \theta \wedge \alpha + (-1)^{|\alpha|}\alpha \wedge \theta\right) \\
&= -\text{tr}\left(\theta \wedge \alpha + (-1)^{|\alpha|}\alpha \wedge \theta\right) \quad \text{by (3.30)} \\
&= 0 \quad \text{by (3.31)}.
\end{aligned}$$

Combining this lemma with the *Bianchi identity*

$$[\nabla, \nabla^2] = 0. \qquad (3.33)$$

leads to closed *Chern–Weil forms*.   □

Proposition 10 For any analytic function f on $\mathbb{C}$, the form $\text{tr}\left(f(\nabla^2)\right)$ is closed with de Rham cohomology class independent of the choice of connection.

Proof It is sufficient to carry out the proof for monomials $f(x) = x^i$, in which case we have:

$$\begin{aligned}
d\,\text{tr}\left(f(\nabla^2)\right) &= [\nabla, \text{tr}]\left(f(\nabla^2)\right) + \text{tr}\left([\nabla, f(\nabla^2)]\right) \\
&= \text{tr}\left([\nabla, \nabla^{2i}]\right) \quad \text{by (3.32)} \\
&= \sum_{j=0}^{i} \text{tr}\left([\nabla, \nabla^2]\nabla^{2(i-1)}\right) \\
&= 0 \quad \text{by (3.33)},
\end{aligned}$$

which proves the closedness of $\text{tr}(f(\nabla^2))$.

Let ∇_t, t in $\mathbb{R}$ be a smooth one-parameter family of connections on E. Its derivative with respect to t is a 1-form $\dot{\nabla}_t = \dot{\theta}_t$ in $\Omega^1(X, \mathcal{A})$. Applying (3.30) to $X = \mathbb{R}$ yields

$$\left[\frac{d}{dt}, \text{tr}\right] = \frac{d}{dt} \circ \text{tr} - \text{tr} \circ \frac{d}{dt} = 0$$

and hence

$$\frac{d}{dt}\left(\text{tr}\left(f(\nabla_t^2)\right)\right) = \text{tr}\left(\frac{d}{dt}\nabla_t^{2i}\right) = \sum_{j=0}^{i} \text{tr}\left(\frac{d}{dt}\left(\nabla_t^2\right) \nabla_t^{2(i-1)}\right)$$

$$= \sum_{j=0}^{i} \text{tr}\left(\left(\nabla_t \dot{\nabla}_t + \dot{\nabla}_t \nabla_t\right) \nabla_t^{2(i-1)}\right) = \sum_{j=0}^{i} \text{tr}\left(\left[\nabla_t, \dot{\nabla}_t\right] \nabla_t^{2(i-1)}\right)$$

$$= \sum_{j=0}^{i} \text{tr}\left(\left[\nabla_t, \dot{\nabla}_t \nabla_t^{2(i-1)}\right]\right) \quad \text{by (3.33)}$$

$$= d \sum_{j=0}^{i} \text{tr}\left(\dot{\nabla}_t \nabla_t^{2(i-1)}\right) \quad \text{by (3.32)}.$$

The variation $\frac{d}{dt}(\text{tr}(f(\nabla_t^2)))$ is therefore exact and the de Rham class of $\text{tr}(f(\nabla_t^2))$ is independent of the parameter t. □

Applying these results to $f(x) = x^j$ yields the following important result.

Corollary 2 *The forms* $\text{tr}(\nabla^{2j})$ *are closed with de Rham cohomology class independent of the choice of connection* ∇, *called the ith Chern class.*

3.9 A class of infinite dimensional vector bundles

We consider a class of infinite dimensional vector bundles inspired from the geometric setup of index theory [B], [BGV], which also encompasses the tangent bundle to the space $C^\infty(M, N)$ of smooth maps from a manifold N to a manifold M. This class consists of Fréchet vector bundles with structure group $\text{Diff}(M, E)$ and typical fibre $C^\infty(M, E)$ for some reference finite rank vector bundle $E \to M$ over a closed manifold M.

Proposition 11 Given a smooth, locally trivial fibre bundle $\pi : \mathbb{M} \to X$ of smooth manifolds with fibre $\mathbb{M}_x$ over a point x in a Fréchet manifold X, and with typical fibre a closed manifold M and structure group $\text{Diff}(M)$, and given a smooth vector bundle $\mathbb{E} \to \mathbb{M}$ over $\mathbb{M}$ modelled on the vector bundle $E \to M$, then the infinite dimensional Fréchet vector bundle $\pi_*\mathbb{E} \to X$ modelled on $C^\infty(M, E)$, with fibre over $x \in X$:

$$(\pi_*\mathbb{E})_x = C^\infty(\mathbb{M}_x, \mathbb{E}_{|_{\mathbb{M}_x}}),$$

where $\mathbb{M}_x := \pi^{-1}(\{x\})$ is the fibre of $\mathbb{M}$ over x, has structure group $\mathrm{Diff}(M, E)$.

Proof Above an open subset U of X such that

$$\mathbb{E}|_{\pi^{-1}(U)} \simeq (U \times E \to U \times M),$$

where E is a finite rank vector bundle corresponding to the model fibre of $\mathbb{E} \to \mathbb{M}$, we have the following isomorphism:

$$C^\infty(U, \pi_*\mathbb{E}) \simeq C^\infty(U, C^\infty(M, E)).$$

A change of local trivialisation of the locally trivial fibre bundle $\mathbb{M}_{|_U} \simeq U \times M$ induces a diffeomorphism $f : M \to M$ in $\mathrm{Diff}(M)$, whereas a change of local trivialisation of the finite rank vector bundle $\mathbb{E}_{|_U \times M} \simeq U \times E$ induces a transformation in $\mathrm{Aut}(E)$. The transition maps therefore have values in the group $\mathrm{Diff}(M, E)$ of vector bundle automorphisms of $E \to M$ over a diffeomorphism of M. Thus, the structure group is $\mathrm{Diff}(M, E)$.   □

Remark 15 When M reduces to a point $\{*\}$, then $\pi_*\mathbb{E} \to X$ reduces to a finite rank vector bundle over X, modelled on some vector space V with transition maps in $\mathrm{Gl}(V)$.

Example 3.23 In the context of the family index theorem (see e.g. [BGV]), $\mathbb{E} = \mathbb{F} \otimes |\Lambda_\pi|$ for some vector bundle $\mathbb{F} \to \mathbb{M}$, where Λ_π is the vertical density bundle which, when restricted to the fibres of $\mathbb{M}$, may be identified with the bundle of densities along the fibre.

Example 3.24 Given a Riemannian manifold N, the space $X := C^\infty(M, N)$ of smooth maps from M to N is a Fréchet manifold with tangent space at point γ given by $C^\infty(M, \gamma^*TN)$. The tangent bundle $TC^\infty(M, N)$ can therefore be realised as $\pi_*\mathbb{E}$ where $\pi : \mathbb{M} \to X$ is the trivial fibre bundle with fibre M and $\mathbb{E}$ is the vector bundle over $X \times M$ with fibre at $(\gamma, m) \in C^\infty(M, N) \times M$ given by the vector space

$$\mathbb{E}_{(\gamma,m)} = \gamma^* T_{\gamma(m)} N \quad \text{so that} \quad \pi_*\mathbb{E}_\gamma = C^\infty(M, \gamma^* T_\gamma N).$$

In passing, note that for maps γ in a connected component of $C^\infty(M, N)$, the bundles γ^*TN are isomorphic, so that the tangent spaces $C^\infty(M, \gamma^*TN)$ are also isomorphic. The structure group reduces here to $\mathrm{Aut}(M, N)$.

Example 3.25 We now specialise to mapping groups $C^\infty(M, G)$ with G a finite dimensional Lie group. The left action $l_g : y \mapsto g \cdot y$ on G induces a left action $L_g : \gamma \mapsto g \cdot \gamma$ on $C^\infty(M, G)$ and a vector field $V(\gamma)$ in $C^\infty(M, \gamma^* T_\gamma G)$ is left-invariant if $(L_g)_* V(\gamma) = V(g \cdot \gamma)$ for all maps γ in $C^\infty(M, G)$. Left-invariant vector fields on $C^\infty(M, G)$ can be identified with elements of the Lie algebra $C^\infty(M, \mathrm{Lie}(G))$. $C^\infty(M, G)$ is a Fréchet–Lie group with structure group $C^\infty(M, \mathrm{Aut}(G))$.

Example 3.26 The group $\mathrm{Diff}(M)$ is a Fréchet manifold as an open subset of the Fréchet space $C^\infty(M, M)$. On a connected component of the identity, its tangent space $f^* TM$ is isomorphic to TM. The tangent bundle $\bigcup_{f \in \mathrm{Diff}(M)} C^\infty(M, f^* TM)$ can then be realised as a bundle $\pi_* \mathbb{E} \to X$ with $X = \mathrm{Diff}(M)$ and where $\mathbb{E}$ is a vector bundle over the trivial fibre bundle $\mathbb{M} = X \times M$ with fibre above (f, M) given by the bundle $f^* TM$.

Definition 3.27 With the notation of Proposition 11, we call an *admissible vector bundle* any Fréchet vector bundle of the type $\pi_* \mathbb{E} \to X$, where $\pi : \mathbb{E} \to \mathbb{M}$ is a vector bundle over a smooth fibration $\mathbb{M} \to X$ of manifolds. A Fréchet manifold whose tangent bundle is admissible is called an admissible manifold.

Note that the vector bundles and manifolds considered in the above examples are all admissible.

3.10 Frame bundles and associated ψdo-algebra bundles

Associated to an admissible vector bundle $\pi_* \mathbb{E} \to X$ is a frame bundle $\mathrm{Diff}(\mathbb{M}, \mathbb{E})$, whose fibre over the point $x \in X$ is given by:

$$\mathrm{Diff}(\mathbb{M}, \mathbb{E})_x := \{\Phi_x = (\alpha_x, f_x), \ \alpha_x : E \to \mathbb{E}_x \quad \text{isomorphism,}$$
$$f_x : M \to \mathbb{M}_x \quad \text{diffeomorphism}\}.$$

This is a principal bundle for the group $\mathrm{Diff}(M, E)$ under the action:

$$(\alpha_x, f_x)(\beta, g) = (\alpha_x \circ \beta, f_x \circ g), \quad \forall (\beta, g) \in \mathrm{Diff}(M, E).$$

We now build an associated bundle of algebras. The group $\mathrm{Diff}(M, E)$ acts on smooth sections $\sigma \in C^\infty(M, E)$ by pull-back:

$$\mathrm{Diff}(M, E) \times C^\infty(M, E) \to C^\infty(M, E)$$
$$(\Phi := (f, \alpha), \sigma) \mapsto \Phi^* \sigma := \alpha \left(f^{-1} \right)^* \sigma = \alpha(\sigma \circ f^{-1}),$$

inducing a smooth action on $\mathrm{Cl}(M, E)$ given by [AS]

$$\mathrm{Diff}(M, E) \times \mathrm{Cl}(M, E) \to \mathrm{Cl}(M, E)$$

$$(\phi, A) \mapsto \Phi_* A := \left(\Phi^{-1}\right)^* \circ A \circ \Phi^*, \tag{3.34}$$

since the algebra of classical pseudodifferential operators is invariant under diffeomorphisms. More precisely, the symbol of the operator $\Phi_* A$ differs from the symbol $\alpha^{-1} \circ f_* \sigma \circ \alpha$, by a classical symbol of lower order. Diffeomorphisms therefore preserve the order, so that this action induces one on $\mathrm{Cl}^a(M, E)$, for any complex number a. They moreover transform the leading symbol covariantly and preserve admissibility of an operator.

We build various fibre bundles corresponding to the sets of classical pseudodifferential operators we considered previously (see e.g. (3.5) and (3.19)):

$$\mathrm{Cl}(\mathbb{M}, \mathbb{E}) := \mathrm{Diff}(\mathbb{M}, \mathbb{E}) \times_{\mathrm{Diff}(M,E)} \mathrm{Cl}(\mathbb{M}, \mathbb{E});$$

$$\mathrm{Cl}^a(\mathbb{M}, \mathbb{E}) := \mathrm{Diff}(\mathbb{M}, \mathbb{E}) \times_{\mathrm{Diff}(M,E)} \mathrm{Cl}^a(\mathbb{M}, \mathbb{E}), \quad a \in \mathbb{C};$$

$$\mathrm{Cl}^{\notin \mathbb{Z}}(\mathbb{M}, \mathbb{E}) := \mathrm{Diff}(\mathbb{M}, \mathbb{E}) \times_{\mathrm{Diff}(M,E)} \mathrm{Cl}^{\notin \mathbb{Z}}(\mathbb{M}, \mathbb{E});$$

$$\mathrm{Cl}^{a,k}(\mathbb{M}, \mathbb{E}) := \mathrm{Diff}(\mathbb{M}, \mathbb{E}) \times_{\mathrm{Diff}(M,E)} \mathrm{Cl}^{a,k}(\mathbb{M}, \mathbb{E}), \quad (a, k) \in \mathbb{C} \times \mathbb{Z}_{\geq 0};$$

$$\mathrm{Cl}^{*,*}(\mathbb{M}, \mathbb{E}) := \mathrm{Diff}(\mathbb{M}, \mathbb{E}) \times_{\mathrm{Diff}(M,E)} \mathrm{Cl}^{*,*}(\mathbb{M}, \mathbb{E}).$$

Remark 16 Since the action of $\mathrm{Diff}(M, E)$ preserves ellipticity and admissibility, it makes sense to consider admissible elliptic operators in $\mathrm{Cl}(\mathbb{M}, \mathbb{E})$.

The leading symbol traces tr_0^τ on $\mathrm{Cl}^0(M, E)$, the residue res on $\mathrm{Cl}(M, E)$ and the canonical trace TR on $\mathrm{Cl}^{\notin \mathbb{Z}}(M, E)$ carry over to bundle morphisms on $\mathrm{Cl}^0(\mathbb{M}, \mathbb{E})$, $\mathrm{Cl}(\mathbb{M}, \mathbb{E})$ and $\mathrm{Cl}^{\notin \mathbb{Z}}(\mathbb{M}, \mathbb{E})$, respectively. Indeed, since the structure group is $\mathrm{Diff}(M, E)$ and the traces which lie at our disposal are diffeomorphism invariant, i.e. invariant under $\mathrm{Diff}(M)$, all we need to check is that they are invariant under the adjoint action of the gauge group $C^\infty(M, \mathrm{Aut}(E))$. It turns out that they are invariant under the adjoint action of the group $\mathrm{Cl}^\times(M, E)$ of invertible ψdos.

Lemma 3.28 *Let $E \to M$ be a finite rank vector bundle over a closed manifold M. For any operator C in $\mathrm{Cl}^\times(M, E)$*

$$\mathrm{res}(C^{-1} A C) = \mathrm{res}(A) \quad \forall A \in \mathrm{Cl}(M, E), \tag{3.35}$$

$$\mathrm{tr}_0^\tau(C^{-1} A C) = \mathrm{tr}_0^\tau(A) \quad \forall A \in \mathrm{Cl}^0(M, E), \tag{3.36}$$

and

$$\mathrm{TR}(C^{-1} A C) = \mathrm{TR}(A) \quad \forall A \in \mathrm{Cl}^{\notin \mathbb{Z}}(M, E). \tag{3.37}$$

Proof These properties easily follow from the cyclicity of the respective traces. □

We can therefore build the corresponding bundle morphisms:

$$\text{res} : \text{Cl}(\mathbb{M}, \mathbb{E}) \to C^\infty(M); \quad \text{tr}_0^\tau : \text{Cl}^0(\mathbb{M}, \mathbb{E}) \to C^\infty(M);$$

$$\text{and} \quad \text{TR} : \text{Cl}^{\notin \mathbb{Z}}(\mathbb{M}, \mathbb{E}) \to C^\infty(M). \tag{3.38}$$

A connection on an admissible vector bundle $\pi_*\mathbb{E} \to X$ is a connection 1-form on the associated frame bundle, i.e. an invariant $\text{Lie}\,(\text{Diff}(M, E))$-valued 1-form on $\text{Diff}(\mathbb{M}, \mathbb{E})$. Locally, above a trivialising open subset U in X it reads:

$$\nabla = d + \theta^U, \quad \theta^U \in \Omega^1(U, C^\infty(M, TM) \times C^\infty(M, \text{End}(M))). \tag{3.39}$$

Bismut gave an explicit construction of a connection [B] on admissible vector bundles $\pi_*\mathbb{E} \to X$ from

- a connection on the fibration $\mathbb{M} \to X$ which yields a horizontal splitting

$$T\mathbb{M} = T^H\mathbb{M} \oplus T(\mathbb{M}/X), \tag{3.40}$$

with $T^H\mathbb{M}$ and $T(\mathbb{M}/X)$ the horizontal and vertical parts of $T\mathbb{M}$, respectively, and where a vector field $V \in TX$ lifts to a horizontal vector field V^H;
- a connection $\nabla^\mathbb{E}$ on the finite rank vector bundle $\mathbb{E} \to \mathbb{M}$ over the fibration $\pi : \mathbb{M} \to X$ of smooth closed manifolds.

The connection $\nabla^{\pi_*\mathbb{E}}$ on $\pi_*\mathbb{E}$ is given by:

$$\nabla^{\pi_*\mathbb{E}}_V := \nabla^\mathbb{E}_{V^H} \quad \forall V \in TX. \tag{3.41}$$

Remark 17 The curvature $R^{\pi_*\mathbb{E}}$ of the connection $\nabla^{\pi_*\mathbb{E}}$ applied to two vector fields V_1 and V_2 on X reads ([B] Proposition 1.11)

$$R^{\pi_*\mathbb{E}}(V_1, V_2) = R^\mathbb{E}(V_1^H, V_2^H) - \nabla^\mathbb{E}_{T(V_1^H, V_2^H)},$$

where $R^\mathbb{E}$ is the curvature of $\nabla^\mathbb{E}$ and where T stands for the torsion of the connection on the fibration $\mathbb{M} \to X$. Hence,

$$R^{\pi_*\mathbb{E}} \in \Omega^2(X, \text{Cl}(\mathbb{M}, \mathbb{E})). \tag{3.42}$$

Example 3.29 The precise geometric setup investigated by Bismut is that of a smooth fibration $\pi : \mathbb{M} \to X$ of spin manifolds, with a vector bundle $\mathbb{E} = \mathbb{S} \otimes \mathbb{W}$ over $\mathbb{M}$, where $\mathbb{W} \to \mathbb{M}$ is an external Hermitian vector bundle on $\mathbb{M}$ and $\mathbb{S} \to X$ is the spinor bundle associated with the fibration.

The connection $\nabla^\mathbb{E}$ is built combining a connection $\nabla^{\mathbb{M}/X}$ on the vertical tangent bundle $T(\mathbb{M}/X)$ with a connection $\nabla^\mathbb{W}$ on the external bundle $\mathbb{W}$

$$\nabla^\mathbb{E} = \nabla^{\mathbb{M}/X} \otimes 1 + 1 \otimes \nabla^\mathbb{W}.$$

Here, we have set

$$\nabla^{\mathbb{M}/X} = P \, \nabla^g \, P,$$

where ∇^g is the Levi-Civita connection on $T\mathbb{M}$ with respect to the metric $g = g_X \oplus g_{\mathbb{M}/X}$ on $T\mathbb{M} \simeq \pi^*TX \oplus T(\mathbb{M}/X)$, with g_X a Riemannian metric on the base and $g_{\mathbb{M}/X}$ a Riemannian metric induced by a Riemannian metric on the fibration $\mathbb{M}$. $P : T\mathbb{M} \to T(\mathbb{M}/X)$ is the projection operator along $T^H\mathbb{M}$.

Even if $\nabla^{\mathbb{E}}$ is unitary for a Hermitian structure on $\mathbb{E}$, the connection $\nabla^{\pi_*\mathbb{E}}$ is not unitary as a result of the variation of the volume of the fibres of $\mathbb{M}/X$. In order to turn it into a unitary connection, one can pick a horizontal vector field $K \in T^H\mathbb{M}$ such that for any $V \in TX$

$$\mathrm{div}_{\mathbb{M}/X} V^H = 2\langle K, V^H \rangle,$$

which combined with $\nabla^{\pi_*\mathbb{E}}$ gives rise to a unitary connection:

$$\widetilde{\nabla}_V^{\pi_*\mathbb{E}} = \nabla_V^{\pi_*\mathbb{E}} + \langle K, V^H \rangle,$$

and yields another admissible connection when viewing the map $\sigma \mapsto \langle K, V^H \rangle \sigma$ as a multiplication operator.

As the following example shows, the Lie algebra $\mathrm{Lie}(\mathrm{Diff}(M, E))$ might turn out to be too small to host some useful connection 1-forms.

Example 3.30 With the notation of Example 3.25, let (M, g) be a closed Riemannian manifold and let G be a Lie group which fulfils the same requirements as in Example 3.21.

Let $Q_0 := (\Delta + \pi_\Delta) \otimes 1_{\mathrm{Lie}(G)}$ where Δ stands for the Laplace–Beltrami operator on M and π_Δ for the orthogonal projection onto the kernel. The operator Q_0 is invertible since Δ is an essentially self-adjoint elliptic operator (see Remark 4). In [F], D. Freed introduces a family of left-invariant 1-forms parametrised by $s \in \mathbb{R}$ on the H^s-Sobolev completion $H^s(M, G)$ of the mapping group $C^\infty(M, G)$ (see formula (1.9) in [F]):

$$\theta_0^s(V) := \frac{1}{2} \left(\mathrm{ad}_V + Q_0^{-s} \mathrm{ad}_V Q_0^s - Q_0^{-s} \mathrm{ad}_{Q_0^{-s}V} \right) \quad \forall V \in C^\infty(M, \mathrm{Lie}(G)).$$

These give rise to left-invariant connections ∇^s on $H^s(M, G)$, which in turn induce connections[20] on the mapping group $C^\infty(M, G)$.

Since θ_0^s lies in $\Omega^1(C^\infty(M, G), \mathrm{Cl}^0(M, \mathrm{Lie}(G)))$, only if $s = 0$ does this 1-form correspond to a multiplication operator.

---

[20] They are weak connections since they are defined by weak metrics on $L^2(M, \mathrm{Lie}(G))$; they are not uniquely determined by the usual six-term formula [F].

Allowing for connections of this type requires enlarging the class of connections locally described by (3.41). We introduce the following class of connections, similar to the ones considered in [P1] and later in [PR1, PR2].

Definition 3.31 We call a connection ∇ on an admissible vector bundle $\pi_* \mathbb{E} \to X$ with structure group $\text{Diff}(M, E)$ a ψdo-*connection* if, above a trivialising open subset U in X, it is of the form

$$\nabla = d + \theta^U, \quad \text{where} \quad \theta^U \in \Omega^1\left(U, \text{Cl}^0(\mathbb{M}, \mathbb{E})|_U\right) \quad \forall V \in T_x X. \qquad (3.43)$$

Remark 18 Since $\text{Cl}^0(M, E)$ is invariant under the action of $\text{Diff}(M, E)$, the local condition (3.43) is invariant under a change of local trivialisation of the bundle.

Example 3.32
1. The connection $\nabla^{\pi_* \mathbb{E}}$ is admissible, since it is locally of the form $d + \theta^U$ where $\theta^U(V)$ is a multiplication operator.
2. The left-invariant 1-form θ_0^s on the loop group, introduced in Example 3.30, lies in $\text{Cl}^0(M, \text{Lie}(G))$, so it gives rise to an admissible connection. Here, X is the Lie group G and E is the tangent bundle to G.

When the fibration $\pi_* \mathbb{E} \to X$ is trivial, the curvature R lies in $\Omega^2(X, \text{Cl}^0(\mathbb{M}, \mathbb{E}))$. However, when the connection is non-trivial, it follows from (3.42) that R lies in $\Omega^2(X, \text{Cl}^1(\mathbb{M}, \mathbb{E}))$.

Remark 19 When M reduces to a point $\{*\}$, then any connection is a ψdo-connection since $\mathbb{E} \to \mathbb{M}$ boils down to a bundle $E \to X$ and locally $\nabla = d + \theta^U$ with $\theta^U \in \Omega^1(U, \text{End}(E))$.

3.11 Logarithms and closed forms

We need a bundle counterpart of Proposition 6, which turns covariantly closed operator-valued forms into closed de Rham forms. The adjoint action $\text{ad}_l : a \mapsto [l, a]$ plays here the role of the derivation δ in Proposition 6.

Proposition 12

1. Let $\mathcal{A}$ be a smooth bundle of topological associative complex algebras over a smooth manifold X, equipped with a connection ∇. Let $\mathcal{T} : \mathcal{A} \to X \times \mathbb{C}$ be a continuous fibrewise trace on $\mathcal{A}$. Then the linear form $\mathcal{T}$ commutes with differentiation, i.e.

$$d \circ \mathcal{T} = \mathcal{T} \circ \nabla. \qquad (3.44)$$

2. We further assume that $\mathcal{A}$ is a subbundle of another bundle $\mathcal{L}$ of topological associative complex algebras over X and that ∇ is induced by a connection on $\mathcal{L}$.

Let l be an element in $C^\infty(X, \mathcal{L})$ such that

$$\mathrm{ad}_l : a \mapsto [l, a]$$

stablises $C^\infty(X, \mathcal{A})$.[21] We further assume that

$$[\nabla, l] \in C^\infty(X, \mathcal{A}) \quad \text{and} \quad \nabla^2 \in \Omega^2(X, \mathcal{A}),$$

and that $T : \mathcal{A} \to X \times \mathbb{C}$ vanishes on the range of ad_l:

$$T([l, a]) = 0 \quad \forall a \in C^\infty(X, \mathcal{A}).$$

Then the 2-forms $T(\nabla^{2j} [\nabla, l])$ are closed for any non-negative integer j:

$$d\left(T(\nabla^{2j} [\nabla, l])\right) = 0. \tag{3.45}$$

Proof 1. Let us first check (3.44). Over a trivialising open subset U of X, the connection ∇ reads $\nabla = d + \theta^U$, where $\theta^U \in \Omega^1(U, \mathcal{L})$. Thus, for any $\alpha \in \Omega(U, \mathcal{L})$, we have

$$dT(\alpha) = T(d\alpha) = T(d\alpha + [\theta^U, \alpha]) = T([\nabla, \alpha]).$$

Hence $d \circ T = T \circ \nabla$.

2. To show that $T(\nabla^{2j} [\nabla, l])$ is closed, we proceed in several steps. Let us first observe that $T(a[a, l]) = 0$ for any $a \in \mathcal{A}$. For $a, b \in C^\infty(X\mathcal{A})$ we have

$$[ab, l] = a[b, l] + [a, l]b \implies T(a[b, l]) = -T([a, l]b).$$

Setting $a = b$ yields the statement.

In particular, we have

$$T\left(\nabla^{2j} [\nabla^2, l]\right) = 0. \tag{3.46}$$

We further observe that for any vector fields U and V on X,

$$[\nabla, [\nabla, l]](V_1, V_2) = [\nabla_{V_1}, [\nabla_{V_2}, l]] - [\nabla_{V_1}, [\nabla_{V_2}, l]] - [\nabla_{[V_1, V_2]}, l]$$

$$= [\nabla_{V_1}, \nabla_{V_2}, l] - [\nabla_{V_{12}} \nabla_{V_1}, l] - [\nabla_{[V_1, V_2]}, l]$$

$$= [\nabla^2(V_1, V_2), l]. \tag{3.47}$$

The Leibniz rule implies that

$$d\,T(\nabla^{2j} [\nabla, l]) = T([\nabla, \nabla^{2j}] [\nabla, l]) + T(\nabla^{2j} [\nabla, [\nabla, l]]) \quad \text{by (3.44)}$$

$$= T(\nabla^{2j} [\nabla, [\nabla, l]]) \quad \text{since } [\nabla, \Omega] = 0 \text{ (Bianchi identity)}$$

$$= T(\nabla^{2j} [\Omega, l]) \quad \text{by (3.47)}$$

$$= 0 \quad \text{by (3.46).} \qquad\qquad \square$$

---

[21] Note that neither al nor la a priori lies in $\mathcal{A}$.

Remark 20 The forms $T(\nabla^{2j} [\nabla, l])$ are not *a priori* exact. Since $\nabla^{2j} l$ does not *a priori* lie in $C^\infty(X, \mathcal{A})$, we cannot implement the straightforward computation $T(\nabla^{2j} [\nabla, l]) = T([\nabla, \nabla^{2j} l]) = d\, T(\nabla^{2j} l)$, whose last step is erroneous.

3.12 Chern–Weil forms in infinite dimensions

We aim at generalising the Chern–Weil formalism to ψdo-vector bundles, defining (when possible) traces of even powers of admissible connections. To build traces of powers of the curvature along the lines of Section 3.8, we can use the two types of (singular) traces at our disposal, as described in Section 3.6:

1. leading symbol traces (3.18) on $\mathrm{Cl}^0(M, E)$,
2. the noncommutative residue (3.17) on $\mathrm{Cl}(M, E)$.

As in the finite dimensional case, from a form $\alpha(x) = A(x)\, dx_1 \wedge \cdots \wedge dx_d$ in $\Omega(X, \mathrm{Cl}(M, E))$ (resp. $\Omega(X, \mathrm{Cl}^0(M, E))$) we can build a form $\mathrm{res}(\alpha)(x) := \mathrm{res}(A(x))\, dx_1 \wedge \cdots \wedge dx_d$ in $\Omega(X)$ (resp. $\mathrm{tr}_0^\tau(\alpha)(x) := \mathrm{tr}_0^\tau(A(x))\, dx_1 \wedge \cdots \wedge dx_d$ in $\Omega(X)$ for any τ in $C^\infty(S^*M)'$). The following theorem provides a generalisation of Proposition 10 to infinite dimensions.

Theorem 3.33 *Let $\mathcal{E} = \pi_* \mathbb{E} \to X$ be an admissible vector bundle with $\mathbb{E} \to \mathbb{M}$ a finite rank vector bundle over a locally trivial fibre bundle $\mathbb{M} \to X$, equipped with a ψdo-connection with curvature Ω. For any non-negative integer j,*

1. *the jth residue Chern–Weil form $\mathrm{res}(\nabla^{2j})$ is closed with de Rham cohomology class independent of the choice of connection,*
2. *if the curvature $R = \nabla^2$ lies in $\Omega^2(X, \mathrm{Cl}^0(M, E))$, then $\mathrm{tr}_0^\tau(\nabla^{2j})$ is also closed with de Rham cohomology class independent of the choice of connection.*

We call these singular Chern–Weil classes.

Proof The proof proceeds as in Proposition 10 with $f(x) = x^j$ and using (3.48). □

Remark 21 Singular Chern–Weil forms are clearly insensitive to smoothing perturbations of the connection.

Unfortunately, singular Chern–Weil classes generally seem too coarse to capture interesting information since most examples lead to vanishing singular Chern–Weil classes.

Example 3.34 Going back to Example 3.30, it was shown in [F] that $W \mapsto R_0^s(W, U)V$, with R_0^s the curvature of ∇_0^s, is conditionally trace-class, i.e. that $\mathrm{tr}_{\mathrm{Lie}(G)}\left(R_0^s(\cdot, U)V\right)$ is trace-class, from which we infer that the residue $\mathrm{res}\left(R_0^s(\cdot, U)V\right)$ vanishes. For the same reason, the leading symbol traces also vanish. Thus, singular Chern–Weil forms vanish in the case of mapping groups as can be expected since the latter are parallelisable.

Remark 22 In [LRST] the authors conjecture that Chern classes built from the noncommutative residue on admissible bundles vanish.

We want to apply Proposition 12 to the bundles $\mathcal{L} = \mathrm{Cl}^{*,*}(\mathbb{M}, \mathbb{E})$ and $\mathcal{A} = \mathrm{Cl}(\mathbb{M}, \mathbb{E})$ with the trace given by the residue $\mathcal{T} = \mathrm{res}$. We need the following result.

Lemma 3.35 *Let ∇ be a ψdo-connection on an admissible bundle $\pi^*\mathbb{E}$. Let Q be a smooth section of $\mathrm{Cl}(\mathbb{M}, \mathbb{E})$, consisting of admissible elliptic operators of constant order and with constant spectral cut θ. Then $[\nabla, \log_\theta Q]$ lies in $\mathbb{C}^\infty(X, \mathrm{Cl}^0(\mathbb{M}, \mathbb{E}))$.*

Proof Over a trivialising open subset of X on which $\nabla = d + \theta^U$, we have

$$[\nabla, \log_\theta Q] = d \log_\theta Q + [\theta^U, \log_\theta Q].$$

We know that $[\theta^U, \log_\theta Q] = \mathrm{ad}_{\theta^U}(\log_\theta Q)$ is classical of zero order. On the other hand, since the difference of two logarithms of two admissible operators of the same order is classical of zero order, for any tangent vector U at point $x \in X$ the differentiated operator:

$$d \log_\theta Q(U) = \lim_{t \to 0} \frac{\log_\theta Q_{x+tU} - \log_\theta Q_x}{t}$$

also lies in $\mathbb{C}^\infty(X, \mathrm{Cl}^0(\mathbb{M}, \mathbb{E}))$. $\qquad\square$

Applying Proposition 12 to a ψdo-connection ∇ on some admissible vector bundle $\pi_*\mathbb{E}$, to $\mathcal{L} = \mathrm{Cl}^{*,*}(\mathbb{M}, \mathbb{E})$ and to

1. the residue seen as a bundle morphism on $\mathcal{A} = \mathrm{Cl}(\mathbb{M}, \mathbb{E})$,
2. the leading symbol traces seen as bundle morphisms on $\mathcal{A} = \mathrm{Cl}^0(\mathbb{M}, \mathbb{E})$,

leads to the following conclusion.

Theorem 3.36 *Let ∇ be a ψdo-connection on an admissible bundle $\pi^*\mathbb{E}$. Let Q be a smooth section of $\mathrm{Cl}(\mathbb{M}, \mathbb{E})$, consisting of admissible elliptic operators of constant order and with constant spectral cut θ.*

1. *The ψdo-covariant differentiation commutes with the residue and the leading symbol traces*

$$d \circ \mathrm{res} = \mathrm{res} \circ \nabla \quad \text{on} \quad C^\infty(X, \mathrm{Cl}(\mathbb{M}, \mathbb{E})); \tag{3.48}$$

$$d \circ \mathrm{tr}_0^\tau = \mathrm{tr}^\tau \circ \nabla \quad \forall \tau \in C^\infty(S^*\mathbb{M}/X) \quad \text{on} \quad C^\infty(X, \mathrm{Cl}^0(\mathbb{M}, \mathbb{E})).$$

2. *The 2-forms*

$$\mathrm{res}\left(\nabla^{2j} [\nabla, \log_\theta Q]\right) \tag{3.49}$$

are closed for any non-negative integer j.

3. *If the fibration* $\pi : \mathbb{M} \to X$ *is trivial, then* ∇^{2j} *lies in* $\Omega^2(X, \mathrm{Cl}(\mathbb{M}, \mathbb{E}))$ *and the 2-forms*

$$\mathrm{tr}_0^\tau \left(\nabla^{2j} \, [\nabla, \log_\theta Q] \right) \tag{3.50}$$

are closed for any non-negative integer j *and for any linear form* $\tau \in (C^\infty(S^*\mathbb{M}/X))'$. *Here as before,* $\mathbb{M}/X$ *is the vertical fibre of* $\mathbb{M} \to X$.

Part 4: Circumventing anomalies

The closed forms (3.49) turn out to be exact. They correspond to the differential of some regularised traces of ∇^{2j} and are seen as discrepancies preventing these regularised traces from being closed. We describe ways of circumventing such discrepancies in a manner similar to the way physicists circumvent anomalies in quantum field theory.

3.13 Weighted Chern–Weil forms; discrepancies

Let us first recall the following identity (see formula (3.20)):

$$\mathrm{Res}_{z=0}\mathrm{TR}\,(C(z)) = -\frac{1}{c'(0)}\mathrm{res}(C), \tag{3.51}$$

where $c(z)$ is the holomorphic order of a holomorphic family $C(z)$ around zero, of classical pseudodifferential operators.

This formula provides ways to measure various defects of weighted traces.

3.13.1 The Hochschild coboundary of a weighted trace

The first defect is an obstruction to its cyclicity. We saw in Theorem 3.17 that the Hochschild coboundary of the Q-weighted trace gives rise to the Radul cocycle

$$\partial\mathrm{tr}_\theta^Q\,(A, B) = -\frac{1}{q}\,\mathrm{res}\,\left(A\,[B, \log_\theta Q]\right) = \frac{1}{q}\,c_{\mathrm{res}}^{Q,\theta}(A, B)$$

introduced in Section 3.4. Since it is a noncommutative residue, it is local as the integral over the cotangent unit sphere of a homogeneous component of a symbol.

3.13.2 Dependence on the weight

Formula (3.51) also provides a way to measure the dependence on the weight Q. We first need a technical lemma.

Lemma 3.37 *Let $Q \in \mathrm{Cl}(M, E)$ be an admissible operator of order $q > 0$ with spectral cut θ and let $A \in \mathrm{Cl}(M, E)$. Then*

$$\mathrm{tr}_{\theta}^{Q_{\theta}^t}(A) = \mathrm{tr}_{\theta}^{Q}(A) \quad \forall t > 0,$$

where the t-th power Q_{θ}^t is taken with respect to the spectral cut θ.

Proof We write $\mathrm{TR}(A \, Q_{\theta}^{-z}) = \frac{a_{-1}}{z} + a_0 + o(z)$ in which case we have

$$\mathrm{TR}\left(A \, \left(Q_{\theta}^t\right)^{-z}\right) = \mathrm{TR}\left(A \, (Q_{\theta})^{-tz}\right) = \frac{a_{-1}}{tz} + a_0 + o(t \, z)$$

so that

$$\mathrm{tr}Q_{\theta}^t(A) = \mathrm{fp}_{z=0}\mathrm{TR}(A \, \left(Q_{\theta}^t\right)^{-z}) = a_0 = \mathrm{tr}_{\theta}^{Q}(A). \qquad \square$$

The following proposition provides a well-known expression of the dependence on the weight [KV1, KV2, Ok].

Proposition 13 Let $Q_1, Q_2 \in \mathrm{Cl}(M, E)$ be two admissible operators with positive orders q_1, q_2 and spectral cuts θ_1, θ_2. For any A in $\mathrm{Cl}(M, E)$

$$\mathrm{tr}_{\theta_1}^{Q_1}(A) - \mathrm{tr}_{\theta_2}^{Q_2}(A) = \mathrm{res}\left(A \, \left(\frac{\log_{\theta_1} Q_2}{q_2} - \frac{\log_{\theta_2} Q_1}{q_1}\right)\right).$$

Proof Applying formula (3.51) to the family $C(z) := \frac{A}{z}\left(Q_1^{-\frac{z}{q_1}} - Q_2^{-\frac{z}{q_2}}\right)$, which is a holomorphic family of classical operators of order $a - z$ with $C(0) = A\left(\frac{\log Q_2}{q_2} - \frac{\log Q_1}{q_1}\right)$, we write

$$\mathrm{tr}^{Q_1}(A) - \mathrm{tr}^{Q_2}(A) = \mathrm{tr}^{Q_1^{\frac{1}{q_1}}}(A) - \mathrm{tr}^{Q_2^{\frac{1}{q_2}}}(A)$$

$$= \mathrm{Res}_{z=0}\mathrm{TR}\left(\frac{A\left(Q_1^{-\frac{z}{q_1}} - Q_2^{-\frac{z}{q_2}}\right)}{z}\right)$$

$$= \mathrm{res}\left(A \, \left(\frac{\log Q_2}{q_2} - \frac{\log Q_1}{q_1}\right)\right). \qquad \square$$

3.13.3 *Exterior differential of a weighted trace*

Proposition 14 [CDMP], [P1] Let $Q \in C^{\infty}(X, \mathrm{Cl}(M, E))$ be a differentiable family (for the topology described in (3.6)) of admissible operators of fixed order

q and spectral cut θ, parametrised by a manifold X. Given A in $\mathrm{Cl}(M, E)$, the trace defect $[d, \mathrm{tr}^Q] := d\,\mathrm{tr}^Q - \mathrm{tr}^Q \circ d$ is local as a noncommutative residue:

$$\left[d, \mathrm{tr}_\theta^Q\right](A) = -\frac{1}{q}\,\mathrm{res}\left(A\,d\,\log_\theta Q\right). \tag{3.52}$$

Proof Again, for convenience, we drop the explicit mention of the spectral cut. Let $h \in C^\infty(X, TX)$ be a smooth vector field. Then by Proposition 13 we have

$$d\mathrm{tr}^Q(A)(h) = \lim_{t \to 0} \frac{\mathrm{tr}^{Q+dQ(th)}(A) - \mathrm{tr}^Q(A)}{t}$$

$$= -\frac{1}{q} \lim_{t \to 0} \frac{\mathrm{res}\left(A\,(\log(Q + dQ(th)) - \log Q)\right)}{t}$$

$$= -\frac{1}{q} \frac{\mathrm{res}\left(\lim_{t \to 0}\left(A\,(\log(Q + dQ(th)) - \log Q)\right)\right)}{t}$$

$$= -\frac{1}{q}\,\mathrm{res}\left(A\,d\,\log Q(h)\right),$$

where we have used the continuity of the noncommutative residue on operators of order $a = \mathrm{ord}A$ since $A\left(\frac{\log(Q+dQ(th))-\log Q}{t}\right)$ has order a for any $t > 0$. □

3.13.4 *Weighted traces extended to admissible fibre bundles*

The covariance property (3.29) generalises to weighted traces as follows.

Lemma 3.38 [P1], [PS2]   *Order, ellipticity, admissibility and spectral cuts are preserved under the adjoint action of* $\mathrm{Cl}^\times(M, E)$ *on* $\mathrm{Cl}(M, E)$*. For any admissible operator* $Q \in \mathrm{Cl}(M, E)$ *with spectral cut* θ*, for any operator* $A \in \mathrm{Cl}(M, E)$*, we have*

$$\mathrm{tr}_\theta^{\mathrm{Ad}_C Q}(\mathrm{Ad}_C A) = \mathrm{tr}_\theta^Q(A) \quad \forall C \in \mathrm{Cl}^\times(M, E), \tag{3.53}$$

where the adjoint map Ad *was defined in* (3.10).

Proof For simplicity, we drop the explicit mention of the spectral cut. Since the leading symbol is multiplicative, we have

$$\sigma_L(\mathrm{Ad}_C Q) = \mathrm{Ad}_{\sigma_L(C)}\sigma_L(Q),$$

from which it follows that order, ellipticity, admissibility and spectral cuts are preserved by the adjoint action.

Let us observe that

$$(\mathrm{Ad}_C Q)^{-z} = \frac{1}{2i\pi} \int_\Gamma \lambda^{-z} (\lambda - \mathrm{Ad}_C Q)^{-1}$$

$$= \frac{1}{2i\pi} \int_\Gamma \lambda^{-z} \mathrm{Ad}_C (\lambda - Q)^{-1}$$

$$= \mathrm{Ad}_C Q^{-z}.$$

Consequently,

$$\mathrm{tr}^{C^{-1}QC}(C^{-1}AC) = \mathrm{fp}_{z=0} \mathrm{TR}\left(C^{-1}AC\,(C^{-1}QC)^{-z}\right)$$

$$= \mathrm{fp}_{z=0} \mathrm{TR}(C^{-1}AC\,C^{-1}Q^{-z}C)$$

$$= \mathrm{fp}_{z=0} \mathrm{TR}(C^{-1}AQ^{-z}C)$$

$$= \mathrm{fp}_{z=0} \mathrm{TR}(A\,Q^{-z})$$

$$= \mathrm{tr}^Q(A),$$

where we have used the fact that the canonical trace vanishes on non-integer order brackets.   □

Since the adjoint action $\mathrm{ad}_C : A \mapsto C^{-1}AC$ of $\mathrm{Cl}^{0,\times}(M, E)$ on $\mathrm{Cl}(M, E)$ preserves the spectrum and the invertibility of the leading symbol, it makes sense to define the subbundle $\mathrm{Ell}^{\mathrm{adm}}(\mathbb{M}, \mathbb{E}))$ of $\mathrm{Cl}(\mathbb{M}, \mathbb{E})$ of fibrewise admissible elliptic ψdos with spectral cut θ. Since the order is also preserved under the adjoint action, we can define $\mathbb{Q}$ to be a smooth admissible elliptic section of order q of $\mathrm{Cl}(\mathbb{M}, \mathbb{E})$, in which case $\mathbb{Q}(x) \in \mathrm{Cl}(\mathbb{M}_x, \mathbb{E}_{|\mathbb{M}_x})$ and we set:

$$\mathrm{tr}_\theta^{\mathbb{Q}}(A)(x) := \mathrm{tr}_\theta^{\mathbb{Q}(x)}(A(x)) \quad \forall A \in C^\infty(X, \mathrm{Cl}(\mathbb{M}, \mathbb{E})), \forall x \in X.$$

$\mathbb{Q}$-weighted traces can further be extended to forms $\alpha(x) = A(x)\,dx_1 \wedge \cdots \wedge dx_d$ in $\Omega(X, \mathrm{Cl}(\mathbb{M}, \mathbb{E}))$ by $\mathrm{tr}_\theta^{\mathbb{Q}}(\alpha)(x) := \mathrm{tr}_\theta^{\mathbb{Q}}(A(x))\,dx_1 \wedge \cdots \wedge dx_d$ and using linearity.

3.13.5 *Obstructions to closedness of weighted Chern–Weil forms*

Theorem 3.39 [CDMP]   *An admissible connection ∇ on an admissible vector bundle $\pi_*\mathbb{E}$ induces a connection $[\nabla, A] := \nabla \circ A - A \circ \nabla$ on $\mathrm{Cl}(\mathbb{M}, \mathbb{E})$. For any form α in $\Omega^p(X, \mathrm{Cl}(\mathbb{M}, \mathbb{E}))$ and any admissible section $\mathbb{Q}$ in $C^\infty(X, \mathrm{Cl}(\mathbb{M}, \mathbb{E}))$ with constant spectral cut θ, constant positive order q, and whose kernel has constant dimension, the trace defect*

$$d\mathrm{tr}_\theta^{\mathbb{Q}}(\alpha) := [\nabla, \mathrm{tr}_\theta^{\mathbb{Q}}] := d\,\mathrm{tr}_\theta^{\mathbb{Q}}(\alpha) - \mathrm{tr}_\theta^{\mathbb{Q}}\left([\nabla, \alpha]\right)$$

is local and given by:

$$[\nabla, \mathrm{tr}_\theta^Q](\alpha) = \frac{(-1)^p}{q} \mathrm{res}\left(\alpha\,[\nabla, \log_\theta Q]\right) \quad \forall \alpha \in \Omega^p(X, \mathrm{Cl}(\mathbb{M}, \mathbb{E})), \quad (3.54)$$

where π_Q stands for the orthogonal projection onto the kernel of Q.

Proof For simplicity, we drop as usual the explicit mention of the spectral cut. Let us prove the result for a zero form; the proof easily extends to higher-order forms. In a local trivialisation of $\mathcal{E}$ over a sufficiently small open subset U of X we write $\nabla = d + \theta$ so that $[\nabla, \cdot] = d + [\theta, \cdot]$. In this local trivialisation we have for any $\in C^\infty(X, \mathrm{Cl}(\mathbb{M}, \mathbb{E}))$:

$$\begin{aligned}
[\nabla, \mathrm{tr}^Q](A) &= d\left(\mathrm{tr}^Q(A)\right) - \mathrm{tr}^Q\left([\nabla, A]\right) \\
&= d\left(\mathrm{tr}^Q(A)\right) - \mathrm{tr}^Q(d\,A) - \mathrm{tr}^Q\left([\theta, A]\right) \\
&= -\frac{1}{q}\mathrm{res}\left(A\,d\log Q\right) - \frac{1}{q}\mathrm{res}\left(A\,[\theta, \log Q]\right) \\
&= -\frac{1}{q}\mathrm{res}\left(A\,[\nabla, \log Q]\right),
\end{aligned}$$

where we have combined (3.22) and (3.52). $\qquad\square$

The obstruction $[\nabla^{\mathrm{ad}}, \mathrm{tr}^Q]$ described in Theorem 3.39 prevents a straightforward generalisation of the Chern–Weil formalism to an infinite dimensional setup where the trace on matrices is replaced by a weighted trace provided the connections are admissible connections.

Corollary 3 *Let ∇ be a ψdo-connection on $\mathcal{E} = \pi_*\mathbb{E} \to X$ with curvature Ω and let Q be an admissible section of $\mathrm{Cl}(\mathbb{M}, \mathbb{E})$ with constant spectral cut θ and positive order q. The de Rham coboundary of the Q-weighted Chern–Weil type form $\mathrm{tr}_\theta^Q(\Omega^i)$ reads:*

$$d\,\mathrm{tr}_\theta^Q(\Omega^i) = -\frac{1}{q}\mathrm{res}\left(\Omega^i\,[\nabla, \log_\theta Q]\right).$$

Proof We follow the finite dimensional proof (see Proposition 10).

$$\begin{aligned}
d\,\mathrm{tr}_\theta^Q(\Omega^i) &= [d\,\mathrm{tr}_\theta^Q](\Omega^i) + \mathrm{tr}_\theta^Q([\nabla, \Omega^i]) \\
&= [d\,\mathrm{tr}_\theta^Q](\Omega^i) \text{ by the Bianchi identity} \\
&= -\frac{1}{q}\mathrm{res}\left(\Omega^i\,[\nabla, \log_\theta Q]\right) \quad \text{using (3.54).} \qquad\square
\end{aligned}$$

3.14 Renormalised Chern–Weil forms on ψdo Grassmannians

In view of Corollary 3, which tells us that a weighted trace of a power of the curvature is generally not closed, it seems hopeless to use weighted traces as a substitute for ordinary traces in order to extend finite dimensional Chern–Weil formalism to infinite dimensions. However, there are different ways to circumvent this difficulty, one of which is to introduce counterterms in order to compensate for the lack of closedness measured in Corollary 3 by a noncommutative residue. Such a renormalisation procedure by the introduction of counterterms can be carried out in a Hamiltonian approach to gauge theory as it was shown in joint work with J. Mickelsson [MP] on which we report here.

Let us first review a finite dimensional situation which will serve as a model for infinite dimensional generalisations.

We consider the finite dimensional Grassmann manifold $\mathrm{Gr}(n, n)$ consisting of rank n projections in $\mathbb{C}^{2n}$, which we parametrise by grading operators $F = 2P - 1$, where P is a finite rank projection.

Lemma 3.40 *The even forms*

$$\omega_{2j} = \mathrm{tr}\,(F(dF)^{2j}), \tag{3.55}$$

where $j = 1, 2, \ldots$, are closed forms on $\mathrm{Gr}(n, n)$.

Remark 23 The cohomology of $\mathrm{Gr}(n, n)$ is known to be generated by even (non-normalised) forms of the type ω_{2j}, $j = 1, \ldots, n$. This follows from the fact (see Proposition 23.2 in [BT]) that the Chern classes of the quotient bundle Q over $\mathrm{Gr}(n, n)$ defined by the exact sequence $0 \to S \to \mathrm{Gr}(n, n) \times \mathbb{C}^n \to Q \to 0$ where S is the universal bundle over $\mathrm{Gr}(n, n)$ with fibre V above the vector space V, generate the cohomology ring $H^*(\mathrm{Gr}(n, n))$. Then the jth Chern class of Q turns out to be proportional to $\mathrm{tr}(F(dF)^{2j})$, where $P(V)$ stands for the orthogonal projection on V.

Proof (of Lemma 3.40) By the traciality of tr we have

$$\begin{aligned}
d\,\omega_{2j} = d\,\mathrm{tr}\,\left(F(dF)^{2j} \right) &= \mathrm{tr}\,\left((dF)^{2j+1} \right) \\
&= \mathrm{tr}\,\left(F^2\,(dF)^{2j+1} \right) \quad \text{since } F^2 = 1 \\
&= -\mathrm{tr}\,\left(F\,(dF)^{2j+1}\,F \right) \quad \text{since } F\,dF = -dF\,F \\
&= -\mathrm{tr}\,\left((dF)^{2j+1}\,F^2 \right) \quad \text{since } \mathrm{tr}\,([A, B]) = 0 \\
&= -\mathrm{tr}\,\left((dF)^{2j+1} \right) = 0. \qquad\qquad\qquad \square \tag{3.56}
\end{aligned}$$

We now want to extend these constructions to ψdo Grassmannians.

Let us consider a finite rank bundle $\mathbb{E}$ over a trivial fibre bundle $\pi : \mathbb{M} = M \times X \to X$ with typical fibre a closed (Riemannian) spin manifold M. Let $D_x \in \mathrm{Cl}(\mathbb{M}, \mathbb{E})$, $x \in X$ be a smooth family of Dirac operators parametrised by X.

On each open subset $U_\lambda := \{x \in X, \lambda \notin \mathrm{spec}(D_x)\} \subset X$ there is a well-defined map

$$F : X \to \mathrm{Cl}^0(M, E)$$

$$x \mapsto F_x := (D_x - \lambda I)/|D_x - \lambda I|.$$

Since $F_x^2 = F_x$, $P_x := \frac{I + F_x}{2}$ is a projection, the range $\mathrm{Gr}(M, E) := \mathrm{Im} F$ of F coincides with the Grassmannian consisting of classical pseudodifferential projections P with kernel and cokernel of infinite rank, acting in the complex Hilbert space $H := L^2(M, E)$ of square-integrable sections of the vector bundle E over the compact manifold M.

Since the map $x \mapsto F_x$ is generally not contractible we want to define, from F_x, cohomology classes on X in the way we built Chern–Weil classes in finite dimensions. This issue usually arises in Hamiltonian quantisation in field theory, when the physical space M is an odd dimensional manifold. In this infinite dimensional setup traces are generally ill-defined, so that we use weighted traces as in the previous section. As expected, there are a priori obstructions to the closedness of the corresponding weighted forms.

Proposition 15 Let $Q \in \mathrm{Cl}(M, E)$ be a fixed admissible elliptic operator with positive order. The exterior differential of the form

$$\omega_{2j}^Q(F) = \mathrm{tr}^Q\left(F(dF)^{2j}\right) \tag{3.57}$$

on $\mathrm{Gr}(M, E)$,

$$d\omega_{2j}^Q = \frac{1}{2q}\mathrm{res}([\log Q, F](dF)^{2k+1}F),$$

is a local expression which only depends on F modulo smoothing operators.

Proof The locality and the dependence on F modulo smoothing operators follow from the expression of the exterior differential in terms of a noncommutative residue. To derive this expression, we mimic the finite dimensional proof, taking into account that this time tr^Q is not cyclic:

$$
\begin{aligned}
d\omega_{2j}^Q &= d\mathrm{tr}^Q\left(F(dF)^{2j}\right) = \mathrm{tr}^{\,Q}\left((dF)^{2j+1}\right) = \mathrm{tr}^Q\left(F^2(dF)^{2j+1}\right) \\
&= -\mathrm{tr}^Q\left(F(dF)^{2j+1}F\right) \quad \text{since } F\,dF = -dF\,F \\
&= \frac{1}{q}\mathrm{res}\left([\log Q, F](dF)^{2j+1}F\right) - \mathrm{tr}^Q\left((dF)^{2j+1}F^2\right) \\
&= \frac{1}{q}\mathrm{res}\left([\log Q, F](dF)^{2j+1}F\right) - \mathrm{tr}^Q\left((dF)^{2j+1}\right),
\end{aligned}
$$

where we have used (3.22) to write

$$
\begin{aligned}
\mathrm{tr}^Q\left([F, (dF)^{2j+1}F]\right) &= -\frac{1}{q}\mathrm{res}\left(F\,[(dF)^{2j+1}F, \log Q]\right) \\
&= \frac{1}{q}\mathrm{res}\left([F, \log Q]\,(dF)^{2j+1}F\right).
\end{aligned}
$$

Hence

$$
\mathrm{tr}^Q\left(F^2(dF)^{2j+1}\right) = \frac{1}{2q}\mathrm{res}\left([\log Q, F](dF)^{2j+1}F\right)
$$

from which the result then follows. $\square$

Let us consider the map

$$
\sigma : X \to \mathrm{Cl}^0(M, E)/\mathrm{Cl}^{-\infty}(M, E)
$$

$$
x \mapsto \bar{F}(x) := p \circ F(x),
$$

where $p : \mathrm{Cl}^0(M, E) \to \mathrm{Cl}^0(M, E)/\mathrm{Cl}^{-\infty}(M, E)$ is the canonical projection map.

The following theorem builds from the original forms ω_{2j}^Q new "renormalised" forms which are closed in contrast with the original ones.

Theorem 3.41 *When $\sigma(X)$ is contractible, there are even forms θ_{2j}^Q such that*

$$
\omega_{2j}^{\mathrm{ren},\,Q} := \omega_{2j}^Q - \theta_{2j}^Q
$$

is closed. The forms θ_{2j}^Q vanish when the order of $(dF)^{2j+1}$ is less than $-\dim M$. This holds in particular if the order of $(dF)^{2j}$ is less than $-\dim M$ in which case $\omega_{2j}^{\mathrm{ren},\,Q} = \omega_{2j}^Q = \mathrm{tr}(F(dF)^{2j})$ is independent of Q.

Proof The form $d\omega_{2j}^Q$, being a noncommutative residue, is insensitive to smoothing perturbations and is therefore a pull-back by the projection map p of a form β_{2j}^Q. The pull-back of β_{2j}^Q with respect to σ is a closed form θ_{2j+1}^Q on X which is exact since σ is contractible. Indeed, selecting a contraction σ_t with $\sigma_1 = \sigma$ and σ_0 a constant map, we have the standard formula $d\theta_{2j}^Q = \theta_{2j+1}^Q$, with

$$\theta_{2j} = \frac{1}{2j+1}\int_0^1 t^{2j}\iota_{\dot\sigma_t}\theta_{2j+1}^Q(\sigma_t)dt, \tag{3.58}$$

where ι_X is the contraction by a vector field X and the dot means differentiation with respect to the parameter t.

When the order of $(dF)^{2j+1}$ is less than $-\dim M$ the correction terms θ_{2j}^Q vanish, and if the order of $(dF)^{2j}$ is less than $-\dim M$, the weighted trace tr^Q coincides with the usual trace so that the naive expression ω_{2j}^Q is a closed form independent of Q. $\qquad\square$

This way, one builds renormalised Chern classes $[\omega_{2j}^{\mathrm{ren},Q}]$. We refer the reader to [MP] for the 2-form case which arises in the quantum field theory gerbe [CMM].

3.15 Regular Chern–Weil forms in infinite dimensions

We describe further geometric setups for which weighted traces actually do give rise to closed Chern–Weil forms.

Mapping groups studied by Freed [F] and later further investigated in e.g. [CDMP], [Mick1], [MRT] provide a first illustration of such a situation.

Going back to Example 3.25, we specialise to the circle $M = S^1$; the Sobolev based loop group $H_e^{\frac{1}{2}}(S^1, G)$ can be equipped with a complex structure and its first Chern form was studied by Freed [F]. We saw in Example 3.34 that the corresponding curvature is conditionally trace-class which leads to the following result. Using (3.28) gives as an alternative description of Freed's conditioned first Chern form.

Proposition 16 ([F] Theorem 2.20; see also [CDMP] Proposition 3) Let Q_0 in $\mathrm{Cl}(S^1)$ be an admissible elliptic operator on S^1 with spectral cut θ and let $\nabla^{\frac{1}{2}}$ be a left-invariant connection on $C^\infty(S^1, G)$ with curvature given by a 2-form $\Omega_0^{\frac{1}{2}}$ as in Example 3.21.The following 2-form

$$\mathrm{tr}_\theta^{Q_0}(\Omega_0^{\frac{1}{2}}) = \mathrm{tr}_{\mathrm{cond}}(\Omega_0^{\frac{1}{2}}) = \mathrm{TR}(\Omega_0^{\frac{1}{2}})$$

is closed and coincides with Freed's conditioned first Chern form.

Remark 24 It was observed by Freed in [F] that this weighted first Chern form $\mathrm{tr}_\theta^{Q_0}(\Omega_0^{\frac{1}{2}})$ relates to the Kähler form on the based loop group $H_e^{\frac{1}{2}}(S^1, G)$. See also [CDMP] for further interpretations of this 2-form.

Another way around the obstructions described previously is to choose a weight $\mathbb{Q}$ and a connection ∇ such that the bracket $[\nabla, \log_\theta \mathbb{Q}]$ vanishes; this can be achieved using superconnections, leading to a second geometric setup in which regularised traces do give rise to closed Chern–Weil forms.

Definition 3.42 A *superconnection* (introduced by Quillen [Q]; see also [B], [BGV]) on an admissible vector bundle $\pi_* \mathbb{E}$, where $\pi : \mathbb{M} \to X$ is a locally trivial fibre bundle of manifolds, adapted to a smooth family $D \in C^\infty \left(X, \mathrm{Cl}^d(\mathbb{M}, \mathbb{E}) \right)$ of formally self-adjoint elliptic ψdos with odd parity, is a linear map $\mathbb{A}$ acting on $\Omega(X, \pi_* \mathbb{E})$ of odd parity with respect to the $\mathbb{Z}_2$-grading such that:

$$\mathbb{A}(\omega \cdot \sigma) = d\omega \wedge \sigma + (-1)^{|\omega|} \omega \wedge \mathbb{A}(\sigma) \quad \forall \omega \in \Omega(X), \quad \forall \sigma \in \Omega(X, \pi_* \mathbb{E})$$

and $\mathbb{A}_{[0]} := D$, where we have written $\mathbb{A} = \sum_{i=0}^{\dim B} \mathbb{A}_{[i]}$ and $\mathbb{A}_{[i]} : \Omega^*(X, \mathcal{E}) \mapsto \Omega^{*+i}(X, \mathcal{E})$.

Example 3.43 A ψdo-connection ∇ as in Definition 3.31 gives rise to a superconnection

$$\mathbb{A} := \nabla + D.$$

The curvature of a superconnection $\mathbb{A}$ is a ψdo-valued form

$$\mathbb{A}^2 \in \Omega^2(X, \mathrm{Cl}(M, \mathbb{E}));$$

it actually is a differential operator valued 2-form.

Following [Sc] and [PS2], we call a ψdo-valued form $\omega = \sum_{i=0}^{\dim B} \omega_{[i]}$ with $\omega_{[i]} : \Omega^*(X, \pi_* \mathbb{E}) \mapsto \Omega^{*+i}(X, \pi_* \mathbb{E})$ elliptic, admissible, or with spectral cut α whenever $\omega_{[0]} \in \mathrm{Cl}(X, \pi_* \mathbb{E})$ has these properties. We refer the reader to [PS2] for detailed explanations on this point.

Since

$$\mathbb{A}^2 = D^2 + \mathbb{A}_{[>0]}^2,$$

where $\mathbb{A}_{[>0]}^2$ is a ψdo-valued form of positive degree, we have $\mathbb{A}_{[0]}^2 = D^2$ so that the ψdo-valued form $\mathbb{A}^2$ is elliptic with spectral cut π. Assuming that $\mathbb{E}$ comes equipped with a Hermitian structure, one can build an L^2-inner product on the fibres $C^\infty(\mathbb{M}_x, \mathbb{E}_x)$ using the Riemannian structure on the compact fibres $\mathbb{M}_x$ and build a smooth family of orthogonal projectors π_{D_x} onto the kernel of D_x parametrised by X. Since $D^2 + \pi_D$ is admissible, it follows that

$$\mathbb{Q} := \mathbb{A}^2 + \pi_{D^2}$$

is an admissible ψdo-valued form. Its complex powers and logarithm can be defined as for ordinary admissible ψdos.

With these conventions, weighted traces associated with locally trivial fibre bundles of ψdo-algebras can be generalised to weights given by admissible ψdo-valued forms such as the curvature $\mathbb{A}^2$ of the superconnection. Along the lines of the proof of the previous theorem one can check that the trace defect $[\mathbb{A}, \text{tr}^Q]$ vanishes:

Proposition 17 [PS2] For any $\omega \in \Omega(X, \text{Cl}(\mathcal{E}))$ we have

$$d \, \text{str}^Q(\omega) = \text{str}^Q([\mathbb{A}, \omega]).$$

Proof Since

$$d \, \text{str}^Q(\omega) = \text{str}^Q([\mathbb{A}, \omega]) + [\mathbb{A}, \text{str}^Q](\omega)$$

this follows from Theorem 3.39 combined with the fact that $[\mathbb{A}, \log(\mathbb{A}^2 + \pi_D)]$ is smoothing. □

Theorem 3.44 *The form* $\text{str}^Q(\mathbb{A}^{2j})$ *defines a closed form called the jth Chern form associated with the superconnection* $\mathbb{A}$ *and a de Rham cohomology class independent of the choice of connection.*

Proof This follows from the Bianchi identity $[\mathbb{A}, \mathbb{A}^{2j}] = 0$ combined with Proposition 17. □

Remark 25 Let $\mathbb{A} = D + \nabla$ be a superconnection associated with a family of Dirac operators D on a trivial fibre bundle of manifolds. The expression

$$\text{tr}^Q(\nabla^{2j}) - \text{tr}^Q(\mathbb{A}^{2j})_{[2j]}$$

compares the naive infinite dimensional analog $\text{tr}^Q(\nabla^{2j})$, where we have set $Q := D^2 + \pi_D$ of the finite dimensional Chern form $\text{tr}(\nabla^{2j})$, with the closed form $\text{tr}^Q(\mathbb{A}^{2j})_{[2j]}$ built from the super connection. As was observed in [MP], it is local in so far as it is insensitive to smoothing perturbations of the connection. The weighted Chern–Weil form $\text{tr}^Q(\mathbb{A}^{2j})_{[2j]}$ is therefore interpreted as a renormalised version of $\text{tr}^Q(\nabla^{2j})$. This is similar to the formula derived in the previous section where a residue correction term was added to the naive weighted form involving the curvature.

If we specialise to a locally trivial fibre bundle $\pi : \mathbb{M} \to X$ of even dimensional closed spin manifolds with the Bismut superconnection $\mathbb{A} := D + \nabla + c(T)$, where c is the Clifford multiplication and T is the curvature of the horizontal distribution on $\mathbb{M}$, we get an explicit description of the jth Chern form $\text{str}^Q(\mathbb{A}^{2j})$. Indeed, as a consequence of the local index theorem for families [B] (see also [BGV]), the component of degree $2j$ of the form $\text{str}^Q(\mathbb{A}^{2j})$ can be expressed in terms of the $\hat{A}$-genus $\hat{A}(\mathbb{M}/X)$ of the vertical tangent bundle $T(\mathbb{M}/X)$ and the relative Chern character $\text{ch}(\mathbb{E}/\mathbb{S})$.

Theorem 3.45 [MP] *Given a family of Dirac operators on even dimensional spin manifolds parametrised by X and the Bismut superconnection $\mathbb{A}$ adapted to this*

Sylvie Paycha

family, we have

$$\text{str}^{\mathbb{Q}}\left(\mathbb{A}^{2j}\right)_{[2j]} = \frac{(-1)^j j!}{(2i\pi)^{\frac{n}{2}}} \left(\int_{M/X} \hat{A}(M/X) \wedge \text{ch}(\mathbb{E}/\mathbb{S})\right)_{[2j]}. \tag{3.59}$$

Proof Along the lines of the heat-kernel proof of the index theorem (see e.g. [BGV]) we introduce the kernel $k_\epsilon(\mathbb{A}^2)$ of $e^{-\epsilon \mathbb{A}^2}$ for some $\epsilon > 0$. Since D is a family of Dirac operators, we have (see e.g. ch. 10 in [BGV])

$$k_\epsilon(\mathbb{A}^2)(x, x) \sim_{\epsilon \to 0} \frac{1}{(4\pi\epsilon)^{\frac{n}{2}}} \sum_{j=0}^{\infty} \epsilon^j k_j(\mathbb{A}^2)(x, x). \tag{3.60}$$

We observe that the jth Chern form associated with $\mathbb{A}$ is given by an integration along the fibre of $\mathbb{M}$:

$$\text{str}^{\mathbb{Q}}\left(\mathbb{A}^{2j}\right) = \frac{(-1)^j j!}{(4\pi)^{\frac{n}{2}}} \int_{M/B} \text{str}(k_{j+\frac{n}{2}}(\mathbb{A}^2))$$

and proceed to compute $\text{str}(k_{j+\frac{n}{2}}(\mathbb{A}^2))$.

Let us introduce Getzler's rescaling, which transforms a homogeneous form $\alpha_{[i]}$ of degree i to the expression

$$\delta_t \cdot \alpha_{[i]} \cdot \delta_t^{-1} = \frac{\alpha_{[i]}}{\sqrt{t}^i},$$

so that a superconnection $\mathbb{A} = \mathbb{A}_{[0]} + \mathbb{A}_{[1]} + \mathbb{A}_{[2]}$ transforms to

$$\tilde{\mathbb{A}}_t = \delta_t \cdot \mathbb{A} \cdot \delta_t^{-1} = \mathbb{A}_{[0]} + \frac{\mathbb{A}_{[1]}}{\sqrt{t}} + \frac{\mathbb{A}_{[2]}}{t}.$$

As in [BGV] par. 10.4, in view of the asymptotic expansion (3.60) we have:

$$\text{ch}(\tilde{\mathbb{A}}_t) = \delta_t x \left(\text{str}(e^{-t\mathbb{A}^2})\right)$$

$$\sim_{t \to 0} (4\pi t)^{-\frac{n}{2}} \sum_j t^j \int_{M/B} \delta_t \left(\text{str}(k_j(\mathbb{A}^2))\right)$$

$$\sim_{t \to 0} (4\pi)^{-\frac{n}{2}} \sum_{j,p} t^{j-(n+p)/2} \left(\int_{M/B} \text{str}\left(k_j(\mathbb{A}^2)\right)\right)_{[p]},$$

so that

$$\text{fp}_{t=0}\text{ch}(\tilde{\mathbb{A}}_t)_{[p]} = (4\pi)^{-\frac{n}{2}} \left(\int_{M/X} \text{str}\left(k_{\frac{p+n}{2}}(\mathbb{A}^2)\right)\right)_{[p]}. \tag{3.61}$$

The family index theorem ([B]; see also Theorem 10.23 in [BGV]) yields the existence of the limit as $t \to 0$ and

$$\lim_{t \to 0} \mathrm{ch}(\mathbb{A}_t) = (2i\pi)^{-\frac{n}{2}} \int_{\mathbb{M}/X} \hat{A}(\mathbb{M}/X) \wedge \mathrm{ch}(\mathbb{E}/\mathbb{S}).$$

Combining these two facts leads to:

$$\left(\int_{\mathbb{M}/B} \mathrm{str}\left(k_{\frac{n+2j}{2}}(\mathbb{A}^2) \right) \right)_{[2j]} = \frac{(4\pi)^{\frac{n}{2}}}{(2i\pi)^{\frac{n}{2}}} \left(\int_{\mathbb{M}/X} \hat{A}(\mathbb{M}/X) \wedge \mathrm{ch}(\mathbb{E}/\mathbb{S}) \right)_{[2j]}.$$

It follows that

$$\mathrm{str}^{\mathbb{Q}}\left(\mathbb{A}^{2j} \right)_{[2j]} = \frac{(-1)^j j!}{(2i\pi)^{\frac{n}{2}}} \left(\int_{\mathbb{M}/X} \hat{A}(\mathbb{M}/X) \wedge \mathrm{ch}(\mathbb{E}/\mathbb{S}) \right)_{[2j]}. \qquad \square$$

As could be expected, $\mathbb{Q}$-weighted Chern forms therefore relate to the Chern character $(2i\pi)^{\frac{n}{2}} \int_{\mathbb{M}/X} \hat{A}(\mathbb{M}/X) \wedge \mathrm{ch}(\mathbb{E}/\mathbb{S})$ of a family of Dirac operators associated with the fibre bundle $\mathbb{M} \to X$ [BGV].

Acknowledgements

I am very grateful to Jouko Mickelsson and Steven Rosenberg for their comments on preliminary versions of this paper. I also thank Frédéric Rochon for his useful observations, Raphaël Ponge for drawing my attention to relevant references, Claude Roger for enlightening comments on certain infinite dimensional Lie algebra extensions and Karl-Hermann Neeb for his very helpful comments on a previous version of this survey. Last but not least, let me express my gratitude to Alexander Cardona, Iván Contreras and Andrés Reyes, coeditors of this volume, for their diligence in preparing these proceedings.

References

[Ad] M. Adler, On a trace functional for formal pseudodifferential operators and the symplectic structure of the Korteweg–de Vries type equation, *Invent. Math.* **50** (1987), 219–248.

[AB] M. F. Atiyah, R. Bott, The Yang–Mills equations over Riemann surfaces, *Phil. Trans. R. Soc. Lond.* **A 308** (1982), 523–615.

[ARS] M. R. Adams, T. Ratiu, R. Schmidt, The Lie group structure of diffeomorphism groups and invertible Fourier integral operators, with applications. In *Infinite-dimensional groups with applications*, (ed. V. Kac), Springer, 1985, pp. 1–69.

[AS] M. F. Atiyah, I. M. Singer, The index of elliptic operators: IV, *Ann. Math.* **93** (1971), 119–149.

[BGV] N. Berline, E. Getzler, M. Vergne, *Heat kernels and Dirac operators*, Grundlehren Math. Wiss. **298**, Springer Verlag, 1996.

[B]   J.-M Bismut, The Atiyah–Singer theorem for families of Dirac operators: two heat equation proofs, *Invent. Math.* **83** (1986), 91–151.

[BW]   B. Booss-Bavnbek, K. Wojciechowski, *Elliptic boundary problems for Dirac operators*, Mathematics: Theory and Applications, Birkhäuser, 1993.

[Bo]   J.-B Bost, Principe d'Oka, K-théorie et systèmes dynamiques non commutatifs, *Invent. Math.* **101** (1990), 261–333.

[Bott] R. Bott, On the Chern–Weil homomorphism and the continuous cohomology of Lie groups, *Adv. Math.* **11** (1973) 289–303.

[BT]   R. Bott, L. W. Tu, *Differential forms in algebraic topology*, Springer Verlag, 1982.

[BGJ] R. Bott, Lectures on characteristic classes and foliations. In *Lectures on algebraic and differential topology* (ed. R. Bott, S. Gitler, I. M. James), Lecture Notes in Math. **279**, Springer, 1972, pp. 1–94.

[BG]   J. L. Brylinski, E. Getzler, The homology of algebras of pseudodifferential symbols and non commutative residues, *K-theory* **1** (1987), 385–403.

[BL]   J. Brüning, M. Lesch, On the eta-invariant of certain nonlocal boundary value problems, *Duke Math. J.* **96**:2 (1999), 425–468.

[Bu]   T. Burak, On spectral projections of elliptic differential operators, *Ann. Scuola Norm. Sup. Pisa* **3**: 22 (1968), 113–132.

[CDP] A. Cardona, C. Ducourtioux, S. Paycha, From tracial anomalies to anomalies in quantum field theory, *Comm. Math. Phys.* **242** (2003), 31–65.

[CFNW] M. Cederwall, G. Ferretti, B. Nilsson, A. Westerberg, Schwinger terms and cohomology of pseudodifferential operators, *Comm. Math. Phys.* **175** (1996), 203–220.

[CDMP] A. Cardona, C. Ducourtioux, J.-P. Magnot, S. Paycha, Weighted traces on algebras of pseudodifferential operators and geometry on loop groups, *Inf. Dim. Anal. Quan. Prob. Rel. Top.* **5** (2002), 1–38.

[CMM] A. Carey, J. Mickelsson, M. Murray, Index theory, Gerbes, and Hamiltonian quantization, *Comm. Math. Phys.* **183** (1997), 707–722.

[C1]   S.-S Chern, A simple intrinsic proof of the Gauss–Bonnet formula for closed Riemannian manifolds, *Ann. Math.* **45** (1944), 747–762.

[C2]   S.-S Chern, *Topics in differential geometry*, Institute for Advanced Study, mimeographed lecture notes (1951).

[CM]   R. Cirelli, A. Manià, The group of gauge transformations as a Schwartz–Lie group, *J. Math. Phys.* **26** (1985), 3036–3041.

[D]   C. Ducourtioux, Weighted traces on pseudodifferential operators and associated determinants, PhD Thesis, Université Blaise Pascal, Clermont-Ferrand, 2001 (unpublished).

[F]   D. Freed, The geometry of loop groups, *J. Diff. Geom.* **28** (1988), 223–276.

[FGLS] B. V. Fedosov, F. Golse, E. Leichtnam, E. Schrohe, The noncommutative residue for manifolds with boundary, *J. Funct. Anal.* **142** (1996), 1–31.

[Gi]   P. Gilkey, *Invariance theory, the heat equation and the Atiyah–Singer index theorem*, Studies in Advanced Mathematics, CRC Press, 1995.

[Gl]   H. Glöckner, Algebras whose groups of units are Lie groups, *Studia Math.* **153** (2002), 147–177.

[GN]   H. Glöckner, K.-H. Neeb, *Introduction to infinite-dimensional Lie groups*, Vol. 1, in preparation.

[GR]   L. Guieu, C. Roger, *L'algèbre et le groupe de Virasoro: Aspects géometriques et algébriques* (French) [*Algebra and the Virasoro group: Geometric and algebraic aspects, generalizations*], Les Publications CRM, 2007.

[Gu]   V. Guillemin, Residue traces for certain algebras of Fourier integral operators, *J. Funct. Anal.* **115**: 2 (1993), 391–417.

[H]   L. Hörmander, *The analysis of linear partial differential operators III. Pseudodifferential operators*, Grundlehren Math. Wiss. **274**, Springer Verlag, 1994.

[Ka]   M. Karoubi, *K-theory (An introduction)*, Grundlehren Math. Wiss. **226**, Springer Verlag, 1978.

[Kas]   Ch. Kassel, Le résidu non commutatif (d'après M. Wodzicki), *Séminaire Bourbaki, Astérisque* **177–178** (1989), 199–229.

[KK]   O. Kravchenko, B. Khesin, Central extension of the Lie lagebra of (pseudo)-differential symbols, *Funct. Anal. Appl.* **25** (1991), 83–85.

[KV1]   M. Kontsevich, S. Vishik, Determinants of elliptic pseudodifferential operators, Max Planck Preprint (1994) (unpublished) arXiv-hep-th/9404046.

[KV2]   M. Kontsevich, S. Vishik, *Geometry of determinants of elliptic operators*. In *Functional analysis on the Eve of the 21st century*, Vol. 1 (ed. S. Gindikin, J. Lepowsky, R. Wilson). Progress in Mathematics **131**. Birkhäuser Boston, 1994, pp. 173–197.

[KM]   A. Kriegel, P. Michor, *The convenient setting of global analysis*, Mathematical Surveys and Monographs **53**, American Mathematical Society, 1997.

[La]   A. Larrain-Hubach, Explicit computations of the symbols of order 0 and -1 of the curvature operator of ΩG, *Lett. Math. Phys.* **89** (2009) 265–275.

[LRST] A. Larrain-Hubach, S. Rosenberg, S. Scott, F. Torres-Ardila, Characteristic classes and zeroth order pseudodifferential operators. In *Spectral theory and geometric analysis* (ed. M. Braverman, L. Friedlander, Th. Kappeler, P. Kuchment, P. Topalov and J. Weitsman), Cont. Math. **532**, American Mathematical Society, 2011, pp. 141–158.

[L]   M. Lesch, On the non commutative residue for pseudodifferential operators with log-polyhomogeneous symbols, *Ann. Global Anal. Geom.* **17** (1998), 151–187.

[Le]   J. Leslie, On a differential structure for the group of diffeomorphisms, *Topology* **6** (1967), 263–271.

[LMR]   E. Langmann, J. Mickelsson, S. Rydh, Anomalies and Schwinger terms in NCG field theory models, *J. Math. Phys.* **42** (2001), 4779.

[LN]   M. Lesch, C. Neira-Jimenez, Classification of traces and hypertraces on spaces of classical pseudodifferential operators, *J. Noncomm. Geom.* in press.

[LP]   J.-M. Lescure, S. Paycha, Traces on pseudo-differential operators and associated determinants, *Proc. Lond. Math. Soc.* **94**: 2 (2007), 772–812.

[Man]   Yu. I. Manin, Aspects algébriques des équations différentielles non linéaires, Itogi Nauk. i Tekhn. Sovrem. Probl. Matematik. **11** (1978) 5–152 (in Russian); Engl. transl. *J. Soviet Math.* **11** (1979) 1–122.

[MRT]   Y. Maeda, S. Rosenberg, F. Torres-Ardila, Riemannian geometry on loop spaces, arXiv:0705.1008 (2007).

[MSS]   L. Maniccia, E. Schrohe, J. Seiler, Uniqueness of the Kontsevich–Vishik trace, *Proc. Amer. Math. Soc.* **136** (2008), 747–752.

[MN]   R. Melrose, N. Nistor, Homology of pseudo-differential operators I. Manifolds with boundary, funct-an/9606005 (1999) (unpublished).

[Mich] P. Michor, *Gauge theory for fiber bundles*, Monographs and Textbooks in Physical Science **19**. Bibliopolis, 1991.

[Mick1] J. Mickelsson, Second quantization, anomalies and group extensions, Lecture notes given at the "Colloque sur les Méthodes Géométriques en physique", C.I.R.M, Luminy, June 1997.

[Mick2] J. Mickelsson, Noncommutative residue and anomalies on current algebras. In *Integrable models and strings* (ed. A. Alekseev *et al.*), Lecture Notes in Physics **436**, Springer Verlag, 1994.

[Mil] J. Milnor, Remarks on infinite dimensional Lie groups. In *Relativity, groups and topology II* In (ed. B. De Witt and R. Stora), North Holland, 1984.

[MP] J. Mickelsson, S. Paycha, Renormalised Chern–Weil forms associated with families of Dirac operators, *J. Geom. Phys.* **57** (2007), 1789–1814.

[MS] J. Milnor, J. Stasheff, *Characteristic classes*, Annals of Mathematics Studies **76**, Princeton University Press, University of Tokyo Press, 1974.

[N1] K.-H. Neeb, Central extensions of infinite-dimensional Lie groups, *Ann. Inst. Fourier* **52** (2002), 1365–1442.

[N2] K.-H. Neeb, Towards a Lie theory of locally convex groups, *Jap. J. Math.* **1** (2006), 291–468.

[Ok] K. Okikiolu, The multiplicative anomaly for determinants of elliptic operators, *Duke Math. J.* **79** (1995), 722–749.

[Om] H. Omori, On the group of diffeomorphisms of a compact manifold. In *Global analysis*, Proc. Sympos. Pure Math. **15**. American Mathematical Society, 1970, pp. 167–183. See also *Infinite dimensional Lie groups*, AMS Translations of Mathematical Monographs **158**, 1997.

[OMYK] H. Omori, Y. Maeda, A. Yoshida, O. Kobayashi, On regular Fréchet–Lie groups: Several basic properties, *Tokyo Math. J.* **6** (1986), 39–64.

[OP] M.-F. Ouedraogo, S. Paycha, The multiplicative anomaly for determinants revisited; locality. *Commun. Math. Anal.* **12** (2012) 28–63.

[P1] S. Paycha, Renormalised traces as a looking glass into infinite-dimensional geometry *Inf. Dim. Anal. Quan. Prob. Rel. Top.* **4** (2001), 221–266.

[P2] S. Paycha, The uniqueness of the Wodzicki residue and the canonical trace in the light of Stokes' and continuity properties, arXiv:0708.0531 (2007).

[P3] S. Paycha, *Regularised traces, integrals and sums; an analytic point of view*, American Mathematical Society University Lecture Notes **59**, American Mathematical Society, 2012.

[PR1] S. Paycha, S. Rosenberg, Curvature on determinant bundles and first Chern forms, *J. Geom. Phys.* **45** (2003), 393–429.

[PR2] S. Paycha, S. Rosenberg, Traces and characteristic classes in loop groups. In *Infinite dimensional groups and manifolds* (ed. T. Wurzbacher), I.R.M.A. Lectures in Mathematical and Theoretical Physics **5**. De Gruyter, 2004, pp. 185–212.

[PS1] S. Paycha, S. Scott, A Laurent expansion for regularised integrals of holomorphic symbols, *Geom. Funct. Anal.*, **17**:2 (2005), 491–536.

[PS2] S. Paycha, S. Scott, Chern–Weil forms associated with superconnections. In *Analysis, geometry and topology of ellipitc operators* (ed. B. Booss-Bavnbeck, S. Klimek, M. Lesch, W. Zhang), World Scientific, 2006, pp. 79–104.

[Po1] R. Ponge, Spectral asymmetry, zeta functions and the noncommutative residue, *Int. J. Math.* **17** (2006), 1065–1090.

[Po2] R. Ponge, Noncommutative residue for the Heisenberg calculus and applications in CR and contact geometry, *J. Funct. Anal.* **252** (2007), 399–463.

[Po3] R. Ponge, Traces on pseudodifferential operators and sums of commutators, arXiv:0707.4265v2 [math.AP] (2008).

[PS] A. Pressley, G. Segal, *Loop groups*, Oxford Mathematical Monographs, Oxford University Press, 1986.

[Q] D. Quillen, Superconnections and the Chern character, *Topology* **24** (1985), 89–95.

[Ra] A. O. Radul, Lie algebras on differential operators, their central extensions, and W-algebras, *Funct. Anal.* **25** (1991), 33–49.

[Ro] F. Rochon, Sur la topologie de l'espace des opérateurs pseudodifférentiels inversible d'ordre 0, *Ann. Inst. Fourier* **58**: 1 (2008), 29–62.

[Rog] C. Roger, Sur les origines du cocycle de Virasoro (2001). Published as a historical appendix in [GR].

[Schm] S. Schmid, Infinite dimensional Lie groups with applications to mathematical physics, *J. Geom. Symm. Phys.* **1** (2004), 1–67.

[Schr] E. Schrohe, Wodzicki's noncommutative residue and traces for operator algebras on manifolds with conical singularities. In *Microlocal analysis and spectral theory* (ed. L. Rodino), Proceedings of the NATO Advanced Study Institute, Il Ciocco, Castelvecchio Pascoli (Lucca), Italy, 1996, NATO ASI Ser. C, Math. Phys. Sci. **490**. Kluwer Academic Publishers, 1997, pp. 227–250.

[Sc] S. Scott, Zeta-Chern forms and the local family index theorem, *Trans. Amer. Math. Soc.* **359**: 5 (2007), 1925–1957.

[Se] R. T. Seeley, Complex powers of an elliptic operator. In *Singular integrals.* (Proc. Symp. Pure Math., Chicago) American Mathematical Society, 1966, pp. 288–307.

[Sh] A. Shubin, *Pseudodifferential operators and spectral theory*, Springer Verlag, 1980.

[T] M. E. Taylor, *Pseudodifferential operators*, Princeton University Press, 1981.

[Tr] F. Trèves, *Introduction to Pseudodifferential and Fourier integral operators*, vol. 1, Plenum Press, 1980.

[W1] M. Wodzicki, Spectral asymmetry and noncommutative residue (in Russian). Habilitation thesis, Steklov Institute (former) Soviet Academy of Sciences, Moscow, 1984.

[W2] M. Wodzicki, *Non commutative residue*, Chapter 1. Fundamentals, *K*-theory, arithmetic and geometry, Springer Lecture Notes **1289**. Springer, 1987, pp. 320–399.

[W3] M. Wodzicki, Report on the cyclic homology of symbols. Preprint, IAS Princeton, Jan. 87, Available online at http://math.berkeley.edu/ wodzicki.

[Woc] Ch. Wockel, Lie group structures on symmetry groups of principal bundles, *J. Funct. Anal.* **251** (2007), 254–288.

4

Introduction to Feynman integrals

STEFAN WEINZIERL

Abstract

In these lecture notes I will give an introduction to Feynman integrals. In the first part I review the basics of the perturbative expansion in quantum field theories. In the second part I will discuss more advanced topics: mathematical aspects of loop integrals related to periods, shuffle algebras and multiple polylogarithms are covered as well as practical algorithms for evaluating Feynman integrals.

4.1 Introduction

In these lecture notes I will give an introduction to perturbation theory and Feynman integrals occurring in quantum field theory. But before embarking onto a journey of integration and special function theory, it is worth recalling the motivation for such an effort.

High-energy physics is successfully described by the Standard Model. The term "Standard Model" has become a synonym for a quantum field theory based on the gauge group $SU(3) \otimes SU(2) \otimes U(1)$. At high energies all coupling constants are small and perturbation theory is a valuable tool to obtain predictions from the theory. For the Standard Model there are three coupling constants, g_1, g_2 and g_3, corresponding to the gauge groups $U(1)$, $SU(2)$ and $SU(3)$, respectively. As all methods which will be discussed below do not depend on the specific nature of these gauge groups and are even applicable to extensions of the Standard Model (like super-symmetry), I will just talk about a single expansion in a single coupling constant. All observable quantities are taken as a power series expansion in the coupling constant, and calculated order by order in perturbation theory.

Geometric and Topological Methods for Quantum Field Theory, ed. Alexander Cardona, Iván Contreras and Andrés F. Reyes-Lega. Published by Cambridge University Press. © Cambridge University Press 2013.

Over the years particle physics has become a field where precision measurements have become possible. Of course, the increase in experimental precision has to be matched with more accurate calculations from the theoretical side. This is the *raison d'être* for loop calculations: a higher accuracy is reached by including more terms in the perturbative expansion. There is even an additional "bonus" that we get from loop calculations: inside the loops we have to take into account all particles which could possibly circle there, even the ones which are too heavy to be produced directly in an experiment. Therefore loop calculations in combination with precision measurements allow us to extend the range of sensitivity of experiments from the region which is directly accessible towards the range of heavier particles which manifest themselves only through quantum corrections. As an example, the mass of the top quark was predicted before the discovery of the top quark from the loop corrections to electro-weak precision experiments. The same experiments currently predict a range for the mass of the yet undiscovered Higgs boson.

It is generally believed that a perturbative series is only an asymptotic series, which will diverge if more and more terms beyond a certain order are included. However, this will be of no concern to us here. We restrict ourselves to the first few terms in the perturbative expansion, with the implicit assumption that the point where the power series starts to diverge is far beyond our computational abilities. In fact, our computational abilities are rather limited. The complexity of a calculation obviously increases with the number of loops, but also with the number of external particles or the number of non-zero internal masses associated to propagators. To give an idea of the state of the art, specific quantities which are just pure numbers have been computed up to an impressive fourth or third order. Examples are the calculation of the four-loop contribution to the quantum chromodynamics (QCD) β-function [1], the calculation of the anomalous magnetic moment of the electron up to three loops [2], and the calculation of the ratio

$$R = \frac{\sigma(e^+ e^- \to \text{hadrons})}{\sigma(e^+ e^- \to \mu^+ \mu^-)} \tag{4.1}$$

of the total cross-section for hadron production to the total cross-section for the production of a $\mu^+ \mu^-$ pair in electron–positron annihilation to order $O(g_3^3)$ (also involving a three-loop calculation) [3]. Quantities which depend on a single variable are known at the best to the third order. Outstanding examples are the computation of the three-loop Altarelli–Parisi splitting functions or the calculation of the two-loop amplitudes for the most interesting $2 \to 2$ processes The complexity of a two-loop computation increases if the result depends on more than one variable. An example of a two-loop calculation whose result depends on two variables is the computation of the two-loop amplitudes for $e^+ e^- \to 3$ jets.

On the other hand the mathematics encountered in these calculations is of interest in its own right and has led in recent years to a fruitful interplay between mathematicians and physicists. Examples are the relation of Feynman integrals to periods, mixed Hodge structures and motives, as well as the occurrence of certain transcendental constants in the result of a calculation [4–23]. Typical transcendental constants which occur in the final results are multiple zeta values. They are obtained from multiple polylogarithms at special values of the arguments. I will discuss these functions in detail.

The outline of this chapter is as follows. In Section 4.2 I review the basics of perturbative quantum field theory and I give a brief outline of how Feynman rules are derived from the Lagrangian of the theory. Issues related to the regularisation of otherwise divergent integrals are treated in Section 4.3. Section 4.4 is devoted to basic techniques, which allow us to exchange the integrals over the loop momenta against integrals over Feynman parameters. In Section 4.5 I discuss how the Feynman parametrisation for a generic scalar l-loop integral can be read off directly from the underlying Feynman graph. The first part of this chapter closes with Section 4.6, which shows how finite results are obtained within perturbation theory. The remaining sections are more mathematical in nature: Section 4.7 states a general theorem which relates Feynman integrals to periods. Shuffle algebras and multiple polylogarithms are treated in Section 4.8 and Section 4.9, respectively. In Section 4.10 we discuss how multiple polylogarithms emerge in the calculation of Feynman integrals. Finally, Section 4.11 provides a summary.

4.2 Basics of perturbative quantum field theory

Elementary particle physics is described by quantum field theory. To begin with let us start with a single field $\phi(x)$. Important concepts in quantum field theory are the Lagrangian, the action and the generating functional. If $\phi(x)$ is a scalar field, a typical Lagrangian is

$$\mathcal{L} = \frac{1}{2}\left(\partial_\mu \phi(x)\right)\left(\partial^\mu \phi(x)\right) - \frac{1}{2}m^2\phi(x)^2 + \frac{1}{4}\lambda\phi(x)^4. \tag{4.2}$$

The quantity m is interpreted as the mass of the particle described by the field $\phi(x)$, the quantity λ describes the strength of the interactions among the particles. Integrating the Lagrangian over Minkowski space yields the action:

$$S[\phi] = \int d^4x \, \mathcal{L}(\phi). \tag{4.3}$$

The action is a functional of the field ϕ. In order to arrive at the generating functional we introduce an auxiliary field $J(x)$, called the source field, and integrate over all field configurations $\phi(x)$:

$$Z[J] = \mathcal{N} \int \mathcal{D}\phi \, e^{i\left(S[\phi] + \int d^4x \, J(x)\phi(x)\right)}. \tag{4.4}$$

The integral over all field configurations is an infinite-dimensional integral. It is called a path integral. The prefactor $\mathcal{N}$ is chosen such that $Z[0] = 1$. The n-point Green function is given by

$$\langle 0|T(\phi(x_1)\ldots\phi(x_n))|0\rangle = \frac{\int \mathcal{D}\phi \, \phi(x_1)\ldots\phi(x_n)e^{iS(\phi)}}{\int \mathcal{D}\phi \, e^{iS(\phi)}}. \tag{4.5}$$

With the help of functional derivatives this can be expressed as

$$\langle 0|T(\phi(x_1)\ldots\phi(x_n))|0\rangle = (-i)^n \left.\frac{\delta^n Z[J]}{\delta J(x_1)\ldots\delta J(x_n)}\right|_{J=0}. \tag{4.6}$$

We are in particular interested in connected Green functions. These are obtained from a functional $W[J]$, which is related to $Z[J]$ by

$$Z[J] = e^{iW[J]}. \tag{4.7}$$

The connected Green functions are then given by

$$G_n(x_1, \ldots, x_n) = (-i)^{n-1} \left.\frac{\delta^n W[J]}{\delta J(x_1)\ldots\delta J(x_n)}\right|_{J=0}. \tag{4.8}$$

It is convenient to go from position space to momentum space by a Fourier transformation. We define the Green functions in momentum space by

$$G_n(x_1, \ldots, x_n)$$
$$= \int \frac{d^4 p_1}{(2\pi)^4} \cdots \frac{d^4 p_n}{(2\pi)^4} e^{-i\sum p_j x_j} (2\pi)^4 \, \delta\left(p_1 + \cdots + p_n\right) \tilde{G}_n(p_1, \ldots, p_n). \tag{4.9}$$

Note that the Fourier transform $\tilde{G}_n$ is defined by explicitly factoring out the δ-function $\delta(p_1 + \cdots + p_n)$ and a factor $(2\pi)^4$. We denote the two-point function in momentum space by $\tilde{G}_2(p)$. In this case we have to specify only one momentum, since the momentum flowing into the Green function on one side has to be equal to the momentum flowing out of the Green function on the other side due to the presence of the δ-function in eq. (4.9). We now are in a position to define the scattering amplitude: in momentum space the scattering amplitude with n external particles is given by the connected n-point Green function multiplied by the inverse two-point function for each external particle:

$$\mathcal{A}_n(p_1, \ldots, p_n) = \tilde{G}_2(p_1)^{-1} \cdots \tilde{G}_2(p_n)^{-1} \tilde{G}_n(p_1, \ldots, p_n). \tag{4.10}$$

The scattering amplitude enters directly the calculation of a physical observable. Let us first consider the scattering process of two incoming particles with four-momenta p_1' and p_2' and $(n-2)$ outgoing particles with four-momenta p_1 to p_{n-2}. Let us assume that we are interested in an observable $O(p_1, \ldots, p_{n-2})$ which depends on the momenta of the outgoing particles. In general the observable depends on the experimental set-up and can be an arbitrary complicated function of the four-momenta. In the simplest case this function is just a constant equal to 1, corresponding to the situation where we count every event with $(n-2)$ particles in the final state. In more realistic situations one takes into account, for example that it is not possible to detect particles close to the beam pipe. The function O would then be zero in these regions of phase space. Furthermore any experiment has a finite resolution. Therefore it will not be possible to detect particles which are very soft or which are very close in angle to other particles. We will therefore sum over the number of final state particles. In order to obtain finite results within perturbation theory we have to require that, in the case where one or more particles become unresolved, the value of the observable O has a continuous limit agreeing with the value of the observable for a configuration where the unresolved particles have been merged into "hard" (or resolved) particles. Observables having this property are called infrared-safe observables. The expectation value for the observable O is given by

$$\langle O \rangle = \frac{1}{2(p_1' + p_2')^2} \sum_n \int d\phi_{n-2} O(p_1, \ldots, p_{n-2}) |\mathcal{A}_n|^2, \qquad (4.11)$$

where $1/2(p_1' + p_2')^2$ is a normalisation factor taking into account the incoming flux. The phase space measure is given by

$$d\phi_n = \frac{1}{n!} \prod_{i=1}^{n} \frac{d^3 p_i}{(2\pi)^3 2E_i} (2\pi)^4 \delta^4 \left(p_1' + p_2' - \sum_{i=1}^{n} p_i \right), \qquad (4.12)$$

where E_i is the energy of particle i:

$$E_i = \sqrt{\vec{p}_i^2 + m_i^2}. \qquad (4.13)$$

We see that the expectation value of O is given by the phase space integral over the observable, weighted by the norm squared of the scattering amplitude. As the integrand can be a rather complicated function, the phase space integral is usually performed numerically by Monte Carlo integration.

Let us now look towards a more realistic theory. As an example I will take quantum chromodynamics (QCD), which describes strong force and which is formulated in terms of quarks and gluons. Quarks and gluons are collectively called partons. There are a few modifications to eq. (4.11). The master formula now reads

$$\langle O \rangle = \sum_{a,b} \int dx_1 f_a(x_1) \int dx_2 f_b(x_2) \tag{4.14}$$

$$\times \frac{1}{2\hat{s} n_s(1) n_s(2) n_c(1) n_c(2)} \sum_n \int d\phi_{n-2} O(p_1, \ldots, p_{n-2}) \sum_{\text{spins,colour}} |\mathcal{A}_n|^2 .$$

The partons have internal degrees of freedom, given by the spin and the colour of the partons. In squaring the amplitude we sum over these degrees of freedom. For the particles in the initial state we would like to average over these degrees of freedom. This is done by dividing by the factors $n_s(i)$ and $n_c(i)$, giving the number of spin degrees of freedom (2 for quarks and gluons) and the number of colour degrees of freedom (3 for quarks, 8 for gluons). The second modification is due to the fact that the particles brought into collision are not partons, but composite particles like protons. At high energies the constituents of the protons interact and we have to include a function $f_a(x)$ giving us the probability of finding a parton a with momentum fraction x of the original proton momentum inside the proton. Here $\hat{s}$ is the centre-of-mass energy squared of the two partons entering the hard interaction. In addition there is a small change in eq. (4.12). The quantity $(n!)$ is replaced by $(\prod n_j!)$, where n_j is the number of times a parton of type j occurs in the final state.

As before, the scattering amplitude $\mathcal{A}_n$ can be calculated once the Lagrangian of the theory has been specified. For QCD the Lagrange density reads:

$$\mathcal{L}_{\text{QCD}} = -\frac{1}{4} F^a_{\mu\nu}(x) F^{a\mu\nu}(x) - \frac{1}{2\xi}(\partial^\mu A^a_\mu(x))^2$$

$$+ \sum_{\text{quarks } q} \bar{\psi}_q(x) \left(i\gamma^\mu D_\mu - m_q \right) \psi_q(x) + \mathcal{L}_{\text{FP}}, \tag{4.15}$$

with

$$F^a_{\mu\nu}(x) = \partial_\mu A^a_\nu(x) - \partial_\nu A^a_\mu(x) + g f^{abc} A^b_\mu(x) A^c_\nu,$$

$$D_\mu = \partial_\mu - i g T^a A^a_\mu(x). \tag{4.16}$$

The gluon field is denoted by $A^a_\mu(x)$, the quark fields are denoted by $\psi_q(x)$. The sum is over all quark flavours. The masses of the quarks are denoted by m_q. There is a summation over the colour indices of the quarks, which is not shown explicitly. The variable g gives the strength of the strong coupling. The generators of the group $SU(3)$ are denoted by T^a and satisfy

$$\left[T^a, T^b \right] = i f^{abc} T^c. \tag{4.17}$$

The quantity $F^a_{\mu\nu}$ is called the field strength, the quantity D_μ is called the covariant derivative. The variable ξ is called the gauge-fixing parameter. Gauge-invariant quantities like scattering amplitudes are independent of this parameter. The term

$\mathcal{L}_{\text{FP}}$ is the Faddeev–Popov term, which arises through the gauge-fixing procedure and which is only relevant for loop amplitudes.

Unfortunately it is not possible to calculate from this Lagrangian the scattering amplitude $\mathcal{A}_n$ exactly. The best that can be done is to expand the scattering amplitude in the small parameter g and to calculate the first few terms. The amplitude $\mathcal{A}_n$ with n external partons has the perturbative expansion

$$\mathcal{A}_n = g^{n-2} \left(\mathcal{A}_n^{(0)} + g^2 \mathcal{A}_n^{(1)} + g^4 \mathcal{A}_n^{(2)} + g^6 \mathcal{A}_n^{(3)} + \cdots \right). \tag{4.18}$$

In principle we could now calculate every term in this expansion by taking the functional derivatives according to eq. (4.8). This is rather tedious and there is a short-cut to arrive at the same result, which is based on Feynman graphs. The recipe for the computation of $\mathcal{A}_n^{(l)}$ is as follows: draw first all Feynman diagrams with the given number of external particles and l loops. Then translate each graph into a mathematical formula with the help of the Feynman rules. $\mathcal{A}_n^{(l)}$ is then given as the sum of all these terms.

In order to derive the Feynman rules from the Lagrangian one proceeds as follows: one first separates the Lagrangian into a part which is bilinear in the fields, and a part where each term contains three or more fields. (A "normal" Lagrangian does not have parts with just one or zero fields.) From the part bilinear in the fields one derives the propagators, while the terms with three or more fields give rise to vertices. As an example we consider the gluonic part of the QCD Lagrange density:

$$\mathcal{L}_{\text{QCD}} = \frac{1}{2} A_\mu^a(x) \left[\partial_\rho \partial^\rho g^{\mu\nu} \delta^{ab} - \left(1 - \frac{1}{\xi} \right) \partial^\mu \partial^\nu \delta^{ab} \right] A_\nu^b(x) \tag{4.19}$$

$$- g f^{abc} \left(\partial_\mu A_\nu^a(x) \right) A^{b\mu}(x) A^{c\nu}(x)$$

$$- \frac{1}{4} g^2 f^{eab} f^{ecd} A_\mu^a(x) A_\nu^b(x) A^{c\mu}(x) A^{d\nu}(x) + \mathcal{L}_{\text{quarks}} + \mathcal{L}_{\text{FP}}.$$

Within perturbation theory we always assume that all fields fall off rapidly enough at infinity. Therefore we can ignore boundary terms within partial integrations. The expression in the first line is bilinear in the fields. The terms in the square bracket in this line define an operator

$$P^{\mu\nu\,ab}(x) = \partial_\rho \partial^\rho g^{\mu\nu} \delta^{ab} - \left(1 - \frac{1}{\xi} \right) \partial^\mu \partial^\nu \delta^{ab}. \tag{4.20}$$

For the propagator we are interested in the inverse of this operator

$$P^{\mu\sigma\,ac}(x) \left(P^{-1} \right)_{\sigma\nu}^{cb} (x - y) = g^\mu_{\ \nu} \delta^{ab} \delta^4(x - y). \tag{4.21}$$

Working in momentum space we are more specifically interested in the Fourier transform of the inverse of this operator:

$$\left(P^{-1} \right)_{\mu\nu}^{ab} (x) = \int \frac{d^4 k}{(2\pi)^4} e^{-ik\cdot x} \left(\tilde{P}^{-1} \right)_{\mu\nu}^{ab} (k). \tag{4.22}$$

The Feynman rule for the propagator is then given by $(\tilde{P}^{-1})^{ab}_{\mu\nu}(k)$ times the imaginary unit. For the gluon propagator one finds the Feynman rule

$$\mu, a \;\text{00000}\; \nu, b = \frac{i}{k^2}\left(-g_{\mu\nu} + (1-\xi)\frac{k_\mu k_\nu}{k^2}\right)\delta^{ab}. \tag{4.23}$$

To derive the Feynman rules for the vertices we look as an example at the first term in the second line of eq. (4.19):

$$\mathcal{L}_{ggg} = -gf^{abc}\left(\partial_\mu A^a_\nu(x)\right) A^{b\mu}(x)A^{c\nu}(x). \tag{4.24}$$

This term contains three gluon fields and will give rise to the three-gluon vertex. We rewrite this term as follows:

$$\mathcal{L}_{ggg} = \int d^4x_1 d^4x_2 d^4x_3 \alpha^{abc\,\mu\nu\lambda}(x, x_1, x_2, x_3)A^a_\mu(x_1)A^b_\nu(x_2)A^c_\lambda(x_3), \tag{4.25}$$

where

$$\alpha^{abc\,\mu\nu\lambda}(x, x_1, x_2, x_3) = gf^{abc}g^{\mu\lambda}\left(\partial^\nu_{x_1}\delta^4(x-x_1)\right)\delta^4(x-x_2)\delta^4(x-x_3). \tag{4.26}$$

Again we are interested in the Fourier transform of this expression:

$$\alpha^{abc\,\mu\nu\lambda}(x, x_1, x_2, x_3)$$
$$= \int \frac{d^4k_1}{(2\pi)^4}\frac{d^4k_2}{(2\pi)^4}\frac{d^4k_3}{(2\pi)^4}e^{-ik_1(x_1-x)-ik_2(x_2-x)-ik_3(x_3-x)}\tilde{\alpha}^{abc\,\mu\nu\lambda}(k_1, k_2, k_3).$$

Working this out we find

$$\tilde{\alpha}^{abc\,\mu\nu\lambda}(k_1, k_2, k_3) = -gf^{abc}g^{\mu\lambda}ik^\nu_1. \tag{4.27}$$

The Feynman rule for the vertex is then given by the sum over all permutations of identical particles of the function $\tilde{\alpha}$ multiplied by the imaginary unit i. (In the case of identical fermions there would also be a minus sign for every odd permutation of the fermions.) We thus obtain the Feynman rule for the three-gluon vertex:

$$\begin{array}{c}\\ k^\mu_1, a\\ \\ k^\lambda_3, c \qquad k^\nu_2, b\end{array} = gf^{abc}\left[g^{\mu\nu}\left(k^\lambda_2 - k^\lambda_1\right) + g^{\nu\lambda}\left(k^\mu_3 - k^\mu_2\right) + g^{\lambda\mu}\left(k^\nu_1 - k^\nu_3\right)\right].$$
$$\tag{4.28}$$

Note that there is momentum conservation at each vertex; for the three-gluon vertex this implies

$$k_1 + k_2 + k_3 = 0. \tag{4.29}$$

Following the procedures outlined above we can derive the Feynman rules for all propagators and vertices of the theory. If an external particle carries spin, we have to associate a factor, which describes the polarisation of the corresponding particle when we translate a Feynman diagram into a formula. Thus, there is a polarisation vector $\varepsilon^\mu(p)$ for each external gauge boson and a spinor $\bar{u}(p)$, $u(p)$, $\bar{v}(p)$ or $v(p)$ for each external fermion.

Furthermore there are a few additional rules. First of all, there is an integration

$$\int \frac{d^4 k}{(2\pi)^4} \tag{4.30}$$

for each internal momentum which is not constrained by momentum conservation. Such an integration is called a "loop integration" and the number of independent loop integrations in a diagram is called the loop number of the diagram. Secondly, each closed fermion loop gets an extra factor of (-1). Finally, each diagram gets multiplied by a symmetry factor $1/S$, where S is the order of the permutation group of the internal lines and vertices leaving the diagram unchanged when the external lines are fixed.

Let us finish this section by listing the remaining Feynman rules for QCD. The quark and the ghost propagators are given by

$$j \underleftarrow{\hspace{2cm}} l = i \frac{\not{k} + m}{k^2 - m^2} \delta_{jl},$$

$$a \underleftarrow{\text{-----}} b = \frac{i}{k^2} \delta^{ab}. \tag{4.31}$$

The Feynman rules for the four-gluon vertex, the quark–gluon vertex and the ghost-gluon vertex are

$$= -ig^2 \left[f^{abe} f^{ecd} \left(g^{\mu\lambda} g^{\nu\rho} - g^{\mu\rho} g^{\nu\lambda} \right) + f^{ace} f^{ebd} g^{\mu\nu} g^{\lambda\rho} \right.$$
$$\left. - f^{ace} f^{ebd} g^{\mu\rho} g^{\lambda\nu} + f^{ade} f^{ebc} \left(g^{\mu\nu} g^{\lambda\rho} - g^{\mu\lambda} g^{\nu\rho} \right) \right],$$

$$= ig\gamma^\mu T^a_{jl}, \tag{4.32}$$

$$= -g f^{abc} k_\mu.$$

The Feynman rules for the electro-weak sector of the Standard Model are similar, but too numerous to list explicitly here.

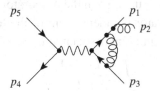

Figure 4.1 A one-loop Feynman diagram contributing to the process $e^+e^- \rightarrow q g \bar{q}$.

Having stated the Feynman rules, let us look at some examples. We have seen that for a given process with a specified set of external particles the scattering amplitude is given as the sum of all Feynman diagrams with this set of external particles. We can order the diagrams by the powers of the coupling factors. In QCD we obtain for each three-particle vertex one power of g, while the four-gluon vertex contributes two powers of g. The leading order result for the scattering amplitude is obtained by taking only the diagrams with the minimal number of coupling factors g into account. These are diagrams which have no closed loops. There are no conceptual difficulties in evaluating these diagrams. However, going beyond the leading order in perturbation theory, loop diagrams appear which involve integrations over the loop momenta. These diagrams are more difficult to evaluate and I will discuss them in more detail. Fig. 4.1 shows a Feynman diagram contributing to the one-loop corrections for the process $e^+e^- \rightarrow q g \bar{q}$. At high energies we can ignore the masses of the electron and the light quarks. From the Feynman rules one obtains for this diagram:

$$-e^2 g^3 C_F T_{jl}^a \frac{\bar{v}(p_4)\gamma^\mu u(p_5)}{p_{123}^2} \int \frac{d^4 k_1}{(2\pi)^4} \frac{1}{k_2^2} \bar{u}(p_1)\displaystyle{\not}\epsilon(p_2)\frac{\displaystyle{\not}p_{12}}{p_{12}^2}\gamma_\nu \frac{\displaystyle{\not}k_1}{k_1^2}\gamma_\mu \frac{\displaystyle{\not}k_3}{k_3^2}\gamma^\nu v(p_3). \quad (4.33)$$

Here, $p_{12} = p_1 + p_2$, $p_{123} = p_1 + p_2 + p_3$, $k_2 = k_1 - p_{12}$, $k_3 = k_2 - p_3$. Further $\displaystyle{\not}\epsilon(p_2) = \gamma_\tau \epsilon^\tau(p_2)$, where $\epsilon^\tau(p_2)$ is the polarisation vector of the outgoing gluon. All external momenta are assumed to be massless: $p_i^2 = 0$ for $i = 1, \dots, 5$. We can reorganise this formula into a part which depends on the loop integration and a part which does not. The loop integral to be calculated is:

$$\int \frac{d^4 k_1}{(2\pi)^4} \frac{k_1^\rho k_3^\sigma}{k_1^2 k_2^2 k_3^2}, \quad (4.34)$$

while the remainder, which is independent of the loop integration, is given by

$$-e^2 g^3 C_F T_{jl}^a \bar{v}(p_4)\gamma^\mu u(p_5)\frac{1}{p_{123}^2 p_{12}^2}\bar{u}(p_1)\displaystyle{\not}\epsilon(p_2)\displaystyle{\not}p_{12}\gamma_\nu \gamma_\rho \gamma_\mu \gamma_\sigma \gamma^\nu v(p_3). \quad (4.35)$$

The loop integral in eq. (4.34) contains in the denominator three propagator factors and in the numerator two factors of the loop momentum. We call a loop integral in

which the loop momentum occurs also in the numerator a "tensor integral". A loop
integral in which the numerator is independent of the loop momentum is called a
"scalar integral". The scalar integral associated to eq. (4.34) is given by

$$\int \frac{d^4k_1}{(2\pi)^4} \frac{1}{k_1^2 k_2^2 k_3^2}. \tag{4.36}$$

It is always possible to reduce tensor integrals to scalar integrals [24, 25]. The
calculation of integrals like the one in eq. (4.36) is the main topic of this chapter.
More information on the basics of perturbation theory and quantum field theory can
be found in one of the many textbooks on quantum field theory, like for example
[26, 27].

4.3 Dimensional regularisation

Before we start with the actual calculation of loop integrals, I should mention one
complication: loop integrals are often divergent! Let us first look at the simple
example of a scalar two-point one-loop integral with zero external momentum:

$$p = 0 \quad = \int \frac{d^4k}{(2\pi)^4} \frac{1}{(k^2)^2} = \frac{1}{(4\pi)^2} \int\limits_0^\infty dk^2 \frac{1}{k^2}$$

$$= \frac{1}{(4\pi)^2} \int\limits_0^\infty \frac{dx}{x}. \tag{4.37}$$

This integral diverges at $k^2 \to \infty$ as well as at $k^2 \to 0$. The former divergence is
called ultraviolet (UV) divergence, the latter is called infrared (IF) divergence. Any
quantity which is given by a divergent integral is of course an ill-defined quantity.
Therefore the first step is to make these integrals well-defined by introducing a
regulator. There are several possibilities for how this can be done, but the method
of dimensional regularisation [28–30] has almost become a standard, as the calculations in this regularisation scheme turn out to be the simplest. Within dimensional
regularisation one replaces the four-dimensional integral over the loop momentum
by a D-dimensional integral, where D is now an additional parameter, which can
be a non-integer or even a complex number. We consider the result of the integration as a function of D and we are interested in the behaviour of this function as
D approaches 4. The D-dimensional integration still fulfils the standard laws for
integration, like linearity, translation invariance and scaling behaviour [31, 32]. If
f and g are two functions, and if a and b are two constants, linearity states that

$$\int d^Dk \left(af(k) + bg(k)\right) = a \int d^Dk f(k) + b \int d^Dk g(k). \tag{4.38}$$

Translation invariance requires that

$$\int d^D k f(k + p) = \int d^D k f(k).$$ (4.39)

for any vector p.

The scaling law states that

$$\int d^D k f(\lambda k) = \lambda^{-D} \int d^D k f(k).$$ (4.40)

The D-dimensional integral has also a rotation invariance:

$$\int d^D k f(\Lambda k) = \int d^D k f(k),$$ (4.41)

where Λ is an element of the Lorentz group $SO(1, D - 1)$ of the D-dimensional vector-space. Here we assumed that the D-dimensional vector-space has the metric diag$(+1, -1, -1, -1, \ldots)$. The integral measure is normalised such that it agrees with the result for the integration of a Gaussian function for all integer values D:

$$\int d^D k \exp\left(\alpha k^2\right) = i \left(\frac{\pi}{\alpha}\right)^{\frac{D}{2}}.$$ (4.42)

We will further assume that we can always decompose any vector into a 4-dimensional part and a $(D - 4)$-dimensional part

$$k^\mu_{(D)} = k^\mu_{(4)} + k^\mu_{(D-4)},$$ (4.43)

and that the 4-dimensional and $(D - 4)$-dimensional subspaces are orthogonal to each other:

$$k_{(4)} \cdot k_{(D-4)} = 0.$$ (4.44)

If D is an integer greater than 4, this is obvious. We postulate that these relations are true for any value of D. One can think of the underlying vector-space as a space of infinite dimension, where the integral measure mimics the one in D dimensions.

In practice we will always arrange things such that every function we integrate over D dimensions is rotational invariant, for example is a function of k^2. In this case the integration over the $(D - 1)$ angles is trivial and can be expressed in a closed form as a function of D. Let us assume that we have an integral, which has a UV-divergence, but no IR-divergences. Let us further assume that this integral would diverge logarithmically if we used a cut-off regularisation instead of dimensional regularisation. It turns out that this integral will be convergent if the real part of D is smaller than 4. Therefore we may compute this integral under the assumption that $\text{Re}(D) < 4$ and we will obtain a function of D as a result. This function can be analytically continued to the whole complex plane. We are mainly interested in what happens close to the point $D = 4$. For a UV-divergent one-loop integral we will find that the analytically continued result will exhibit a pole at $D = 4$. It should be mentioned that there are also integrals which are quadratically divergent, if a cut-off regulator is used. In this case we can repeat the argument above with the replacement $\text{Re}(D) < 2$.

Similarly, we can consider an IR-divergent integral, which has no UV-divergence. This integral will be convergent if $\text{Re}(D) > 4$. Again, we can compute the integral in this domain and continue the result to $D = 4$. Here we find that each IR-divergent loop integral can lead to a double pole at $D = 4$.

We will use dimensional regularisation to regulate both the ultraviolet and infrared divergences. The attentive reader may ask how this goes together, as we argued above that UV-divergences require $\text{Re}(D) < 4$ or even $\text{Re}(D) < 2$, whereas IR-divergences are regulated by $\text{Re}(D) > 4$. Suppose for the moment that we use dimensional regularisation just for the UV-divergences and that we use a second regulator for the IR-divergences. For the IR-divergences we could keep all external momenta off-shell, or introduce small masses for all massless particles or even raise the original propagators to some power ν. The exact implementation of this regulator is not important, as long as the IR-divergences are screened by this procedure. We then perform the loop integration in the domain where the integral is UV-convergent. We obtain a result which we can analytically continue to the whole complex D-plane, in particular to $\text{Re}(D) > 4$. There we can remove the additional regulator and the IR-divergences are now regulated by dimensional regularisation. Then the infrared divergences will also show up as poles at $D = 4$.

There is one more item which needs to be discussed in the context of dimensional regularisation. Let us look again at the example in eqs. (4.33) to (4.35). We separated the loop integral from the remainder in eq. (4.35), which is independent of the loop integration. In this remainder the following string of Dirac matrices occurs:

$$\gamma_\nu \gamma_\rho \gamma_\mu \gamma_\sigma \gamma^\nu. \tag{4.45}$$

If we anti-commute the first Dirac matrix, we can achieve that the two Dirac matrices with index ν are next to each other:

$$\gamma_\nu \gamma^\nu. \tag{4.46}$$

In four dimensions this equals 4 times the unit matrix. What is the value within dimensional regularisation? The answer depends on how we treat the Dirac algebra. Does the Dirac algebra remain in four dimensions or do we also continue the Dirac algebra to D dimensions? There are several schemes on the market which treat this issue differently. To discuss these schemes it is best to look how they treat the momenta and the polarisation vectors of observed and unobserved particles. Unobserved particles are particles circulating inside loops or emitted particles not resolved within a given detector resolution. The most commonly used schemes are the conventional dimensional regularisation scheme (CDR) [32], where all momenta and all polarisation vectors are taken to be in D dimensions, the 't Hooft–Veltman scheme (HV) [28, 33], where the momenta and the helicities of the unobserved particles are D-dimensional, whereas the momenta and the helicities of the observed particles are four-dimensional, and the four-dimensional helicity scheme (FD) [34–36], where all polarisation vectors are kept in four dimensions, as well as the momenta of the observed particles. Only the momenta of the unobserved particles are continued to D dimensions.

The conventional scheme is mostly used for an analytical calculation of the interference of a one-loop amplitude with the Born amplitude by using polarisation sums corresponding to D dimensions. For the calculation of one-loop helicity amplitudes the 't Hooft–Veltman scheme and the four-dimensional helicity scheme are possible choices. All schemes have in common that the propagators appearing in the denominator of the loop integrals are continued to D dimensions. They differ in how they treat the algebraic part in the numerator. In the 't Hooft–Veltman scheme the algebraic part is treated in D dimensions, whereas in the FD scheme the algebraic part is treated in four dimensions. It is possible to relate results obtained in one scheme to another scheme, using simple and universal transition formulae [37–39]. Therefore, if we return to the example above, we have

$$\gamma_\nu \gamma^\nu = \begin{cases} D \cdot \mathbf{1}, & \text{in the CDR and HV scheme,} \\ 4 \cdot \mathbf{1}, & \text{in the FD scheme.} \end{cases} \tag{4.47}$$

To summarise, we are interested into loop integrals regulated by dimensional regularisation. As a result we seek the Laurent expansion around $D = 4$. It is common practice to parametrise the deviation of D from 4 by

$$D = 4 - 2\varepsilon. \tag{4.48}$$

Divergent loop integrals will therefore have poles in $1/\varepsilon$. In an l-loop integral ultraviolet divergences will lead to poles $1/\varepsilon^l$ at the worst, whereas infrared divergences can lead to poles up to $1/\varepsilon^{2l}$.

4.4 Loop integration in D dimensions

In this section I will discuss how to perform the D-dimensional loop integrals. It would be more correct to say that we exchange them for some parameter integrals. As an example we take the one-loop integral of eq. (4.36):

$$I = \int \frac{d^D k_1}{i\pi^{D/2}} \frac{1}{(-k_1^2)(-k_2^2)(-k_3^2)}. \tag{4.49}$$

The integration is now in D dimensions. In eq. (4.49) there are some overall factors, which I inserted for convenience: the integral measure is now $d^D k/(i\pi^{D/2})$ instead of $d^D k/(2\pi)^D$, and each propagator is multiplied by (-1). The reason for doing this is that the final result will be simpler.

As already discussed above, the only functions we really want to integrate over D dimensions are the ones which depend on the loop momentum only through k^2. The integrand in eq. (4.49) is not yet in such a form. To bring the integrand into this form, we first convert the product of propagators into a sum. We can do this with the Feynman parameter technique. In its full generality it is also applicable to cases where each factor in the denominator is raised to some power ν. The formula reads:

$$\prod_{i=1}^{n} \frac{1}{(-P_i)^{\nu_i}} = \frac{\Gamma(\nu)}{\prod\limits_{i=1}^{n} \Gamma(\nu_i)} \int_0^1 \left(\prod_{i=1}^{n} dx_i \, x_i^{\nu_i-1} \right) \frac{\delta\left(1 - \sum\limits_{i=1}^{n} x_i\right)}{\left(-\sum\limits_{i=1}^{n} x_i P_i\right)^\nu}, \tag{4.50}$$

with $\nu = \sum_{i=1}^{n} \nu_i$. The proof of this formula can be found in many textbooks and is not repeated here. $\Gamma(x)$ is Euler's gamma function, $\delta(x)$ denotes Dirac's delta function. The price we have to pay for converting the product into a sum are $(n-1)$ additional integrations. Let us look at the example from eq. (4.36):

$$\frac{1}{(-k_1^2)(-k_2^2)(-k_3^2)} = 2 \int_0^1 dx_1 \int_0^1 dx_2 \int_0^1 dx_3 \frac{\delta(1 - x_1 - x_2 - x_3)}{\left(-x_1 k_1^2 - x_2 k_2^2 - x_3 k_3^2\right)^3}. \tag{4.51}$$

In the next step we complete the square and shift the loop momentum, so that the integrand becomes a function of k^2. With $k_2 = k_1 - p_{12}$ and $k_3 = k_2 - p_3$ we have

$$-x_1 k_1^2 - x_2 k_2^2 - x_3 k_3^2$$
$$= -(k_1 - x_2 p_{12} - x_3 p_{123})^2 - x_1 x_2 s_{12} - x_1 x_3 s_{123}, \tag{4.52}$$

where $s_{12} = (p_1 + p_2)^2$ and $s_{123} = (p_1 + p_2 + p_3)^2$. We can now define

$$k_1' = k_1 - x_2 p_{12} - x_3 p_{123} \tag{4.53}$$

and using translational invariance our loop integral becomes

$$I = 2 \int \frac{d^D k_1'}{i\pi^{D/2}} \int_0^1 dx_1 \int_0^1 dx_2 \int_0^1 dx_3 \frac{\delta(1 - x_1 - x_2 - x_3)}{\left(-k_1'^2 - x_1 x_2 s_{12} - x_1 x_3 s_{123}\right)^3}. \tag{4.54}$$

The integrand is now a function of $k_1'^2$, which we can relabel as k^2.

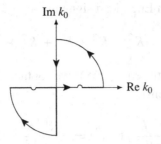

Figure 4.2 Integration contour for the Wick rotation. The little circles along the real axis exclude the poles.

Having succeeded in rewriting the integrand as a function of k^2, we then perform a Wick rotation, which transforms Minkowski space into a Euclidean space. Remember that k^2 written out in components in D-dimensional Minkowski space reads

$$k^2 = k_0^2 - k_1^2 - k_2^2 - k_3^2 - \cdots \qquad (4.55)$$

(Here k_j denotes the jth component of the vector k, in contrast to the previous notation, where we used the subscript to label different vectors k_j. It should be clear from the context what is meant.) Furthermore, when integrating over k_0, we encounter poles which are avoided by Feynman's $i\delta$-prescription. In the complex k_0-plane we consider the integration contour shown in Fig. 4.2. Since the contour does not enclose any poles, the integral along the complete contour is zero:

$$\oint dk_0 f(k_0) = 0. \qquad (4.56)$$

If the quarter-circles at infinity give a vanishing contribution (it can be shown that this is the case) we obtain

$$\int_{-\infty}^{\infty} dk_0 f(k_0) = - \int_{i\infty}^{-i\infty} dk_0 f(k_0). \qquad (4.57)$$

We now make the following change of variables:

$$k_0 = i K_0,$$
$$k_j = K_j, \quad \text{for } 1 \leq j \leq D - 1. \qquad (4.58)$$

As a consequence we have

$$k^2 = -K^2, \quad d^D k = i d^D K, \qquad (4.59)$$

where K^2 is now given with Euclidean signature:

$$K^2 = K_0^2 + K_1^2 + K_2^2 + K_3^2 + \cdots \tag{4.60}$$

Combining eq. (4.57) with eq. (4.58) we obtain for the integration of a function $f(k^2)$ over D dimensions

$$\int \frac{d^D k}{i\pi^{D/2}} f(-k^2) = \int \frac{d^D K}{\pi^{D/2}} f(K^2), \tag{4.61}$$

whenever there are no poles inside the contour of Fig. 4.2 and the arcs at infinity give a vanishing contribution. The integral on the right-hand side is now over D-dimensional Euclidean space. Equation (4.61) justifies our conventions to introduce a factor i in the denominator and a minus sign for each propagator in eq. (4.49). These conventions are just such that after Wick rotation we have simple formulae.

We now have an integral over D-dimensional Euclidean space, where the integrand depends only on K^2. It is therefore natural to introduce spherical coordinates. In D dimensions they are given by

$$K_0 = K \cos\theta_1,$$

$$K_1 = K \sin\theta_1 \cos\theta_2,$$

$$\vdots$$

$$K_{D-2} = K \sin\theta_1 \ldots \sin\theta_{D-2} \cos\theta_{D-1},$$

$$K_{D-1} = K \sin\theta_1 \ldots \sin\theta_{D-2} \sin\theta_{D-1}. \tag{4.62}$$

In D dimensions we have one radial variable K, $D-2$ polar angles θ_j (with $1 \le j \le D-2$) and one azimuthal angle θ_{D-1}. The measure becomes

$$d^D K = K^{D-1} dK d\Omega_D, \quad d\Omega_D = \prod_{i=1}^{D-1} \sin^{D-1-i}\theta_i \, d\theta_i. \tag{4.63}$$

Integration over the angles yields

$$\int d\Omega_D = \int_0^\pi d\theta_1 \sin^{D-2}\theta_1 \ldots \int_0^\pi d\theta_{D-2} \sin\theta_{D-2} \int_0^{2\pi} d\theta_{D-1} = \frac{2\pi^{D/2}}{\Gamma\left(\frac{D}{2}\right)}. \tag{4.64}$$

Note that the integration on the left-hand side of eq. (4.64) is defined for any natural number D, whereas the result on the right-hand side is an analytic function of D, which can be continued to any complex value.

It is now the appropriate place to say a few words on Euler's gamma function. The gamma function is defined for $\text{Re}(x) > 0$ by

$$\Gamma(x) = \int_0^\infty e^{-t} t^{x-1} dt. \qquad (4.65)$$

It fulfils the functional equation

$$\Gamma(x+1) = x \, \Gamma(x). \qquad (4.66)$$

For positive integers n it takes the values

$$\Gamma(n+1) = n! = 1 \cdot 2 \cdot 3 \cdot \ldots \cdot n. \qquad (4.67)$$

For integers n we have the reflection identity

$$\frac{\Gamma(x-n)}{\Gamma(x)} = (-1)^n \frac{\Gamma(1-x)}{\Gamma(1-x+n)}. \qquad (4.68)$$

The gamma function $\Gamma(x)$ has poles located on the negative real axis at $x = 0, -1, -2, \ldots$. Quite often we will need the expansion around these poles. This can be obtained from the expansion around $x = 1$ and the functional equation. The expansion around $\varepsilon = 1$ is given by

$$\Gamma(1+\varepsilon) = \exp\left(-\gamma_E \varepsilon + \sum_{n=2}^\infty \frac{(-1)^n}{n} \zeta_n \varepsilon^n\right), \qquad (4.69)$$

where γ_E is Euler's constant

$$\gamma_E = \lim_{n \to \infty} \left(\sum_{j=1}^n \frac{1}{j} - \ln n\right) = 0.577\,215\,664\,9\ldots \qquad (4.70)$$

and ζ_n is given by

$$\zeta_n = \sum_{j=1}^\infty \frac{1}{j^n}. \qquad (4.71)$$

For example, we obtain for the Laurent expansion around $\varepsilon = 0$

$$\Gamma(\varepsilon) = \frac{1}{\varepsilon} - \gamma_E + O(\varepsilon). \qquad (4.72)$$

We are now in a position to perform the integration over the loop momentum. Let us discuss again the example from eq. (4.54). After Wick rotation we have

$$
\begin{aligned}
I &= \int \frac{d^D k_1}{i\pi^{D/2}} \frac{1}{(-k_1^2)(-k_2^2)(-k_3^2)} \\
&= 2 \int \frac{d^D K}{\pi^{D/2}} \int d^3 x \frac{\delta(1 - x_1 - x_2 - x_3)}{\left(K^2 - x_1 x_2 s_{12} - x_1 x_3 s_{123}\right)^3}.
\end{aligned}
\tag{4.73}
$$

Introducing spherical coordinates and performing the angular integration, this becomes

$$
I = \frac{2}{\Gamma\left(\frac{D}{2}\right)} \int_0^\infty dK^2 \int d^3 x \frac{\delta(1 - x_1 - x_2 - x_3)\left(K^2\right)^{\frac{D-2}{2}}}{\left(K^2 - x_1 x_2 s_{12} - x_1 x_3 s_{123}\right)^3}.
\tag{4.74}
$$

For the radial integration we have after the substitution $t = K^2/(-x_1 x_2 s_{12} - x_1 x_3 s_{123})$

$$
\begin{aligned}
&\int_0^\infty dK^2 \frac{\left(K^2\right)^{\frac{D-2}{2}}}{\left(K^2 - x_1 x_2 s_{12} - x_1 x_3 s_{123}\right)^3} \\
&= (-x_1 x_2 s_{12} - x_1 x_3 s_{123})^{\frac{D}{2} - 3} \int_0^\infty dt \frac{t^{\frac{D-2}{2}}}{(1 + t)^3}.
\end{aligned}
\tag{4.75}
$$

The remaining integral is a standard integral and yields

$$
\int_0^\infty dt \frac{t^{\frac{D-2}{2}}}{(1 + t)^3} = \frac{\Gamma\left(\frac{D}{2}\right) \Gamma\left(3 - \frac{D}{2}\right)}{\Gamma(3)}.
\tag{4.76}
$$

Putting everything together and setting $D = 4 - 2\varepsilon$ we obtain

$$
I = \Gamma(1 + \varepsilon) \int d^3 x \, \delta(1 - x_1 - x_2 - x_3) \, x_1^{-1-\varepsilon} (-x_2 s_{12} - x_3 s_{123})^{-1-\varepsilon}.
\tag{4.77}
$$

Therefore we succeeded in performing the integration over the loop momentum k at the expense of introducing a two-fold integral over the Feynman parameters. We will learn techniques for how to perform the Feynman parameter integrals later in these lecture notes. Let me, however, already state the final result:

$$
I = -\frac{1}{s_{123} - s_{12}} \left[\left(\frac{1}{\varepsilon} - \gamma_E - \ln(-s_{123})\right) \ln x - \frac{1}{2} \ln^2 x \right] + \mathcal{O}(\varepsilon),
\tag{4.78}
$$

with $x = \frac{-s_{12}}{-s_{123}}$. The result has been expanded in the regularisation parameter ε up to the order $\mathcal{O}(\varepsilon)$. We see that the result has a term proportional to $1/\varepsilon$. Poles in ε

in the final (regularised) result reflect the original divergences in the unregularised integral. In this example the pole corresponds to a collinear singularity.

4.5 Multi-loop integrals

As the steps discussed in the previous section always occur in any loop integration we can combine them into a master formula. Let us consider a scalar Feynman graph G with m external lines and n internal lines. We denote by I_G the associated scalar l-loop integral. For each internal line j the corresponding propagator in the integrand can be raised to an integer power ν_j. Therefore the integral will depend also on the numbers $\nu_1, \ldots, \nu_n$.

$$I_G = \int \prod_{r=1}^{l} \frac{d^D k_r}{i\pi^{\frac{D}{2}}} \prod_{j=1}^{n} \frac{1}{(-q_j^2 + m_j^2)^{\nu_j}}. \tag{4.79}$$

The independent loop momenta are labelled $k_1, \ldots, k_l$. The momenta flowing through the propagators are then given as a linear combination of the external momenta p and the loop momenta k with coefficients $-1, 0$ or 1:

$$q_i = \sum_{j=1}^{l} \lambda_{ij} k_j + \sum_{j=1}^{m} \sigma_{ij} p_j, \quad \lambda_{ij}, \sigma_{ij} \in \{-1, 0, 1\}. \tag{4.80}$$

We can repeat for each loop integration the steps of the previous section. Doing so, we arrive at the following Feynman parameter integral:

$$I_G = \frac{\Gamma(\nu - lD/2)}{\prod_{j=1}^{n} \Gamma(\nu_j)} \int_{x_j \geq 0} d^n x \, \delta\left(1 - \sum_{i=1}^{n} x_i\right) \left(\prod_{j=1}^{n} dx_j \, x_j^{\nu_j - 1}\right) \frac{\mathcal{U}^{\nu - (l+1)D/2}}{\mathcal{F}^{\nu - lD/2}}. \tag{4.81}$$

The functions $\mathcal{U}$ and $\mathcal{F}$ depend on the Feynman parameters x_j. If one expresses

$$\sum_{j=1}^{n} x_j (-q_j^2 + m_j^2) = -\sum_{r=1}^{l} \sum_{s=1}^{l} k_r M_{rs} k_s + \sum_{r=1}^{l} 2k_r \cdot Q_r + J, \tag{4.82}$$

where M is an $l \times l$ matrix with scalar entries and Q is an l-vector with four-vectors as entries, one obtains

$$\mathcal{U} = \det(M), \quad \mathcal{F} = \det(M)\left(J + QM^{-1}Q\right). \tag{4.83}$$

As an example let us look at the two-loop double box graph of Fig. 4.3. In Fig. 4.3 there are two independent loop momenta. We can choose them to be $k_1 = q_3$ and

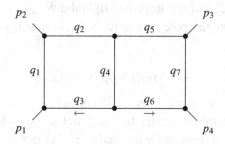

Figure 4.3 The "double box" graph: A two-loop Feynman diagram with four external lines and seven internal lines. The momenta flowing out along the external lines are labelled $p_1, \ldots, p_4$; the momenta flowing through the internal lines are labelled $q_1, \ldots, q_7$.

$k_2 = q_6$. Then all other internal momenta are expressed in terms of k_1, k_2 and the external momenta $p_1, \ldots, p_4$:

$$q_1 = k_1 - p_1, \qquad q_2 = k_1 - p_1 - p_2, \quad q_4 = k_1 + k_2,$$
$$q_5 = k_2 - p_3 - p_4, \quad q_7 = k_2 - p_4. \tag{4.84}$$

We will consider the case

$$p_1^2 = 0, \quad p_2^2 = 0, \quad p_3^2 = 0, \quad p_4^2 = 0,$$
$$m_1 = m_2 = m_3 = m_4 = m_5 = m_6 = m_7 = 0. \tag{4.85}$$

We define

$$s = (p_1 + p_2)^2 = (p_3 + p_4)^2, \quad t = (p_2 + p_3)^2 = (p_1 + p_4)^2. \tag{4.86}$$

We have

$$\sum_{j=1}^{7} x_j \left(-q_j^2\right) = -(x_1 + x_2 + x_3 + x_4)\, k_1^2 - 2x_4 k_1 \cdot k_2$$

$$-(x_4 + x_5 + x_6 + x_7)\, k_2^2 + 2\left[x_1 p_1 + x_2\left(p_1 + p_2\right)\right] \cdot k_1$$
$$+2\left[x_5\left(p_3 + p_4\right) + x_7 p_4\right] \cdot k_2 - (x_2 + x_5)\, s. \tag{4.87}$$

In comparing with eq. (4.82) we obtain

$$M = \begin{pmatrix} x_1 + x_2 + x_3 + x_4 & x_4 \\ x_4 & x_4 + x_5 + x_6 + x_7 \end{pmatrix},$$

$$Q = \begin{pmatrix} x_1 p_1 + x_2 \left(p_1 + p_2\right) \\ x_5 \left(p_3 + p_4\right) + x_7 p_4 \end{pmatrix},$$

$$J = (x_2 + x_5)(-s). \tag{4.88}$$

Plugging this into eq. (4.83) we obtain the graph polynomials as

$$\mathcal{U} = (x_1 + x_2 + x_3)(x_5 + x_6 + x_7) + x_4(x_1 + x_2 + x_3 + x_5 + x_6 + x_7),$$

$$\mathcal{F} = [x_2 x_3 (x_4 + x_5 + x_6 + x_7) + x_5 x_6 (x_1 + x_2 + x_3 + x_4) + x_2 x_4 x_6$$

$$+ x_3 x_4 x_5](-s) + x_1 x_4 x_7 (-t). \tag{4.89}$$

There are several other ways to obtain the two polynomials $\mathcal{U}$ and $\mathcal{F}$ [40]. Let me mention one method, where the two polynomials can be read off directly from the topology of the graph G. We consider first connected tree graphs T, which are obtained from the graph G by cutting l lines. The set of all such trees (or 1-trees) is denoted by T_1. The Feynman parameters corresponding to the cut lines define a monomial of degree l, and $\mathcal{U}$ is the sum over all such monomials. Cutting one more line of a 1-tree leads to two disconnected trees (T_1, T_2), or a 2-tree; T_2 is the set of all such pairs. The cut lines define monomials of degree $l + 1$. Each 2-tree of a graph corresponds to a cut defined by cutting the lines which connected the two now disconnected trees in the original graph. The square of the sum of momenta through the cut lines of one of the two disconnected trees T_1 or T_2 defines a Lorentz invariant

$$s_{(T_1, T_2)} = \left(\sum_{j \notin (T_1, T_2)} q_j \right)^2. \tag{4.90}$$

The function $\mathcal{F}_0$ is the sum over all such monomials times minus the corresponding invariant. The function $\mathcal{F}$ is then given by $\mathcal{F}_0$ plus an additional piece involving the internal masses m_j. In summary, the functions $\mathcal{U}$ and $\mathcal{F}$ are obtained from the graph as follows:

$$\mathcal{U} = \sum_{T \in T_1} \left[\prod_{j \notin T} x_j \right], \tag{4.91}$$

$$\mathcal{F}_0 = \sum_{(T_1, T_2) \in T_2} \left[\prod_{j \notin (T_1, T_2)} x_j \right] (-s_{(T_1, T_2)}), \quad \mathcal{F} = \mathcal{F}_0 + \mathcal{U} \sum_{j=1}^{n} x_j m_j^2.$$

4.6 How to obtain finite results

We have already seen in eq. (4.78) that the final result of a regularised Feynman integral may contain poles in the regularisation parameter ε. These poles reflect the original ultraviolet and infrared singularities of the unregularised integral. What shall we do with these poles? The answer has to come from physics and we distinguish again the case of UV-divergences and IR-divergences. The UV-divergences are removed through renormalisation. Ultraviolet divergences are absorbed into a redefinition of the parameters. As an example we consider the renormalisation of the coupling:

$$\underbrace{g}_{\text{divergent}} = \underbrace{Z_g}_{\text{divergent}} \underbrace{g_r}_{\text{finite}} . \tag{4.92}$$

The renormalisation constant Z_g absorbs the divergent part. However Z_g is not unique: one may always shift a finite piece from g_r to Z_g or vice versa. Different choices for Z_g correspond to different renormalisation schemes. Two different renormalisation schemes are always connected by a finite renormalisation. Note that different renormalisation schemes give numerically different answers. Therefore one always has to specify the renormalisation scheme. Some popular renormalisation schemes are the on-shell scheme, where the renormalisation constants are defined by conditions at a scale where the particles are on-shell. A second widely used scheme is modified minimal subtraction. In this scheme one always absorbs the combination

$$\Delta = \frac{1}{\varepsilon} - \gamma_E + \ln 4\pi \tag{4.93}$$

into the renormalisation constants. One proceeds similarly with all other quantities appearing in the original Lagrangian. For example:

$$A_\mu^a = \sqrt{Z_3} A_{\mu,r}^a, \quad \psi_q = \sqrt{Z_2} \psi_{q,r}, \quad g = Z_g g_r, \quad m = Z_m m_r, \quad \xi = Z_\xi \xi_r. \tag{4.94}$$

The fact that square roots appear for the field renormalisation is just convention. Let us look a little bit closer into the coupling renormalisation within dimensional regularisation and the $\overline{\text{MS}}$-renormalisation scheme. Within dimensional regularisation the renormalised coupling g_r is a dimensionful quantity. We define a dimensionless quantity g_R by

$$g_r = g_R \mu^\varepsilon, \tag{4.95}$$

where μ is an arbitrary mass scale. From a one-loop calculation one obtains

$$Z_g = 1 - \frac{1}{2} \beta_0 \frac{g_R^2}{(4\pi)^2} \Delta + \mathcal{O}(g_R^4), \quad \beta_0 = \frac{11}{3} N_c - \frac{2}{3} N_f, \tag{4.96}$$

where N_c is the number of colours and N_f the number of light quarks. The quantity g_R will depend on the arbitrary scale μ. To derive this dependence one first notes that the unrenormalised coupling constant g is of course independent of μ:

$$\frac{d}{d\mu} g = 0. \tag{4.97}$$

Substituting $g = Z_g \mu^\varepsilon g_R$ into this equation one obtains

$$\mu \frac{d}{d\mu} g_R = -\varepsilon g_R - \left(Z_g^{-1} \mu \frac{d}{d\mu} Z_g \right) g_R. \tag{4.98}$$

From eq. (4.96) one obtains

$$Z_g^{-1} \mu \frac{d}{d\mu} Z_g = \beta_0 \frac{g_R^2}{(4\pi)^2} + \mathcal{O}(g_R^4). \tag{4.99}$$

Instead of g_R one often uses the quantity $\alpha_s = g_R^2/(4\pi)$, Going to $D = 4$ one arrives at

$$\mu^2 \frac{d}{d\mu^2} \frac{\alpha_s}{4\pi} = -\beta_0 \left(\frac{\alpha_s}{4\pi} \right)^2 + \mathcal{O} \left(\frac{\alpha_s}{4\pi} \right)^3. \tag{4.100}$$

This differential equation gives the dependence of α_s on the renormalisation scale μ. At leading order (LO) the solution is given by

$$\frac{\alpha_s(\mu)}{4\pi} = \frac{1}{\beta_0 \ln \left(\frac{\mu^2}{\Lambda^2} \right)}, \tag{4.101}$$

where Λ is an integration constant. The quantity Λ is called the QCD scale parameter. For QCD β_0 is positive and $\alpha_s(\mu)$ decreases with larger μ. This property is called asymptotic freedom: the coupling becomes smaller at high energies. In quantum electrodynamics (QED) β_0 has the opposite sign and the fine-structure constant $\alpha(\mu)$ increases with larger μ. The electromagnetic coupling becomes weaker when we go to smaller energies.

Let us now look at the infrared divergences. We first note that any detector has a finite resolution. Therefore two particles which are sufficiently close to each other in phase space will be detected as one particle. Now let us look again at eqs. (4.11) and (4.18). The next-to-leading order (NLO) term will receive contributions from the interference term of the one-loop amplitude $\mathcal{A}_n^{(1)}$ with the leading-order amplitude $\mathcal{A}_n^{(0)}$, both with $(n-2)$ final state particles. This contribution is of order g^{2n-2}. Of the same order is the square of the leading-order amplitude $\mathcal{A}_{n+1}^{(0)}$ with $(n-1)$ final state particles. This contribution we have to take into account whenever our detector resolves only n particles. It turns out that the phase-space integration over the regions where one or more particles become unresolved is also divergent and, when performed in D dimensions, leads to poles with the opposite sign to the one encountered in the loop amplitudes. Therefore the sum of the two contributions is finite. The Kinoshita–Lee–Nauenberg theorem [41, 42] guarantees that all infrared divergences cancel, when summed over all degenerate physical states. As an example we consider the NLO corrections to $\gamma^* \to 2$ jets. The interference term of the one-loop amplitude with the Born amplitude is given by

$$2\mathrm{Re}\,\mathcal{A}_3^{(0)\,*}\,\mathcal{A}_3^{(1)} = \frac{\alpha_s}{\pi}C_F\left(-\frac{1}{\varepsilon^2} - \frac{3}{2\varepsilon} - 4 + \frac{7}{12}\pi^2\right)S_\varepsilon\left|\mathcal{A}_3^{(0)}\right|^2 + \mathcal{O}(\varepsilon). \quad (4.102)$$

$S_\varepsilon = (4\pi)^\varepsilon e^{-\varepsilon\gamma_E}$ is the typical phase-space volume factor in $D = 4 - 2\varepsilon$ dimensions. For simplicity we have set the renormalisation scale μ equal to the centre-of-mass energy squared, s. The square of the Born amplitude is given by

$$\left|\mathcal{A}_3^{(0)}\right|^2 = 16\pi N_c\alpha\,(1 - \varepsilon)\,s. \quad (4.103)$$

This is independent of the final state momenta and the integration over the phase space can be written as

$$\int d\phi_2\left(2\,\mathrm{Re}\,\mathcal{A}_3^{(0)\,*}\,\mathcal{A}_3^{(1)}\right)$$

$$= \frac{\alpha_s}{\pi}C_F\left(-\frac{1}{\varepsilon^2} - \frac{3}{2\varepsilon} - 4 + \frac{7}{12}\pi^2\right)S_\varepsilon\int d\phi_2\left|\mathcal{A}_3^{(0)}\right|^2 + \mathcal{O}(\varepsilon). \quad (4.104)$$

The real corrections are given by the leading-order matrix element for $\gamma^* \to q g\bar{q}$ and read

$$\left|\mathcal{A}_4^{(0)}\right|^2 = 128\pi^2\alpha\alpha_s C_F N_c(1 - \varepsilon)$$

$$\times\left[\frac{2}{x_1 x_2} - \frac{2}{x_1} - \frac{2}{x_2} + (1 - \varepsilon)\frac{x_2}{x_1} + (1 - \varepsilon)\frac{x_1}{x_2} - 2\varepsilon\right], \quad (4.105)$$

where $x_1 = s_{12}/s_{123}$, $x_2 = s_{23}/s_{123}$ and $s_{123} = s$ is again the centre-of-mass energy squared. For this particular simple example we can write the three-particle phase space in D dimensions as

$$d\phi_3 = d\phi_2 d\phi_{\text{unres}},$$

$$d\phi_{\text{unres}} = \frac{(4\pi)^{\varepsilon-2}}{\Gamma(1 - \varepsilon)}s_{123}^{1-\varepsilon}d^3x\delta(1 - x_1 - x_2 - x_3)(x_1 x_2 x_3)^{-\varepsilon}. \quad (4.106)$$

Integration over the phase space yields

$$\int d\phi_3\left|\mathcal{A}_4^{(0)}\right|^2 = \frac{\alpha_s}{\pi}C_F\left(\frac{1}{\varepsilon^2} + \frac{3}{2\varepsilon} + \frac{19}{4} - \frac{7}{12}\pi^2\right)S_\varepsilon\int d\phi_2\left|\mathcal{A}_3^{(0)}\right|^2 + \mathcal{O}(\varepsilon).$$

$$(4.107)$$

We see that in the sum the poles cancel and we obtain the finite result

$$\int d\phi_2\left(2\mathrm{Re}\,\mathcal{A}_3^{(0)\,*}\,\mathcal{A}_3^{(1)}\right) + \int d\phi_3\left|\mathcal{A}_4^{(0)}\right|^2 = \frac{3}{4}C_F\frac{\alpha_s}{\pi}\int d\phi_2\left|\mathcal{A}_3^{(0)}\right|^2 + \mathcal{O}(\varepsilon).$$

$$(4.108)$$

In this example we have seen the cancellation of the infrared (soft and collinear) singularities between the virtual and the real corrections according to the Kinoshita–Lee–Nauenberg theorem. We integrated over the phase space of all final state particles. In practice one is often interested in differential distributions. In these cases the cancellation is technically more complicated, as the different contributions live on phase spaces of different dimensions and one integrates only over restricted regions of phase space. Methods to overcome this obstacle are known under the name "phase-space slicing" and "subtraction method" [43–50].

The Kinoshita–Lee–Nauenberg theorem is related to the finite experimental resolution in detecting final state particles. In addition we have to discuss initial state particles. Let us go back to eq. (4.14). The differential cross-section can be written schematically as

$$d\sigma_{H_1 H_2} = \sum_{a,b} \int dx_1 f_{H_1 \to a}(x_1) \int dx_2 f_{H_2 \to b}(x_2) d\sigma_{ab}(x_1, x_2), \quad (4.109)$$

where $f_{H \to a}(x)$ is the parton distribution function, giving us the probability of finding a parton of type a in a hadron of type H carrying a fraction x to $x + dx$ of the hadron's momentum, and $d\sigma_{ab}(x_1, x_2)$ is the differential cross-section for the scattering of partons a and b. Now let us look at the parton distribution function $f_{a \to b}$ of a parton inside another parton. At leading order this function is trivially given by $\delta_{ab}\delta(1 - x)$, but already at the next order a parton can radiate off another parton and thus lose some of its momentum and/or convert to another flavour. One finds in D dimensions

$$f_{a \to b}(x, \varepsilon) = \delta_{ab}\delta(1 - x) - \frac{1}{\varepsilon}\frac{\alpha_s}{4\pi} P^0_{a \to b}(x) + O(\alpha_s^2), \quad (4.110)$$

where $P^0_{a \to b}$ is the lowest order Altarelli–Parisi splitting function. To calculate a cross-section $d\sigma_{H_1 H_2}$ at NLO involving parton densities one first calculates the cross-section $d\hat{\sigma}_{ab}$ where the hadrons H_1 and H_2 are replaced by partons a and b to NLO:

$$d\hat{\sigma}_{ab} = d\hat{\sigma}^0_{ab} + \frac{\alpha_s}{4\pi}d\hat{\sigma}^1_{ab} + O(\alpha_s^2). \quad (4.111)$$

The hard scattering part $d\sigma_{ab}$ is then obtained by inserting the perturbative expansions for $d\hat{\sigma}_{ab}$ and $f_{a \to b}$ into the factorisation formula:

$$d\hat{\sigma}^0_{ab} + \frac{\alpha_s}{4\pi}d\hat{\sigma}^1_{ab}$$

$$= d\sigma^0_{ab} + \frac{\alpha_s}{4\pi}d\sigma^1_{ab} - \frac{1}{\varepsilon}\frac{\alpha_s}{4\pi}\sum_c \int dx_1 P^0_{a \to c} d\sigma^0_{cb} - \frac{1}{\varepsilon}\frac{\alpha_s}{4\pi}\sum_d \int dx_2 P^0_{b \to d} d\sigma^0_{ad}.$$

One therefore obtains for the LO and the NLO terms of the hard scattering part

$$d\sigma_{ab}^0 = d\hat{\sigma}_{ab}^0$$

$$d\sigma_{ab}^1 = d\hat{\sigma}_{ab}^1 + \frac{1}{\varepsilon}\sum_c \int dx_1 P_{a\to c}^0 d\hat{\sigma}_{cb}^0 + \frac{1}{\varepsilon}\sum_d \int dx_2 P_{b\to d}^0 d\hat{\sigma}_{ad}^0. \quad (4.112)$$

The last two terms remove the collinear initial state singularities in $d\hat{\sigma}_{ab}^1$.

4.7 Feynman integrals and periods

In the previous section we have seen how all divergences disappear in the final result. However, in intermediate steps of a calculation we will in general have to deal with expressions which contain poles in the regularisation parameter ε. Let us go back to our general Feynman integral as in eq. (4.81). We multiply this integral by $e^{l\gamma_E\varepsilon}$, which avoids the occurrence of Euler's constant in the final result:

$$\hat{I}_G = e^{l\gamma_E\varepsilon}\frac{\Gamma(\nu - lD/2)}{\prod\limits_{j=1}^n \Gamma(\nu_j)}\int\limits_{x_j\geq 0} d^n x\, \delta\left(1 - \sum_{i=1}^n x_i\right)\left(\prod_{j=1}^n dx_j\, x_j^{\nu_j-1}\right)\frac{\mathcal{U}^{\nu-(l+1)D/2}}{\mathcal{F}^{\nu-lD/2}}.$$

$$(4.113)$$

This integral has a Laurent series in ε. For a graph with l loops the highest pole of the corresponding Laurent series is of power $(2l)$:

$$\hat{I}_G = \sum_{j=-2l}^{\infty} c_j \varepsilon^j. \quad (4.114)$$

We see that there are three possibilities for how poles in ε can arise from the integral in eq. (4.113):

First of all the gamma function $\Gamma(\nu - lD/2)$ of the prefactor can give rise to a (single) pole if the argument of this function is close to zero or to a negative integer value. This divergence is called the overall ultraviolet divergence.

Secondly, we consider the polynomial $\mathcal{U}$. Depending on the exponent $\nu - (l + 1)D/2$ of $\mathcal{U}$ the vanishing of the polynomial $\mathcal{U}$ in some part of the integration region can lead to poles in ε after integration. From the definition of $\mathcal{U}$ in eq. (4.91) one sees that each term of the expanded form of the polynomial $\mathcal{U}$ has coefficient $+1$, therefore $\mathcal{U}$ can only vanish if some of the Feynman parameters are equal to zero. In other words, $\mathcal{U}$ is non-zero (and positive) inside the integration region, but may vanish on the boundary of the integration region. Poles in ε resulting from the vanishing of $\mathcal{U}$ are related to ultraviolet sub-divergences.

Thirdly, we consider the polynomial $\mathcal{F}$. In an analytic calculation one often considers the Feynman integral in the Euclidean region. The Euclidean region is defined as the region where all invariants $(p_{i_1} + p_{i_2} + \cdots + p_{i_k})^2$ are negative or zero, and all internal masses are positive or zero. The result in the physical region is then obtained by analytic continuation. It can be shown that in the Euclidean region the polynomial $\mathcal{F}$ is also non-zero (and positive) inside the integration region. Therefore, under the assumption that the external kinematics is within the Euclidean region, the polynomial $\mathcal{F}$ can only vanish on the boundary of the integration region, in a similar way to what has been observed for the the polynomial $\mathcal{U}$. Depending on the exponent $\nu - lD/2$ of $\mathcal{F}$ the vanishing of the polynomial $\mathcal{F}$ on the boundary of the integration region may lead to poles in ε after integration. These poles are related to infrared divergences.

Now let us consider the integral in the Euclidean region and let us further assume that all values of kinematical invariants and masses are given by rational numbers. Then it can shown that all coefficients c_j in eq. (4.114) are periods [23]. I should first state what a period actually is. There are several equivalent definitions for a period, but probably the most accessible definition is the following [51]: a period is a complex number whose real and imaginary parts are values of absolutely convergent integrals of rational functions with rational coefficients over domains in $\mathbb{R}^n$ given by polynomial inequalities with rational coefficients. The number of periods is a countable set. Any rational and algebraic number is a period, but there are also transcendental numbers which are periods. An example is the number π, which can be expressed through the integral

$$\pi = \iint_{x^2+y^2 \leq 1} dx \, dy. \tag{4.115}$$

The integral on the right-hand side clearly shows that π is a period. On the other hand, it is conjectured that the basis of the natural logarithm e and Euler's constant γ_E are not periods. Although there are uncountably many numbers, which are not periods, only very recently an example for a number which is not a period has been found [52].

The proof that all coefficients in eq. (4.114) are periods is constructive [23] and based on sector decomposition [53–58]. The method can be used to compute numerically each coefficient of the Laurent expansion. This is a very reliable method, but unfortunately also a little bit slow.

4.8 Shuffle algebras

Before we continue the discussion of loop integrals, it is first useful to discuss shuffle algebras and generalisations thereof from an algebraic viewpoint. Consider a set of letters A. The set A is called the alphabet. A word is an ordered sequence of letters:

$$w = l_1 l_2 \ldots l_k. \tag{4.116}$$

The word of length zero is denoted by e. Let K be a field and consider the vector space of words over K. A shuffle algebra $\mathcal{A}$ on the vector space of words is defined by

$$(l_1 l_2 \ldots l_k) \cdot (l_{k+1} \ldots l_r) = \sum_{\text{shuffles } \sigma} l_{\sigma(1)} l_{\sigma(2)} \ldots l_{\sigma(r)}, \tag{4.117}$$

where the sum runs over all permutations σ, which preserve the relative order of $1, 2, \ldots, k$ and of $k+1, \ldots, r$. The name "shuffle algebra" is related to the analogy of shuffling cards: if a deck of cards is split into two parts and then shuffled, the relative order within the two individual parts is conserved. A shuffle algebra is also known under the name "mould symmetral" [59]. The empty word e is the unit in this algebra:

$$e \cdot w = w \cdot e = w. \tag{4.118}$$

A recursive definition of the shuffle product is given by

$$(l_1 l_2 \ldots l_k) \cdot (l_{k+1} \ldots l_r) = l_1 \left[(l_2 \ldots l_k) \cdot (l_{k+1} \ldots l_r) \right]$$
$$+ l_{k+1} \left[(l_1 l_2 \ldots l_k) \cdot (l_{k+2} \ldots l_r) \right]. \tag{4.119}$$

It is a well-known fact that the shuffle algebra is actually a (non-cocommutative) Hopf algebra [60]. In this context let us briefly review the definitions of a coalgebra, a bialgebra and a Hopf algebra, which are closely related. First note that the unit in an algebra can be viewed as a map from K to A and that the multiplication can be viewed as a map from the tensor product $A \otimes A$ to A (e.g. one takes two elements from A, multiplies them and gets one element out).

A coalgebra has instead of multiplication and unit the dual structures: a comultiplication Δ and a counit $\bar{e}$. The counit is a map from A to K, whereas comultiplication is a map from A to $A \otimes A$. Note that comultiplication and counit go in the reverse direction compared with multiplication and unit. We will always assume that the comultiplication is coassociative. The general form of the coproduct is

$$\Delta(a) = \sum_i a_i^{(1)} \otimes a_i^{(2)}, \tag{4.120}$$

where $a_i^{(1)}$ denotes an element of A appearing in the first slot of $A \otimes A$ and $a_i^{(2)}$ correspondingly denotes an element of A appearing in the second slot. Sweedler's notation [61] consists in dropping the dummy index i and the summation symbol:

$$\Delta(a) = a^{(1)} \otimes a^{(2)}. \tag{4.121}$$

The sum is implicitly understood. This is similar to Einstein's summation convention, except that the dummy summation index i is also dropped. The superscripts (1) and (2) indicate that a sum is involved.

A bialgebra is an algebra and a coalgebra at the same time, such that the two structures are compatible with each other. Using Sweedler's notation, the compatibility between the multiplication and comultiplication is expressed as

$$\Delta (a \cdot b) = \left(a^{(1)} \cdot b^{(1)}\right) \otimes \left(a^{(2)} \cdot b^{(2)}\right). \tag{4.122}$$

A Hopf algebra is a bialgebra with an additional map from A to A, called the antipode S, which fulfils

$$a^{(1)} \cdot S\left(a^{(2)}\right) = S\left(a^{(1)}\right) \cdot a^{(2)} = e \cdot \bar{e}(a). \tag{4.123}$$

With this background at hand we can now state the coproduct, the counit and the antipode for the shuffle algebra. The counit $\bar{e}$ is given by:

$$\bar{e}(e) = 1, \quad \bar{e}(l_1 l_2 \ldots l_n) = 0. \tag{4.124}$$

The coproduct Δ is given by:

$$\Delta (l_1 l_2 \ldots l_k) = \sum_{j=0}^{k} \left(l_{j+1} \ldots l_k\right) \otimes \left(l_1 \ldots l_j\right). \tag{4.125}$$

The antipode S is given by:

$$S (l_1 l_2 \ldots l_k) = (-1)^k \, l_k l_{k-1} \ldots l_2 l_1. \tag{4.126}$$

The shuffle algebra is generated by the Lyndon words. If one introduces a lexicographic ordering on the letters of the alphabet A, a Lyndon word is defined by the property

$$w < v \tag{4.127}$$

for any sub-words u and v such that $w = uv$.

An important example of a shuffle algebra is iterated integrals. Let $[a, b]$ be a segment of the real line and $f_1, f_2, \ldots$ functions on this interval. Let us define the following iterated integrals:

$$I(f_1, f_2, \ldots, f_k; a, b) = \int_a^b dt_1 \, f_1(t_1) \int_a^{t_1} dt_2 \, f_2(t_2) \ldots \int_a^{t_{k-1}} dt_k \, f_k(t_k). \tag{4.128}$$

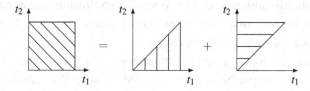

Figure 4.4 Sketch of the proof for the shuffle product of two iterated integrals. The integral over the square is replaced by two integrals over the upper and lower triangle.

For fixed a and b we have a shuffle algebra:

$$I(f_1, f_2, \ldots, f_k; a, b) \cdot I(f_{k+1}, \ldots, f_r; a, b)$$

$$= \sum_{\text{shuffles } \sigma} I(f_{\sigma(1)}, f_{\sigma(2)}, \ldots, f_{\sigma(r)}; a, b), \qquad (4.129)$$

where the sum runs over all permutations σ, which preserve the relative order of $1, 2, \ldots, k$ and of $k+1, \ldots, r$. The proof is sketched in Fig. 4.4. The two outermost integrations are recursively replaced by integrations over the upper and lower triangle.

We now consider generalisations of shuffle algebras. Assume that for the set of letters we have an additional operation

$$(.,.) : A \otimes A \to A,$$

$$l_1 \otimes l_2 \to (l_1, l_2), \qquad (4.130)$$

which is commutative and associative. Then we can define a new product of words recursively through

$$(l_1 l_2 \ldots l_k) * (l_{k+1} \ldots l_r) = l_1 [(l_2 \ldots l_k) * (l_{k+1} \ldots l_r)] + l_{k+1}[(l_1 l_2 \ldots l_k)$$

$$* (l_{k+2} \ldots l_r)] + (l_1, l_{k+1})[(l_2 \ldots l_k) * (l_{k+2} \ldots l_r)].$$

$$(4.131)$$

This product is a generalisation of the shuffle product and differs from the recursive definition of the shuffle product in eq. (4.119) through the extra term in the last line. This modified product is known under the names quasi-shuffle product [62], mixable shuffle product [63], stuffle product [64] or mould symmetrel [59]. Quasi-shuffle algebras are Hopf algebras. Comultiplication and counit are defined as for the shuffle algebras. The counit $\bar{e}$ is given by:

$$\bar{e}(e) = 1, \quad \bar{e}(l_1 l_2 \ldots l_n) = 0. \qquad (4.132)$$

The coproduct Δ is given by:

$$\Delta(l_1 l_2 \ldots l_k) = \sum_{j=0}^{k} (l_{j+1} \ldots l_k) \otimes (l_1 \ldots l_j). \qquad (4.133)$$

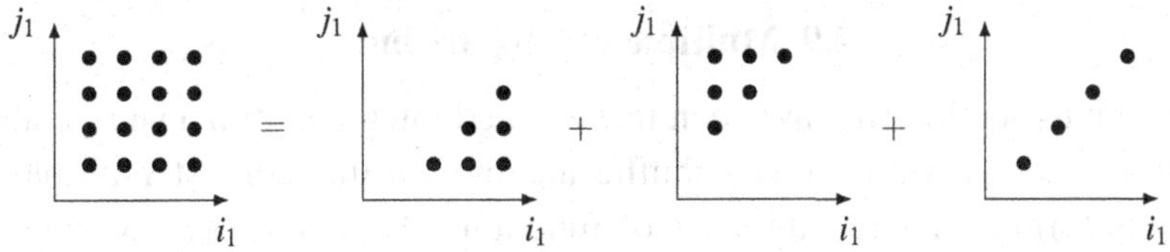

Figure 4.5 Sketch of the proof for the quasi-shuffle product of nested sums. The sum over the square is replaced by the sum over the three regions on the right-hand side.

The antipode S is recursively defined through

$$S\left(l_1 l_2 \ldots l_k\right) = -l_1 l_2 \ldots l_k - \sum_{j=1}^{k-1} S\left(l_{j+1} \ldots l_k\right) * \left(l_1 \ldots l_j\right). \quad (4.134)$$

An example of a quasi-shuffle algebra is nested sums. Let n_a and n_b be integers with $n_a < n_b$ and let $f_1, f_2, \ldots$ be functions defined on the integers. We consider the following nested sums:

$$S(f_1, f_2, \ldots, f_k; n_a, n_b) = \sum_{i_1=n_a}^{n_b} f_1(i_1) \sum_{i_2=n_a}^{i_1-1} f_2(i_2) \ldots \sum_{i_k=n_a}^{i_{k-1}-1} f_k(i_k). \quad (4.135)$$

For fixed n_a and n_b we have a quasi-shuffle algebra:

$$S(f_1, f_2, \ldots, f_k; n_a, n_b) * S(f_{k+1}, \ldots, f_r; n_a, n_b)$$

$$= \sum_{i_1=n_a}^{n_b} f_1(i_1) \, S(f_2, \ldots, f_k; n_a, i_1 - 1) * S(f_{k+1}, \ldots, f_r; n_a, i_1 - 1)$$

$$+ \sum_{j_1=n_a}^{n_b} f_k(j_1) \, S(f_1, f_2, \ldots, f_k; n_a, j_1 - 1) * S(f_{k+2}, \ldots, f_r; n_a, j_1 - 1)$$

$$+ \sum_{i=n_a}^{n_b} f_1(i) f_k(i) \, S(f_2, \ldots, f_k; n_a, i - 1) * S(f_{k+2}, \ldots, f_r; n_a, i - 1).$$

$$(4.136)$$

Note that the product of two letters corresponds to the point-wise product of the two functions:

$$(f_i, f_j)(n) = f_i(n) f_j(n). \quad (4.137)$$

The proof that nested sums obey the quasi-shuffle algebra is sketched in Fig. 4.5. The outermost sums of the nested sums on the left-hand side of (4.136) are split into the three regions indicated in Fig. 4.5.

4.9 Multiple polylogarithms

In the previous section we have seen that iterated integrals form a shuffle algebra, while nested sums form a quasi-shuffle algebra. In this context multiple poly-logarithms form an interesting class of functions. They have a representation as iterated integrals as well as nested sums. Therefore multiple polylogarithms form a shuffle algebra as well as a quasi-shuffle algebra. The two algebra structures are independent. Let us start with the representation as nested sums. The multiple polylogarithms are defined by [65–68]

$$\mathrm{Li}_{m_1,\dots,m_k}(x_1,\dots,x_k) = \sum_{i_1>i_2>\dots>i_k>0} \frac{x_1^{i_1}}{i_1^{m_1}} \cdots \frac{x_k^{i_k}}{i_k^{m_k}}. \tag{4.138}$$

The multiple polylogarithms are generalisations of the classical polylogarithms $\mathrm{Li}_n(x)$, whose most prominent examples are

$$\mathrm{Li}_1(x) = \sum_{i_1=1}^{\infty} \frac{x^{i_1}}{i_1} = -\ln(1-x), \quad \mathrm{Li}_2(x) = \sum_{i_1=1}^{\infty} \frac{x^{i_1}}{i_1^2}, \tag{4.139}$$

as well as Nielsen's generalised polylogarithms [69]

$$S_{n,p}(x) = \mathrm{Li}_{n+1,1,\dots,1}(x, \underbrace{1,\dots,1}_{p-1}), \tag{4.140}$$

and the harmonic polylogarithms [70, 71]

$$H_{m_1,\dots,m_k}(x) = \mathrm{Li}_{m_1,\dots,m_k}(x, \underbrace{1,\dots,1}_{k-1}). \tag{4.141}$$

In addition, multiple polylogarithms have an integral representation. To discuss the integral representation it is convenient to introduce for $z_k \neq 0$ the following functions

$$G(z_1,\dots,z_k; y) = \int_0^y \frac{dt_1}{t_1-z_1} \int_0^{t_1} \frac{dt_2}{t_2-z_2} \cdots \int_0^{t_{k-1}} \frac{dt_k}{t_k-z_k}. \tag{4.142}$$

In this definition one variable is redundant owing to the following scaling relation:

$$G(z_1,\dots,z_k; y) = G(xz_1,\dots,xz_k; xy). \tag{4.143}$$

If one further defines

$$g(z; y) = \frac{1}{y-z}, \tag{4.144}$$

then one has

$$\frac{d}{dy}G(z_1, \ldots, z_k; y) = g(z_1; y)G(z_2, \ldots, z_k; y) \tag{4.145}$$

and

$$G(z_1, z_2, \ldots, z_k; y) = \int_0^y dt \; g(z_1; t)G(z_2, \ldots, z_k; t). \tag{4.146}$$

One can slightly enlarge the set and define $G(0, \ldots, 0; y)$ with k zeros for z_1 to z_k to be

$$G(0, \ldots, 0; y) = \frac{1}{k!}(\ln y)^k. \tag{4.147}$$

This permits us to allow trailing zeros in the sequence $(z_1, \ldots, z_k)$ by defining the function G with trailing zeros via (4.146) and (4.147). To relate the multiple polylogarithms to the functions G it is convenient to introduce the following shorthand notation:

$$G_{m_1, \ldots, m_k}(z_1, \ldots, z_k; y) = G(\underbrace{0, \ldots, 0}_{m_1-1}, z_1, \ldots, z_{k-1}, \underbrace{0 \ldots, 0}_{m_k-1}, z_k; y). \tag{4.148}$$

Here, all z_j for $j = 1, \ldots, k$ are assumed to be non-zero. One then finds

$$\mathrm{Li}_{m_1, \ldots, m_k}(x_1, \ldots, x_k) = (-1)^k G_{m_1, \ldots, m_k}\left(\frac{1}{x_1}, \frac{1}{x_1 x_2}, \ldots, \frac{1}{x_1 \ldots x_k}; 1\right). \tag{4.149}$$

The inverse formula reads

$$G_{m_1, \ldots, m_k}(z_1, \ldots, z_k; y) = (-1)^k \, \mathrm{Li}_{m_1, \ldots, m_k}\left(\frac{y}{z_1}, \frac{z_1}{z_2}, \ldots, \frac{z_{k-1}}{z_k}\right). \tag{4.150}$$

Equation (4.149) together with (4.148) and (4.142) defines an integral representation for the multiple polylogarithms.

Up to now we have treated multiple polylogarithms from an algebraic point of view. Equally important are the analytical properties, which are needed for an efficient numerical evaluation. As an example we first discuss the numerical evaluation of the dilogarithm [72]:

$$\mathrm{Li}_2(x) = -\int_0^x dt \frac{\ln(1-t)}{t} = \sum_{n=1}^{\infty} \frac{x^n}{n^2}. \tag{4.151}$$

The power series expansion can be evaluated numerically, provided $|x| < 1$. Using the functional equations

$$\text{Li}_2(x) = -\text{Li}_2\left(\frac{1}{x}\right) - \frac{\pi^2}{6} - \frac{1}{2}\left(\ln(-x)\right)^2,$$

$$\text{Li}_2(x) = -\text{Li}_2(1-x) + \frac{\pi^2}{6} - \ln(x)\ln(1-x), \quad (4.152)$$

any argument of the dilogarithm can be mapped into the region $|x| \leq 1$ and $-1 \leq \text{Re}(x) \leq 1/2$. The numerical computation can be accelerated by using an expansion in $[-\ln(1-x)]$ and the Bernoulli numbers B_i:

$$\text{Li}_2(x) = \sum_{i=0}^{\infty} \frac{B_i}{(i+1)!}\left(-\ln(1-x)\right)^{i+1}. \quad (4.153)$$

The generalisation to multiple polylogarithms proceeds along the same lines [73]: using the integral representation eq. (4.142) one transforms all arguments into a region, where one has a converging power series expansion. In this region eq. (4.138) may be used. However, it is advantageous to speed up the convergence of the power series expansion. This is done as follows. The multiple polylogarithms satisfy the Hölder convolution [64]. For $z_1 \neq 1$ and $z_w \neq 0$ this identity reads

$$G(z_1, \ldots, z_w; 1) \quad (4.154)$$

$$= \sum_{j=0}^{w} (-1)^j \, G\left(1 - z_j, 1 - z_{j-1}, \ldots, 1 - z_1; 1 - \frac{1}{p}\right) G\left(z_{j+1}, \ldots, z_w; \frac{1}{p}\right).$$

The Hölder convolution can be used to accelerate the convergence for the series representation of the multiple polylogarithms.

4.10 From Feynman integrals to multiple polylogarithms

In Section 4.5 we saw that the Feynman parameter integrals depend on two graph polynomials $\mathcal{U}$ and $\mathcal{F}$, which are homogeneous functions of the Feynman parameters. In this section we will discuss how multiple polylogarithms arise in the calculation of Feynman parameter integrals. We will discuss two approaches. In the first approach one uses a Mellin–Barnes transformation and sums residues. This leads to the sum representation of multiple polylogarithms. In the second approach one first derives a differential equation for the Feynman parameter integral, which is then solved by an ansatz in terms of the iterated integral representation of multiple polylogarithms.

Let us start with the first approach. Assume for the moment that the two graph polynomials $\mathcal{U}$ and $\mathcal{F}$ are absent from the Feynman parameter integral. In this case we have

$$\int\limits_0^1 \left(\prod_{j=1}^n dx_j\, x_j^{\nu_j-1} \right) \delta\left(1 - \sum_{i=1}^n x_i \right) = \frac{\prod\limits_{j=1}^n \Gamma(\nu_j)}{\Gamma(\nu_1 + \cdots + \nu_n)}. \qquad (4.155)$$

With the help of the Mellin–Barnes transformation we now reduce the general case to eq. (4.155). The Mellin–Barnes transformation reads

$$(A_1 + A_2 + \cdots + A_n)^{-c} = \frac{1}{\Gamma(c)} \frac{1}{(2\pi i)^{n-1}} \int\limits_{-i\infty}^{i\infty} d\sigma_1 \ldots \int\limits_{-i\infty}^{i\infty} d\sigma_{n-1}$$

$$\times \Gamma(-\sigma_1)\ldots\Gamma(-\sigma_{n-1})\Gamma(\sigma_1 + \cdots + \sigma_{n-1} + c)$$

$$\times A_1^{\sigma_1} \ldots A_{n-1}^{\sigma_{n-1}} A_n^{-\sigma_1-\ldots-\sigma_{n-1}-c}. \qquad (4.156)$$

Each contour is such that the poles of $\Gamma(-\sigma)$ are to the right and the poles of $\Gamma(\sigma + c)$ are to the left. This transformation can be used to convert the sum of monomials of the polynomials $\mathcal{U}$ and $\mathcal{F}$ into a product, such that all Feynman parameter integrals are of the form of eq. (4.155). As this transformation converts sums into products it is the "inverse" of Feynman parametrisation. Equation (4.156) is derived from the theory of Mellin transformations: let $h(x)$ be a function which is bounded by a power law for $x \to 0$ and $x \to \infty$, e.g.

$$|h(x)| \le K x^{-c_0} \qquad \text{for } x \to 0,$$

$$|h(x)| \le K' x^{c_1} \qquad \text{for } x \to \infty. \qquad (4.157)$$

Then the Mellin transform is defined for $c_0 < \operatorname{Re} \sigma < c_1$ by

$$h_{\mathcal{M}}(\sigma) = \int\limits_0^\infty dx\, h(x)\, x^{\sigma-1}. \qquad (4.158)$$

The inverse Mellin transform is given by

$$h(x) = \frac{1}{2\pi i} \int\limits_{\gamma-i\infty}^{\gamma+i\infty} d\sigma\, h_{\mathcal{M}}(\sigma)\, x^{-\sigma}. \qquad (4.159)$$

The integration contour is parallel to the imaginary axis and $c_0 < \operatorname{Re} \gamma < c_1$. As an example of the Mellin transform we consider the function

$$h(x) = \frac{x^c}{(1+x)^c} \qquad (4.160)$$

with Mellin transform $h_{\mathcal{M}}(\sigma) = \Gamma(-\sigma)\Gamma(\sigma + c)/\Gamma(c)$. For $\mathrm{Re}(-c) < \mathrm{Re}\,\gamma < 0$ we have

$$\frac{x^c}{(1+x)^c} = \frac{1}{2\pi i} \int\limits_{\gamma-i\infty}^{\gamma+i\infty} d\sigma \, \frac{\Gamma(-\sigma)\Gamma(\sigma+c)}{\Gamma(c)} \, x^{-\sigma}. \qquad (4.161)$$

From eq. (4.161) one obtains with $x = B/A$ the Mellin–Barnes formula

$$(A+B)^{-c} = \frac{1}{2\pi i} \int\limits_{\gamma-i\infty}^{\gamma+i\infty} d\sigma \, \frac{\Gamma(-\sigma)\Gamma(\sigma+c)}{\Gamma(c)} \, A^{\sigma} B^{-\sigma-c}. \qquad (4.162)$$

Equation (4.156) is then obtained by repeated use of eq. (4.162).

With the help of eq. (4.155) and eq. (4.156) we may exchange the Feynman parameter integrals for multiple contour integrals. A single contour integral is of the form

$$I = \frac{1}{2\pi i} \int\limits_{\gamma-i\infty}^{\gamma+i\infty} d\sigma \, \frac{\Gamma(\sigma+a_1)\ldots\Gamma(\sigma+a_m)\,\Gamma(-\sigma+b_1)\ldots\Gamma(-\sigma+b_n)}{\Gamma(\sigma+c_2)\ldots\Gamma(\sigma+c_p)\,\Gamma(-\sigma+d_1)\ldots\Gamma(-\sigma+d_q)} \, x^{-\sigma}.$$

$$(4.163)$$

If $\max(\mathrm{Re}(-a_1), \ldots, \mathrm{Re}(-a_m)) < \min(\mathrm{Re}(b_1), \ldots, \mathrm{Re}(b_n))$ the contour can be chosen as a straight line parallel to the imaginary axis with

$$\max(\mathrm{Re}(-a_1), \ldots, \mathrm{Re}(-a_m)) \; < \; \mathrm{Re}\,\gamma \; < \; \min(\mathrm{Re}(b_1), \ldots, \mathrm{Re}(b_n)); \quad (4.164)$$

otherwise the contour is indented, such that the residues of $\Gamma(\sigma+a_1), \ldots,$ $\Gamma(\sigma+a_m)$ are to the right of the contour, whereas the residues of $\Gamma(-\sigma+b_1), \ldots,$ $\Gamma(-\sigma+b_n)$ are to the left of the contour. The integral eq. (4.163) is most conveniently evaluated with the help of the residue theorem by closing the contour to the left or to the right. To sum up all residues which lie inside the contour it is useful to know the residues of the gamma function:

$$\mathrm{res}\,(\Gamma(\sigma+a), \sigma = -a-n) = \frac{(-1)^n}{n!},$$

$$\mathrm{res}\,(\Gamma(-\sigma+a), \sigma = a+n) = -\frac{(-1)^n}{n!}. \qquad (4.165)$$

In general there are multiple contour integrals, and as a consequence one obtains multiple sums. In particularly simple cases the contour integrals can be performed in closed form with the help of two lemmas of Barnes. Barnes' first lemma states that

$$\frac{1}{2\pi i} \int\limits_{-i\infty}^{i\infty} d\sigma \, \Gamma(a+\sigma)\Gamma(b+\sigma)\Gamma(c-\sigma)\Gamma(d-\sigma)$$

$$= \frac{\Gamma(a+c)\Gamma(a+d)\Gamma(b+c)\Gamma(b+d)}{\Gamma(a+b+c+d)}, \qquad (4.166)$$

if none of the poles of $\Gamma(a+\sigma)\Gamma(b+\sigma)$ coincides with the ones from $\Gamma(c-\sigma)\Gamma(d-\sigma)$. Barnes' second lemma reads

$$\frac{1}{2\pi i} \int\limits_{-i\infty}^{i\infty} d\sigma \, \frac{\Gamma(a+\sigma)\Gamma(b+\sigma)\Gamma(c+\sigma)\Gamma(d-\sigma)\Gamma(e-\sigma)}{\Gamma(a+b+c+d+e+\sigma)}$$

$$= \frac{\Gamma(a+d)\Gamma(b+d)\Gamma(c+d)\Gamma(a+e)\Gamma(b+e)\Gamma(c+e)}{\Gamma(a+b+d+e)\Gamma(a+c+d+e)\Gamma(b+c+d+e)}. \qquad (4.167)$$

Although the Mellin–Barnes transformation has been known for a long time, the method has seen a revival in applications in recent years [22, 74–88].

Having collected all residues, one obtains multiple sums. The task is then to expand all terms in the dimensional regularisation parameter ε and to re-express the resulting multiple sums in terms of known functions. It depends on the form of the multiple sums whether this can be done systematically. The following types of multiple sums occur often and can be evaluated further systematically:

Type A:

$$\sum_{i=0}^{\infty} \frac{\Gamma(i+a_1)}{\Gamma(i+a'_1)} \cdots \frac{\Gamma(i+a_k)}{\Gamma(i+a'_k)} x^i.$$

Up to prefactors the hyper-geometric functions $_{J+1}F_J$ fall into this class.

Type B:

$$\sum_{i=0}^{\infty} \sum_{j=0}^{\infty} \frac{\Gamma(i+a_1)}{\Gamma(i+a'_1)} \cdots \frac{\Gamma(i+a_k)}{\Gamma(i+a'_k)} \frac{\Gamma(j+b_1)}{\Gamma(j+b'_1)} \cdots \frac{\Gamma(j+b_l)}{\Gamma(j+b'_l)} \frac{\Gamma(i+j+c_1)}{\Gamma(i+j+c'_1)}$$

$$\cdots \frac{\Gamma(i+j+c_m)}{\Gamma(i+j+c'_m)} x^i y^j.$$

An example of a function of this type is the first Appell function F_1.

Type C:

$$\sum_{i=0}^{\infty} \sum_{j=0}^{\infty} \binom{i+j}{j} \frac{\Gamma(i+a_1)}{\Gamma(i+a'_1)} \cdots \frac{\Gamma(i+a_k)}{\Gamma(i+a'_k)} \frac{\Gamma(i+j+c_1)}{\Gamma(i+j+c'_1)} \cdots \frac{\Gamma(i+j+c_m)}{\Gamma(i+j+c'_m)} x^i y^j.$$

Here, an example is the Kampé de Fériet function S_1.

Type D:

$$\sum_{i=0}^{\infty}\sum_{j=0}^{\infty}\binom{i+j}{j}\frac{\Gamma(i+a_1)}{\Gamma(i+a_1')}\cdots\frac{\Gamma(i+a_k)}{\Gamma(i+a_k')}\frac{\Gamma(j+b_1)}{\Gamma(j+b_1')}$$

$$\cdots\frac{\Gamma(j+b_l)}{\Gamma(j+b_l')}\frac{\Gamma(i+j+c_1)}{\Gamma(i+j+c_1')}\cdots\frac{\Gamma(i+j+c_m)}{\Gamma(i+j+c_m')}x^i y^j.$$

An example of a function of this type is the second Appell function F_2.

Note that in these examples there are always as many gamma functions in the numerator as in the denominator. We assume that all a_n, a_n', b_n, b_n', c_n and c_n' are of the form "integer + const $\cdot \varepsilon$". The generalisation towards the form "rational number + const $\cdot \varepsilon$" is discussed in [89]. The task is now to expand these functions systematically into a Laurent series in ε. We start with the formula for the expansion of the gamma function:

$$\Gamma(n+\varepsilon) = \Gamma(1+\varepsilon)\Gamma(n)$$

$$\times \left[1 + \varepsilon Z_1(n-1) + \varepsilon^2 Z_{11}(n-1) + \varepsilon^3 Z_{111}(n-1)\right.$$

$$\left. + \cdots + \varepsilon^{n-1} Z_{11\ldots1}(n-1)\right], \tag{4.168}$$

where $Z_{m_1,\ldots,m_k}(n)$ are Euler–Zagier sums defined by

$$Z_{m_1,\ldots,m_k}(n) = \sum_{n\geq i_1>i_2>\ldots>i_k>0}\frac{1}{i_1^{m_1}}\cdots\frac{1}{i_k^{m_k}}. \tag{4.169}$$

This motivates the following definition of a special form of nested sums, called Z-sums [89–92]:

$$Z(n;m_1,\ldots,m_k;x_1,\ldots,x_k) = \sum_{n\geq i_1>i_2>\ldots>i_k>0}\frac{x_1^{i_1}}{i_1^{m_1}}\cdots\frac{x_k^{i_k}}{i_k^{m_k}}. \tag{4.170}$$

k is called the depth of the Z-sum and $w = m_1 + \cdots + m_k$ is called the weight. If the sums go to infinity ($n = \infty$) the Z-sums are multiple polylogarithms:

$$Z(\infty;m_1,\ldots,m_k;x_1,\ldots,x_k) = \mathrm{Li}_{m_1,\ldots,m_k}(x_1,\ldots,x_k). \tag{4.171}$$

For $x_1 = \cdots = x_k = 1$ the definition reduces to the Euler–Zagier sums [93–97]:

$$Z(n;m_1,\ldots,m_k;1,\ldots,1) = Z_{m_1,\ldots,m_k}(n). \tag{4.172}$$

For $n = \infty$ and $x_1 = \cdots = x_k = 1$ the sum is a multiple ζ-value [64, 98]:

$$Z(\infty;m_1,\ldots,m_k;1,\ldots,1) = \zeta_{m_1,\ldots,m_k}. \tag{4.173}$$

The usefulness of the Z-sums lies in the fact that they interpolate between multiple polylogarithms and Euler–Zagier sums. The Z-sums form a quasi-shuffle algebra. In this approach multiple polylogarithms appear through eq. (4.171).

As an example let us look again at eq. (4.49). Setting $D = 4 - 2\varepsilon$ we obtain:

$$I = \int \frac{d^{4-2\varepsilon}k_1}{i\pi^{2-\varepsilon}} \frac{1}{(-k_1^2)} \frac{1}{(-k_2^2)} \frac{1}{(-k_3^2)}$$

$$= (-s_{123})^{-1-\varepsilon} \frac{\Gamma(-\varepsilon)\Gamma(1-\varepsilon)}{\Gamma(1-2\varepsilon)} \sum_{n=1}^{\infty} \frac{\Gamma(n+\varepsilon)}{\Gamma(n+1)} (1-x)^{n-1}, \qquad (4.174)$$

with $x = (-s_{12})/(-s_{123})$. The simplest way to arrive at the sum representation is to use the following Feynman parametrisation:

$$I = (-s_{123})^{-1-\varepsilon} \Gamma(1+\varepsilon) \int_0^1 da \int_0^1 db \, b^{-\varepsilon-1}(1-b)^{-\varepsilon}[1-a(1-x)]^{-1-\varepsilon}.$$

$$(4.175)$$

One then expands $[1 - a(1-x)]^{-1-\varepsilon}$ according to

$$(1-z)^{-c} = \frac{1}{\Gamma(c)} \sum_{n=0}^{\infty} \frac{\Gamma(n+c)}{\Gamma(n+1)} z^n. \qquad (4.176)$$

We continue with eq. (4.174). Expanding $\Gamma(n+\varepsilon)$ according to eq. (4.168), one obtains:

$$I = \frac{\Gamma(-\varepsilon)\Gamma(1-\varepsilon)\Gamma(1+\varepsilon)}{\Gamma(1-2\varepsilon)} \frac{(-s_{123})^{-1-\varepsilon}}{1-x} \sum_{n=1}^{\infty} \varepsilon^{n-1} H_{\underbrace{1,\ldots,1}_{n}}(1-x).$$

In this special case all harmonic polylogarithms can be expressed in terms of powers of the standard logarithm:

$$H_{\underbrace{1,\ldots,1}_{n}}(1-x) = \frac{(-1)^n}{n!} (\ln x)^n. \qquad (4.177)$$

This particular example is very simple and one recovers the well-known all-order result

$$\frac{\Gamma(1-\varepsilon)^2 \Gamma(1+\varepsilon)}{\Gamma(1-2\varepsilon)} \frac{(-s_{123}^2)^{-1-\varepsilon}}{\varepsilon^2} \frac{1-x^{-\varepsilon}}{1-x}, \qquad (4.178)$$

which (for this simple example) can also be obtained by direct integration. If we expand this result in ε we recover eq. (4.78).

An alternative approach to the computation of Feynman parameter integrals is based on differential equations [71, 99–104]. To evaluate these integrals within this approach one first finds for each master integral a differential equation which this master integral has to satisfy. The derivative is taken with respect to an external scale, or a ratio of two scales. An example for a one-loop four-point function is given by eq. (4.179).

$$(4.179)$$

The two-point functions on the right-hand side are simpler and can be considered to be known. This equation is solved iteratively by an ansatz for the solution as a Laurent expression in ε. Each term in this Laurent series is a sum of terms, consisting of basis functions times some unknown (and to be determined) coefficients. This ansatz is inserted into the differential equation and the unknown coefficients are determined order by order from the differential equation. The basis functions are taken as a subset of multiple polylogarithms. In this approach the iterated integral representation of multiple polylogarithms is the most convenient form. This is immediately clear from the simple formula for the derivative as in eq. (4.145).

4.11 Conclusions

In these lecture notes I have discussed Feynman integrals. After an introduction to the basic techniques, the lecture notes focused on the computation of Feynman parameter integrals, with an emphasis on the mathematical structures underlying these computations. One encounters iterated structures as nested sums or iterated integrals, which form a Hopf algebra with a shuffle or quasi-shuffle product. Of particular importance are multiple polylogarithms. The algebraic properties of these functions are very rich: they form at the same time a shuffle algebra and a quasi-shuffle algebra. Based on these algebraic structures, I discussed algorithms which evaluate Feynman integrals to multiple polylogarithms.

References

[1] T. van Ritbergen, J. A. M. Vermaseren and S. A. *Larin, Phys. Lett.* **B400**, 379, (1997), hep-ph/9701390.
[2] S. Laporta and E. Remiddi, *Phys. Lett.* **B379** 283 (1996), hep-ph/9602417.
[3] S. G. Gorishnii, A. L. Kataev and S. A. Larin, *Phys. Lett.* **B259** 144 (1991).
[4] P. Belkale and P. Brosnan, *Int. Math. Res. Not.* 2655 (2003).
[5] S. Bloch, H. Esnault and D. Kreimer, *Comm. Math. Phys.* **267** 181 (2006), math.AG/0510011.
[6] S. Bloch and D. Kreimer, *Commun. Num. Theor. Phys.* **2** 637 (2008), arXiv:0804.4399.
[7] S. Bloch, (2008), arXiv:0810.1313.
[8] F. Brown, *Commun. Math. Phys.* **287** 925 (2008), arXiv:0804.1660.
[9] F. Brown, (2009), arXiv:0910.0114.
[10] F. Brown and K. Yeats, (2009), arXiv:0910.5429.
[11] O. Schnetz, (2008), arXiv:0801.2856.
[12] O. Schnetz, (2009), arXiv:0909.0905.
[13] P. Aluffi and M. Marcolli, *Commun. Num. Theor. Phys.* **3** 1 (2009), arXiv:0807.1690.
[14] P. Aluffi and M. Marcolli, (2008), arXiv:0811.2514.
[15] P. Aluffi and M. Marcolli, (2009), arXiv:0901.2107.
[16] P. Aluffi and M. Marcolli, (2009), arXiv:0907.3225.
[17] C. Bergbauer, R. Brunetti and D. Kreimer, (2009), arXiv:0908.0633.
[18] S. Laporta, *Phys. Lett.* **B549** 115 (2002), hep-ph/0210336.
[19] S. Laporta and E. Remiddi, *Nucl. Phys.* **B704** 349 (2005), hep-ph/0406160.
[20] S. Laporta, *Int. J. Mod. Phys.* **A23** 5007 (2008), arXiv:0803.1007.
[21] D. H. Bailey, J. M. Borwein, D. Broadhurst and M. L. Glasser, (2008), arXiv:0801.0891.
[22] I. Bierenbaum and S. Weinzierl, *Eur. Phys. J.* **C32** 67 (2003), hep-ph/0308311.
[23] C. Bogner and S. Weinzierl, *J. Math. Phys.* **50** 042302 (2009), arXiv:0711.4863.
[24] O. V. Tarasov, *Phys. Rev.* **D54** 6479 (1996), hep-th/9606018.
[25] O. V. Tarasov, *Nucl. Phys.* **B502** 455 (1997), hep-ph/9703319.
[26] M. E. Peskin and D. V. Schroeder, *An Introduction to Quantum Field Theory* (Perseus Books, 1995).
[27] C. Itzykson and J. B. Zuber, *Quantum Field Theory* (McGraw-Hill, 1980).
[28] G. 't Hooft and M. J. G. Veltman, *Nucl. Phys.* **B44** 189 (1972).
[29] C. G. Bollini and J. J. Giambiagi, *Nuovo Cim.* **B12** 20 (1972).
[30] G. M. Cicuta and E. Montaldi, *Nuovo Cim. Lett.* **4** 329 (1972).
[31] K. G. Wilson, *Phys. Rev.* **D7** 2911 (1973).
[32] J. Collins, *Renormalization* (Cambridge University Press, 1984).
[33] P. Breitenlohner and D. Maison, *Commun. Math. Phys.* **52** 11 (1977).
[34] Z. Bern and D. A. Kosower, *Nucl. Phys.* **B379** 451 (1992).
[35] S. Weinzierl, (1999), hep-ph/9903380.
[36] Z. Bern, A. De Freitas, L. Dixon and H. L. Wong, *Phys. Rev.* **D66** 085002 (2002), hep-ph/0202271.

[37] Z. Kunszt, A. Signer and Z. Trocsanyi, *Nucl. Phys.* **B411** 397 (1994), hep-ph/9305239.

[38] A. Signer, PhD thesis, Diss. ETH Nr. 11143 (1995).

[39] S. Catani, M. H. Seymour and Z. Trocsanyi, *Phys. Rev.* **D55** 6819 (1997), hep-ph/9610553.

[40] C. Bogner and S. Weinzierl, (2010), arXiv:1002.3458.

[41] T. Kinoshita, *J. Math. Phys.* **3** 650 (1962).

[42] T. D. Lee and M. Nauenberg, *Phys. Rev.* **133** B1549 (1964).

[43] W. T. Giele and E. W. N. Glover, *Phys. Rev.* **D46** 1980 (1992).

[44] W. T. Giele, E. W. N. Glover and D. A. Kosower, *Nucl. Phys.* **B403** 633 (1993), hep-ph/9302225.

[45] S. Keller and E. Laenen, *Phys. Rev.* **D59** 114004 (1999), hep-ph/9812415.

[46] S. Frixione, Z. Kunszt and A. Signer, *Nucl. Phys.* **B467** 399 (1996), hep-ph/9512328.

[47] S. Catani and M. H. Seymour, *Nucl. Phys.* **B485** 291 (1997), hep-ph/9605323.

[48] S. Dittmaier, *Nucl. Phys.* **B565** 69 (2000), hep-ph/9904440.

[49] L. Phaf and S. Weinzierl, *JHEP* **04** 006 (2001), hep-ph/0102207.

[50] S. Catani, S. Dittmaier, M. H. Seymour and Z. Trocsanyi, *Nucl. Phys.* **B627** 189 (2002), hep-ph/0201036.

[51] M. Kontsevich and D. Zagier, in B. Engquis and W. Schmid (eds), *Mathematics Unlimited – 2001 and Beyond* (Springer, 2001), 771.

[52] M. Yoshinaga, (2008), arXiv:0805.0349.

[53] K. Hepp, *Commun. Math. Phys.* **2** 301 (1966).

[54] M. Roth and A. Denner, *Nucl. Phys.* **B479** 495 (1996), hep-ph/9605420.

[55] T. Binoth and G. Heinrich, *Nucl. Phys.* **B585** 741 (2000), hep-ph/0004013.

[56] T. Binoth and G. Heinrich, *Nucl. Phys.* **B680** 375 (2004), hep-ph/0305234.

[57] C. Bogner and S. Weinzierl, *Comput. Phys. Commun.* **178** 596 (2008), arXiv:0709.4092.

[58] A. V. Smirnov and M. N. Tentyukov, *Comput. Phys. Commun.* **180** 735 (2009), arXiv:0807.4129.

[59] J. Ecalle, (2002), (available at http://www.math.u-psud.fr/ biblio/ppo/2002/ppo2002-23.html).

[60] C. Reutenauer, *Free Lie Algebras* (Clarendon Press, 1993).

[61] M. Sweedler, *Hopf Algebras* (Benjamin, 1969).

[62] M. E. Hoffman, *J. Algebraic Combin.* **11** 49 (2000), math.QA/9907173.

[63] Guo and W. Keigher, *Adv. in Math.* **150** 117 (2000), math.RA/0407155.

[64] J. M. Borwein, D. M. Bradley, D. J. Broadhurst and P. Lisonek, *Trans. Amer. Math. Soc.* **353:3** 907 (2001), math.CA/9910045.

[65] A. B. Goncharov, *Math. Res. Lett.* **5** 497 (1998), (available at http://www.math.uiuc.edu/K-theory/0297).

[66] H. M. Minh, M. Petitot and J. van der Hoeven, *Discrete Math.* **225:1–3** 217 (2000).

[67] P. Cartier, *Séminaire Bourbaki* 885 (2001).

[68] G. Racinet, *Publ. Math. Inst. Hautes Études Sci.* **95** 185 (2002), math.QA/0202142.

[69] N. Nielsen, *Nova Acta Leopoldina* (Halle) **90** 123 (1909).

[70] E. Remiddi and J. A. M. Vermaseren, *Int. J. Mod. Phys.* **A15** 725 (2000), hep-ph/9905237.

[71] T. Gehrmann and E. Remiddi, *Nucl. Phys.* **B601** 248 (2001), hep-ph/0008287.

[72] G. 't Hooft and M. J. G. Veltman, *Nucl. Phys.* **B153** 365 (1979).

[73] J. Vollinga and S. Weinzierl, *Comput. Phys. Commun.* **167** 177 (2005), hep-ph/0410259.

[74] E. E. Boos and A. I. Davydychev, *Theor. Math. Phys.* **89** 1052 (1991).

[75] A. I. Davydychev, *J. Math. Phys.* **32** 1052 (1991).
[76] A. I. Davydychev, *J. Math. Phys.* **33** 358 (1992).
[77] V. A. Smirnov, *Phys. Lett.* **B460** 397 (1999), hep-ph/9905323.
[78] V. A. Smirnov and O. L. Veretin, *Nucl. Phys.* **B566** 469 (2000), hep-ph/9907385.
[79] J. B. Tausk, *Phys. Lett.* **B469** 225 (1999), hep-ph/9909506.
[80] V. A. Smirnov, *Phys. Lett.* **B491** 130 (2000), hep-ph/0007032.
[81] V. A. Smirnov, *Phys. Lett.* **B500** 330 (2001), hep-ph/0011056.
[82] V. A. Smirnov, *Phys. Lett.* **B567** 193 (2003), hep-ph/0305142.
[83] G. Heinrich and V. A. Smirnov, *Phys. Lett.* **B598** 55 (2004), hep-ph/0406053.
[84] S. Friot, D. Greynat, and E. De Rafael, *Phys. Lett.* **B628** 73 (2005), hep-ph/0505038.
[85] Z. Bern, L. J. Dixon, and V. A. Smirnov, *Phys. Rev.* **D72** 085001 (2005), hep-th/0505205.
[86] C. Anastasiou and A. Daleo, *JHEP* **10** 031 (2006), hep-ph/0511176.
[87] M. Czakon, *Comput. Phys. Commun.* **175** 559 (2006), hep-ph/0511200.
[88] J. Gluza, K. Kajda and T. Riemann, *Comput. Phys. Commun.* **177** 879 (2007), arXiv:0704.2423.
[89] S. Weinzierl, *J. Math. Phys.* **45** 2656 (2004), hep-ph/0402131.
[90] S. Moch, P. Uwer and S. Weinzierl, *J. Math. Phys.* **43** 3363 (2002), hep-ph/0110083.
[91] S. Weinzierl, *Comput. Phys. Commun.* **145** 357 (2002), math-ph/0201011.
[92] S. Moch and P. Uwer, *Comput. Phys. Commun.* **174** 759 (2006), math-ph/0508008.
[93] L. Euler, *Novi Comm. Acad. Sci. Petropol.* **20** 140 (1775).
[94] D. Zagier, *First European Congress of Mathematics*, Vol. II (Birkhauser, 1994), 497.
[95] J. A. M. Vermaseren, *Int. J. Mod. Phys.* **A14** 2037 (1999), hep-ph/9806280.
[96] J. Blümlein and S. Kurth, *Phys. Rev.* **D60** 014018 (1999), hep-ph/9810241.
[97] J. Blümlein, *Comput. Phys. Commun.* **159** 19 (2004), hep-ph/0311046.
[98] J. Blümlein, D. J. Broadhurst, and J. A. M. Vermaseren, *Comput. Phys. Commun.* **181** 582 (2010), arXiv:0907.2557.
[99] A. V. Kotikov, *Phys. Lett.* **B254** 158 (1991).
[100] A. V. Kotikov, *Phys. Lett.* **B267** 123 (1991).
[101] E. Remiddi, *Nuovo Cim.* **A110** 1435 (1997), hep-th/9711188.
[102] T. Gehrmann and E. Remiddi, *Nucl. Phys.* **B580** 485 (2000), hep-ph/9912329.
[103] T. Gehrmann and E. Remiddi, *Nucl. Phys.* **B601** 287 (2001), hep-ph/0101124.
[104] M. Argeri and P. Mastrolia, *Int. J. Mod. Phys.* **A22** 4375 (2007), arXiv:0707.4037.

5

Iterated integrals in quantum field theory

FRANCIS BROWN

Abstract

These notes are based on a series of lectures given to a mixed audience of mathematics and physics students at Villa de Leyva in Colombia in 2009. The first half is an introduction to iterated integrals and polylogarithms, with emphasis on the case $\mathbb{P}^1 \backslash \{0, 1, \infty\}$. The second half gives an overview of some recent results connecting them with Feynman diagrams in perturbative quantum field theory.

5.1 Introduction

The theory of iterated integrals was first invented by K. T. Chen in order to construct functions on the (infinite-dimensional) space of paths on a manifold, and has since become an important tool in various branches of algebraic geometry, topology and number theory. It turns out that this theory makes contact with physics in (at least) the following ways:

1. the theory of Dyson series,
2. conformal field theory and the KZ equation,
3. the Feynman path integral and calculus of variations,
4. Feynman diagram computations in perturbative quantum field theory (QFT).

The relation between Dyson series and Chen's iterated integrals is more or less tautological. The relationship with conformal field theory is well-documented, and we discuss a special case of the KZ equation in these notes. The relationship with the Feynman path integral is perhaps the deepest and most mysterious, and we say nothing about it here. Our belief is that a complete understanding of the path integral

Geometric and Topological Methods for Quantum Field Theory, ed. Alexander Cardona, Iván Contreras and Andrés F. Reyes-Lega. Published by Cambridge University Press. © Cambridge University Press 2013.

will only be possible via the perturbative approach, and by first understanding the relationship with (2) and (4). Thus the first goal of these notes is to try to explain why iterated integrals should occur in perturbative quantum field theory.

Our main example is the thrice punctured Riemann sphere $M = \mathbb{P}^1 \backslash \{0, 1, \infty\}$. The iterated integrals on M (Section 5.2) can be written in terms of multiple polylogarithms, which go back to Poincaré and Lappo-Danilevskyy and are defined for integers $n_1, \ldots, n_r \in \mathbb{N}$ by

$$\mathrm{Li}_{n_1,\ldots,n_r}(z) = \sum_{0 < k_1 < \cdots < k_r} \frac{z^{k_r}}{k_1^{n_1} \ldots k_r^{n_r}}. \tag{5.1}$$

This sum converges for $|z| < 1$ and has an analytic continuation to a multivalued function on $\mathbb{C} \backslash \{0, 1\}$. The monodromy of these functions can be expressed in terms of multiple zeta values, which were first discovered by Euler and are given by

$$\zeta(n_1, \ldots, n_r) = \mathrm{Li}_{n_1,\ldots,n_r}(1) = \sum_{0 < k_1 < \cdots < k_r} \frac{1}{k_1^{n_1} \ldots k_r^{n_r}}, \tag{5.2}$$

where now $n_r \geq 2$ to ensure convergence of the sum. Polylogarithms, and especially multiple zeta values, have undergone a huge renewal of interest in recent years owing to their appearance in many branches of geometry and number theory, but especially in particle physics. The remarkable fact is that (5.1) and (5.2) suffice to express the Feynman amplitudes (Section 5.7) for a vast number of different processes at low loop orders, and the particle physics literature is filled with complicated expressions involving them. One can therefore say that the iterated integrals on the *single space M* generate a class of numbers and functions which are sufficient to express almost all perturbative quantum field theory at low loop orders.

It is unfortunate that (perhaps owing to a lack of communication) various subclasses and variants of the functions (5.1) were rediscovered in their own right by physicists in an ad hoc manner, and go by the name of classical, Nielsen or harmonic polylogarithms, amongst others. However, the general theory of iterated integrals gives a single class of functions which are universal in a certain sense (the salient property is having unipotent monodromy), contain all these varaints, and have better properties. Thus the second goal of these notes is to present a systematic treatment of polylogarithms from this more general viewpoint, which we hope may be of use to working physicists.

In the second half (Section 5.7) of these notes, we give an overview of the mysterious appearance of multiple zeta values in calculations in perturbative quantum field theory, and suggest that the reason for this, at least at low loop orders, comes from this same unipotency property. However, the general number-theoretic content of perturbative quantum field theories is very far from being understood, and we hope that an account of this might also be of interest to mathematicians.

Finally, it is important to mention that we have made very little use of cohomology or the theory of motives, for reasons of space. A more sophisticated approach would require a detailed explanation of the theory of mixed Tate motives, and also the close relation between algebraic geometry and Feynman integrals, as in [2]. However, it turns out *a posteriori* that the mixed Tate motives which occur lie in the subcategory of mixed Tate motives generated by the fundamental group of $\mathbb{P}^1 \backslash \{0, 1, \infty\}$ [12], so in fact they can be cut out of the story altogether in a first approximation.

5.1.1 Overview of the lectures

Sections 5.2–5.6 approximately correspond to one lecture each and give a standard and purely mathematical account of iterated integrals. Since it is probably impossible to improve on the many excellent survey articles (e.g. [8, 9, 14, 15]) on this topic, we have tried to shift the emphasis to the specific example of the punctured Riemann sphere $\mathbb{P}^1 \backslash \{0, 1, \infty\}$ and relate the general structures to the theory of polylogarithms. Another reason for this is that the iterated integrals on punctured Riemann surfaces of higher genus are still not known, and the genus 0 case is one of the very few examples of manifolds where all the constructions can be made explicit.

The final section, Section 5.7, is an expanded version of various talks on numbers and periods in quantum field theory given at Villa de Leyva, Durham, Berlin and Paris and is more or less independent from the previous lectures. Owing to the rapidly expanding nature of this topic, a complete survey is both inappropriate at this point in time and would be very lengthy so the presentation is highly biased by our own recent work in this direction.

5.2 Definition and first properties of iterated integrals

We motivate the definition of iterated integrals by recalling Picard's method for solving a system of ordinary linear differential equations by successive approximation. We then state some first properties of iterated integrals.

5.2.1 Picard integration

Let $A(t)$ be an $n \times n$ matrix of continuous functions defined on an open subset U of $\mathbb{R}$, and consider the system of linear differential equations:

$$\frac{d}{dt} X(t) = A(t) X(t), \quad t \in U, \tag{5.3}$$

with some initial condition $X(t_0) = X_0$, where $t_0 \in U$, and X_0 is some $n \times n$ matrix. The differential equation (5.3) is equivalent to the integral equation

$$X(t) - X_0 = \int_{t_0}^{t} A(s)X(s)ds. \tag{5.4}$$

Picard's method for solving this differential system is by successive approximation. We denote by $X_0(t)$ the constant function $t \mapsto X_0$, and define

$$X_{n+1}(t) = X_0(t) + \int_{t_0}^{t} A(s)X_n(s)ds \quad \text{for } n \geq 0,$$

where $t_0 \leq t \in U$. If the limit $\lim_{n \to \infty} X_n(t)$ were to exist, then it would give a solution to the original problem (5.3). The first two terms are:

$$X_1(t) = X_0 + \int_{t_0}^{t} A(s)ds \, X_0,$$

$$X_2(t) = X_0 + \int_{t_0}^{t} A(s)ds \, X_0 + \int_{t_0}^{t} A(s) \int_{t_0}^{s} A(s')ds'ds \, X_0.$$

Assuming $t_0 < t$, the second term in the previous equation can be written as

$$\int_{t_0 \leq s_1 \leq s_2 \leq t} A(s_2)A(s_1)ds_1 ds_2 \, X_0.$$

Continuing in the same way, we can formally write the limit

$$X(t) = \lim_{n \to \infty} X_n(t)$$

as $X(t) = T(t, t_0)X_0$, where $T(t, t_0)$ is given explicitly by:

$$T(t, t_0) = 1_n + \sum_{n \geq 1} \int_{t_0 \leq s_1 \leq \cdots \leq s_n \leq t} A(s_n)A(s_{n-1}) \ldots A(s_1) \, ds_1 \ldots ds_n \tag{5.5}$$

and 1_n is the identity $n \times n$ matrix. The right-hand side is an infinite sum of what are nowadays called iterated integrals (see below), and the quantity $T(t, t_0)$ is known as the transport of the equation (5.3).

Now the sum (5.5) converges absolutely on compacta $K \subset U$ by the bound

$$\left| \int_{t_0 \leq s_1 \leq \cdots \leq s_n \leq t} A(s_n) \ldots A(s_1) \, ds_1 \ldots ds_n \right| \leq \sup_{s \in K} ||A(s)||^n \frac{(t - t_0)^n}{n!},$$

where the second factor is the volume of the bounded simplex

$$\Delta_n(t_0, t) = \{(s_1, \ldots, s_n) \in \mathbb{R}^n : t_0 \leq s_1 \leq \cdots \leq s_n \leq t\}$$

for $t_0 < t$. Therefore setting $X(t) = T(t, t_0)X_0$ where $T(t, t_0)$ is given by (5.5) does indeed define the desired solution to (5.3) on U.

There are two special cases which are of interest. Suppose first of all that the matrices $A(s)$, $A(s')$ commute for all s, s'. Then we can rearrange the order of integration in each integrand of (5.5) and rewrite it as an exponential series

$$T(t, t_0) = \sum_{n \geq 0} \frac{1}{n!} \int_{[t_0, t]^n} A(t_1)A(t_2) \ldots A(t_n) \, dt_1 \ldots dt_n = \sum_{n \geq 0} \frac{1}{n!} \left(\int_{t_0}^t A(t) dt \right)^n.$$

This also follows from a special case of the shuffle product formula, which we discuss below. In this case we can write the full solution in the form

$$X(t) = e^{\int_{t_0}^t A(s)ds} X_0.$$

This formula should be familiar to physicists under the name of Dyson series, and is the first point of contact between iterated integrals and quantum field theory.

The second case of interest is when any product $A(s_1) \ldots A(s_N)$ vanishes for sufficiently large N. This occurs, for example, when $A(t)$ is strictly upper triangular. In this case the series (5.5) and hence $X(t)$ is a finite sum of iterated integrals, and the solution $X(t)$ is simply called an iterated integral. Thus we should expect iterated integrals to appear whenever there are differential equations of this type (the condition is that they should have unipotent monodromy; see below).

5.2.2 Iterated integrals

Let k be the real or complex numbers, and let M be a smooth manifold over k. Let $\gamma : [0, 1] \to M$ be a piecewise smooth path on M, and let $\omega_1, \ldots, \omega_n$ be smooth k-valued 1-forms on M. We write

$$\gamma^*(\omega_i) = f_i(t)dt$$

for the pull-back of the forms ω_i to the interval $[0, 1]$. Recall that the ordinary line integral is given by

$$\int_\gamma \omega_1 = \int_{[0,1]} \gamma^*(\omega_1) = \int_0^1 f_1(t_1)dt_1$$

and does not depend on the choice of parameterization of γ.

Definition 5.2.1 The iterated integral of $\omega_1, \ldots, \omega_n$ along γ is defined by

$$\int_\gamma \omega_1 \ldots \omega_n = \int_{0 \leq t_1 \leq \cdots \leq t_n \leq 1} f_1(t_1)dt_1 \ldots f_n(t_n)dt_n.$$

More generally, an iterated integral is any k-linear combination of such integrals. The empty iterated integral (when $n = 0$) is defined to be the constant function 1.

Proposition 5.2.2 Iterated integrals satisfy the following first properties:

(i) The iterated integral $\int_\gamma \omega_1 \ldots \omega_n$ does not depend on the choice of parametrization of the path γ.

(ii) If $\gamma^{-1}(t) = \gamma(1 - t)$ denotes the reversal of the path γ, then

$$\int_{\gamma^{-1}} \omega_1 \ldots \omega_n = (-1)^n \int_\gamma \omega_n \ldots \omega_1.$$

(iii) If $\alpha, \beta : I \to M$ are two paths such that $\beta(0) = \alpha(1)$, then let $\alpha\beta$ denote the composed path obtained by traversing first α and then β. Then

$$\int_{\alpha\beta} \omega_1 \ldots \omega_n = \sum_{i=0}^{n} \int_\alpha \omega_1 \ldots \omega_i \int_\beta \omega_{i+1} \ldots \omega_n,$$

where we recall that the empty iterated integral ($n = 0$) is just the constant function 1.

(iv) The shuffle product formula holds:

$$\int_\gamma \omega_1 \ldots \omega_r \int_\gamma \omega_{r+1} \ldots \omega_{r+s} = \sum_{\sigma \in \Sigma(r,s)} \int_\gamma \omega_{\sigma(1)} \ldots \omega_{\sigma(r+s)}, \qquad (5.6)$$

where $\Sigma(r, s)$ is the set (r, s)-shuffles:

$$\Sigma(r, s)$$

$$= \{\sigma \in \Sigma(r + s) : \sigma^{-1}(1) < \cdots < \sigma^{-1}(r) \text{ and } \sigma^{-1}(r + 1) < \cdots < \sigma^{-1}(r + s)\},$$

and $\Sigma(n)$ is the set of permutations on $\{1, \ldots, n\}$.

Proof (i) and (ii) are left as exercises. The identities (iii) and (iv) for $n = 2$ can be seen from the following two pictures:

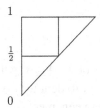 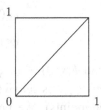

which show that $[0, 1] \times [0, 1] = \{0 \le t_1 \le t_2 \le 1\} \cup \{0 \le t_2 \le t_1 \le 1\}$ (right) and $\{0 \le t_1 \le t_2 \le 1\} = \{0 \le t_1 \le t_2 \le \frac{1}{2}\} \cup \{0 \le t_1 \le \frac{1}{2} \le t_2 \le 1\} \cup \{\frac{1}{2} \le t_1 \le t_2 \le 1\}$ (left). In general (iii) follows from the formula:

$$\Delta_n(0, 1) \cong \bigcup_{i=0}^{n} \Delta_i(0, \tfrac{1}{2}) \times \Delta_{n-i}(\tfrac{1}{2}, 1)$$

plus the fact that all overlaps are of codimension at least 2 and do not contribute to the integral. Likewise, (iv) follows from the formula for the decomposition of a product of simplices into smaller simplices:

$$\Delta_m(0, 1) \times \Delta_n(0, 1) = \bigcup_{\sigma \in \Sigma(r,s)} \sigma_* \Delta_{m+n}(0, 1),$$

where all overlaps again do not contribute to the integral. □

One way to see the iterated integral as an ordinary integral is to notice that a smooth path $\gamma : [0, 1] \to M$ gives rise to a map

$$\gamma_n = \underbrace{\gamma \times \cdots \times \gamma}_{n} : [0, 1]^n \longrightarrow \underbrace{M \times \cdots \times M}_{n}.$$

If $pr_i : M^{\times n} \to M$ denotes projection onto the ith component, the product $\Omega = pr_1^*(\omega_1) \wedge \ldots \wedge pr_n^*(\omega_n)$ defines a differential form on $M^{\times n}$. It is pulled back by the map γ_n^* to the hypercube $[0, 1]^n$. The iterated integral along γ can be written as an ordinary integral

$$\int_{\gamma} \omega_1 \ldots \omega_n = \int_{\Delta_n(0,1)} \gamma_n^*(\Omega),$$

and all the above properties follow from purely combinatorial properties of the simplices $\Delta_n(0, 1)$.

5.2.3 Homotopy functionals

Two continuous paths $\gamma_0, \gamma_1 : [0, 1] \to M$ such that $\gamma_0(0) = \gamma_1(0) = x_0$ and $\gamma_0(1) = \gamma_1(1) = x_1$ are said to be homotopic relative to their endpoints x_0, x_1 if there exists a continuous map $\phi : [0, 1] \times [0, 1] \to M$ such that

$$\phi(0, t) = \gamma_0(t),$$
$$\phi(1, t) = \gamma_1(t)$$

for all $0 \le t \le 1$, and $\phi(s, 0) = x_0, \phi(s, 1) = x_1$ for all $0 \le s \le 1$. This defines an equivalence relation on paths, and we write $\gamma_0 \sim \gamma_1$ to denote two homotopic paths (relative to their endpoints). We pass freely between piecewise continuous and smooth paths, since one can show that any piecewise continuous path is homotopic to a smooth one.

Definition 5.2.3 Let PM denote the set of piecewise smooth paths $\gamma : [0, 1] \to M$. A function $F : PM \to k$ is called a homotopy functional if

$$\gamma_0 \sim \gamma_1 \qquad \Rightarrow \qquad F(\gamma_0) = F(\gamma_1).$$

Example 5.2.4 Let $M = \mathbb{R}^2$ and, for any $r, s > 0$, let $\gamma_{r,s} : [0, 1] \to M$ be the path defined by $\gamma(t) = (t^r, t^s)$, whose endpoints are $(0, 0)$ and $(1, 1)$. All such paths are homotopic relative to their endpoints. Let x, y be the standard coordinates on $\mathbb{R}^2$, and consider the exact 1-forms $\omega_1 = dx$, $\omega_2 = dy$. Then the iterated integral

$$\int_\gamma \omega_1 \omega_2 = \int_{0 \le t_1 \le t_2 \le 1} r\, t_1^{r-1} dt_1\, s t_2^{s-1} dt_2 = \frac{s}{r+s}$$

clearly depends on the paths $\gamma_{r,s}$. Thus the general iterated integral is *not* a homotopy functional. The reason for this, as we shall see later, is that the form $\omega_1 \wedge \omega_2$ is non-zero.

Remark 5.2.5 It is proved in [11], Theorem 4.2, that if $\omega_1, \ldots, \omega_k$ are 1-forms on M which span the cotangent bundle at every point of M, and γ_1, γ_2 are smooth paths in M which have the same initial point, then

$$\int_{\gamma_1} \omega_{i_1} \ldots \omega_{i_n} = \int_{\gamma_2} \omega_{i_1} \ldots \omega_{i_n}$$

for all $n \ge 1$, and $i_1, \ldots, i_n \in \{1, \ldots, k\}$, if and only if γ_1 and γ_2 are two different parametrizations of the same path. In other words, iterated integrals define functions on the infinite-dimensional space of paths on a manifold, and are sufficient to separate points on it. This is perhaps the first hint that iterated integrals might be related to the Feynman path integral.

From now on, we will only be interested in iterated integrals which do give rise to homotopy functionals. We will give necessary and sufficient conditions for an iterated integral to be a homotopy functional in Section 5.6.

5.2.4 Multivalued functions and monodromy

Let M be a connected and locally simply connected topical space and let $\pi : \widetilde{M} \to M$ be a universal covering. A multivalued function on M (see [8], §1.3 for a historical discussion) will refer to a continuous function f on $\widetilde{M}$. If, on some simply connected open set $U \subset M$, one chooses a continuous section $s : M \to \widetilde{M}$ of π, then the function $f \circ s$ defines a local branch of the multivalued function f.

Suppose that F is a homotopy functional on M. Let us fix a point $x_0 \in M$ and allow $x_1 \in M$ to move around. Then we can consider

$$F(x_1) := F(\gamma, \text{ for any piecewise smooth } \gamma \text{ such that } \gamma(0) = x_0, \gamma(1) = x_1).$$

Thus any homotopy functional defines a multivalued function on M. To see this, let $\widetilde{M}_{x_0}$ denote the universal covering space of M based at x_0 and let $\pi : \widetilde{M}_{x_0} \to M$ be the covering map. Let $x \in \widetilde{M}_{x_0}$ be such that $\pi(x) = x_0$. Let y be any point on

$\widetilde{M}_{x_0}$ and let $\widetilde{\gamma} : [0, 1] \rightarrow \widetilde{M}_x$ be a path from x to y. Its image $\gamma = \pi(\widetilde{\gamma})$ is a path in M from x to $\pi(y)$, and we can define $F(y)$ to be $F(\gamma)$. It is well-defined, since any two paths $\widetilde{\gamma}_1, \widetilde{\gamma}_2$ from x to y are homotopic (because $\widetilde{M}_{x_0}$ is simply connected), and hence so are $\gamma_1 = \pi(\widetilde{\gamma}_1)$ and $\gamma_2 = \pi(\widetilde{\gamma}_2)$, so $F(\gamma_1) = F(\gamma_2)$.

Example 5.2.6 Let $M = \mathbb{C}\backslash\{0\}$, and let $x_0 \in M$. Let $\omega_0 = \frac{dz}{z}$, and let $\gamma : [0, 1] \rightarrow M$ be a smooth path such that $\gamma(0) = x_0$ and $\gamma(1) = z$. We have

$$\int_\gamma \omega_0 = \log(z) - \log(x_0).$$

It follows from the shuffle product formula (5.6) that the iterated integral of ω_0 can be expressed in terms of the logarithm:

$$\int_\gamma \underbrace{\omega_0 \ldots \omega_0}_{n} = \frac{1}{n!}\left(\int_\gamma \omega_0\right)^n = \frac{1}{n!}(\log(z) - \log(x_0))^n,$$

and therefore clearly only depends on the homotopy class of γ and the endpoints x_0, z.

Example 5.2.7 Let $M = \mathbb{C}\backslash\{0, 1\}$, and let $x_0 \in M$. Consider the closed 1-forms:

$$\omega_0 = \frac{dz}{z}, \quad \omega_1 = \frac{dz}{z-1},$$

whose cohomology classes give a basis for the de Rham cohomology $H^1_{dR}(M; \mathbb{C})$. Let $\gamma : I \rightarrow M$ be a smooth path such that $\gamma(0) = x_0$ and $\gamma(1) = z$, and consider the family of iterated integrals

$$I(\omega_{i_1} \ldots \omega_{i_k}, \gamma) = \int_\gamma \omega_{i_1} \ldots \omega_{i_n},$$

where $i_k \in \{0, 1\}$. We shall see in Section 5.3 that these integrals only depend on the homotopy class of γ and the endpoints x_0, z, and will write branches of the corresponding multivalued functions explicitly in terms of polylogarithms.

5.2.5 Digression on tangential base points

It is often the case that one wants to integrate from a point which is not in fact in the space M, and for this one requires tangential base points. Suppose that $M = \mathbb{C}\backslash\{\sigma_1, \ldots, \sigma_N\}$, where $\sigma_i \in \mathbb{C}$ are distinct, and let γ be a smooth path in $\mathbb{C}\backslash\{\sigma_2, \ldots, \sigma_N\}$ from σ_1 to any point $z \in M$. Let

$$\omega_1, \ldots, \omega_m \in \left\{\frac{dz}{z - \sigma_i}, i = 1, \ldots, N\right\}.$$

Then one shows that, for $0 < \varepsilon \ll 1$, the iterated integral on $[\varepsilon, 1]$ has an expansion

$$\int_\varepsilon^1 \gamma^*(\omega_1) \ldots \gamma^*(\omega_m) = a_0(\varepsilon) + a_1(\varepsilon) \log \varepsilon + \cdots + a_m(\varepsilon) \log^m \varepsilon, \qquad (5.7)$$

where $a_i(\varepsilon)$ are analytic functions at $\varepsilon = 0$. One defines its regularized value by formally setting $\log \varepsilon$ to zero, and taking the limit at $\varepsilon = 0$:

$$\mathrm{Reg}_\varepsilon \int_\varepsilon^1 \gamma^*(\omega_1) \ldots \gamma^*(\omega_m) = a_0(0). \qquad (5.8)$$

Finally, note that this depends not on the choice of coordinate ε, but only on the tangent vector $\partial/\partial\varepsilon$. To see this, consider a change of variables $\varepsilon' = \lambda_1 \varepsilon + \lambda_2 \varepsilon^2 + \cdots$, where $\lambda_1 \neq 0$. Then

$$\log \varepsilon' = \log(\lambda_1 \varepsilon + \lambda_2 \varepsilon^2 + \cdots) = \log \lambda_1 + \log \varepsilon + \log(1 + \lambda_2 \lambda_1^{-1} \varepsilon + \cdots).$$

Since the last term $\log(1 + \lambda_2 \lambda_1^{-1} \varepsilon + \cdots)$ is of order ε, and since, for $m \geq 0$, $\varepsilon \log^m \varepsilon$ tends to zero as ε goes to zero, we deduce that the regularization only depends on $\lambda_1 = \partial\varepsilon'/\partial\varepsilon$. In particular, if $\lambda_1 = 1$, then $\mathrm{Reg}_\varepsilon = \mathrm{Reg}_{\varepsilon'}$. It follows that if we define the regularized value of an iterated integral along such a path γ to be

$$\int_\gamma \omega_1 \ldots \omega_m := \mathrm{Reg}_\varepsilon \int_\varepsilon^1 \gamma^*(\omega_1) \ldots \gamma^*(\omega_m)$$

then it only depends on the tangent vector $\gamma'(0)$ of the path γ.

One defines a *tangential base point* at σ_1 to be the choice of a tangent vector t at the point σ_1 in $\mathbb{C}$: a path γ which starts at this tangential base point is by definition a path satisfying $\gamma(0) = \sigma_1$ and $\gamma'(0) = t$. There is a corresponding notion of homotopy: one considers homotopy classes of paths γ such that $\gamma(0), \gamma(1)$ and $\gamma'(0)$ are fixed (and possibly $\gamma'(1)$ if the endpoint is also a tangential base point), and all the properties of Proposition 5.2.2 remain true for tangential basepoints.

Example 5.2.8 In Example 5.2.6, one often wants to replace the base point x_0 with the tangential base point defined by the tangent vector 1 at the point $0 \in \mathbb{C}$. If γ is a path from 0 to z whose interior is contained in $\mathbb{C}\backslash\{0\}$ and which satisfies $\gamma'(0) = 1$, then with the above definitions

$$\int_\gamma \underbrace{\omega_0 \ldots \omega_0}_{n} = \frac{1}{n!} \log^n(z), \qquad (5.9)$$

as an equality of multivalued functions. When one must specify a branch, then we take the simply connected open subset $U = \mathbb{C}\backslash(-\infty, 0] \subset M$, and define $\log(z)$ to be the principal branch which vanishes at $z = 1$.

The same applies to Example 5.2.7. Let γ be a path in $\mathbb{C}$ whose interior is contained in $\mathbb{C}\backslash\{0, 1\}$, such that $\gamma(0) = 0$, $\gamma'(0) = 1$ and $\gamma(1) = z$, where $z \in \mathbb{C}\backslash\{0, 1\}$. Then the iterated integral of $\omega_0 \ldots \omega_0$ along γ is again given by (5.9), and one can verify that for all other iterated integrals we have

$$\lim_{\varepsilon \to 0} \int_\varepsilon^1 \gamma^*(\omega_{i_1}) \ldots \gamma^*(\omega_{i_n}) < \infty \tag{5.10}$$

if at least one $i_k \neq 0$ and the integral actually converges. In other words, in the expansion (5.7), all terms $a_i(0) = 0$ for $i \geq 1$, and there is no need to regularize. Thus, in this case, we can skirt around the issue of taking tangential base points by simply taking care to separate the two different cases where the i_k are all equal to 0 (equation (5.9)) or not (equation (5.10)). This is what we will do in Section 5.3.

5.3 The case $\mathbb{P}^1\backslash\{0, 1, \infty\}$ and polylogarithms

In this section and the next, we study the set of iterated integrals on $\mathbb{C}\backslash\{0, 1\}$ and relate them to polylogarithms. Because of their historical and mathematical importance, we first consider the special case of the classical polylogarithms.

5.3.1 The classical polylogarithms

Let $n \geq 1$. The classical polylogarithms are over 250 years old and are defined by the series

$$\mathrm{Li}_n(z) = \sum_{k=1}^\infty \frac{z^k}{k^n}, \tag{5.11}$$

which converges absolutely for $|z| < 1$ and therefore defines a holomorphic function in a neighbourhood of the origin. These functions are generalizations of the logarithm $\mathrm{Li}_1(z) = -\log(1 - z)$ and satisfy the differential equations

$$\frac{d}{dz}\mathrm{Li}_n(z) = \frac{1}{z}\mathrm{Li}_{n-1}(z)$$

for all $n \geq 2$. It follows that, for $n \geq 2$, we can also define

$$\mathrm{Li}_n(z) = \int_\gamma \mathrm{Li}_{n-1}(t)\frac{dt}{t}, \tag{5.12}$$

where γ is a smooth path from 0 to z in $\mathbb{C}\backslash\{0, 1\}$. This integral formula proves by induction that $\mathrm{Li}_n(z)$ has an analytic continuation to a multivalued function on $M = \mathbb{C}\backslash\{0, 1\}$. Recall that we had 1-forms $\omega_0 = \frac{dz}{z}$ and $\omega_1 = \frac{dz}{1-z}$ on M. It follows from (5.12) and the definition of iterated integrals that

$$\mathrm{Li}_n(z) = \int_\gamma \omega_1 \underbrace{\omega_0 \ldots \omega_0}_{n} = \int_{0 \leq t_1 \leq \ldots \leq t_n \leq 1} \frac{z \, dt_1}{1 - z t_1} \frac{dt_2}{t_2} \cdots \frac{dt_n}{t_n}.$$

Another way to see this is to do an expansion of the first differential form $\frac{z \, dt_1}{1 - z t_1}$ as the geometric series $\sum_{k \geq 1} z^k t_1^{k-1} \, dt_1$. Integrating from 0 to t_2 yields $\sum_{k \geq 1} \frac{1}{k} z^k t_2^k$. Multiplying by $\frac{dt_2}{t_2}$, and integrating again from 0 to t_3, gives $\sum_{k \geq 1} \frac{1}{k^2} z^k t_3^k$. Continuing in this way, one obtains after n integrations

$$\sum_{k \geq 1} \frac{1}{k^n} z^k t_n^k,$$

which is evaluated at $t_n = 1$ to give $\mathrm{Li}_n(z)$.

Note that the lower bound of integration 0 is not in the space M, but this in fact poses no problem here, since ω_1 has no pole at 0 and so the integral converges. The representation as functions also shows that these iterated integrals are homotopy functionals. This will follow from the general results of Section 5.6 as a consequence of the fact that $d\omega_0 = d\omega_1 = \omega_0 \wedge \omega_1 = 0$.

Thus we see that the classical polylogarithms are special cases of iterated integrals on M. For an exposition of some their numerous applications in number theory and geometry see the survey paper [17].

5.3.2 Monodromy

To describe the monodromy of the classical polylogarithms, we need a base point x on M. If one wishes, one can take $x = \frac{1}{2}$ throughout this section.[1] The fundamental group $\pi_1(M, x)$ is then the free group generated by the homotopy classes of two loops γ_0 and γ_1 which wind once around 0 and 1, respectively:

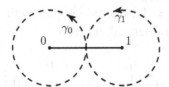

For $i = 0, 1$, let $\mathcal{M}_i$ (monodromy around the point i) denote the operator which associates to a local branch of a multivalued function $f(z)$ on M its analytic continuation along the path γ_i. It follows from the properties of analytic continuation

---

[1] Since this choice is not canonical, it is better to take the real interval $x = (0, 1)$ as a base point. One can verify that all the usual properties one requires of a base point are satisfied, because $(0, 1)$ is contractible. From this point of view, there are three canonical base points given by the three connected components $(-\infty, 0)$, $(0, 1)$ and $(1, \infty)$ of the set of real points $\mathbb{R} \setminus \{0, 1\}$.

that the operators $\mathcal{M}_i$ are multiplicative and commute with differentiation: i.e. for any multivalued functions f, g on $\mathbb{C}\backslash\{0, 1\}$, and $i \in \{0, 1\}$,

$$\mathcal{M}_i\left(\frac{d}{dz}f(z)\right) = \frac{d}{dz}(\mathcal{M}_i f(z)), \qquad (5.13)$$

$$\mathcal{M}_i(f(z)g(z)) = (\mathcal{M}_i f(z))(\mathcal{M}_i g(z)).$$

We will show in Section 5.4.2 that:

$$\mathcal{M}_0 \mathrm{Li}_n(z) = \mathrm{Li}_n(z), \qquad (5.14)$$

$$\mathcal{M}_1 \mathrm{Li}_n(z) = \mathrm{Li}_n(z) + \frac{2i\pi}{(n-1)!}\log^{n-1}(z).$$

Note that the local branch of $\mathrm{Li}_n(z)$ defined by (5.11) is holomorphic at the origin, but its Riemann surface is ramified there, because after doing an analytic continuation around the point 1 it acquires a term $\log^{n-1}(z)$ which has a singularity at the origin.

5.3.3 Matrix representation for the monodromy

Before passing to more general iterated integrals on M, we relate the above, in the case of the dilogarithm $\mathrm{Li}_2(z)$, to Picard's method (Section 5.2.1). Let $U \subset M$ and define

$$A(z) = \begin{pmatrix} 0 & 0 & 0 \\ \frac{1}{z} & 0 & 0 \\ 0 & \frac{1}{1-z} & 0 \end{pmatrix} \quad \text{and} \quad \Omega(z) = A(z)dz = \begin{pmatrix} 0 & 0 & 0 \\ \omega_0 & 0 & 0 \\ 0 & \omega_1 & 0 \end{pmatrix}. \qquad (5.15)$$

Since $A(z)$ is nilpotent, applying Picard's method to the equation $X'(z) = A(z)X(z)$ with initial condition $X(z_0) = 1_2$, where 1_2 is the identity 2×2 matrix, on U gives a finite expansion in terms of iterated integrals

$$X(z) = \begin{pmatrix} 1 & 0 & 0 \\ \int_\gamma \omega_0 & 1 & 0 \\ \int_\gamma \omega_1\omega_0 & \int_\gamma \omega_1 & 1 \end{pmatrix}.$$

It is customary to rescale this integral by multiplying the nth column by $(2\pi i)^n$. Equivalently, consider the trivial bundle $\mathbb{C}^3 \times M \to M$. On it we have a connection $\nabla f = df - \Omega f$ for $f : M \to \mathbb{C}$ smooth. It extends via the map $M \hookrightarrow \mathbb{C}$ to a connection on $\mathbb{C}^3 \times \mathbb{C} \to \mathbb{C}$ which has regular singularities at $z = 0$ and $z = 1$. In other words, $dV(z) = \Omega V(z)$ is a Fuchsian differential equation and has the solution:

$$V_2 = \begin{pmatrix} 1 & 0 & 0 \\ \mathrm{Li}_1(z) & 2i\pi & 0 \\ \mathrm{Li}_2(z) & 2i\pi \log(z) & (2i\pi)^2 \end{pmatrix}.$$

The monodromy of the dilogarithm $\mathrm{Li}_2(z)$ given by (5.14) can be conveniently encoded as follows. We have $\mathcal{M}_0 V_2 = V_2 M_0$ and $\mathcal{M}_1 V_2 = V_2 M_1$, where

$$M_0 = \begin{pmatrix} 1 & 0 & 0 \\ 0 & 1 & 0 \\ 0 & 1 & 1 \end{pmatrix} \quad \text{and} \quad M_1 = \begin{pmatrix} 1 & 0 & 0 \\ 1 & 1 & 0 \\ 0 & 0 & 1 \end{pmatrix}.$$

Consider, therefore, the unipotent group

$$G = \begin{pmatrix} 1 & 0 & 0 \\ * & 1 & 0 \\ * & * & 1 \end{pmatrix}$$

of lower triangular matrices. Let $G(R)$ denote the corresponding group with its entries $*$ in any ring R. The monodromy gives a representation

$$\rho : \pi_1(M, x) \longrightarrow G(\mathbb{Z})$$

$$\gamma_0, \gamma_1 \mapsto M_0, M_1.$$

Thus we obtain a commutative diagram:

$$\begin{array}{ccc} \widetilde{M} & \longrightarrow & G(\mathbb{C}) \\ \downarrow & & \downarrow \\ M & \longrightarrow & G(\mathbb{C})/G(\mathbb{Z}) \end{array} \tag{5.16}$$

where the horizontal map along the top associates the matrix $V_2(z)$ to a point $z \in \widetilde{M}$, and the lower horizontal map takes a point $z \in M$ to $V_2(z) \mod G(\mathbb{Z})$. In this sense, the dilogarithm has unipotent monodromy.

One can write down similar matrices A_n for each of the classical polylogarithms $\mathrm{Li}_n(z)$ (exercise), and we shall see later how to generalize this picture to include all the iterated integrals on M, using an infinite-dimensional bundle.

5.3.4 Multiple polylogarithms in one variable

We now describe all the iterated integrals on $\mathbb{C}\backslash\{0, 1\}$ by writing down the functions which they represent. These are known as multiple polylogarithms (in one variable), which were studied in great detail by Lappo-Danilevsky, and partly rediscovered by physicists more recently.

Definition 5.3.1 Let $n_1, \ldots, n_r \in \mathbb{N}$ and define the multiple polylogarithm in one variable (also frequently called hyperlogarithm)

$$\mathrm{Li}_{n_1,\ldots,n_r}(z) = \sum_{1 \le k_1 < \cdots < k_r} \frac{z^{k_r}}{k_1^{n_1} \ldots k_r^{n_r}}.$$

This converges absolutely for $|z| < 1$ and defines a germ of a holomorphic function in the neighbourhood of the origin.

It follows immediately from the definitions that

$$\frac{d}{dz}\mathrm{Li}_{n_1,\ldots,n_r}(z) = \begin{cases} \frac{1}{z}\mathrm{Li}_{n_1,\ldots,n_r-1}(z) & \text{if } n_r > 1, \\ \frac{1}{1-z}\mathrm{Li}_{n_1,\ldots,n_{r-1}}(z) & \text{if } n_r = 1. \end{cases} \qquad (5.17)$$

As in the case of the classical polylogarithms it follows from these equations that we have an iterated integral representation

$$\mathrm{Li}_{n_1,\ldots,n_r}(z) = \int_\gamma \omega_1\omega_0^{n_1-1} \ldots \omega_1\omega_0^{n_r-1}, \qquad (5.18)$$

where $\gamma : [0, 1] \to \mathbb{C}$ is a smooth path whose interior is contained in $\mathbb{C}\backslash\{0, 1\}$, such that $\gamma(0) = 0$ and $\gamma(1) = z$. This formula also defines an analytic continuation to the whole of M, and thus the multiple polylogarithms are multivalued functions on M. All these iterated integrals begin with ω_1.

It remains to describe the iterated integrals consisting of words in ω_0, ω_1 which begin with ω_0. One way to do this is to use the regularisation at 0 defined in equation (5.9) and use the shuffle product. It turns out that all remaining iterated integrals can be obtained from the ones we have already along with $\int_\gamma \omega_0$, which is defined to be $\log z$. For example, we have from Proposition 5.2.2 (iii) that

$$\int_\gamma \omega_0 \int_\gamma \omega_1 = \int_\gamma \omega_0\omega_1 + \int_\gamma \omega_1\omega_0$$

and so we are obliged to define $\int_\gamma \omega_0\omega_1 := \int_\gamma \omega_1\omega_0 - \int_\gamma \omega_0 \int_\gamma \omega_1 = \mathrm{Li}_2(z) - \log(z)\mathrm{Li}_1(z)$. Therefore if we fix the value of $\int_\gamma \omega_0$ then the shuffle product determines all the iterated integrals uniquely. We formalize this in a slightly more algebraic setting below.

5.3.5 Algebraic representation

Let $X = \{x_0, x_1\}$ denote an alphabet in two letters, and let $X^\times$ denote the free monoid generated by X, i.e. the set of words in the letters x_0, x_1 along with the empty word e. Let $\mathbb{Q}\langle x_0, x_1 \rangle$ denote the vector space generated by the words in X, equipped with the shuffle product

$$x_{i_1} \dots x_{i_r} \, \text{\scriptsize III} \, x_{i_{r+1}} \dots x_{i_{r+s}} = \sum_{\sigma \in \Sigma(r,s)} x_{\sigma(1)} \dots x_{\sigma(r+s)},$$

and where $e \, \text{\scriptsize III} \, w = w \, \text{\scriptsize III} \, e = w$ for all $w \in X^*$. To every word $w \in X^*$ we associate a multivalued function $\mathrm{Li}_w(z)$, as follows:

(i) Firstly, if $w \in X^\times x_1$, then we can write $w = x_0^{n_1-1} x_1 \dots x_0^{n_r-1} x_1$, and we set

$$\mathrm{Li}_w(z) = \int_0^z \omega_1 \omega_0^{n_1-1} \omega_1 \dots \omega_0^{n_r-1} = \mathrm{Li}_{n_1,\dots,n_k}(z).$$

Note the reversal of the letters in the iterated integral. By the shuffle product formula (Proposition 5.2.2 (iii)), we have

$$\mathrm{Li}_w(z)\mathrm{Li}_{w'}(z) = \mathrm{Li}_{w \, \text{\scriptsize III} \, w'}(z), \tag{5.19}$$

where the notation Li is extended by linearity: $\mathrm{Li}_{\sum w_i}(z) := \sum \mathrm{Li}_{w_i}(z)$.

(ii) We can extend the definition of $\mathrm{Li}_w(z)$ to all words $w \in X^\times$ by setting:

$$\mathrm{Li}_{x_0^n}(z) = \int_\gamma \underbrace{\omega_0 \dots \omega_0}_{n} = \frac{1}{n!} \log^n(z). \tag{5.20}$$

Exercise 5.3.2 Every word $w \in X^\times$ is a unique sum of shuffles

$$w = \sum_{i=0}^k x_0^i \, \text{\scriptsize III} \, w_i,$$

where $w_i \in X^* x_1$ are convergent.

Setting $\mathrm{Li}_w(z) = \sum_{i=0}^k \mathrm{Li}_{x_0^i}(z)\mathrm{Li}_{w_i}(z)$ completes the definition of the functions $\mathrm{Li}_w(z)$. With this definition, the shuffle relations (5.19) are valid for all words $w \in X^*$.

5.4 The KZ equation and the monodromy of polylogarithms

Recall that we defined multiple polylogarithm functions $\mathrm{Li}_w(z)$, indexed by the set of words in X, as multivalued functions on $M = \mathbb{C}\backslash\{0, 1\}$ which satisfy:

$$d\,\mathrm{Li}_{x_i w}(z) = \omega_i \mathrm{Li}_w(z) \quad \text{for } i = 0, 1, \tag{5.21}$$

where $\omega_0 = dz/z$ and $\omega_1 = dz/(1 - z)$. Furthermore, we had

$$\lim_{z \to 0} \mathrm{Li}_w(z) = 0 \quad \text{for all words} w \neq x_0^n, \tag{5.22}$$

$$\mathrm{Li}_{x_0^n}(z) = \frac{1}{n!} \log(z).$$

Since the properties (5.22) fix all the constants of integration in the differential equations (5.21), this determines the functions $\mathrm{Li}_w(z)$ uniquely. Recall that the general shuffle product formula for iterated integrals also implied the shuffle product for multiple polylogarithms: $\mathrm{Li}_w(z)\mathrm{Li}_{w'}(z) = \mathrm{Li}_{w \, \text{ш} \, w'}(z)$.

In order to study the monodromy of the functions $\mathrm{Li}_w(z)$ it is helpful to consider their generating series:

$$L(z) = \sum_{w \in X^*} w \, \mathrm{Li}_w(z).$$

Formally, let

$$\mathbb{C}\langle\langle X \rangle\rangle = \left\{ \sum_{w \in X^*} S_w w : S_w \in \mathbb{C} \right\}$$

denote the ring of non-commutative formal power series in the words X^*, equipped with the concatenation product ($w.w' = ww'$). The function $L(z)$ defines a multi-valued function on M taking values in $\mathbb{C}\langle\langle X \rangle\rangle$, and satisfies the differential equation:

$$\frac{d}{dz} L(z) = \left(\frac{x_0}{z} + \frac{x_1}{z - 1} \right) L(z). \tag{5.23}$$

Thus $L(z)$ should be viewed as a flat section of the corresponding connection on the infinite-dimensional trivial bundle $\mathbb{C}\langle\langle X \rangle\rangle \times M \to M$. Term by term, (5.23) is equivalent to the differential equations (5.21). This equation is known as the Knizhnik–Zamolodchikov (KZ) equation (in the one-dimensional case). The conditions (5.22) can be written

$$L(z) \sim \exp(x_0 \log z) \quad \text{as } z \to 0.$$

This notation means that there exists a $\mathbb{C}\langle\langle X \rangle\rangle$-valued function $h(z)$, which is holomorphic in the neighbourhood of the origin, such that

$$L(z) = h(z) \exp(x_0 \log z) \quad \text{for } z \text{ near } 0, \tag{5.24}$$

and $h(0)$ is the constant series $1 \in \mathbb{C}\langle\langle X \rangle\rangle$. As before, the condition (5.24) uniquely determines the solution $L(z)$ to (5.23).

5.4.1 Drinfel'd associator and multiple zeta values

By the same argument (or by symmetry $z \mapsto 1 - z$), there exists another solution $L^1(z)$ to (5.23) which satisfies:

$$L^1(z) \sim \exp(x_1 \log(1 - z)) \quad \text{as } z \to 1. \tag{5.25}$$

The series is invertible in $\mathbb{C}\langle\langle X \rangle\rangle$ since its leading term (the coefficient of the empty word) is 1. In this situation, it is usual to consider the parallel transport:

$$\Phi(z) = \left(L^1(z)\right)^{-1} L(z), \tag{5.26}$$

which relates the two solutions. It follows from differentiating $L^1(z)\Phi(z) = L(z)$ and equation (5.23) that $L^1(z)d\Phi(z) = 0$. It follows that $d\Phi(z) = 0$, and so $\Phi(z)$ is a constant series denoted $\Phi(x_0, x_1) \in \mathbb{C}\langle\langle X \rangle\rangle$ and known as Drinfel'd's associator. It follows from the asymptotics of the solution $L^1(z)$ near 1 (equations (5.25) and (5.26)) that we can write:

$$\Phi(x_0, x_1) = \lim_{z \to 1^-} \left(\exp(-x_1 \log(1 - z))L(z)\right). \tag{5.27}$$

In this sense, $\Phi(x_0, x_1)$ is a regularized limit of $L(z)$ as $z \to 1^-$ along the real axis. More precisely, it follows from (5.25) that every multiple polylogarithm $\text{Li}_w(z)$ has a canonical branch for $z \in (0, 1)$ which can be written in the form

$$\text{Li}_w(z) = a_0(z) + a_1(z) \log(1 - z) + \cdots + a_{|w|}(z) \log^{|w|}(1 - z),$$

where $a_i(z)$ is holomorphic in a neighbourhood of $z = 1$. The regularized value at $z = 1$ can therefore be defined as[2]

$$\text{Reg}_{z=1} \text{Li}_w(z) = a_0(1),$$

and $\Phi(x_0, x_1)$ is the generating series of these regularized values. In particular, it has real coefficients, i.e. $\Phi(x_0, x_1) \in \mathbb{R}\langle\langle X \rangle\rangle$. One can determine the coefficients completely in terms of multiple zeta values.

Definition 5.4.1 Let $n_1, \ldots, n_r \in \mathbb{N}$, such that $n_r \geq 2$. The multiple zeta value is defined by the absolutely convergent sum:

$$\zeta(n_1, \ldots, n_r) = \sum_{0 < k_1 < \cdots < k_r} \frac{1}{k_1^{n_1} \ldots k_r^{n_r}} \in \mathbb{R}.$$

---

[2] Comparing with Section 5.2.5, this means we are taking tangential base points: the tangent vector 1 at 0, and the tangent vector -1 at 1.

Lemma 5.4.2 There is a unique function $\zeta : \mathbb{Q}\langle X \rangle \longrightarrow \mathbb{R}$ such that

$$\zeta(x_0) = \zeta(x_1) = 0,$$

$$\zeta(x_0^{n_r-1} x_1 \ldots x_0^{n_1-1} x_1) = \zeta(n_1, \ldots, n_r),$$

$$\zeta(w)\zeta(w') = \zeta(w \, \text{ш} \, w').$$

The coefficients of $\Phi(x_0, x_1)$ are exactly the $\zeta(w)$, i.e.

$$\Phi(x_0, x_1) = \sum_{w \in X^*} w \, \zeta(w).$$

Proof First of all, for all words $w \in x_0 X^* x_1$, $\text{Li}_w(z)$ converges at the point $z = 1$, and we have $\zeta(w) = \text{Li}_w(1)$. The definition of the series $\text{Li}_w(z)$ gives the formula for $\zeta(w)$ as a nested sum. Furthermore, the shuffle product for the functions $\text{Li}_w(z)$ implies the corresponding formula $\zeta(w)\zeta(w') = \zeta(w \, \text{ш} \, w')$ for all words $w, w' \in x_0 X^* x_1$. It is then an exercise to show that every word $w \in X$ can be uniquely written as a sum of shuffles of words of the form x_0^i, x_1^j and $w \in x_0 X^* x_1$. It follows that there is a unique way to extend ζ by linearity to all words $w \in X^*$ after fixing the values of $\zeta(x_0)$ and $\zeta(x_1)$ such that the three properties are satisfied. This proves the first part. To prove that the coefficients of $\Phi(x_0, x_1)$ are given by the $\zeta(w)$, one verifies from (5.27) that the coefficient of w is $\text{Reg}_{z=1}\text{Li}_w(z)$, which satisfies the shuffle relations for convergent w, and vanishes for $w = x_0, x_1$, since $\text{Li}_{x_0}(z) = \log z$ and $\text{Li}_{x_1}(z) = -\log(1-z)$. $\qquad\square$

The fact that the coefficients of $\Phi(x_0, x_1)$ are multiple zeta values was first observed by Kontsevich. Explicitly, one can write

$$\Phi(x_0, x_1) = 1 + \zeta(2)[x_0, x_1] + \zeta(3)\big([x_0, [x_0, x_1]] - [[x_0, x_1], x_1]\big) + \cdots .$$

5.4.2 Monodromy

We can now proceed with the calculation of the monodromy of the multiple polylogarithms. In this section we write Φ instead of $\Phi(x_0, x_1)$.

Proposition 5.4.3 The action of the monodromy operators $\mathcal{M}_i$, $i = 0, 1$, on the generating series $L(z)$ is given by:

$$\mathcal{M}_0 L(z) = L(z) \exp(2i\pi x_0), \tag{5.28}$$

$$\mathcal{M}_1 L(z) = L(z) \, \Phi^{-1} \exp(2i\pi x_1)\Phi.$$

Proof The first line follows immediately from the asymptotics of $L(z)$ (5.24). Using the fact that $L^1(z)\Phi = L(z)$, and that the series Φ is invertible in $\mathbb{C}\langle\langle X \rangle\rangle$, we have

$$\mathcal{M}_1 L(z) = \mathcal{M}_1 L^1(z)\Phi = L^1(z) \exp(2i\pi x_1)\Phi = L(z) \, \Phi^{-1} \exp(2i\pi x_1)\Phi. \quad\square$$

Corollary 5.4.4 The monodromy (or holonomy) of $\mathbb{P}^1 \setminus \{0, 1, \infty\}$ can be expressed in terms of multiple zeta values and $2i\pi$.

We can easily deduce the monodromy of the classical polylogarithms from (5.28). Since $\mathrm{Li}_n(z)$ corresponds to $\mathrm{Li}_{x_0^{n-1}x_1}(z)$, it suffices to compute the coefficient of $x_0^{n-1}x_1$ in $\mathcal{M}_1 L(z)$. Since words which contain two or more x_1s do not contribute:

$$L(z)\Phi^{-1} \exp(2i\pi x_1)\Phi = L(z)\left(1 + \Phi^{-1} 2i\pi x_1 \Phi + \cdots\right),$$

where the coefficient of $x_0^{n-1}x_1$ is

$$\mathrm{Li}_{x_0^{n-1}x_1}(z) + \sum_{i+j=n-1} \mathrm{Li}_{x_0^i}(z)\Phi_{x_0^j}^{-1} 2i\pi$$

and $\Phi_{x_0^j}^{-1}$ is the coefficient of x_0^j in Φ^{-1}. But the coefficient of x_0^j, for $j \geq 1$, in Φ is just $\zeta(x_0^j) = 0$, and this implies that the coefficient of x_0^j in Φ^{-1} is also 0. The previous formula therefore reduces to $\mathrm{Li}_n(z) + \frac{2i\pi}{(n-1)!} \log^{n-1}(z)$, as promised.

5.4.3 A (pro-)unipotent group

Now consider the non-commutative algebra $\mathbb{C}\langle X\rangle$ equipped with the concatenation product. For $n \geq 1$, the set I_n of words of length $\geq n$ is a (two-sided) ideal in $\mathbb{C}\langle X\rangle$, and the quotient $W_n = \mathbb{C}\langle X\rangle / I_n$ is isomorphic as a vector space to the $\mathbb{C}$-vector space spanned by the set of all words of length $< n$. There are two nilpotent operators, $X_0, X_1 : W_n \to W_n$ given by left multiplication by x_0, x_1, respectively.

By passing to the quotient $\mathbb{C}\langle X\rangle \to W_n$, (5.23) defines a connection on W_n for all n, which are compatible with the quotient maps $W_n \to W_{n-1}$. The monodromy of this connection is unipotent (i.e. can be written as a lower-triangular matrix in a suitable basis of W_n), and is given explicitly by (5.28). This is the promised generalization of the dilogarithm variation of Section 5.3.3.

5.4.4 Structure of iterated integrals on $\mathbb{C}\setminus\{0, 1\}$

In summary, we have the following description of the iterated integrals on $\mathbb{C}\setminus\{0, 1\}$.

Theorem 5.4.5 The map

$$\left(\mathbb{C}\langle X\rangle, \text{\cyr{sh}}\right) \longrightarrow \{\text{homotopy invariant iterated integrals on } \mathbb{C}\setminus\{0, 1\}\}$$

$$w \mapsto \mathrm{Li}_w(z) \tag{5.29}$$

is an isomorphism. In other words, every homotopy-invariant iterated integral on $\mathbb{C}\setminus\{0, 1\}$ is a linear combination of the $\mathrm{Li}_w(z)$, and the $\mathrm{Li}_w(z)$ are linearly independent.

In fact, one can show more: every multivalued function on $\mathbb{C}\backslash\{0, 1\}$ with unipotent monodromy and satisfying a polynomial growth condition at 0, 1 and ∞ is a linear combination of $\mathrm{Li}_w(z)$ with coefficients in $\mathbb{C}(z)$, and furthermore, the only algebraic relations between the $\mathrm{Li}_w(z)$ over $\mathbb{C}(z)$ are given by the shuffle product. Thus the multiple polylogarithms are the universal unipotent functions on the thrice punctured projective line.

The multiple polylogarithms also inherit a Hopf algebra structure from the previous theorem. This is because the shuffle algebra $\mathbb{Q}\langle X\rangle$ is itself a graded commutative Hopf algebra for the deconcatenation coproduct defined by:

$$\Delta : \mathbb{Q}\langle X\rangle \longrightarrow \mathbb{Q}\langle X\rangle \otimes_{\mathbb{Q}} \mathbb{Q}\langle X\rangle,$$

$$\Delta(x_{i_1}\ldots x_{i_n}) = \sum_{k=0}^n x_{i_1}\ldots x_{i_k} \otimes x_{i_{k+1}}\ldots x_{i_n},$$

where the grading is given by the number of letters. As we shall see in Section 5.6, this is a general feature of iterated integrals.

A different way to state the universality of the multiple polylogarithms as a class of functions on $\mathbb{P}^1\backslash\{0, 1, \infty\}$ is as follows. Let $\mathcal{O} = \mathbb{Q}[z, \frac{1}{z}, \frac{1}{1-z}]$ denote the ring of regular functions on $\mathbb{P}^1\backslash\{0, 1, \infty\}$. It is a differential algebra with respect to the operator $\frac{\partial}{\partial z}$. Consider the algebra

$$L = \mathcal{O}\langle \mathrm{Li}_w(z) : w \in X^*\rangle$$

consisting of linear combinations of $\mathrm{Li}_w(z)$ with coefficients in $\mathcal{O}$. It too is a differential algebra with respect to the operator $\frac{\partial}{\partial z}$.

Theorem 5.4.6 Every element in L has a primitive, which is unique up to a constant.

In other words, for every $f \in L$, there exists $F \in L$ such that $\frac{\partial F}{\partial z} = f$, and L is the smallest extension of $\mathcal{O}$ with this property. Thus L is the smallest class of polylogarithm functions on $\mathbb{P}^1\backslash\{0, 1, \infty\}$ which are stable under multiplication and taking primitives. This is the key property which will be used in Section 5.7 in relation to Feynman integrals.

5.5 A brief overview of multiple zeta values

5.5.1 Single zeta values

First consider the values of the Riemann zeta function $\zeta(n)$, for n an integer $n \geq 2$. It was shown by Euler in 1735 that $\zeta(2) = \frac{\pi^2}{6}$ (this had been conjectured by Mengoli in 1644), and subsequently that

$$\zeta(2n) = -\frac{(2\pi i)^{2n} B_{2n}}{2\,(2n)!},$$

where the Bernoulli numbers B_m are given by the generating series

$$\frac{x}{1 - e^{-x}} = \sum_{m=0}^{\infty} B_m \frac{x^m}{m!}.$$

In particular, $\zeta(2n)$ is an explicit rational multiple of π^{2n}. It is expected that no such formula should exist for the odd zeta values.

Conjecture 1 The numbers $\pi, \zeta(3), \zeta(5), \dots$ are algebraically independent over $\mathbb{Q}$.

Surprisingly little is known about this conjecture. What is known is that π is transcendental (Lindemann 1882), that $\zeta(3)$ is irrational (Apéry 1978), that infinitely many values of $\zeta(2n + 1)$ are irrational (Ball-Rivoal 2000) and various refinements in this direction. However, it is still not known whether $\zeta(5)$ is irrational or not.

5.5.2 *Multiple zeta values*

Multiple zeta values, on the other hand, satisfy a huge number of algebraic relations over $\mathbb{Q}$. The so-called 'standard relations' or 'double shuffle relations' are linear and quadratic relations which we briefly summarize below. Recall that $X = \{x_0, x_1\}$ is an alphabet on two letters, and let $\mathbb{Q}\langle X \rangle$ denote the free non-commutative algebra on the symbols x_0 and x_1 with the concatenation product, i.e. the vector space generated by all words in X including the empty word 1. We have already defined a map $\zeta : x_0 X^* x_1 \longrightarrow \mathbb{R}$,

$$\zeta(x_0^{n_r-1} x_1 \dots x_0^{n_0-1} x_1) = \zeta(n_1, \dots, n_r),$$

which we can extend by linearity to the subspace $x_0 \mathbb{Q}\langle X \rangle x_1 \subset \mathbb{Q}\langle X \rangle$ of convergent words. The weight of $\zeta(n_1, \dots, n_r)$ is defined to be the quantity $n_1 + \dots + n_r$. This defines a filtration on the vector space of multiple zeta values. The standard relations can be described in terms of two different product structures on $x_0 \mathbb{Q}\langle X \rangle x_1$.

- *Shuffle product.* Recall that we had a commutative and associative product

$$\text{ш} : \mathbb{Q}\langle X \rangle \times \mathbb{Q}\langle X \rangle \to \mathbb{Q}\langle X \rangle$$

which is uniquely determined by the properties (exercise)

$$w \, \text{ш} \, 1 = w, \quad 1 \, \text{ш} \, w = w \quad \text{for all } w \in X^*,$$

$$x_i w \, \text{ш} \, x_j w' = x_i(w \, \text{ш} \, x_j w') + x_j(x_i w \, \text{ш} \, w')$$

for all $x_i, x_j \in X$, $w, w' \in X^*$. The subspace $\mathbb{Q}1 \oplus x_0 \mathbb{Q}\langle X \rangle x_1$ of convergent words is a subalgebra of $\mathbb{Q}\langle X \rangle$ with respect to ш. It follows immediately from the corresponding shuffle product formula for the functions $\mathrm{Li}_w(z)$ that

$$\zeta(w)\zeta(w') = \zeta(w \shuffle w') \quad \text{for all } w, w' \in x_0 X^* x_1.$$

This gives, for example, $\zeta(x_0x_1)\zeta(x_0x_1) = 2\,\zeta(x_0x_1x_0x_1) + 4\,\zeta(x_0x_0x_1x_1)$, that is $\zeta(2)\zeta(2) = 2\zeta(2, 2) + 4\zeta(1, 3)$.

- *The stuffle (or quasi-shuffle) product.* The stuffle product comes from the representation of multiple zetas as nested sums. We will not prove the general case (see e.g. [21, 23]), but only illustrate in the simplest case. Decomposing the domain $\mathbb{N} \times \mathbb{N}$ of summation into three regions in the following gives:

$$\sum_{k \geq 1} \frac{1}{k^m} \sum_{\ell \geq 1} \frac{1}{\ell^n} = \left(\sum_{k < \ell} + \sum_{\ell < k} + \sum_{k = \ell} \right) \frac{1}{k^m \ell^n},$$

$$\zeta(m)\zeta(n) = \zeta(m, n) + \zeta(n, m) + \zeta(m + n).$$

The general case is similar and gives a formula relating the product of any multiple zetas as a linear combination of other multiple zetas of the same weight. This product can also be encoded symbolically with words as follows. For each $i \geq 1$, we write $y_i = x_0^{i-1} x_1$. Then $\mathbb{Q}1 \oplus \mathbb{Q}\langle X \rangle x_1 \cong \mathbb{Q}\langle Y \rangle$, where $Y = \{y_n, n \in \mathbb{N}\}$. The stuffle product, written

$$\star : \mathbb{Q}\langle Y \rangle \times \mathbb{Q}\langle Y \rangle \to \mathbb{Q}\langle Y \rangle,$$

is defined inductively as follows:

$$w \star 1 = w, \quad 1 \star w = w \quad \text{for all } w \in Y^*,$$

$$y_i w \star y_j w' = y_i(w \star y_j w') + y_j(y_i w \star w') + y_{i+j}(w \star w'),$$

for all $i, j \geq 1$ and $w, w' \in Y^*$. The stuffle relation is then:

$$\zeta(w)\zeta(w') = \zeta(w \star w') \quad \text{for all } w, w' \in x_0 X^* x_1.$$

For example, the relation we derived earlier by decomposing the domain of summation corresponds to $y_m \star y_n = y_m y_n + y_n y_m + y_{m+n}$.

- *Regularization relation.* Let $w \in x_0 X^* x_1$ be a convergent word. One can prove that $x_1 \star w - x_1 \shuffle w \in x_0 X^* x_1$ is also a linear combination of convergent words. The regularization, or Hoffman, relation is given by:

$$\zeta(x_1 \star w - x_1 \shuffle w) = 0.$$

Applying this identity to $w = x_0 x_1$, for example, yields the relation

$$\zeta(3) = \zeta(1, 2), \tag{5.30}$$

which was first proved by Euler.

It is conjectured that all algebraic relations over $\mathbb{Q}$ satisfied by the multiple zeta values are generated by the previous three identities (to get a sense of this, prove that $\zeta(4)$ is a multiple of $\zeta(2)^2$ using the standard identities only).

One can then try to write down a minimal basis for the multiple zeta values in each weight by solving these equations. What one finds is the following table, to weight 8:

Weight	1	2	3	4	5	6	7	8
		$\zeta(2)$	$\zeta(3)$	$\zeta(4)$	$\zeta(5)$	$\zeta(6)$	$\zeta(7)$	$\zeta(8)$
					$\zeta(2)\zeta(3)$	$\zeta(3)^2$	$\zeta(2)\zeta(5)$	$\zeta(3)\zeta(5)$
							$\zeta(3)\zeta(4)$	$\zeta(3)^2\zeta(2)$
								$\zeta(3,5)$
dim	0	1	1	1	2	2	3	4

For example, in depth 3 there are *a priori* two multiple zeta values, $\zeta(3)$ and $\zeta(1, 2)$, but the identity (5.30) tells us that they coincide. So $\dim_{\mathbb{Q}}(\mathbb{Q}\zeta(3) + \mathbb{Q}\zeta(1, 2)) = 1$. We could have replaced $\zeta(2n)$ with $\zeta(2)^n$ in the table, by Euler's theorem. Note that, up to weight 7, all multiple zeta values are spanned by products of ordinary zeta values $\zeta(n)$. In weight 8, something interesting happens for the first time, and there is a new quantity $\zeta(3, 5)$ which is irreducible, in the sense that it (conjecturally, at least) cannot be expressed as a polynomial in the $\zeta(n)$.

The dimensions at the bottom, which we denote by d_k in weight k, mean the following. Take the $\mathbb{Q}$-vector space spanned by the symbols $\zeta(w)$, where w ranges over the set of convergent words of length k, and take the quotient by all linear relations $\zeta(\eta \sqcup \eta') - \zeta(\eta \star \eta')$ deduced from the standard relations. This gives an upper bound for the dimension of the vector space spanned by the actual values $\zeta(w) \in \mathbb{R}$.

Conjecture 2 (Zagier) $d_k = d_{k-2} + d_{k-3}$.

This conjecture is purely algebraic, but in fact Zagier also conjectured that the dimension of the *actual* zeta values in weight k should be d_k. This transcendental part of the conjecture is already completely unknown in weight 5 and is equivalent to

$$\frac{\zeta(5)}{\zeta(2)\zeta(3)} \notin \mathbb{Q}.$$

What is known in the algebraic direction is the following:

Theorem 5.5.1 (Goncharov, Terasoma)   Let $D_0 = 1$, $D_1 = 0$, $D_2 = 1$, and define D_k for $k \geq 2$ by $D_k = D_{k-2} + D_{k-3}$. Then

$$\dim_{\mathbb{Q}} \langle \zeta(k) \text{ of weight } k \rangle \leq D_k.$$

The proof of this theorem is one of the most striking applications of the theory of mixed Tate motives. What is not known, however, is whether the standard relations are enough to span all $\mathbb{Q}$-relations satisfied by the multiple zeta values, and there is no known algorithm to reduce a given multiple zeta value into a given basis using the above relations.

There is also a more precise conjecture for the number of zeta values of given weight and depth (the depth of $\zeta(n_1, \ldots, n_r)$ being the number r) which is due to Broadhurst and Kreimer [3].

5.5.3 Hopf algebra interpretation

Some of the previous conjectures can be reformulated in terms of a certain Hopf algebra. Let $\mathcal{L}$ denote the free Lie algebra generated by one element σ_{2n+1} in every odd degree $-2n - 1$ for $n \geq 1$. The underlying graded vector space is generated by, in decreasing weight:

$$\sigma_3, \sigma_5, \sigma_7, [\sigma_3, \sigma_5], \sigma_9, [\sigma_3, \sigma_7], \sigma_{11}, [\sigma_3, [\sigma_5, \sigma_3]], \ldots$$

Let $\mathcal{UL}$ be its universal enveloping algebra, and let $\mathbb{M}'$ be its graded dual, which is a commutative Hopf algebra. Concretely, $\mathbb{M}'$ is the set of all non-commutative words in letters f_{2n+1} in degree $2n + 1$ dual to σ_{2n+1}, equipped with the shuffle product ɯ. Finally, set

$$\mathbb{M} = \mathbb{Q}[f_2] \otimes_{\mathbb{Q}} (\mathbb{Q}\langle f_3, f_5, f_7 \rangle \ldots, \text{ɯ})$$

where f_2 is a new generator of degree 2 which commutes with all the others. The generators in each weight up to 8 are precisely:

$$f_2; \; f_3; \; f_2^2; \; f_5, f_3 f_2; \; f_3^2, f_2^3; \; f_7, f_5 f_2, f_3 f_2^2; \; f_2^4, f_3^2 f_2, f_3 \text{ ɯ } f_5, f_3 f_5,$$

which matches the table for multiple zeta values above. In weight 8 the basis can also be written $f_3 f_5$, $f_5 f_3$, f_2^4, $f_3^2 f_2$. The following conjecture, due to Goncharov [13], is a more precise version of Conjecture 2.

Conjecture 3 The algebra spanned by the multiple zetas over $\mathbb{Q}$ is isomorphic to $\mathbb{M}$.

This conjecture comes from the theory of mixed Tate motives over $\mathbb{Z}$ and has several consequences.

- Firstly, since $\mathbb{M}$ is graded by the weight, it implies that there should exist no algebraic relations between any multiple zeta values of different weights.
- Secondly, it implies Conjecture 2. To see this, let $V = \bigoplus_{k \geq 0} V_k$ be any graded vector space such that the V_k are finite-dimensional, and let

$$\chi(V)(t) = \sum_k \dim_k(V_k)\, t^k$$

denote the generating series of its graded dimensions. For any two such graded vector spaces V, V' we have $\chi(V \otimes_{\mathbb{Q}} V') = \chi(V)\chi(V')$. Then

$$\chi(\mathbb{Q}\langle f_3, f_5, \ldots \rangle)(t) = \frac{1}{1 - t^3 - t^5 - t^7 - \cdots} = \frac{1 - t^2}{1 - t^2 - t^3}$$

and since $\mathbb{M} = \mathbb{Q}[f_2] \otimes_{\mathbb{Q}} \mathbb{Q}\langle f_3, f_5, \ldots, \rangle$ we deduce that

$$\chi(\mathbb{M})(t) = \Big(\frac{1}{1 - t^2}\Big)\Big(\frac{1 - t^2}{1 - t^2 - t^3}\Big) = \frac{1}{1 - t^2 - t^3} = \sum_{k \geq 0} D_k t^k,$$

where the numbers D_k satisfy $D_k = D_{k-2} + D_{k-3}$.
- Since $\mathbb{M}'$ is a Hopf algebra, it implies that the coproduct $\Delta : \mathbb{M}' \to \mathbb{M}' \otimes \mathbb{M}'$ should also exist on the level of multiple zeta values. The coproduct on $\mathbb{M}'$ is given by the deconcatenation coproduct

$$\Delta(f_{2i_1+1} \cdots f_{2i_r+1}) = f_{2i_1+1} \cdots f_{2i_r+1} \otimes 1 + 1 \otimes f_{2i_1+1} \cdots f_{2i_r+1}$$

$$+ \sum_{k=1}^{r-1} f_{2i_1+1} \cdots f_{2i_k+1} \otimes f_{2i_{k+1}+1} \cdots f_{2i_r+1}.$$

It makes no sense to define a coproduct on the level of the numbers $\zeta(w)$ because of the inaccessibility of the transcendence conjectures. However, one can lift the ordinary zeta values to "motivic multiple zeta values" $\zeta^M(w)$, which generate a sub-Hopf algebra of $\mathbb{M}'$ (conjecturally equal to it). The objects $\zeta^M(w)$ satisfy the standard relations together with the equation $\zeta^M(2) = 0$. Goncharov has computed the corresponding coproduct on the level of the motivic multiple zetas [13] and, for example, gives:

$$\Delta \zeta^M(2n + 1) = 1 \otimes \zeta^M(2n + 1) + \zeta^M(2n + 1) \otimes 1,$$

$$\Delta \zeta^M(3, 5) = 1 \otimes \zeta^M(3, 5) - 5\,\zeta^M(3) \otimes \zeta^M(5) + \zeta^M(3, 5) \otimes 1.$$

Note, however, that the coproduct on the motivic multiple zeta values is very complicated in general and shows that $\zeta^M(n_1, \ldots, n_r)$ does not correspond to an element of $\mathbb{M}'$ (i.e. word in the fs) in any straightforward way.

Remark 5.5.2 The (conjectural) existence of a coproduct on the multiple zeta values is a genuinely new feature given by the motivic theory, and it seems it has not yet been exploited by physicists. We believe that it should have much relevance to perturbative quantum field theories: many Feynman amplitudes should be certain linear combinations of multiple zeta values which are 'simple' with respect to the coproduct Δ (in other words, they should be filtered in an interesting way by the coradical filtration).

5.6 Iterated integrals and homotopy invariance

We now return to the general theory of iterated integrals and state necessary and sufficient conditions for an iterated integral to be a homotopy functional. We then state Chen's π_1-de Rham theorem, and relate it to the previous discussions.

5.6.1 Homotopy functionals

For motivation we begin by constructing some simple examples of homotopy-invariant iterated integrals. Let M be a smooth manifold, ω a smooth 1-form on M and $\gamma : [0, 1] \to M$ a smooth path.

Lemma 5.6.1 The line integral $\int_\gamma \omega$ is a homotopy functional if and only if ω is closed.

Proof Since the result is well known, we only sketch the argument. Suppose that the integral is a homotopy functional. Then for every closed loop $\gamma : [0, 1] \to M$ which bounds a small disk $D \subset M$, the integral of ω along γ is zero. By Stokes' theorem

$$\oint_\gamma \omega = \int_D d\omega.$$

Since this vanishes for all small disks centred at every point of M, we conclude that $d\omega = 0$. In the converse direction, by the Poincaré lemma, a closed form is locally exact, and so $d\omega = 0$ implies that the integral around any small loop is zero. $\quad\square$

Now suppose that ω_1, ω_2 are closed 1-forms. By the previous lemma, the line integrals $\int \omega_i$, for $i = 1, 2$, are homotopy functionals. Therefore if $\gamma : [0, 1] \to M$ is a smooth path with fixed initial point $\gamma(0)$ but variable endpoint $z = \gamma(1)$ then $F_2(z) = \int_\gamma \omega_2$ defines a multivalued function on M which satisfies $dF_2(z) = \omega_2$ by the fundamental theorem of calculus. Consider the iterated integral

$$I = \int_\gamma \omega_1 \omega_2 + \omega_{12}.$$

Recall from the definition of the iterated integrals that $I = \int_\gamma \omega_1 \wedge F_2(z) + \omega_{12}$. By the lemma, it is a homotopy functional if and only if the integrand is closed, i.e.

$$d\left(\omega_1 \wedge F_2(z) + \omega_{12}\right) = 0.$$

By Leibniz' rule, this gives: $d\omega_1 \wedge F_2(z) + \omega_1 \wedge \omega_2 + d\omega_{12} = 0$. Thus I defines a homotopy functional if and only if

$$\omega_1 \wedge \omega_2 + d\omega_{12} = 0. \tag{5.31}$$

More generally, suppose that ω_i are closed 1-forms. A similar calculation shows that

$$\int \sum_{i,j} \omega_i \omega_j + \omega_k$$

is a homotopy functional if and only if $\sum_{i,j} \omega_i \wedge \omega_j + d\omega_k = 0$.

Example 5.6.2 Let $\omega_1, \omega_2, \omega_3$ be closed 1-forms on M, and suppose that $\omega_1 \wedge \omega_2$ and $\omega_2 \wedge \omega_3$ are exact. Then we can find ω_{12}, ω_{13} such that $d\omega_{12} = -\omega_1 \wedge \omega_2$ and $d\omega_{23} = -\omega_2 \wedge \omega_3$. It follows that $\omega_1 \wedge \omega_{23} + \omega_{12} \wedge \omega_3$ is closed. Now suppose that there is a 1-form ω_{123} (known as a Massey triple product of $\omega_1, \omega_2, \omega_3$) such that

$$d\omega_{123} = \omega_1 \wedge \omega_{23} + \omega_{12} \wedge \omega_3.$$

Then $\int \omega_1 \omega_2 \omega_3 + \omega_{12} \omega_3 + \omega_1 \omega_{23} + \omega_{123}$ is a homotopy functional.

Exercise 5.6.3 Consider the trivial bundle $k^3 \times M \to M$. The following matrix of 1-forms

$$\Omega = \begin{pmatrix} 0 & 0 & 0 \\ \omega_1 & 0 & 0 \\ \omega_{12} & \omega_2 & 0 \end{pmatrix}$$

defines a connection on it which is integrable if and only if $d\Omega = \Omega \wedge \Omega$, which is exactly the requirement $d\omega_{12} = \omega_1 \wedge \omega_2$ (5.31).

5.6.2 The bar construction

The general condition for the homotopy invariance of iterated integrals can be stated in terms of the bar construction. Let M be connected, let $A^*(M)$ be the complex of C^∞ forms on M, and let us suppose that $X \subset A^*(M)$ is a connected model for M, i.e, such that $X^0 \cong k$ and the map

$$X \longrightarrow A^*(M) \tag{5.32}$$

induces an isomorphism $H^*(X) \to H^*(\mathcal{A}^*(M)) = H^*(M)$ (in fact it suffices to be an isomorphism on H^1 and injective on H^2). When considering the tensor product $(X^1)^{\otimes n}$, it is customary in this context to use the bar notation and write $[\omega_1| \cdots |\omega_n]$ for $\omega_1 \otimes \cdots \otimes \omega_n$. Consider the map:

$$D : (X^1)^{\otimes n} \to (\mathcal{A}^\bullet(M))^{\otimes n}$$

$$D([\omega_1| \cdots |\omega_n]) = \sum_{i=1}^{n} [\omega_1| \cdots |\omega_{i-1}|d\omega_i|\omega_{i+1}| \cdots |\omega_n]$$

$$+ \sum_{i=1}^{n-1} [\omega_1| \cdots |\omega_{i-1}|\omega_i \wedge \omega_{i+1}| \cdots |\omega_n].$$

Let us define

$$B_n(M) = \left\{ \xi = \sum_{\ell=0}^{n} \sum_{i_1,\dots,i_\ell} [\omega_{i_1}| \cdots |\omega_{i_\ell}] \text{ such that } D\xi = 0, \text{ where } \omega_{i_j} \in X^1 \right\}.$$

We call such an element ξ satisfying $D\xi = 0$ an integrable word in X^1. The limit $B(M) = \sum_{n \geq 0} B_n(M)$ is a vector space over k, and the index n defines a filtration on it which is called the length filtration. One verifies that $B(M)$ is a commutative algebra for the shuffle product

$$\text{ш} : B(M) \otimes_k B(M) \longrightarrow B(M)$$

(for this, one must check that D defined above satisfies the Leibniz rule with respect to the shuffle product on $X^{\otimes n}$) and is in fact a commutative Hopf algebra for the deconcatenation coproduct:

$$\Delta : B(M) \longrightarrow B(M) \otimes_k B(M),$$

$$[\omega_{i_1}| \cdots |\omega_{i_\ell}] \mapsto \sum_{r=0}^{\ell} [\omega_{i_1}| \cdots |\omega_{i_r}] \otimes [\omega_{i_{r+1}}| \cdots |\omega_{i_\ell}]. \tag{5.33}$$

In conclusion, $B(M)$ is a filtered, commutative Hopf algebra. It follows from the definition of D and (5.32) that in length 1 we have:

$$B_1(M) \cong k \oplus H^1_{dR}(M; k).$$

To any element ξ in $B_n(M)$, we can associate the corresponding iterated integral

$$\sum_{\ell=0}^{n} \sum_{i_1,\dots,i_\ell} [\omega_{i_1}| \cdots |\omega_{i_\ell}] \mapsto \sum_{\ell=0}^{n} \sum_{i_1,\dots,i_\ell} \int_\gamma \omega_{i_1} \dots \omega_{i_\ell} \tag{5.34}$$

where $\gamma : [0, 1] \rightarrow M$ is a smooth path in M. Chen's theorem states that this iterated integral is a homotopy functional and, furthermore, that all such homotopy functionals arise in this way.

Theorem 5.6.4 (Chen) The integration map (5.34) gives an isomorphism:

$$B_n(M) \longrightarrow \{\text{homotopy invariant iterated integrals of length } \leq n\}.$$

In particular, if we consider paths with fixed $\gamma(0) = z_0$ and let $\gamma(1) = z$, then the iterated integrals of elements $\xi \in B_n(M)$ along γ define multivalued functions on M. One can check that their differential is given by

$$d\Big(\sum_{\ell=0}^{n} \sum_{i_1,\ldots,i_\ell} \int_\gamma \omega_{i_1} \ldots \omega_{i_\ell}\Big) = \sum_{\ell=0}^{n} \sum_{i_1,\ldots,i_\ell} \omega_{i_\ell} \int_\gamma \omega_{i_1} \ldots \omega_{i_{\ell-1}},$$

which is induced from the $(\ell - 1, 1)$-part of the coproduct Δ.

Remark 5.6.5 $B_n(M)$ is quasi-isomorphic to $H_0(\overline{B}_n(\mathcal{A}^\bullet(M)))$, where $\overline{B}_n$ denotes Chen's reduced bar construction.

One can generalize the definition of iterated integrals for forms of any degree. Such an iterated integral defines a form on the space of paths of M. The H_0 refers to locally constant functions on the space of paths, i.e. the homotopy invariant functions on M.

5.6.3 Chen's theorem

Let M be a smooth manifold, and let $\pi_1(M, x)$ be the fundamental group of M based at a point $x \in M$. One can extend the definition of homotopy-invariant iterated integrals by linearity to the group ring

$$\mathbb{Q}[\pi_1(M, x)]$$

as follows. For any combination of paths $g = \sum_{i=1}^{n} a_i \gamma_i \in \mathbb{Q}[\pi_1(M, x)]$, and any 1-forms $\omega_1, \ldots, \omega_r$ on M, one can define the iterated integral along g by:

$$\int_g \omega_1 \ldots \omega_r = \sum_{i=1}^{n} a_i \int_{\gamma_i} \omega_1 \ldots \omega_r.$$

Recall that, for ordinary line integrals, we have the formula

$$\int_{(\gamma_1-1)(\gamma_2-1)} \omega = 0,$$

where 1 is the constant path at x. By expanding $(\gamma_1 - 1)(\gamma_2 - 1)$, this is equivalent to the composition of paths formula (Proposition 5.2.2 (iii)):

$$\int_{\gamma_1 \gamma_2} \omega = \int_{\gamma_1} \omega + \int_{\gamma_2} \omega.$$

The generalization for iterated integrals is the following.

Lemma 5.6.6 Let $\omega_1, \ldots, \omega_r$ be smooth 1-forms on M, and let $\gamma_1, \ldots, \gamma_s$ be loops on M based at x. If $s > r$ then

$$\int_{(\gamma_1 - 1)\ldots(\gamma_s - 1)} \omega_1 \ldots \omega_r = 0.$$

Proof The formula for the composition of paths (Proposition 5.2.2 (iii)), remains valid for linear combinations of paths. Note that the composition of paths is associative up to reparametrization, but iterated integrals do not depend on the choice of parametrization of the path. Recall that, in that formula, the empty iterated integral ($r = 0$) over a path γ is just the constant function 1. It follows that the empty iterated integral over $(\gamma_1 - 1)$ is zero. The general case follows by induction. Setting $\beta = (\gamma_2 - 1) \cdots (\gamma_s - 1) \in \mathbb{Q}[\pi_1(M, x)]$, we can write $(\gamma_1 - 1) \ldots (\gamma_s - 1) = \gamma_1 \beta - \beta$. Therefore by the composition of paths formula,

$$\int_{(\gamma_1 - 1)\beta} \omega_1 \ldots \omega_r = \sum_{i=0}^{r} \int_{\gamma_1} \omega_1 \ldots \omega_i \int_{\beta} \omega_{i+1} \ldots \omega_r - \int_{\beta} \omega_1 \ldots \omega_r$$

$$= \sum_{i=1}^{r} \int_{\gamma_1} \omega_1 \ldots \omega_i \int_{\beta} \omega_{i+1} \ldots \omega_r.$$

But the second term in the last line is an integral of at most $r - 1$ forms over a composition $\beta = (\gamma_2 - 1) \ldots (\gamma_s - 1)$, which vanishes by induction since $s - 1 > r - 1$. $\square$

Consider the augmentation map

$$\varepsilon : \mathbb{Q}[\pi(M, x)] \longrightarrow \mathbb{Q},$$

$$\gamma \mapsto 1,$$

and let $J = \ker \varepsilon \subset \mathbb{Q}[\pi_1(M, x)]$ be the augmentation ideal. It is generated by the elements of the form $\gamma - 1$, with $\gamma \in \pi_1(M, x)$. The truncated group ring is defined by

$$V_n = \mathbb{Q}\pi_1(M, x)/J^{n+1}.$$

Recall that the integration map gives a pairing

$$\mathbb{Q}[\pi_1(M, x)] \otimes_\mathbb{Q} B_n(M) \longrightarrow \mathbb{C}, \qquad (5.35)$$

$$g \otimes [\omega_1| \cdots |\omega_n] \mapsto \int_g \omega_1 \ldots \omega_n.$$

Lemma 5.6.6 states that this map vanishes on J^m if $m \geq n + 1$. From this we deduce a map

$$B_n(M) \otimes_k \mathbb{C} \xrightarrow{\phi} \mathrm{Hom}_\mathbb{Q}(V_{n+1}, \mathbb{C}) \qquad (5.36)$$

$$\xi \mapsto \left(g \mapsto \int_g \xi\right).$$

Theorem 5.6.7 (Chen's π_1-de Rham theorem)   The map ϕ is an isomorphism.

References for the proofs of the above theorems can be found in [15], [10], [14].

Example 5.6.8 Recall that we had $B_1(M) \cong H^1_{DR}(M) \oplus k$. On the other hand

$$V_2 = \mathbb{Q}[\pi_1(M, x)]/J^2 \cong H_1(M) \oplus \mathbb{Q}$$

by Hurwitz' theorem. Therefore, in the case $n = 1$, Chen's theorem is equivalent to saying that the integration map

$$H^1_{DR}(M; \mathbb{C}) \longrightarrow \mathrm{Hom}(H_1(M), \mathbb{C})$$

is an isomorphism. This is exactly de Rham's theorem in degree 1.

5.6.4 Example: multiple polylogarithms in one variable

Let $M = \mathbb{P}^1 \backslash \{0, 1, \infty\}$. In this case we can take $X \subset \mathcal{A}^1(M)$ to be the $\mathbb{C}$-vector space spanned by $\omega_0 = \frac{dz}{z}$ and $\omega_1 = \frac{dz}{1-z}$ since these form a basis for $H^1_{dR}(M)$. Since ω_0, ω_1 are closed and satisfy $\omega_i \wedge \omega_j = 0$ for all $i, j \in \{0, 1\}$ (M is one-dimensional) the integrability condition $D\xi = 0$ is trivially satisfied for any word ξ in the forms ω_0, ω_1. Thus

$$B_n(M) = \{[\omega_{i_1}| \cdots |\omega_{i_k}] : i_j \in \{0, 1\}, \quad k \leq n\}$$

is the set of words in $\{\omega_0, \omega_1\}^*$ with at most n letters, and the bar construction is isomorphic to the shuffle algebra

$$B(M) \cong \mathbb{C}\langle\{\omega_0, \omega_1\}\rangle$$

equipped with the deconcatenation coproduct; Chen's theorem boils down to Theorem 5.4.5. In this case, it follows from the fact that M is defined over $\mathbb{Q}$ that $B(M)$ can also be defined over $\mathbb{Q}$, simply by taking $\mathbb{Q}$-linear combinations of words in ω_0, ω_1. Note that it is this $\mathbb{Q}$-structure that defines the $\mathbb{Q}$-algebra of multiple zeta values.

5.6.5 Example: multiple polylogarithms in several variables

The previous example generalizes to the family of manifolds $M = \mathfrak{M}_{0,n+3}$, where $n \geq 1$, defined by the complement of hyperplanes

$$\mathfrak{M}_{0,n+3}(\mathbb{C}) = \{(t_1, \ldots, t_n) \in \mathbb{C}^n : t_i \neq t_j, \ t_i \neq 0, 1\}.$$

The notation $\mathfrak{M}_{0,n+3}$ comes from the fact that these are isomorphic to the moduli spaces of curves of genus 0 with $n + 3$ ordered marked points. The previous example concerns the iterated integrals on $\mathfrak{M}_{0,4}(\mathbb{C}) \cong \mathbb{C}\backslash\{0, 1\}$. In order to write down the bar construction, we need a model for the de Rham complex on $\mathfrak{M}_{0,n}$. Therefore consider the set of differential forms

$$\Omega = \left\{ \frac{dt_i - dt_j}{t_i - t_j}, \frac{dt_i}{t_i}, \frac{dt_i}{1 - t_i} \right\}$$

and denote its elements by $\omega_1, \ldots, \omega_N$, where $N = n(n - 3)/2$. Let $X = \bigoplus_{n \geq 0} X^n$ denote the $\mathbb{Q}$-subalgebra of regular differential forms on $\mathfrak{M}_{0,n}$ spanned by Ω and graded by the degree. It is connected, i.e. $X^0 \cong \mathbb{Q}$.

Theorem 5.6.9 (Arnold) The map $X \otimes_{\mathbb{Q}} \mathbb{C} \to \mathcal{A}^*(\mathfrak{M}_{0,n+3})$ is a quasi-isomorphism, i.e. it induces an isomorphism on cohomology.

This means that the space $\mathfrak{M}_{0,n+3}$ is formal and implies that there are no Massey products in this case. Furthermore, X has a natural $\mathbb{Q}$ structure and so we henceforth work over $\mathbb{Q}$. The set of integrable words of length n in $B(\Omega)$ can be written

$$\xi = \sum_{I=(i_1,\ldots,i_n)} c_I[\omega_{i_1}|\cdots|\omega_{i_n}] \quad c_I \in \mathbb{Q}$$

such that

$$\omega_{i_1} \otimes \cdots \otimes \omega_{i_k} \wedge \omega_{i_{k+1}} \otimes \cdots \otimes \omega_{i_n} = 0 \in (\Omega^1)^{\otimes k-1} \otimes \Omega^2 \otimes (\Omega^1)^{\otimes n-k-1}.$$

In particular, the bar construction $B(\mathfrak{M}_{0,n+3})$ is graded by the length.

The description of the iterated integrals on $\mathfrak{M}_{0,n+3}$ requires the definition of multiple polylogarithms in several variables.

Definition 5.6.10 (Goncharov) Let $n_1, \ldots, n_r \in \mathbb{N}$ and define the multiple polylogarithm in several variables by the nested sum

$$\mathrm{Li}_{n_1,\ldots,n_r}(x_1, \ldots, x_r) = \sum_{0 < k_1 < \cdots < k_r} \frac{x_1^{k_1} \ldots x_r^{k_r}}{k_1^{n_1} \ldots k_r^{n_r}},$$

which converges absolutely on compacta in the polydisc $|x_i| < 1$.

In order to make the connection with $\mathfrak{M}_{0,n+3}$ consider the functions

$$I_{n_1,\ldots,n_r}(t_1,\ldots,t_n) = \mathrm{Li}_{n_1,\ldots,n_r}\left(\frac{t_1}{t_2},\ldots,\frac{t_{r-1}}{t_r},t_r\right)$$

which can be shown to have an analytic continuation as multivalued functions on the whole of $\mathfrak{M}_{0,r+3}$, and have unipotent monodromy. The converse is also true:

Theorem 5.6.11 Every (homotopy-invariant) iterated integral on $\mathfrak{M}_{0,n+3}$ can be expressed (non-uniquely) as a sum of products of the functions $\log(t_i)$, $1 \leq i \leq n$, and the multiple polylogarithms $I_{n_1,\ldots,n_r}(\frac{s_1}{s_2},\ldots,\frac{s_{r-1}}{s_r},s_r)$ where $s_i \in \{1, t_1, \ldots, t_n\}$.

A basis for the iterated integrals on $\mathfrak{M}_{0,n+3}$ is given in [4].

Example 5.6.12 Consider the function $I_{1,1}(t_1, t_2)$. It follows from the definition as a nested sum that it satisfies a differential equation of the form:

$$dI_{1,1}(t_1, t_2) = dI_1\left(\frac{t_1}{t_2}\right)I_1(t_2) + dI_1(t_2)I_1(t_1) - dI_1\left(\frac{t_2}{t_1}\right)I_1(t_1).$$

Since $I_1(t) = -\log(1 - t)$, one deduces that the following word

$$\xi = \left[d\log\left(1 - \frac{t_1}{t_2}\right)|d\log(1 - t_2)\right] + \left[d\log(1 - t_2)|d\log(1 - t_1)\right]$$

$$- \left[\log\left(1 - \frac{t_2}{t_1}\right)|\log(1 - t_1)\right]$$

is integrable, and $I_{1,1}(t_1, t_2)$ is the iterated integral:

$$I_{1,1}(t_1, t_2) = \int_\gamma \xi,$$

where γ is a path from $(0, 0)$ to (t_1, t_2). In particular, the expression for ξ gives a formula for the coproduct of $I_{1,1}(t_1, t_2)$.

Remark 5.6.13 By computing the integable word for $I_{n_1,\ldots,n_r}(t_1,\ldots,t_r)$, and then taking the appropriate limit of its coproduct as $t_1,\ldots,t_r \to 1$, one can retrieve the correct expression for the motivic coproduct for $\zeta^M(n_1,\ldots,n_r)$ mentioned earlier (Section 5.5.2).

The multivariable analogue of Theorem 5.4.6 is the following. Let $\mathcal{O}(\mathfrak{M}_{0,n+3}) = \mathbb{Q}[t_1,\ldots,t_n,t_i^{-1},(1-t_i)^{-1},(t_i-t_j)^{-1}]$ denote the ring of regular functions on $\mathfrak{M}_{0,n+3}$ and let L denote the $\mathcal{O}(\mathfrak{M}_{0,n+3})$-algebra spanned by the functions in Theorem 5.6.11.

Theorem 5.6.14 The following complex

$$0 \to L \to L \otimes_\mathbb{Q} X^1 \to L \otimes_\mathbb{Q} X^2 \to \cdots \to L \otimes_\mathbb{Q} X^n \to 0$$

is exact in all degrees ≥ 1.

In other words, every closed form of degree ≥ 1 on $\mathfrak{M}_{0,n+3}$ with coefficients in L is exact. Concretely, any closed differential form of degree ≥ 1 whose coefficients are multiple polylogarithms and rational functions in $\mathcal{O}(\mathfrak{M}_{0,n+3})$ always has a primitive of the same form. As in the single-variable case $\mathfrak{M}_{0,4}$, it is this property which provides the connection with Feynman diagram computations (see Section 5.7).

Remark 5.6.15 Even though the picture in genus 0 is quite complete, the analogous numbers and functions are not known in the case of curves of higher genus.

5.7 Feynman integrals

5.7.1 Very short introduction to perturbative quantum field theory

A quantum field theory can be represented by certain families of vertices and edges of different types. By putting them together, one obtains Feynman diagrams, such as on the right below:

The graph elements on the left are taken from quantum electrodynamics: the solid line represents an electron moving in the direction of the arrow (or a positron moving in the opposite direction), the wiggly line represents a photon, and the three-valent vertex represents their interaction. A graph such as the one on the right represents a process involving these elementary particles: in this case, the exchange of a photon between two electrons. The same diagram has other interpretations, depending on the direction of the time axis (if the time axis goes from left to right, it shows an electron and positron annihilating to give a photon, which in turn decays into an electron and positron). To this graph one associates a "probability" or Feynman amplitude, which is an integral determined from the graph by the Feynman rules. From the mathematician's point of view, a Feynman graph can be thought of as a compact way to encode an integral. In order to obtain a physical prediction for a process, one must then sum over all possible Feynman amplitudes which contribute to the given process. The diagrams with higher numbers of loops represent successive approximations – the higher the loop order diagrams one can calculate, the more accurate the theoretical prediction. But this comes at great cost, since the number of diagrams grows very fast at increasing loop orders, and the integrals themselves become extremely hard to compute. Indeed, a vast amount of effort in the modern physics literature is devoted to the calculation of Feynman integrals, and this is the main way in which theoretical predictions for particle collider experiments are currently obtained.

In the next section, we try to justify the (rather arduous, but well-understood [16], [19]) passage from realistic quantum field theories to the simplified scalar Feynman diagrams we shall consider in what follows. The reader only interested in the connection with number theory can skip straight to Section 5.7.3.

5.7.2 Parametric forms for scalar integrals

The reduction of (momentum space) Feynman integrals down to a convenient convergent parametric form is a long and involved process involving several steps, which are well documented and which we do not wish to reproduce here. Roughly speaking, there are three main stages (as well as a Wick rotation to reduce to the case of Euclidean space–time):

1. reduction to scalar integrals,
2. change of variables to parametric form (via the Schwinger or Feynman trick),
3. regularization and renormalization of divergent subgraphs (e.g. BPHZ [16]).

Here are some comments on each step. In this discussion we shall only consider scalar integrals. The general effect of tensor structures only affects the numerators of the resulting parametric integrals, so it follows that the mathematical structure, and number theory content, of the general case is very similar. It also turns out that the parametric form of Feynman integrals (2), which goes back to the early days of quantum field theory, is not the most frequently used by practitioners at present, but turns out to be the closest to algebraic geometry, and hence most convenient for our purposes. It is likely that much of what follows could also be translated to the coordinate space setting. The renormalization of (ultraviolet) divergences (3) is the most delicate point. In recent calculations, a common approach appears to be the following: first, one regularizes divergent integrals using dimensional regularization (working in dimensions $D = 4 - 2\varepsilon$, and expanding integrals as Laurent series in ε); secondly, counter-terms are subtracted according to the BPHZ formula to retrieve the finite convergent part. In this discussion, we shall mainly consider subdivergence-free graphs for which no renormalization is required. However, the methods surely carry over to the general case. For this, one must completely reconsider the approach to step 3, and perform the subtraction of counter-terms directly on the level of the parametric integral. At the end one can write the renormalized Feynman integral as a single absolutely convergent parametric integral and apply the methods below. Again, the mathematics is not so dissimilar from the subdivergence-free case (this will be discussed in a forthcoming paper with D.

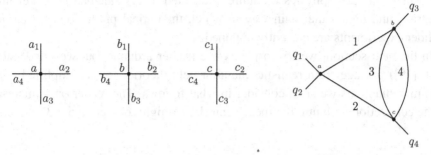

Figure 5.1 The dunce's cap, with edges 1, 2, 3, 4 and external momenta $q_1, \ldots, q_4$. Conservation of momentum gives $q_1 + q_2 = q_3 + q_4$. The graph is obtained from the three corollas on the left by gluing the half-edges $1 = \{a_3, b_4\}$, $2 = \{a_4, c_3\}$, $3 = \{b_1, c_1\}$, $4 = \{b_2, c_2\}$. The external edges $\{a_1, a_2, b_3, c_4\}$ carry momenta q_1, q_2, q_3, q_4, respectively.

Kreimer). In conclusion, the reduction to scalar, subdivergence-free parametric integrals does not forfeit much and the mathematical structure that we consider below captures much of the general case.

5.7.3 The first and second Symanzik polynomials

A Feynman graph G is a collection of *corollas* of degree $n \geq 3$ (a single vertex v surrounded by n half-edges $v_1, \ldots, v_n$) and a collection of *internal edges* E_{int} which are unordered pairs $e = \{v_i, w_j\}$ of half-edges, where v, w are (possibly the same) corolla. A half-edge may occur in at most one internal edge, and the set of half-edges E_{ext} which occur in no internal edge are called *external edges*. An internal edge $\{v_i, v_j\}$ which consists of two half-edges from the same corolla is called a *tadpole*. Each internal edge $e \in E_{\text{int}}$ has associated to it a mass m_e, and each external edge $e \in E_{\text{ext}}$ has a momentum q_e, which is a vector in $\mathbb{R}^4$. We shall always assume that G is connected. Finally, we also require that the total momentum entering G adds up to zero (conservation of momentum). For example, consider the graph on the right in Fig. 5.1, with edges numbered 1–4, which is obtained from the three corollas of degree 4 on the left.

To such a Feynman graph G, one associates two polynomials as follows. For each internal edge $e \in E_{\text{int}}$, we associate a variable α_e known as the Schwinger parameter. The *graph polynomial*, or first Symanzik polynomial, of G does not depend in any way on the external edges and is the polynomial defined by

$$\Psi_G = \sum_{T \subset G} \prod_{e \notin E_T} \alpha_e,$$

where the sum is over all spanning trees T of G and the product is over all internal edges of G which are not in T. A spanning tree is a subgraph of G which is connected, has no loops, and passes through every vertex of G. In the example above, the set of spanning trees is $\{1, 2\}, \{1, 3\}, \{1, 4\}, \{2, 3\}, \{3, 4\}$, so it follows that

$$\Psi_G = \alpha_3\alpha_4 + \alpha_2\alpha_4 + \alpha_2\alpha_3 + \alpha_1\alpha_4 + \alpha_1\alpha_2.$$

The degree of Ψ_G is equal to h_G, the loop number (or first Betti number) of G.

The second Symanzik polynomial of G (which we will not actually require in what follows) is a function of the external momenta and is defined by

$$\Phi_G = \sum_S \prod_{e \notin S} \alpha_e (q^S)^2$$

where the sum is over spanning 2-trees $S = T_1 \cup T_2$ and q^S is the total momentum entering either T_1 or T_2 (which is the same, by conservation of momentum). A spanning 2-tree is defined to be a subgraph S with exactly two connected components T_1, T_2, each of which is a tree (which can reduce to a single vertex), and such that S contains every vertex of G. In the example above, the spanning 2-trees are $\{\{1\}, \{c\}\}, \{\{2\}, \{b\}\}, \{\{a\}, \{3\}\}$ and $\{\{a\}, \{4\}\}$, so

$$\Phi_G = q_4^2 \alpha_2\alpha_3\alpha_4 + q_3^2 \alpha_1\alpha_3\alpha_4 + (q_1 + q_2)^2 (\alpha_1\alpha_2\alpha_4 + \alpha_1\alpha_2\alpha_3),$$

where we recall that $(q_1 + q_2)^2 = (q_3 + q_4)^2$. Up to (omitted) Γ-factors, the general shape of the unregularized (divergent) parametric Feynman integral of G in d space–time dimensions is:

$$I_G(m, q) = \int_{[0,\infty]^{E_{\mathrm{int}}}} \frac{\Psi_G^{N_G - (h_G+1)d/2}}{(\Psi_G \sum_{e \in E_{\mathrm{int}}} m_e^2 \alpha_e - \Phi_G)^{N_G - h_G d/2}} \, \delta\left(\sum_e \alpha_e - 1\right), \quad (5.37)$$

which is a function of the masses m_e and external momenta q_i. Here we only consider the case when all masses are equal to zero, there is a single external momentum q, and the momentum dependence is trivial, i.e. the momentum dependence factors out of the integral (5.37). This is the case, for example, for the graph below on the left, known as the master 2-loop diagram, which has an external particle entering on the left with momentum q:

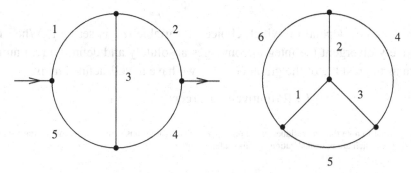

For such graphs, one can show from an identity between the two Symanzik polynomials Φ_G and Ψ_G that (5.37) reduces, up to trivial factors, to the Feynman integral of the graph on the right given by closing up the external edges:

$$I_G = \int_{[0,\infty]^{E_{\text{int}}}} \frac{1}{\Psi_G^{d/2}} \, \delta\left(\sum_e \alpha_e - 1\right). \tag{5.38}$$

In summary, one can reduce massless Feynman integrals with trivial momentum dependence by a series of tricks to the apparently unphysical-looking graphs such as the wheel with three spokes graph above on the right, which have no external edges. In so doing, the second Symanzik polynomial drops out altogether, and one obtains a single number, the *residue* of G which we study below. Although these graphs now bear little resemblance to the bona-fide Feynman integrals of realistic quantum field theories, they will capture much of their number-theoretic content, and indeed most of the difficulty in practical Feynman diagram computations reduces to the problem of computing such so-called master integrals.

5.7.4 Massless ϕ^4 theory

Having performed all the above reductions, we can simply restrict ourselves to looking at graphs G with no external edges. We work in $d = 4$ space–time dimensions. In order to ensure convergence, one says that a graph G is *primitively divergent*[3] if

- $N_G = 2h_G$,
- $N_\gamma > 2h_\gamma$ for all strict subgraphs $\gamma \subset G$.

Furthermore, we say that G is in ϕ^4 theory if the valency of every vertex is at most 4. The dunce's cap graph pictured above is not primitively divergent, since it contains a subgraph $\{3, 4\}$ which has one loop and two edges, and therefore violates the second condition of primitive divergence. A convenient way to remove the δ-function in (5.38) is simply to write it as an affine integral:

$$I_G = \int_{[0,\infty]^{N-1}} \frac{d\alpha_1 \dots d\alpha_{N-1}}{\Psi_G^2}\bigg|_{\alpha_N=1}$$

which does not depend on which choice of variable α_N is set to 1. When G is primitively divergent the integral converges absolutely and defines a real number known as the residue of the graph G. Thus we have a well-defined map

$$I : \{\text{Primitively divergent } G\} \longrightarrow \mathbb{R}$$

---

[3] It is important to note that the residues of the primitively divergent graphs give contributions to the β-function of ϕ^4 theory which are renormalization-scheme independent.

which is entirely determined by the combinatorics of each graph. Unfortunately, the map I is very difficult to evaluate at present, and is known analytically in only a handful of cases. We give a brief survey of known results below.

Numerology of massless ϕ^4

We first state the main operations on massless primitively divergent graphs and their effect on the residue, before giving a brief list of some known residue computations in Section 5.7.5.

Two-vertex join. Let G, G' be two primitively divergent graphs and choose edges e, e' with endpoints v_1, v_2 and v'_1, v'_2, in G and G' respectively. The *two, vertex join* $G{\bullet\atop\bullet}G'$ of G, G' is the graph obtained by gluing $G\backslash e$ and $G'\backslash e'$ by identifying the vertices v_i with v'_i, $i = 1, 2$:

The two-vertex join is also primitively divergent. One can show that

$$I_{G{\bullet\atop\bullet}G'} = I_G I_{G'}, \tag{5.39}$$

i.e. the residue is multiplicative with respect to the two-vertex join.

Completion. Let G be a primitively divergent graph in ϕ^4 theory with at least two loops. It is easy to show from the definition of primitive divergence that there are exactly four vertices which have valency 3. Let $\widehat{G}$ denote the completed graph obtained by adding a new vertex to G, which is joined to each of the four 3-valent vertices in G. The graph $\widehat{G}$ is 4-regular, i.e. every vertex has valency exactly 4. Now it can happen that two distinct primitively divergent graphs G, G' have the same completion (see below). In this case, they have the same residue:

$$I_G = I_{G'} \quad \text{if } \widehat{G} = \widehat{G}'. \tag{5.40}$$

This identity is a well-known consequence of conformal symmetries of Feynman integrals, but this elegant combinatorial interpretation is due to O. Schnetz.

Planar duality. Let G be a planar graph. Then for every planar embedding of G there is a well-defined dual graph G' obtained by placing a vertex in the interior of each face of G and connecting any two vertices by an edge whenever the corresponding faces are neighbouring. Then one shows that the residues coincide:

$$I_G = I_{G'}. \tag{5.41}$$

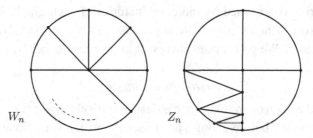

Figure 5.2 The wheels with spokes (left), and zig-zags (right) are the only two families of primitively divergent graphs for which a conjectural formula for the residue exists.

There are more sophisticated identities between residues obtained by more complex operations on graphs (see [18]). Note also that a sequence of the above operations can take one outside the class of ϕ^4 graphs (or primitively divergent graphs, or even graphs altogether), but result in an identity between actual primitively divergent graphs. There is currently no conjectural complete explanation to determine when $I_G = I_{G'}$ for non-isomorphic graphs G, G', let alone when $\sum_i n_i I_{G_i} = 0$, where $n_i \in \mathbb{Z}$.

5.7.5 Numerology

First consider the two families of graphs depicted in Fig. 5.2: the wheels with n spokes W_n, and the zig-zag graphs Z_n ($n \geq 3$). The index n refers to the number of loops of the graph. Both families W_n and Z_n are primitively divergent, but only W_3, W_4 and the family Z_n is in ϕ^4, since W_n has a vertex of valency n. One expects that the corresponding resides satisfy (see [3], [18]):

$$I_{W_n} = \binom{2n-2}{n-1}\zeta(2n-3), \quad I_{Z_n} = 4\frac{(2n-2)!}{n!(n-1)!}\left(1 - \frac{1-(-1)^n}{2^{2n-3}}\right)\zeta(2n-3).$$

The statement for I_{W_n} can be proved by Gegenbauer polynomial techniques, but no 'parametric' or algebro-geometric proof is known for general n. The statement for I_{z_n} is a conjecture, and is proved for small values of n only.

The smallest non-trivial primitively divergent graph is the wheel with three spokes W_3, and it is the unique such graph at this loop order. At four loops, the unique primitively divergent graph is W_4. Next, there are only three primitively divergent graphs at five loops, shown in Fig. 5.3. The one on the far left is the two-vertex join of W_3, and so its residue is the product $I_{W_3}^2$. One can check that the non-planar graph on the right satisfies

$$\widehat{NP_5} \cong \widehat{W_3 \bullet W_3}$$

so this implies that its residue is also equal to $I_{W_3}^2$.

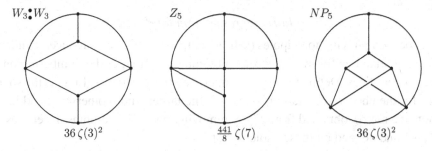

Figure 5.3 Primitively divergent graphs with five loops.

At higher loop orders, the number of primitively divergent graphs grows very fast, and the exact analytic results peter out very quickly. Indeed, at seven and higher loops, there remain graphs whose residue cannot be determined to a single significant digit. The census [18] gives an excellent survey of what is known on this topic. There are a few remarks which need to be made in this context:

1. As first observed by Broadhurst and Kreimer [3], the single zeta values do not suffice to express the residues I_G starting from six loops. The complete bipartite graph $K_{3,4}$, for example, is primitively divergent with six loops and its residue involves $\zeta(3, 5)$, which is conjecturally irreducible. There are also several examples which are known (numerically) to evaluate to multiple zeta values of depth 3, so it is likely that multiple zeta values of all depths occur.
2. As discussed earlier, the multiple zeta values are filtered (conjecturally, graded) by their weight, so it makes sense to ask what the transcendental weight of a Feynman diagram is. One can show that the generic weight, for a primitively divergent graph G with ℓ loops which evaluates to multiple zeta values, is $2\ell - 3$. However, for many graphs the transcendental weight drops (at five loops, for example, the graphs $W_3 \bullet \atop \bullet W_3$ and NP_5 but not Z_5 have a drop in the weight). Some combinatorial criteria for a weight drop to occur are given in [6], but a complete answer to the weight problem is not known.
3. The holy grail of the motivic approach to Feynman integrals would be to compute a motivic coproduct for the residue I_G in terms of the combinatorics of G, but it is not known how to do this for a single non-trivial G. Were such a formula available, one could intuit the value of a Feynman graph just by looking at its combinatorics, a dream first expressed by Kreimer.

5.7.6 *Properties of graph polynomials*

In order to give some idea why multiple zeta values and polylogarithms should occur in massless scalar quantum field theories, we must look at some properties of graph polynomials.

Matrix representation for Ψ_G

Let G be a graph with no tadpoles (self-edges), and let $\mathcal{E}_G$ be its reduced incidence matrix, obtained as follows. Choose an orientation on G, and a numbering on its edges and vertices. Define the $e_G \times v_G$ matrix $\mathcal{E}'_G(e_i, v_j)$ to be -1 if e_i is the source of v_j for the chosen orientation, $+1$ if it is the target, and 0 otherwise, and let $\mathcal{E}_G$ denote the matrix obtained from $\mathcal{E}'_G$ by removing one of its columns. Its entries are $0, 1, -1$ and depend on these choices. Let

$$M_G = \left(\begin{array}{c|c} \begin{matrix} \alpha_1 & & \\ & \ddots & \\ & & \alpha_{e_G} \end{matrix} & \mathcal{E}_G \\ \hline -^T\mathcal{E}_G & 0 \end{array} \right).$$

It follows from the Matrix-Tree theorem that $\Psi_G = \det M_G$.

Example 5.7.1 Consider the following oriented graph G. In the middle is its incidence matrix $\mathcal{E}'_G$, and on the right the matrix M_G obtained by deleting column v_3.

	v_1	v_2	v_3
e_1	-1	1	0
e_2	0	1	-1
e_3	-1	0	1

$$\begin{pmatrix} \alpha_1 & 0 & 0 & -1 & 1 \\ 0 & \alpha_2 & 0 & 0 & 1 \\ 0 & 0 & \alpha_3 & -1 & 0 \\ 1 & 0 & 1 & 0 & 0 \\ -1 & -1 & 0 & 0 & 0 \end{pmatrix}$$

We have $\Psi_G = \det(M_G) = \alpha_1 + \alpha_2 + \alpha_3$ in this case.

Dodgson polynomials

For a graph G as above, let us fix a choice of matrix M_G. For any subsets of edges $I, J, K \subset \{1, \ldots N\}$ of G such that $|I| = |J|$,

$$\Psi_{G,K}^{I,J} = \det M_G(I, J)\Big|_{\alpha_k = 0, k \in K}$$

where $M_G(I, J)$ denotes the matrix M_G with rows I and columns J removed. We call $\Psi_{G,K}^{I,J}$ the Dodgson polynomials of G. There also exists a formula for $\Psi_{G,K}^{I,J}$ in terms of spanning trees, which shows that it is a sum of monomials in the α_i with a coefficient of $+1$ or -1. Changing the choice of M_G only modifies all the $\Psi_{G,K}^{I,J}$ by a sign.

Algebraic relations

The key to computing the Feynman integrals is to exploit the many identities between the polynomials $\Psi_{G,K}^{I,J}$. We have

- The *contraction-deletion formula*. For any edge $e \notin I \cup J \cup K$, it follows from the shape of the matrix M_G that the polynomial $\Psi_{G,K}^{I,J}$ is of degree at most 1 in the Schwinger variable α_e. We can therefore write

$$\Psi_{G,K}^{I,J} = \Psi_{G,K}^{Ie,Je} \alpha_e + \Psi_{G,Ke}^{I,J}.$$

  The contraction-deletion relations state that $\Psi_{G,K}^{Ie,Je} = \pm \Psi_{G\backslash e,K}^{I,J}$ where $G\backslash e$ denotes the graph obtained by deleting the edge e and identifying its endpoints, and $\Psi_{G,Ke}^{I,J} = \pm \Psi_{G/e,K}^{I,J}$, where G/e denotes the graph obtained by contracting the edge e. Note that if deleting the edge e disconnects the graph G, then $\Psi_{G\backslash e,K}^{I,J} = 0$ and if e is a tadpole (a self-edge) then $\Psi_{G/e,K}^{I,J} = 0$.

- Generally, the minors of a matrix satisfy *determinantal identities*. This yields quadratic identities such as the following identity:

$$\Psi_{G,Kabx}^{I,J} \Psi_{G,K}^{Iax,Jbx} - \Psi_{G,Kab}^{Ix,Jx} \Psi_{G,Kx}^{Ia,Jb} = \Psi_{G,Kb}^{Ia,Jx} \Psi_{G,Ka}^{Ix,Jb}$$

  or linear Plücker-type identities such as

$$\Psi_{G,K}^{ij,kl} - \Psi_{G,K}^{ik,jl} + \Psi_{G,K}^{il,jk} = 0$$

  when $i < j < k < l$. These classes of identities are true for certain general families of matrices (the first is due to Dodgson), and hold irrespective of the particular combinatorics of the graph G.

- *Graph-specific identities*. If, for example, K contains a loop, then $\Psi_{G,K}^{I,J} = 0$. Likewise, one can show that if E is the set of all edges which meet a given vertex of G, then $\Psi_{G,K}^{I,J} = 0$ whenever $E \subset I$ or $E \subset J$. In general, the local structure of the graph can cause certain Dodgson polynomials to vanish, which will in turn induce new relations between other Dodgson polynomials via the quadratic equations mentioned above.

5.7.7 A naive integration method

A basic idea is to try to compute the Feynman integral in parametric form by integrating out one variable at a time.

To illustrate this we can attempt to compute the residue:

$$I_G = \int_{[0,\infty]^{N-1}} \frac{d\alpha_1 \dots d\alpha_{N-1}}{\Psi_G^2}\bigg|_{\alpha_N=1}.$$

For more general Feynman diagrams, the numerator will be a polynomial in α_i, $\log \alpha_i$ and $\log \Psi_G$. This won't affect the method significantly.

By the contraction-deletion formula, we can write $\Psi = \Psi^{1,1}\alpha_1 + \Psi_1$. We will henceforth drop the Gs from the notation. Therefore

$$I_G = \int_0^\infty \frac{d\alpha_1 \ldots d\alpha_{N-1}}{\Psi^2}$$

can be written

$$\int_0^\infty \frac{d\alpha_1 \ldots d\alpha_{N-1}}{(\Psi^{1,1}\alpha_1 + \Psi_1)^2} = \int_0^\infty \frac{d\alpha_2 \ldots d\alpha_{N-1}}{\Psi^{1,1}\Psi_1}.$$

By contraction-deletion, the polynomials $\Psi^{1,1}$ and Ψ_1 are linear in the variable α_2:

$$\Psi^{1,1} = \Psi^{12,12}\alpha_2 + \Psi_2^{1,1},$$

$$\Psi_1 = \Psi_1^{2,2}\alpha_2 + \Psi_{12}.$$

We can write the previous integral as

$$\int \frac{1}{\Psi^{1,1}\Psi_1} = \int_0^\infty \frac{d\alpha_2 \ldots d\alpha_{N-1}}{(\Psi^{12,12}\alpha_2 + \Psi_2^{1,1})(\Psi_1^{2,2}\alpha_2 + \Psi_{12})}.$$

By decomposing into partial fractions, one can then integrate out α_2. This leaves an integrand of the form

$$\frac{\log \Psi_2^{1,1} + \log \Psi_1^{2,2} - \log \Psi^{12,12} - \log \Psi_{12}}{\Psi_2^{1,1}\Psi_1^{2,2} - \Psi^{12,12}\Psi_{12}}.$$

At this point, we should be stuck since the denominator is quadratic in every variable. One would expect to have to take a square root at the next stage of integration, but, miraculously, we can use the quadratic identities between Dodgson polynomials to get a factorization:

$$\Psi_2^{1,1}\Psi_1^{2,2} - \Psi^{12,12}\Psi_{12} = (\Psi^{1,2})^2.$$

So after two integrations we have

$$\int \frac{d\alpha_1 d\alpha_2}{\Psi^2} = \frac{\log \Psi_2^{1,1} + \log \Psi_1^{2,2} - \log \Psi^{12,12} - \log \Psi_{12}}{(\Psi^{1,2})^2}.$$

We can then write $\Psi^{1,2} = \Psi^{13,23}\alpha_3 + \Psi_3^{1,2}$ and keep integrating out the variables $\alpha_3, \alpha_4, \ldots$ successively.

As long as we can find a Schwinger coordinate α_i in which all the terms in the integrand are *linear*, then we can always perform the next integration using iterated integrals (this is Theorem 5.6.14). This requires choosing a good order on the edges of G; it can be the case that for some choices of orderings the integration process terminates, but for others one is blocked by polynomials which are of degree > 1 in every remaining variable.

In this case, the integral is expressible as multiple polylogarithms:

$$\mathrm{Li}_{n_1,\ldots,n_r}(x_1, \ldots, x_r) = \sum_{1 \le k_1 < \cdots < k_r} \frac{x_1^{k_1} \ldots x_r^{k_r}}{k_1^{n_1} \ldots k_r^{n_r}},$$

where the arguments are quotients of Dodgson polynomials $\Psi_{G,K}^{I,J}$ (this follows from the explicit description of iterated integrals on the moduli spaces $\mathfrak{M}_{0,n}$ in Section 5.6). When this process terminates, the Feynman integral is expressed as values of multiple polylogarithms evaluated at 1 (or roots of unity).

We say that G is *linearly reducible* if this integration process terminates, i.e. we can find a variable with respect to which all the arguments are linear. The conclusion is that if G is linearly reducible, the residue of G is computable in terms of multiple zeta values, or similar numbers. Likewise, when there are masses or momenta in the integrand, one would in this case obtain multiple polylogarithms in certain rational functions of the kinematic variables. Note that it can happen at a certain stage of the integration that we obtain a quadratic function of a Schwinger coordinate, which factorizes precisely when one introduces a square root of its discriminant. In this way one can also obtain polylogarithms evaluated at certain functions involving square roots of rational functions of the external kinematic variables. Many examples of this type are already well known in perturbative quantum field theory.

5.7.8 The five-invariant

Most of the terms which occur are of the form $\Psi_{G,K}^{I,J}$ which are linear in every variable. However, this is not always the case. The first obstruction which can occur is the *five-invariant*, defined for any five edges i, j, k, l, m in G:

$$^5\Psi(i, j, k, l, m) = \pm \det \begin{pmatrix} \Psi_m^{ij,kl} & \Psi^{ijm,klm} \\ \Psi_m^{ik,jl} & \Psi^{ikm,jlm} \end{pmatrix}.$$

It is not obvious that this is well defined, but one can show that if one permutes the five indices i, j, k, l, m in the definition, the five-invariant only changes by a sign. Thus it does give an invariant of the set of edges $\{i, j, k, l, m\} \subset G$. In the general case, the five-invariant is of degree > 1 in its variables and does not factorize. In other words, we run out of identities in the generic case.

But if, for example, i, j, k, l, m contains a triangle (or if three of the edges meet at a 3-valent vertex) then one of the matrix entries, say $\Psi_m^{ik,jl}$, vanishes, and

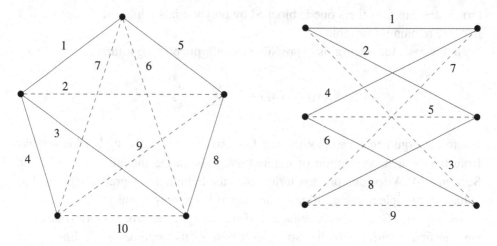

Figure 5.4 The non-planar graphs K_5 and $K_{3,3}$.

$^5\Psi(i, j, k, l, m)$ factorizes into a product of Dodgson polynomials

$$\Psi^{ijm,klm}\Psi^{ik,jl}_m,$$

which are linear in each variable, and so we can keep on going. Thus the *generic* graph is not at all linearly reducible – it is the local combinatorial structure of the graph which generates sufficiently many identities of type to ensure linear reducibility. One can ask which are the smallest graphs which have a non-trivial five-invariant at all. These are the non-planar graphs K_5 (fewest vertices, left in Fig. 5.4) and $K_{3,3}$ (fewest edges, right in Fig. 5.4).

For example, the five-invariant $^5\Psi_{K_{3,3}}(1, 2, 4, 6, 8)$ for the graph on the right is given by:

$$\alpha_5\alpha_9^2 + \alpha_3\alpha_5\alpha_9 + \alpha_5\alpha_7\alpha_9 + \alpha_3\alpha_5\alpha_7 - \alpha_3\alpha_7\alpha_9.$$

One can see that this polynomial is in fact of degree 1 in all its variables except α_9; it turns out that these two graphs are still linearly reducible; one must only choose a more intelligent order in which to integrate out the edges.

The first *serious* obstruction to the integration method of Section 5.7.7 occur at eight loops.

5.7.9 General results on the periods

One can ask if there is a simple combinatorial criterion to ensure linear reducibility, and this requires the following definition.

Definition 5.7.2 Let $\mathcal{O}$ be an ordering on the edges of G. It gives rise to a filtration

$$\emptyset = G_0 \subset G_1 \ldots \subset G_{N-1} \subset G_N = G$$

of subgraphs of G, where G_i has exactly i edges. To any such sequence we obtain a sequence of integers $v_i^{\mathcal{O}}$ = number of vertices of $G_i \cap (G \backslash G_i)$. We say that G has *vertex width* at most n if there exists an ordering $\mathcal{O}$ such that $v_i^{\mathcal{O}} \leq n$ for all i.

Example 5.7.3 Consider the following graph, with the ordering on its edges as shown. Set $G_i = \{1, \ldots, i\}$, and let W_i be the set of vertices in $G_i \cap (G \backslash G_i)$.

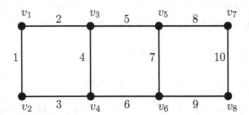

We have $W_1 = \{v_1, v_2\}$, $W_2 = \{v_2, v_3\}$, $W_3 = \{v_3, v_4\}$, $W_4 = \{v_3, v_4\}$, $W_5 = \{v_4, v_5\}$ and so on. This shows that the vertex width is at most 2.

Bounding the vertex width is a very strong constraint on a graph: the set of planar graphs have arbitrarily high vertex width.

Theorem 5.7.4 If G has vertex width at most 3, then G is linearly reducible.

In particular, one can prove that, for such graphs, the residue I_G evaluates to multiple zeta values (or perhaps alternating sums, which are values of multiple polylogarithms evaluated at $x_i = \pm 1$). The zig-zags and wheels are examples of families of graphs with vertex width 3.

5.7.10 Counting points over finite fields

Let us now consider the (at first sight) unrelated problem of counting the points of graph hypersurfaces over finite fields. Let G be a graph, and consider the graph hypersurface

$$X_G = \{(\alpha_1, \ldots, \alpha_N) \in \mathbb{A}^N : \Psi_G(\alpha_1, \ldots, \alpha_N) = 0\}$$

defined to be the zero locus of the polynomial Ψ_G, viewed in affine space $\mathbb{A}^N$. Since Ψ_G has integer coefficients, it makes sense to reduce the equation modulo q, where q is any power of a prime p. This gives a function:

$$[X_G] : \{\text{Prime powers } q\} \longrightarrow \mathbb{N}, \qquad (5.42)$$

$$q \mapsto |X_G(\mathbb{F}_q)|,$$

which to q associates the number of solutions of $\Psi_G = 0$ in $\mathbb{F}_q^N$. Indeed such a counting function exists for any polynomial, or system of polynomials, defined over $\mathbb{Z}$. The question, first asked by Kontsevich in 1997, is whether $[X_G]$ is a polynomial in q.

Example 5.7.5 Consider the primitively divergent graphs in ϕ^4 theory up to five loops. They are the wheels W_3, W_4, and at five loops we have the zig-zag graph Z_5, the two-vertex join $W_3 \bullet W_3$, and the non-planar graph $N P_5$. One can compute:

$$[W_3] = q^2(q^3 + q - 1),$$

$$[W_4] = q^2(q^5 + 3q^3 - 6q^2 + 4q - 1),$$

$$[Z_5] = q^2(q^7 + 5q^5 - 10q^4 + 7q^3 - 4q^2 + 3q - 1),$$

$$[W_3 \bullet W_3] = q^3(q^6 + q^5 + q^4 - 3q^3 - q^2 + 3q - 1),$$

$$[N P_5] = q^5(q^4 + 4q^2 - 7q + 3).$$

One way to prove that a given function $[X_G]$ is a polynomial in q is to use the following inductive argument, due to Stembridge [20]. For any set of polynomials $f_1, \ldots, f_k$ in $\mathbb{Z}[\alpha_1, \ldots, \alpha_N]$, denote the point count of the intersection $\cap_k \{f_k = 0\}$ by

$$[f_1, \ldots, f_k] = \left|\{(\alpha_1, \ldots, \alpha_N) \in \mathbb{F}_q^N : f_1 = \cdots = f_k = 0\}\right|.$$

Consider a polynomial $f = f^1 \alpha_1 + f_1$ which is linear in α_1, where $f^1, f_1 \in \mathbb{Z}[\alpha_2, \ldots, \alpha_N]$. For f to vanish, either f^1 is invertible in $\mathbb{F}_q$ (in which case we can solve for α_1), or both f^1 and f_1 vanish (in which case α_1 can take any value in $\mathbb{F}_q$). Hence

$$[f^1 \alpha_1 + f_1] = q[f^1, f_1] + q^{N-1} - [f_1]. \qquad (5.43)$$

Similarly, if we have two polynomials f, g linear in α_1, one checks in a similar way that

$$[f^1 \alpha_1 + f_1, g^1 \alpha_1 + g_1] = q[f_1, g_1, f^1, g^1] + [f^1 g_1 - g^1 f_1] - [f^1, g^1]. \quad (5.44)$$

By a computer implementation of this method, Stembridge proved that, for all graphs G with at most 12 edges, the point counts $[X_G]$ are polynomials in q. Previously, Stanley had shown that certain graphs obtained by deleting edges from complete graphs have this property, and gave explicit formulae for the point counts in these cases.

A landmark result due to Belkale and Brosnan showed that this is false in general.

Theorem 5.7.6 [1]   The point-counting functions $[X_G]$ are, in a certain sense, of general type, as G ranges over the set of all graphs.

Concretely, this means that, given any set of polynomials $f_1, \ldots, f_n$ with integer coefficients as above, one can construct a set of graphs $G_1, \ldots, G_M$ such that

$$[f_1, \ldots, f_n] = \sum_{i=1}^{M} p_i(q)[G_i],$$

where $p_i(q) \in \mathbb{Z}[q, q^{-1}, (q-1)^{-1}]$. In other words, graph hypersurfaces are universal in the sense of motives. However, until recently no explicit non-polynomial example was known, and the method seems to produce counter-examples with very large numbers of edges. It is also not clear in this approach if the conditions of physicality, for example, being primitively divergent and in ϕ^4 theory, are enough to ensure polynomiality, since the graphs G_i are hard to control.

Now, by applying Stembridge's method to the graph polynomial Ψ_G, one sees at the first stage that there are terms Ψ^1, Ψ_1 and at the second stage that we obtain a term $\Psi^{12}\Psi_{12} - \Psi_2^1\Psi_1^2 = (\Psi^{1,2})^2$. In general, one obtains iterated resultants of Dodgson polynomials $\Psi_{G,K}^{I,J}$, as in the previous integration method. One can show the following.

Theorem 5.7.7 [7]   If G has vertex width 3 then there exists a polynomial P_G such that $[X_G](q) = P_G(q)$ for all q not divisible by 2.

It is possible that the condition that q be divisible by 2 can be dropped. This is the case for the wheels and zig-zags whose polynomials can be computed explicitly [7]. It was recently shown independently by D. Doryn and O. Schnetz that some primitively divergent graphs in ϕ^4 at seven loops have 'quasi-polynomial' point counts (more precisely, are given by more than one polynomial according to whether small primes divide q or not). At eight loops, one finds the following.

Theorem 5.7.8 [7]   There exists a primitively divergent graph G in ϕ^4 theory such that

$$[X_G](q) \equiv a_q^2 q^2 \qquad \mathrm{mod}\ q^3$$

for all prime powers q, where $q + 1 - a_q$ is the number of points over $\mathbb{F}_q$ of the (complex multiplication) elliptic curve defined by the equation

$$y^2 + xy = x^3 - x^2 - 2x + 1. \tag{5.45}$$

This graph has vertex width 4.

The a_q are given by the coefficients of a certain modular form, and the theorem proves in particular that $[X_G]$ is not a quasi-polynomial. There is another interpretation of the term $[X_G](q) \mod q^3$ in terms of the denominator in the integration method of the previous section. This strongly suggests that the residue I_G of this graph should not be a multiple zeta value, but possibly a period of the fundamental group of the elliptic curve (5.45) with punctures.

5.7.11 Geometric interpretation

Consider the graph hypersurface $X_G \subset \mathbb{A}^N$, and let $B = \bigcup_{i=1}^N \{\alpha_i = 0\}$ denote the union of the coordinate axes. Consider the projection

$$\pi_i : \mathbb{A}^N \longrightarrow \mathbb{A}^{N-i}, \tag{5.46}$$

$$(\alpha_1, \ldots, \alpha_N) \mapsto (\alpha_{i+1}, \ldots, \alpha_N).$$

For each i, one can show that there exists an algebraic subvariety $L_i \subset \mathbb{A}^{N-i}$ (the discriminant), where L_i is a union of irreducible hypersurfaces, such that

$$\pi_i : \mathbb{A}^N \backslash (X_G \cup B \cup \pi_i^{-1}(L_i)) \longrightarrow \mathbb{A}^{N-i} \backslash L_i \tag{5.47}$$

is locally trivial in the sense of stratified varieties: in other words, the fibers over every complex point of $\mathbb{C}^N \backslash L_i(\mathbb{C})$ are topologically constant. In this situation, the partial Feynman integral

$$I_G^i(\alpha_{i+1}, \ldots, \alpha_n) = \int_{[0,\infty]^i} \frac{d\alpha_1 \ldots d\alpha_i}{\Psi_G^2} \tag{5.48}$$

obtained by integrating in the fibre is a multivalued function of the parameters $\alpha_{i+1}, \ldots, \alpha_n$, and has singularities contained in L_i. Thus the variety L_i corresponds to what physicists refer to as a Landau variety.

The main result of [5] is that when G is of vertex width ≤ 3, then, with the natural numbering of its edges, one can compute the L_i inductively and show that

$$L_i \subseteq \{\Psi_{G,K}^{I,J} = 0 : I \cup J \cup K = \{1, \ldots, i\}\}.$$

This involves an inductive argument using the fact that discriminants of one-dimensional projections can be written as resultants, and the fact that the local topology of the graph G generates enough identities to ensure that these resultants always factorize into products of polynomials $\Psi_{G,K}^{I,J}$. Since each polynomial $\Psi_{G,K}^{I,J}$ is of degree ≤ 1 in each variable, one can show that the partial integral $I_G^i(\alpha_{i+1}, \ldots, \alpha_n)$ has unipotent monodromy. By the universality of multiple polylogarithms, we conclude that (5.48) is expressible in terms of multiple polylogarithms with singularities along $\Psi_{G,K}^{I,J} = 0$ only. This ultimately explains the appearance of multiple zeta values (or alternating sums) in the Feynman integral calculations.

On the other hand, the fibrations (5.47) can also be used to count points over finite fields. For example, the map $\pi_1 : \mathbb{A}^N \backslash (X_G \cup B \cup \pi_1^{-1}(L_1)) \longrightarrow \mathbb{A}^{N-1} \backslash L_1$ is a fibration, whose fibers are a complement of a finite number of points in $\mathbb{A}^1$. Since $L_1 = \{\Psi_G^{1,1} \Psi_{G,1} = 0\}$ we can use the fibration to count points over $\mathbb{F}_q$, which yields a version of (5.43), since for a locally trivial fibration the number of points in the total space is simply the number of points in the base multiplied by the number of points in the fiber. Continuing in this way yields all the terms in the Stembridge reduction algorithm, and ultimately Theorem 5.7.7.

5.7.12 Conclusion

The appearance of multiple zeta values in massless ϕ^4 theory at low loop orders can be explained as follows. For certain classes of graphs, the particular algebraic properties of graph polynomials imply that the partial Feynman integrals (5.48) have unipotent monodromy, which by Chen's theory can be expressed in terms of iterated integrals. The explicit description of iterated integrals on moduli spaces of curves of genus 0 allows one to express these functions in terms of multiple polylogarithms. The multiple zeta values or Euler sums then appear as the special values (or holonomy) of these functions. This argument works for all graphs up to six or seven loops, and a similar argument should certainly also work for a suitable class of Feynman graphs with subdivergences, and non-trivial kinematic variables. Instead of obtaining numbers, the amplitudes will be now expressible as multiple polylogarithm functions whose arguments depend on the masses and momenta.

At higher loop orders, the presence of an elliptic curve strongly suggests that the Feynman integrals will fail to be multiple zeta values. Together with the work of Belkale and Brosnan, this points to the fact that ϕ^4 theory is mathematically far richer than previously thought, and confirms David Broadhurst's philosophy that one should expect totally new phenomena to occur in quantum field theories at every new loop order.

Acknowledgements

Very many thanks to João-Pedro dos Santos for detailed comments and corrections. Many thanks also to Pacho Marrón for interesting discussions, and the University of los Andes for hospitality.

References

[1] P. Belkale, P. Brosnan: Matroids, motives, and a conjecture of Kontsevich, *Duke Math. J.* **116**: 1(2003), 147–188.

[2] S. Bloch, H. Esnault, D. Kreimer: On motives associated to graph polynomials, *Comm. Math. Phys.* **267**: 1(2006), 181–225.

[3] D. Broadhurst, D. Kreimer: Knots and numbers in ϕ^4 theory to 7 loops and beyond, *Int. J. Mod. Phys.* C **6** (1995), 519.

[4] F. C. S. Brown: Multiple zeta values and periods of moduli spaces $\mathfrak{M}_{0,n}$, *Ann. Scient. Éc. Norm. Sup.* 4e série, t. 42, 373–491, math.AG/0606419.

[5] F. C. S. Brown: On the periods of some Feynman integrals, arXiv:0910.0114v1 (2009), 1–69.

[6] F. C. S. Brown, K. Yeats: Spanning forest polynomials and the transcendental weight of Feynman graphs, arXiv:0910.5429 (2009), 1–30.

[7] F. C. S. Brown, O. Schnetz: A K3 in ϕ^4, *Duke Math. J.* **161**(2012), 1817–1862.

[8] P. Cartier: Jacobiennes généralisées, monodromie unipotente et intégrales itérées, *Séminaire Bourbaki, 1987–1988*, exp. no. 687, pp. 31–52.

[9] P. Cartier: Polylogarithmes, polyzetas et groupes pro-unipotents, *Séminaire Bourbaki, 2000–2001*; *Asterisque* **282** (2002) 137–173.

[10] K. T. Chen: Iterated path integrals, *Bull. Amer. Math. Soc.* **83** (1977), 831–879.

[11] K. T. Chen: Integration of paths-a faithful representation of paths by non-commutative formal power series, *Trans. Amer. Math. Soc.* **89** (1958), 395–407.

[12] P. Deligne, A. Goncharov: Groupes fondamentaux motiviques de Tate mixte, *Ann. Sci. École Norm. Sup.* (4) **38**: 1(2005), 1–56.

[13] A. B. Goncharov: Multiplc polylogarithms and mixed Tate motives, preprint (2001), arXiv:math.AG/0103059v4.

[14] R. M. Hain: The geometry of the mixed Hodge structure on the fundamental group, *Proc. Sympos. Pure Math.* **46** (1987), 247–282.

[15] R. M. Hain: Classical polylogarithms, *Proc. Sympos. Pure Math.* **55** (1994), Part 2.

[16] C. Itzykson, J. B. Zuber: *Quantum field theory*, International Series in Pure and Applied Physics. New York: McGraw-Hill, 1980.

[17] J. Oesterlé: Polylogarithmes, *Séminaire Bourbaki 1992–1993*, exp. no. 762, *Astérisque* **216** (1993), 49–67.

[18] O. Schnetz: Quantum periods: A census of ϕ^4 transcendentals, arXiv:0801.2856 (2008).

[19] V. A. Smirnov: *Evaluating Feynman integrals*, Springer Tracts in Modern Physics 211. Berlin: Springer, 2004.

[20] J. Stembridge: Counting points on varieties over finite fields related to a conjecture of Kontsevich, *Ann. Combin.* **2** (1998), 365–385.

[21] M. Waldschmidt: Valeurs zêtas multiples: une introduction, *J. Théorie Nomb. Bordeaux* **12** (2002), 581–595.

[22] S. Weinzierl: The art of computing loop integrals, arXiv:hep-ph/0604068v1 (2006).

[23] W. Zudilin: Algebraic relations for multiple zeta values, *Russ. Math. Surveys* **58**: 1(2003), 1–29.

6

Geometric issues in quantum field theory
and string theory

LUIS J. BOYA

Abstract

These lecture notes begin with an introduction to the basic notions of
differential geometry, suitable for physicists. After the introduction of
holonomy groups, various geometric structures relevant in quantum field
theory and string theory are explained, including compactification and
its relation to M-theory and F-theory.

6.1 Differential geometry for physicists

This section is a survey of the basic differential-geometric constructions that appear
in theoretical physics. A thorough discussion of these notions can be found in [6].
For a reference more adapted to physicists see [8].

6.1.1 Manifolds and vector fields

An *n-dimensional manifold* M is a Hausdorff topological space covered by open
sets U_i, each one homeomorphic to an open set V_i in $\mathbb{R}^n$. If $\phi_i : U_i \to V_i$ denotes
such a homeomorphism, let $\phi_i(P) = \{x^1, x^2, \ldots, x^n\}$ denote the coordinates of
a point $P \in M$ in the coordinate system $\{U_i, \phi_i\}$. If $P \in U_i \cap U_j$, the functions
$\phi_i \circ \phi_j^{-1}$ are supposed to be C^∞ as maps from $\mathbb{R}^n$ to $\mathbb{R}^n$. Let us recall that M is
called *compact* if any open covering $\{U_i\}$ has a finite sub-covering, M is called
connected if one can go continuously from any point to another and M is called
simply connected if any loop (= closed path) drawn from any point is contractible
(shrinks to the point continuously).

Geometric and Topological Methods for Quantum Field Theory, ed. Alexander Cardona, Iván Contreras and
Andrés F. Reyes-Lega. Published by Cambridge University Press. © Cambridge University Press 2013.

Let us denote by $\mathfrak{E}(M)$ the set of C^∞ functions $f : M \to \mathbb{R}$; it is a commutative, infinite-dimensional $\mathbb{R}$-algebra. A vector v at $P \in M$ is a derivation $v : f \to \partial f/\partial n|_p$, where $\partial f/\partial n$, the *directional derivative* along v, can be expressed as $\sum_{i=1}^n a_i \partial/\partial x^i$. The set of vectors in P build up the *tangent space* to M at P, denoted $T_P = T_P(M)$. The union of all T_P, for P in M, define the (total space of) the *tangent vector bundle* $T(M)$:

$$\tau : \mathbb{R}^n \hookrightarrow T(M) \to M. \tag{6.1}$$

This is an example of a *vector bundle*, as the fibers $F(\cong \mathbb{R}^n)$ are vector spaces, M is called the *base* manifold, and $TM = T(M)$ the *total space*. If $\{e_i\}_P$ is a *frame* at P (a basis of T_P), the totality of frames in P for all P build up the *frame bundle*, which is an example of what is called a *principal bundle*, as the group $\mathrm{GL}_n(\mathbb{R})$ acts freely on the fibers:

$$
\begin{array}{ccccc}
\mathrm{GL}_n(\mathbb{R}) & \longrightarrow & B & \longrightarrow & M \qquad \text{Principal bundle} \\
\downarrow & & & \| & \\
\mathbb{R}^n & \longrightarrow & TM & \longrightarrow & M \quad \text{Associated vector bundle}
\end{array}
\tag{6.2}
$$

A *(cross) section* s in a bundle $\pi : E \to M$ is a map sending any point P in M to a point u in the fiber $\pi^{-1}(P)$ over P. Let us denote by $\Gamma(\tau)$ the set of sections of TM. A section defines a vector field X which in coordinates can be written as

$$X = \xi^m \frac{\partial}{\partial x^m}, \tag{6.3}$$

where the ξ^m are functions in $\mathfrak{E}(M)$ and we use the summation convention on repeated indexes. A vector field X in a manifold M defines a *flow*, or set of curves tangent to X: if $\gamma = x^\mu(t)$ is such a parameterized curve, the *system* of differential equations of the flow are

$$\frac{dx^\mu}{dt} = \xi^\mu, \qquad \mu = 1, 2, \ldots, n. \tag{6.4}$$

The set of all vector fields $\mathfrak{L} = \Gamma(TM)$ has a rich structure: on the one hand it is an infinite-dimensional *Lie algebra* with respect to commutation, $[X, Y] = XY - YX$ (notice that the second-order derivative terms cancel) and, on the other hand, it is also a *derivation algebra* of the commutative algebra of functions $\mathfrak{E}(M)$, as Leibniz' rule holds, $X(fh) = (Xf)h + f(Xh)$. Notice also that $\mathfrak{L}(M)$ is, as any type of tensor, a $\mathfrak{E}(M)$-module.

6.1.2 Differential forms, tensors, metrics and de Rham cohomology

Let $T_P^*(M)$ denote the dual vector space to $T_P(M)$ and consider $T^*M := \bigcup_P T_P^*(M)$, the union over all points P in M of the dual spaces $T_P^*(M)$, called the *cotangent bundle* over M. The cotangent bundle over M is a $\mathbb{R}^n$-vector bundle, and sections $\theta \in \Gamma(T^*(M))$ on it are called *1-forms*. Given a vector field X on M the contraction $\theta(X)$ is a function in $\mathfrak{E}(M)$.

The map $\mathrm{d}: \mathfrak{E}(M) \to \Gamma(T^*(M))$ defined by

$$\mathrm{d}f(X) = Xf \tag{6.5}$$

is called the *differential* of the function f. Applying d to the coordinate functions $f = x^\mu$ we obtain 1-forms satisfying

$$\mathrm{d}x^\mu \left(\frac{\partial}{\partial x^\nu} \right) = \delta^\mu_\nu \tag{6.6}$$

so $\mathrm{d}x^\mu$ is the *dual base* for the base $\partial/\partial x^\mu$ of $T_P(M)$. Hence, $\theta := p_\mu \mathrm{d}x^\mu$ is the general form of a dual field of a vector field, i.e. a *1-form*. The space of 1-forms on M will be denoted by $\Omega^1(M)$.

Dual to the role of vector fields X as derivatives, 1-forms serve to *integrate*: let $\gamma : (t_0, t_1) \to M$ be a path on M (i.e. a 1-dimensional submanifold of M), then

$$\int_\gamma \theta := \int_{t_0}^{t_1} p_\mu(x^\nu(t)) \frac{\mathrm{d}x^\mu}{\mathrm{d}t} \mathrm{d}t \tag{6.7}$$

is the integral of the 1-form θ on the curve γ; it does not depend on parameterization $x^\nu(t)$ for γ.

As we did for vector fields and 1-forms on M, we can define many other tensor fields, point by point over M, using the standard operations of linear algebra on tangent spaces. If V is a vector space, call $T(V)$ the tensor algebra of V. Thus, T^0_0 are the *scalars* (field numbers), T^1_0 are the vectors in V, T^0_1 the dual space V^*, T^0_2 the bilinear forms on V, T^0_n include the volume forms, etc. The union of the tensor vector spaces on each point on M will give rise to the *tensor bundle* $\mathfrak{T}M$ on the manifold M:

$$
\begin{array}{ccc}
\mathrm{GL}_n(\mathbb{R}) & \longrightarrow B \longrightarrow & M \\
\downarrow & & \| \\
T(V) & \longrightarrow \mathfrak{T}M \longrightarrow & M
\end{array}
\tag{6.8}
$$

whose sections define *tensor fields* on the manifold, of paramount importance in physics and mathematics. We shall consider in what follows two special types of tensors, the *p-forms* and the *metrics*. Each space of tensors, say T^p_q, is a $\mathfrak{E}(M)$-module.

The antisymmetric tensor product, or wedge product, defined on 1-forms in a point by $a \wedge b := (a \otimes b - b \otimes a)/2$, gives rise to an element in $\wedge T^0_2$ and, generalizing this construction, to elements in $\wedge T^0_p$, which are called *p-forms*. For example, a 2-form ω on a manifold M can be expressed in local coordinates as

$$\omega = p_{\mu\nu}dx^{\mu} \wedge dx^{\nu}. \tag{6.9}$$

If the manifold M has dimension n, the space of differential forms of any degree is an algebra of $\sum_p \binom{n}{p} = 2^n$ dimensions. The p-forms (as tensor fields) are closed under the wedge product, and the *exterior differential operator* d defined in (6.5) can be extended to forms in any degree to give

$$d : \Omega^p(M) \rightarrow \Omega^{p+1}(M), \tag{6.10}$$

which is an *antiderivation* of the wedge product, i.e. satisfies

$$d(\omega_1 \wedge \omega_2) = (d\omega_1) \wedge \omega_2 + (-1)^{\deg \omega_1}\omega_1 \wedge d\omega_2$$

and, more importantly, $d^2 = 0$. Notice that the wedge algebra of *even* p-forms on M is commutative.

Notice that functions lying in $\Omega^0(M)$, i.e. 0-forms on M, "integrate" over points (through evaluation) $\int_P f \equiv f(P)$, while 1-forms integrate over curves as we saw in (6.7), 2-forms integrate on surfaces, and so on; n-forms integrate on n-volumes in M. For a submanifold N of dimension p in M, with boundary ∂N, Stokes' theorem shows that, if $\omega \in \Omega^{p-1}(M)$,

$$\int_{\partial N} \omega = \int_N d\omega. \tag{6.11}$$

A p-form ω on M is called *closed* if $d\omega = 0$, and it is called *exact* if $\omega = d\theta$ for some $\theta \in \Omega^{p-1}(M)$ (of course if exact, it is closed, as $d^2 = 0$). In $\mathbb{R}^n$ any closed form is exact (Poincaré's lemma) but, in general, the existence of closed forms on a manifold M which are *not* exact give topological information about the manifold M; this gives rise to the *de Rham cohomology* of the manifold M. Let us denote by $Z^p = Z^p(M)$ the *p-cocycles* (closed forms) on M, by $B^p = B^p(M)$ the *p-coboundaries* (exact forms) on M, and define the quotient $H^p(M, \mathbb{R}) = Z^p/B^p$. It follows that the space $H^p(M, \mathbb{R})$, called the pth (real) cohomology group of the manifold M, is finite-dimensional for any $p = 1, 2, \ldots, n = \dim M$ [6],

$$\dim H^p(M, \mathbb{R}) = b_p(M), \tag{6.12}$$

and we call the dimensions $b_p(M)$ of $H^p(M, \mathbb{R})$ the *Betti numbers* of M. Such numbers turn out to be *topological invariants* of the manifold: de Rham theory, suggested by É. Cartan, is an "access to the topology of manifolds via exterior forms". It is called *co*-homology because it is the dual to the *homology* theory, whose corresponding spaces are denoted using sub-indexes $H_*(M, \mathbb{R})$, and makes use of cycles and boundaries as given by a *triangulation* of the manifold M. Examples of Betti numbers for some well-known manifolds are: For $\mathbb{R}^n$, $b_0 = 1$

and all other Betti numbers are 0. For the n-dimensional sphere S^n all the Betti numbers are zero, except $b_0 = b_n = 1$. For the n-torus, $b_k = \binom{n}{k}$. In general b_0 measures the number of connected components of the manifold and, if the manifold is *compact and oriented*, $b_k = b_{n-k}$ (Poincaré duality).

Metrics

In a vector space V, a bilinear form b defines a T_2^0 tensor. Such a tensor is called symmetric/antisymmetric if $b(v, w) = \pm b(w, v)$, for vectors $v, w \in V$, and it is called *regular* (or non-degenerate) if the induced map $b^* : V \to V^{\text{dual}}$ is an isomorphism (equivalently, if in any base for V it becomes a matrix (b) with $\det(b) \neq 0$). A *Riemannian manifold* (B. Riemann, 1854) is a manifold endowed with a symmetric regular bilinear field, called the metric field, which is positive-definite and in local coordinates can be written as

$$g|_U = \mathrm{d}s^2 = g_{\mu\nu}\mathrm{d}x^\mu\mathrm{d}x^\nu. \tag{6.13}$$

Metric tensors serve to define length of curves, areas of surfaces and volumes on manifolds. Positivity, $g_P(u, u) \geq 0$, is relaxed in the Lorentzian case. The existence of Riemannian metrics on manifolds can be related to possible reductions of the structure group of their corresponding frame bundle [5]. Actually, any manifold admits a Riemannian metric on it: let $N = \dim GL_n(\mathbb{R}) - \dim O(n) = n^2 - n(n - 1/2) = n(n + 1)/2$ and consider the bundles

$$O(n) \tag{6.14}$$

$$\downarrow$$

$$GL_n(\mathbb{R}) \longrightarrow B(M) \longrightarrow M$$

$$\downarrow$$

$$\mathbb{R}^N \dashrightarrow E \dashrightarrow M$$

Since the $\mathbb{R}^N$ space is *contractible*, the associated bundle – is trivial, hence the GL-bundle reduces to the O-bundle and, as $O(n)$ is the isotropy group of a symmetric regular *positive* bilinear form, this shows that any manifold can be endowed with a Riemannian metric. However, not every manifold admits a metric with signature $\neq 0$.

Riemann spaces were introduced in physics by Einstein, 1915 (with signature $(n - 1, 1)$, where $n = 4$ is the dimension of the space–time manifold). The idea (anticipated by Riemann and considered also by Clifford) is that geometry of our space–time is not fixed *a priori*, but determined by the matter/energy content of the universe. Metrics with non-positive definite g are as easy to handle as Riemannian metrics; the crucial property is *regularity*, i.e. $\det(g) \neq 0$.

A *symplectic manifold* (M, ω) is a manifold M endowed with a regular or non-degenerate 2-form ω. It is easy to see that any symplectic manifold is even-dimensional and orientable, since ω^n is a volume form. The typical domain of symplectic geometry is *classical mechanics*. Let us recall some basic notions in *Hamiltonian dynamics*. If $H = H(p, q)$ denotes the Hamilton function for a dynamical system, where (p, q) denote canonical coordinates (generalized coordinates q and associated momenta p), let $\omega = dp \wedge dq$ be the symplectic form. Taking the exterior derivative of the function H one obtains a 1-form which, using the non-degeneracy of ω, is the dual of a vector field X_H on the manifold modeling the phase space of the system. Then, it turns out that the integral curves of this vector field satisfy Hamilton's equations of motion. There is a *restricted* inverse process: 1-parametric groups of symmetries τ_t generate *constants of motion* (see [8]).

6.1.3 Connections and curvature

Curvature as something *intrinsic* (i.e. embedding independent) was introduced for surfaces by Gauss (1827) and generalized by Riemann (1854); it was seen as consequence of *connections*, rather than of metrics, by Levi-Civita and Weyl (ca. 1917); and understood as an operation in general bundles by Ehresman (1950). We shall define connections on *vector* bundles; but more generally, they are defined on *principal* bundles [6].

Let $\xi : V \to E \to M$ be a vector bundle. A *connection* ∇ is a *linear* map from sections in ξ to sections in the tensor product $\otimes$ of ξ with the cotangent bundle T^*M, which is a *derivation*: if $f \in \mathcal{E}(M)$, $fs = s'$ makes sense, as $\Gamma(\xi)$ is a $\mathcal{E}(M)$-module, and the defining properties of the connection are

$$\nabla : \Gamma(\xi) \to \Gamma(\xi \otimes T^*M); \quad \nabla(fs) = (df)s + f\nabla s. \qquad (6.15)$$

∇ is also called the *covariant differential* (of the connection). The covariant *derivative* with respect to a vector field X is the *contraction* with the vector field X:

$$\nabla_X s := X \lrcorner \nabla s. \qquad (6.16)$$

For instance, for $f \in \mathcal{E}(M)$ and $s \in \Gamma(\xi)$ we have:

$$\nabla_X(fs) = X(f)s + f\nabla_X s.$$

So the covariant differential *increases* the indices by adding a covariant one, whereas the covariant derivative *conserves* the index. A connection in coordinates is a matrix of 1-forms: if $\{s_i\}$ is a frame, $\{s_i\} = \epsilon$ (in some coordinate patch

U), $\nabla s_i = \omega_{ij} s_j$, where ω_{ij} is an $n \times n$ (when n is the dimension of the fiber) matrix of 1-forms in M, so for short we write $\nabla \epsilon = \omega \epsilon$. Connection components transform *inhomogeneously*; they are *not* tensorial: take another frame $\epsilon' = g\epsilon$, where g is unique, and define

$$\nabla \epsilon' := \omega' \epsilon'; \quad \text{then} \nabla \epsilon' = \nabla(g\epsilon) = (dg)\epsilon + g\omega\epsilon = \omega'g\epsilon,$$

$$\omega' = dg \cdot g^{-1} + g \cdot \omega \cdot g^{-1}. \tag{6.17}$$

What about curvature? Let us write $\wedge^2 M$ for the (total space of) the bundle of 2-forms, and try to extend ∇ as *antiderivation*:

$$\nabla \circ \nabla := \nabla^2 : \Gamma(\xi) \to \Gamma(\xi \otimes \wedge^2 M),$$

$$\nabla(\theta s) := d\theta \cdot s - \theta \wedge \nabla s. \tag{6.18}$$

One then shows [7] that $Ks := \nabla^2 s$ is a *tensor*, i.e. it satisfies $K(fs) = f K(s)$, and it is called the *curvature* of the connection. The connection is called *flat* when $K = 0$. In terms of the 1-form matrix ω, we have

$$\nabla^2 s = Ks = \nabla(\omega s) = d\omega \cdot s - \omega \wedge \nabla s = d\omega \cdot s - (\omega \wedge \omega)s,$$

from which we obtain

$$K = d\omega - \omega \wedge \omega, \tag{6.19}$$

a formula familiar to physicists ("$F = dA + A \wedge A$" in the context of Yang–Mills field theories). Here K clearly (in the frame s) is an $n \times n$ matrix of 2-forms ("field strength" in physics).

Applying d (not ∇) to (6.19), we then have

$$dK = -(d\omega) \wedge \omega + \omega \wedge (d\omega) = -(K + \omega \wedge \omega) \wedge \omega + \omega \wedge (K + \omega \wedge \omega)$$

$$= -\omega \wedge K + K \wedge \omega.$$

With the natural definition $\nabla K := dK + \omega \wedge K - K \wedge \omega$ this amounts to the following statement (*Bianchi identity*):

$$\nabla K = 0. \tag{6.20}$$

For all this see [7].

Take now as bundle the tangent bundle TM; there is an extra tensor, the so-called *torsion* of the connection, $T = T^\nabla$:

$$T(X, Y) := Z = \nabla_X Y - \nabla_Y X - [X, Y]. \tag{6.21}$$

It is a tensor, namely

$$T(fX, Y) = fT(X, Y) = T(X, fY). \tag{6.22}$$

The Levi-Civita connection

If (M, g) is a Riemannian manifold, there is a unique connection, characterized by being torsionless ($\Leftrightarrow$ *symmetric*, $T = T^{\nabla} = 0$) and isometric ($\nabla g = 0$). The theorem is *constructive*, in the sense that the explicit form of the connection is computable easily, and allows for signature; for the positive ($g > 0$) case *or not*, the components of the connection are given by the classical Christoffel formulas:

$$[\mu\nu, \lambda] := \frac{1}{2} \left(g_{\nu\lambda,\mu} - g_{\mu\nu,\lambda} + g_{\lambda\mu,\nu} \right), \tag{6.23}$$

$$\Gamma^{\lambda}_{\mu\nu} = g^{\lambda\rho}[\mu\nu, \rho]. \tag{6.24}$$

The curvature in TM can be defined as a an 2-*form operator on vector fields*, or an *2-form: endomorphism-valued*

$$R(X, Y) = [\nabla_X, \nabla_Y] - \nabla_{[X,Y]}. \tag{6.25}$$

Hence $R(X, Y)Z = W$, or in coordinates $\partial_{\lambda} = \partial/\partial x^{\lambda}$

$$R(\partial_{\mu}, \partial_{\nu})\partial_{\lambda} := R^{\rho}_{\lambda\mu\nu}\partial_{\rho}. \tag{6.26}$$

The tensor so defined is called the Riemann tensor, Riem. It has the following symmetries (consider four lower indices $R_{\sigma\lambda\mu\nu}$, sometimes called the *Riemann–Christoffel symbol tensor*, $R_{\sigma\lambda\mu\nu} = g_{\sigma\rho}R^{\rho}_{\lambda\mu\nu}$):

1. It is antisymmetric in $\mu\nu$ and in $\sigma\lambda$ independently.
2. It is symmetric under exchange of both pairs $\mu\nu \Leftrightarrow \sigma\lambda$.
3. Its fully antisymmetric part vanishes (from the Bianchi identity[1]).

Hence the Young tableau symmetry of the Riemann tensor, Riem, is $[2^2]$, with

$$\dim \text{Riem} = n(n + 1)(n - 1)n\frac{2}{4!} = \frac{n^2(n^2 - 1)}{12}, \tag{6.27}$$

which gives $\dim \text{Riem} = 0, 1, 6, 20, 50$ for $n = 1, 2, 3, 4, 5$. Properties 1 and 2 mean the freedom is given by $[1^2] \vee [1^2] = [2^2] + [1^4]$; but property 3 eliminates $[1^4]$. The Riemann tensor Riem, a T^1_3 tensor, has two contractions:

$$\text{Ric}_{\lambda\mu} := R^{\rho}_{\lambda\mu\rho}, \quad \text{which is symmetric;} \tag{6.28}$$

$$R = R_{\text{sc}} \equiv \text{Tr}(g^{-1}\text{Ric}) = R^m_m. \tag{6.29}$$

So in general Riem $\approx$ Weyl $+$ Ric(traceless) $+ R_{\text{sc}}$ where Weyl is the traceless part of Riem. We have:

$$\dim \text{Weyl} = \dim(\text{Riem} - \text{Ric}) = n(n + 1)(n + 2)(n - 3)/12. \tag{6.30}$$

---

[1] Sometimes the Bianchi identity is reserved for a cyclic derivative property of Riem (see [6], Vol. I, p.135).

Let us consider the lowest dimensional cases:

- If dim $M = 1$, $g = ds^2 = f(x)dx^2$; with $y = \int \sqrt{f}dx$, this becomes $g = dy^2$: *any curve is rectifiable.* Curvature requires biplanes, $x \perp y$, and there are none in one dimension.
- If dim $M = 2$, a single curvature suffices, as there is only a biplane; indeed

$$K \equiv \text{Curv(Gauss)} = R_{sc}/2. \tag{6.31}$$

For example Ric $= 1/2g R_{sc}$. Similarly, for Riem itself, we have:

$$R_{\sigma\lambda\mu\nu} = \frac{1}{2}\left(g_{\sigma\nu}g_{\lambda\nu} - g_{\sigma\mu}g_{\lambda\nu}\right)R_{sc}. \tag{6.32}$$

In the two-dimensional case a new choice of coordinates $(u, v) \to (u', v')$ allows a general metric g given in terms of the matrix components E, F, G of the first fundamental form $(ds^2 = E(u, v)du^2 + 2F(u, v)du\, dv + G(u, v)dv^2)$ to be put in *geodesic* (6.33a) or *isothermal* or *conformal* (6.33b) forms:

$$ds^2 = du^2 + G(u, v)dv^2, \tag{6.33a}$$

$$ds^2 = \exp(2\sigma(u, v))(du^2 + dv^2). \tag{6.33b}$$

For dim $M = 3$, we have: dim Riem $= 6 =$ dim Ric, hence Weyl $\equiv 0$. In three dimensions, this means g can be put in orthogonal or *diagonal form:*

$$G = ds^2 = E(u, v, w)du^2 + F(u, v, w)dv^2 + G(u, v, w)dw^2. \tag{6.34}$$

It is only in dimensions ≥ 4 that the Riemann tensor, Riem, exhibits all its grandeur; for example, in dimension 4, dim Weyl $=$ dim Ric $= 10$, so dim Riem $= 10 + 10 = 20$. Gravitation in three dimensions is "conic", with *no* propagating modes; gravitation in four dimensions has the same degrees in the *gravistatic* part as in the *radiating* part. Also, as $2 = 4/2$, the curvature of four-dimensional manifolds can be (anti-)*selfdual.*

One shows that Riem is the obstruction to flatness (Christoffel): any (M, g) manifold can have a metric sum of squares if and only if Riem $= 0$.

One shows likewise that the Weyl tensor W is the obstruction to *conformal flatness* (Gauss in two dimensions; Weyl in general). Any manifold (M, g) with $\dim(M) \geq 4$ can have a metric sum of squares but for a common factor, that is, $ds^2 = \exp(2\sigma(x_i))(dx_1^2 + \cdots + dx_n^2)$ if and only if Weyl $= 0$ (Weyl, 1917); somehow *conformal* is associated to *traceless.*

In dimension 1, Riem $= 0$; in dimension 2, Ric(traceless) $= 0$; in dimension 3, Riem is traceless. In dimension 2, any metric is conformally flat; in dimension 3, any metric is orthogonal (see above). The obstruction to conformal flatness in three dimensions is measured by a different tensor: the *Cotton* (1899) tensor.

6.1.4 Homotopy groups

The reader is perhaps familiar with the *fundamental group* $\pi_1(X)$ or first homotopy group of a topological space X. Denoting by Ω_P the set of *loops* (continuous closed paths) starting and ending in $P \in X$, we define a *composition* rule in the following way. If α and γ are two loops in Ω_P we define $\gamma \circ \alpha$ to be the loop obtained by first following γ and then matching with α. This defines a new loop, as α starts where γ ends. Declare now two loops to be *equivalent* if they are continuously deformable onto each other. Then, $\Omega_P/(\text{equiv. relation}) \equiv \pi_1(X, P)$ becomes a group, the fundamental group of the manifold (H. Poincaré, 1896). The identity Id is the class of contractible loops (i.e. those loops which are continuously deformable, or "shrinkable" to a point). If the space is *arcwise* connected (all connected manifolds are) the P-dependence is spurious, and one talks only of $\pi_1(X)$. Some examples of fundamental groups are:

$$\pi_1(S^1) = \mathbb{Z}, \qquad\qquad \pi_1(S^2) = 0, \qquad\qquad (6.35)$$

$$\pi_1(SU(2) = S^3) = 0, \qquad\qquad \pi_1(SO(3) = \mathbb{R}P^3) = \mathbb{Z}_2. \qquad (6.36)$$

Hurewicz generalized these in 1935 (in 1940 he died from an accident while visiting the Maya ruins in Mexico) to higher homotopy groups.

The $\pi_0(X)$ set is the set of maps: $S^0 \to X$; as S^0 is just two points, it just measures the distinct connected pieces of X.

$$X \text{ connected} \Leftrightarrow \text{card } \pi_0(X) = 1. \qquad\qquad (6.37)$$

If $X = G$ is a Lie group, then $\pi_0(G)$ is a group; for example, $SO(3, 1)$ has four components, and

$$\pi_0(\text{Lorentz (3, 1) group}) = \mathbb{Z}_2 \times \mathbb{Z}_2 = V. \qquad\qquad (6.38)$$

The fundamental group of X can be seen as a group (of equivalence classes) of maps from the (pointed) circle S^1 to the (pointed) space X. Then the generalization is obvious: $\pi_n(X), n = 0, 1, 2, \ldots$, are the classes of maps from $S^n \to X$ with suitably defined equivalence relations; we omit details (see e.g. [8], [9]). All these are important topological invariants. J. P. Serre found (ca. 1952) that the homotopy groups of spheres were (generally) finite, and computable. $\pi_n(S^n) = \mathbb{Z}$ is an old result, but $\pi_3(S^2) = \mathbb{Z}_2$ is a surprise! Some general facts about homotopy groups are:

1. $\pi_0(X)$ is just a discrete set; if $X = G$, a Lie group, $\pi_0(G)$ is a group.
2. $\pi_1(X)$ is, in general, a *non-Abelian* group; but if $X = G$, a Lie group, it is Abelian. For example, $\pi_1(\Sigma_g)$ is non-Abelian for $g > 1$, where Σ_g is a Riemann

surface of genus g: a sphere, torus and "pretzel" for $g = 0, 1, 2$, respectively. For the torus $T^2 = S^1 \times S^1$, we have $\pi_1(T^2) = \mathbb{Z} \times \mathbb{Z}$.

3. Any connected space with non-trivial $\pi_1(X)$ has a unique universal covering space $\widetilde{X}$, with a natural *onto* map $\widetilde{X} \to X$ with inverse images equal to $\pi_1(X)$:

$$\pi_1(X) \hookrightarrow \widetilde{X} \to X. \tag{6.39}$$

4. $\pi_n(X)$ is an Abelian group for $n > 1$.

5. If G is a (finite-dimensional) Lie group, then

$$\pi_2(G) = 0. \tag{6.40}$$

(there is no simple proof!)

6. The homotopy groups exist for any n, regardless of the dimension of the space X; for example, $\pi_4(S^3) = \mathbb{Z}_2$. In this it differs from (co-)homology.

The homotopy groups of a space X are naturally topological *invariants*; let us include the following result of Hurewicz:

The *first nonnull* homology group of a manifold, $H_k(X, \mathbb{Z})$ coincides with the Abelianization of the first nonnull homotopy group, $\pi_k(X)$ (recall that, for any group G, $\mathrm{Ab}(G)$, the Abelianization of G, is $G/\Omega G$, where ΩG is the *commutator* or first derived group of G). For example, $b_1(\text{pretzel}) = 4$, and $\pi_1(\text{pretzel})$ is generated by four elements a, b, c, d subject *only* to the relation $abcda^{-1}b^{-1}c^{-1}d^{-1} = 1$: the Abelianized H_1 (pretzel;$\mathbb{Z}$) is clearly $\mathbb{Z} + \mathbb{Z} + \mathbb{Z} + \mathbb{Z}$.

To conclude this brief survey of homotopy, let us "see" the homotopy solution to the problem of the types of *principal bundles over spheres*:

$$G \hookrightarrow P \to S^n. \tag{6.41}$$

How many principal bundles P are there? The spheres are covered with two charts, skipping e.g. the poles, as $S^n \setminus \{0\} \approx \mathbb{R}^n$. Then, as the two charts overlap at the equator, one shows the total possible spaces Ps are given by homotopy classes of maps from the equator in S^n to the group G; indeed the precise result is (see [9]):

$$G\text{-bundles over } S^n \approx \pi_{n-1}(G). \tag{6.42}$$

6.1.5 Spin groups, Spin(n)

Consider now the real orthogonal group $O(n) = O(n, \mathbb{R})$, $n = 0, 1, 2, \ldots$; the $n \times n$ matrices $M \in O(n)$ verify $M^T M = 1$: hence $\det M = \pm 1$. This implies that the orthogonal group has two connected components. Call $SO(n)$ the connected component of the identity; it is called the *rotation group*. We have:

$$O(1) = \mathbb{Z}_2 \approx S^0, \quad \text{hence } SO(1) = \text{Id}. \tag{6.43}$$

$SO(2) = U(1) = S^1$, the circle. We have $\pi_1(S^1) = \mathbb{Z}$. All higher homotopy groups of the circle S^1 are zero, as follows from the "exact homotopy sequence" where $\mathbb{R}$ is the universal covering of the circle [9],

$$\mathbb{Z} \hookrightarrow \mathbb{R} \to SO(2) \approx S^1, \tag{6.44}$$

the reason being that $\mathbb{R}^n$ is contractible (for a contractible space all homotopy groups vanish, except for π_0).

$SO(3) \approx \mathbb{R}P^3$: any 3D rotation has an axis and an angle, so $SO(3) \approx$ solid 3D ball with radius π, but with antipodal points in the boundary 2-sphere identified; this is clearly homeomorphic to the 3D real projective space, $\mathbb{R}P^3 \approx S^3/\mathbb{Z}_2$, where the quotient is by the antipodal map. Hence $\pi_1(SO(3)) = \mathbb{Z}_2$ and this results holds for all higher rotation groups:

$$\pi_1(SO(n)) = \mathbb{Z}_2, n \geq 3 \Leftrightarrow \mathbb{Z}_2 \to \text{Spin}(n) \to SO(n). \tag{6.45}$$

Hence, for $n \geq 3$ there is a *unique double covering of the rotation group*, which is clearly also a group, called the Spin (n) group. The reader should see that $\widetilde{SO}(3) = S^3$, and also that $\widetilde{SO}(3) = SU(2)$, the form often preferred in quantum mechanics. There are also other coincidences, namely:

$$\text{Spin}(1) = \mathbb{Z}_2; \quad \text{Spin}(2) \approx U(1); \quad \text{Spin}(3) = SU(2); \quad \text{Spin}(4) = \left[\text{Spin}(3)\right]^2 \tag{6.46}$$

$$\text{Spin}(5) = Sq(2); \quad \text{Spin}(6) = SU(4); \quad \text{Spin}(7, 8, 9) \text{ related to octonions}. \tag{6.47}$$

For $n = 1, 2$, the definition of Spin $(1, 2)$ is just a *double covering*; it is not the universal one, but it is well defined (through Clifford algebras).

The orthogonal groups in 2, 4 and 8 dimensions are special. In dimension 2, the rotation part is Abelian and divisible, $SO(2)/\mathbb{Z}_n \approx SO(2)$; the existence of the complex numbers and the regular plane polygons are related to this fact.

Dimension 4: The Spin (4) group factorizes, Spin $(4) = \left[\text{Spin}(3)\right]^2$. The Lie algebra is given by 2-forms in $\mathbb{R}^4$, hence split into self dual and anti-selfdual. The existence of the quaternion numbers $\mathbb{H}$ and of the special regular polytopes in four and three dimensions (e.g. the 24-cell in four dimensions, the icosahedron in three dimensions) is a consequence of this fact.

Dimension 8: Spin (8) exhibits *triality*, that is Spin (8) has an outer automorphism group S_3 (the permutation group of three elements, of order 6). It is at the base of *supersymmetry* in physics! It starts with the two chiral and the vector representation of dimension 8. These three representations are permuted by this S_3. For all this see e.g. [2].

Finally, the homotopy of orthogonal groups exhibit Bott's periodicity: for a generic spin (n big enough), the *stable* homotopy groups are:

$$\pi_{01234567}(O(n), (n \gg)) = \mathbb{Z}_2, \mathbb{Z}_2, 0, \mathbb{Z}, 0, 0, 0, \mathbb{Z}, \quad (6.48)$$

whereas for the symplectic groups there is a shift by 4:

$$\pi_{01234567}(Sq(n), (n \gg)) = 0, 0, 0, \mathbb{Z}, \mathbb{Z}_2, \mathbb{Z}_2, 0, \mathbb{Z}. \quad (6.49)$$

For completeness we add the periodicity 2 for unitary groups:

$$\pi_{01234567}(SU(n), (n \gg)) = 0, \mathbb{Z}, 0, \mathbb{Z}, 0, \mathbb{Z}, 0, \mathbb{Z}. \quad (6.50)$$

Spin(n) groups appear often in physics through representations; for the rotation group $SO(n)$ the set of irreducible (unitary) *linear* representations, *irreps*, are given, starting from the vector, dimension n, by the *traceless* Young tableaux. For example, for $SO(5)$, the first *irreps* are of dimension 1 (identity); 5 (vector); $5 \cdot 6/2 - 1 = 14$, type [2]; $5 \cdot 4/2 = 10$, type $[1]^2$, etc.

Spin $(2\nu + 1)$ has a primitive *irrep* of dimension 2^ν, call it Δ. But Spin (2ν) has two, of different chirality, called $\Delta_{L,R}$ and dimension again $2^{\nu-1}$. All (true, linear) *irreps* of $SO(n)$ are real, of course, but the character (*type*) of the (one or two) spin *irreps* varies:

$\Delta(n)$, and $\Delta_{L,R}(n)$ types: *real* for $n = 0, \pm 1$; complex for dimensions 2, 6: quasireal (or quaternionic, or pseudoreal) for $n = 4, 4 \pm 1$.

The Spin groups realize *projective* representations of the rotation group. This is why they are important in physics, as quantum mechanics seeks projective (or *ray*) representations of physical symmetries; the ubiquitous appearance of $SU(2)$ in quantum mechanics is because all projective *irreps* of $SO(3)$, the physical rotation group, come from the linear *irreps* of the covering group, $SU(2) = \text{Spin}(3)$.

6.2 Holonomy

6.2.1 Parallel transport and holonomy

When there is a connection ∇ (in any bundle, in our case in the tangent bundle), vectors (or sections) are propagated along curves; in particular, *geodesics* (in the *affine* conception) are defined by the flow of auto-parallel vector fields,

$$\nabla_X X = 0, \quad (6.51)$$

which in coordinates becomes the conventional geodesic equation

$$\frac{d^2 x^\lambda}{ds^2} + \Gamma^\lambda_{\mu\nu} \frac{dx^\mu}{ds} \frac{dx^\nu}{ds} = 0. \tag{6.52}$$

As it is a second order equation, we have two Cauchy data: there is a geodesic starting at any point, for any given direction (e.g. meridians in the sphere).

In a Riemannian space (M, g) there is a *metric* definition of geodesics, minimizing the distance function

$$\delta \int ds = 0, \tag{6.53}$$

which yields the same equation as (6.52).

A vector field Y is *parallel translated* along the flow of another vector field X; the first order equation is (X and Y_0 known; find $Y = Y(t)$)

$$\nabla_X Y = 0 \tag{6.54}$$

with a unique solution from a particular value $Y_0 = Y(P_0) = v$. We know that if the connection is *flat*, there is no curvature; how do we measure curvature from parallel transport? Through *holonomy*: consider equation (6.54) for a *frame* (a base) ϵ in P, and make it run through a loop γ; at the end it becomes another frame ϵ' in the same point P. As any two frames *at the same point* are related by an isometry (when the connection, as in our case, is the Levi-Civita connection $\nabla g = 0$), we have an element o of the orthogonal group $O(n)$ depending on the loop γ: $o(\gamma) \in O(n)$. It is easy to see how all loops starting/ending from a point P_0 compose, and the isometries make up a subgroup of $O(n)$, called the *holonomy group* of the connection, $\mathrm{Hol}(\nabla) = \mathrm{Hol}(g)$ (in our case) (Cartan, 1926).

One expects that $\mathrm{Hol} = \mathrm{Id}$ for flat connections, but there is a constructive counter-result, *the Ambrose–Singer theorem*, 1953: the Lie algebra of the holonomy group is generated by the curvature (see e.g. [6], Ch. 2; [8] p. 343).

It is a reasonable result, as the curvature $R^\rho_{\lambda\mu\nu}$ is a 2-form (μ, ν) Lie algebra-valued (ρ, λ): antisymmetry in the second pair implies holonomy lies inside the orthogonal group (whose Lie algebra consists of $A^T = -A$ antisymmetric matrices).

For *contractible* loops the holonomy group has to lie in $SO(n)$, the connected part of $O(n)$. In fact, this is the general case for orientable manifolds; let us look at this in detail. A manifold is *orientable* if the transition functions between charts U_i and U_j (see Section 6.1) can be chosen in GL^+, the connected part of GL (GL has det > 0 or < 0: two connected components). Let us see the obstruction:

$$
\begin{array}{ccccc}
\mathrm{GL}^{+} & \longrightarrow & & & \\
\downarrow & & & & \\
\mathrm{GL} & \longrightarrow & B(M) & \longrightarrow & M \\
\downarrow & & \downarrow & & \| \\
\mathbb{Z}_2 & \longrightarrow & B'(M) & \longrightarrow & M
\end{array}
\tag{6.55}
$$

The middle row generates the lower row, which *defines* the first Stiefel–Whitney (S–W) class w_1 (of the tangent bundle of) the manifold M. The middle row lifts to an upper row $\mathrm{GL}^{+} \to B'' \to M$ *iff* this class is zero:

$$
M \text{ orientable} \Leftrightarrow w_1(M) = 0.
\tag{6.56}
$$

This is a nice example of measuring properties by absence of obstruction: the obstruction of orientability is given by the first S–W class.

By a simple extension, if one asks when an oriented manifold admits a *spin structure*, it is to ask when the tangent bundle, with group $SO(n)$, lifts to the *spin bundle*, with group $\mathrm{Spin}(n)$:

$$
\begin{array}{ccccc}
\mathbb{Z}_2 & & & & \\
\downarrow & & & & \\
\mathrm{Spin}(n) & \cdots\!\!\to & B_0^{\wedge} & \cdots\!\!\to & M \\
\downarrow & & \downarrow & & \big| \\
SO(n) & \longrightarrow & B_0(M) & \longrightarrow & M
\end{array}
\tag{6.57}
$$

We have the result [8]: the lower row lifts to the middle row, that is, the manifold admits a spin structure, *iff* the *second* Stiefel–Whitney class of the tangent bundle is zero, $w_2(TM) = 0$. For spheres it is easy to show that all are *spinable*,

$$
w_2(TS^n) = 0.
\tag{6.58}
$$

For oriented surfaces Σ_g, of genus g ($g = 0, 1, 2$ for sphere, torus, pretzel, respectively) there is *no* obstruction to spin structures: $w_2(\Sigma_g) = 0$. Let us look at it for the ordinary 2-sphere S^2: the principal bundle of the tangent bundle is

$$
SO(2) \to SO(3) \to S^2.
\tag{6.59}
$$

But this has a "square root", because $SU(2)$ covers $SO(3)$ universally: hence $SO(2)$ lifts to a double covering, and this is the spin bundle. Alternatively, the Euler "class" carries to the top S–W class, and as $\chi(S^2) = 2$, it is 0 under mod 2, so $w_2(S^2) = 0$; compare [7], p. 99.

Once a manifold admits a spin structure, how many (inequivalent) spin structure does it admit? The answer is easy to see: as many as there are elements in $H^1(M, \mathbb{Z}_2)$. This result bears on string theory, as one is supposed to sum over all possible spin structures; in fact $H^1(\Sigma_g, \mathbb{Z}_2) = 2^{2g}$, which is therefore the number of spin structures on a "worldsheet" of genus g (see [3], Vol. 2, p. 278).

(Any n-dimensional real vector bundle has $w_1, w_2, \ldots, w_n$ Stiefel–Whitney classes, which take values in the $\mathbb{Z}_2$ cohomology, $w_i \in H^i(M, \mathbb{Z}_2)$; see [7].)

In relation to connections, we also have the *reduction theorem* (see [6], Ch. 2):

Reduction theorem The structure group of the (vector) bundle reduces to the holonomy group.

This means that the transition functions can be taken in the holonomy (sub-)group. The theorem is again reasonable, because we can arrange the transition functions from the parallel transport, so they transform among themselves with the holonomy group.

As a consequence, for an orientable manifold the holonomy group lies in $SO(n)$; but it can be smaller, of course, when the connection conserves some particular objects (some tensor fields, for example); see below.

6.2.2 Classes of holonomy groups

It is interesting to see when the holonomy group is smaller than $O(n)$ or SO, but not Id. This problem was dealt with by M. Berger in 1955, with nearly complete results; the outcome is related to which subgroups of the orthogonal group still act *transitively* on spheres.

Digression A Lie group *acts* "differentiably" in a manifold M if $g \cdot x = y$ is a diffeomorphism (g in G; x, y in M); the action is *effective* if $g \cdot x = x$ for all $x \Rightarrow g = $ Id. Otherwise there is a *kernel* K, and G/K operates effectively. The action is *transitive (trans)*, if $G \cdot x = M$, from any x. In general the images $G \cdot x$ are called "the orbit of G through x". The *isotropy* group (called *little group* (Wigner) in physics) is G_0, when $G_0 \cdot x = x$: points x, y in the same orbit, $G \cdot x = G \cdot y$ have conjugate isotropy groups. When G acts transitively in M, the quotient space $G/H = M$ has a well-defined sense as a manifold, and M is called a *homogeneous space*; *symmetric spaces* are an enhancement of homogeneous spaces [4] when there is geodesic reflection symmetry.

Returning to the issue of when the holonomy group is smaller than $O(n)$, but not Id, it turns out that the possible holonomy groups of any Riemannian manifold are related to the four division algebras $\mathbb{R}$, $\mathbb{C}$, $\mathbb{H}$ and $\mathbb{O}$. They are also, as mentioned, related to transitive actions of groups on spheres, and thirdly they depend on Riemannian spaces leaving some other object (besides the metric), invariant under parallel displacement (for example, the holonomy group of an oriented manifold leaves a volume form invariant).

Let us consider (M, g) an *irreducible nonsymmetric* Riemannian manifold; the possible holonomy groups are 2×3 series for the fields (or skew-fields, $\mathbb{R}$, $\mathbb{C}$ and $\mathbb{H}$) plus two exceptional cases (for Octonions $\mathbb{O}$). Product manifolds have product holonomy groups and all symmetric spaces are known, so they are excluded from the list by convention.

Let us comment on Table 6.1. Holonomy reflects the "parallel" tensors $\nabla T = 0$; as mentioned, orientable manifolds maintain a fixed volume, hence $\mathrm{Hol}(\nabla) = O(n) \cap SL(n) = SO(n)$. This is the most common case, as most manifolds are required to be orientable (for integration, etc.). Recall that a complex manifold (M, J) can acquire a Hermitian metric: the "g" part is automatic (i.e. always possible) in any Riemannian manifold, and ω is concocted from $J : g(J) = \omega$. The natural Levi-Civita connection leaves only g invariant; but the manifold is called Kähler if it also leaves ω fixed, i.e. $\nabla \omega = 0$, equivalent to ω closed, $d\omega = 0$. Hence the holonomy of Kähler manifolds is $O(2n) \cap Sp(2n, \mathbb{R}) = U(n)$. If the holonomy descends to $SU(n)$, the manifold is called Calabi–Yau (for historical reasons).

Quaternionic and hyperkähler manifolds similarly preserve some object related to the skew-field of the quaternions. As $Sq(n) \subset SU(2n) \subset SO(4n)$, hyperkähler manifolds are at least Calabi–Yau (CY), Kähler and orientable. Quaternionic manifolds need not to be CY nor Kähler.

There are also two singular cases related to the octonions $\mathbb{O}$.

[Recall that the complex numbers $\mathbb{C}$ are a 2-dimensional composition and division algebra over $\mathbb{R}$ with the unit $(0,1) \equiv i$ fulfilling $i^2 = -1$; the *quaternions* (4-dimensional over $\mathbb{R}$) have two independent units, i and j, with $i^2 = j^2 = (ij \equiv k)^2 = ijk = -1$. For the *octonions* $\mathbb{O}$ there are three independent units: i, j, k, anti-involutive and anticommuting, and the same for the products $(ij, jk, ki, (ij)k)$, which forces *alternativity*, in lieu of associativity, namely $(ij)k = -i(jk)$. The octonions (8-dimensional over $\mathbb{R}$) are a division algebra; but one cannot proceed beyond this: there are no division or composition $\mathbb{R}$-algebras except in dimensions 1, 2, 4, 8: (Hurwitz' theorem, 1895; see e.g. [2]).]

The descent from $U(n)$ to $SU(n)$ is similar to the commented $O(n)$ to $SO(n)$: the Det map generates the bundle $U(1) = U(n)/SU(n) \to B'(M) \to M$, which defines *the first Chern class* (of the tangent bundle $c_1(TM)$) of the complex manifold M; M Kähler becomes Calabi–Yau (CY_n, Calabi–Yau n-fold, real dim $2n$) *iff* $c_1(M) \equiv c_1(TM) = 0$.

[For complex vector bundles η, Chern classes c_i (i 1 to dim η) take values in $H^{2i}(M, \mathbb{Z})$: there is a "transgression" from $U(1)$-bundles to H^2 $\mathbb{Z}$-cohomology

Table 6.1 *Table of holonomy groups*

$O(n)$	$SO(n)$	$\mathbb{R}$
Generic	Orientable, $w_1 = 0$	
$U(n)$	$SU(n)$	$\mathbb{C}$
Kähler, $\nabla\omega = 0$	Calabi–Yau, $c_1 = 0$	
$q(n)$	$Sq(n)$	$\mathbb{H}$
Quaternionic	Hyperkähler	
Spin(7)	G_2	$\mathbb{O}$
dim $M = 8$	dim $M = 7$	

because of the *resolution*, alluded to above, $\mathbb{Z} \to \mathbb{R} \to S^1 = U(1)$, and similarly for higher Chern classes.]

The Hermitian metric with quaternionic entries admits the isometry group $Sq(n)$ (sometimes written $Sp(n)$: it is the *real compact* form of Cartan's simple complex Lie algebras C_n). As $Sq(n) \subset SU(2n)$, hyperkähler manifolds are also CY, with first Chern class = 0 (they are oriented, of course, because $U(n)$, $SU(n)$, $q(n)$ and $Sq(n)$ are connected and all lie in some $SO(N)$).

The group $q(n)$ (the notation is not universal!) is defined as

$$q(n) = Sq(n) \times /_2 Sq(1) \qquad (6.60)$$

and the quotient is by the common centre $\mathbb{Z}_2$. We have in particular

$$Sq(1) = SU(2) = \text{Spin}(3) \approx S^3, \quad q(1) = SU(2)^2/\mathbb{Z}_2 = SO(4). \qquad (6.61)$$

6.2.3 Octonions and octonion-related groups

Finally, as the octonions $\mathbb{O}$ are not associative, there are only five groups related to them: Cartan's exceptions G_2 (dimension 14), F_4 (dimension 52), E_6 (dimension 78), E_7 (dimension 133) and E_8 (dimension 248). In some ways the spin groups Spin (7,8 and 9) can be considered *also* as octonionic groups, and indeed G_2 as well as Spin (7) can act as exceptional holonomy groups (see [5]). Note that both types of exceptional holonomy manifolds are Ricci flat (but neither is Calabi–Yau). In particular, a manifold with holonomy G_2 preserves a *generic* 3-form in 7-dimensions (check: $7^2 - \binom{7}{3} = 49 - 35 = 14 = \dim G_2$), whereas one with Spin (7) holonomy preserves a particular 4-form called a *Cayley four-fold*; recall that the spin representation(s) of $SO(7)$ acts in $\mathbb{R}^8$: $2^{(7-1)/2} = 8$, real type, so Spin (7)$\subset SO(8)$ in a natural manner.

So the G_2 and Spin (7)-manifolds have a *Ricci flat* metric, i.e. the Riemann tensor is given just by the traceless part, the Weyl tensor; this makes the construction of these manifolds difficult (see [5]).

The five exceptional groups G_2, F_4, $E_{6,7,8}$ are related to octonions; in particular $G_2 = \text{Aut}(\mathbb{O})$ acts transitively on the 6-dimensional sphere of unit imaginary octonions, S^6, so the defining representation has dimension 7. The isotropy group is $SU(3)$ $(14 - 6 = 8)$, acting on the diameter $S^5 \subset \mathbb{R}^6$ through the real irreducible $3 \oplus \bar{3}$ representation; there is another suggestion that perhaps the group $SU(3)$ appearing in physics (as color and flavour group) might also be connected with the octonions!

F_4 acts in the 3×3 Hermitian traceless Jordan octonionic matrices $J(3)$, with real dimension $(3 \cdot 8 + (3 - 1)) = 26$, which is the defining representation of F_4. There is a famous octonionic projective plane (it should be called the Moufang plane, not the Cayley plane), $\mathbb{O}P^2$: the isometry group is again F_4, and $\mathbb{O}P^2 = F_4/\text{Spin}(9)$, $2 \cdot 8 = 2 \cdot 26 - 9 \cdot 8/2 \Rightarrow 16 = 52 - 36$, and in fact there is a natural inclusion $\mathbb{O}P^2 \subset J(3)$, as a class of idempotent elements. A noncompact form of the E_6 group also acts on $\mathbb{O}^2$ as projective transformations; E_7 and E_8 have a more complicated definition (Freundenthal, 1955; J. Tits). We expect that our understanding of these two final exceptional groups will increase in the future; for example, there is a mysterious relation between the four exceptional groups (besides G_2) and spin groups, in the following sense (Adams):

F_4 can be formed from $O(9)$ and the spin representation: $36 + 16 = 52$.

$$\text{This is based on the } \mathbb{O}P^2 \text{ space of above.} \tag{6.62}$$

E_6 can be formed from $O(10)$ and the spin representation $+U(1)$:

$$10 \cdot 9/2 + 2(2^{10/2-1}) + 1 = 78 = \dim E_6. \tag{6.63}$$

E_7 is formed starting from $O(12)$, the spin representation and $Sq(1)$:

$$12 \cdot 11/2 + 2(2^{12/2-1}) + 3 = 66 + 64 + 3 = 133. \tag{6.64}$$

E_8 is concocted from $O(16)$ with one of the spin representations:

$$16 \cdot 15/2 + (2^{16/2-1}) = 120 + 128 = 248. \tag{6.65}$$

The two intermediate cases need an extra ingredient, $U(1)$ for E_6 and $Sq(1)$ for E_7; this is well understood, as E_6 is complex type, and E_7 is quaternionic type.

E_8 is the most spectacular of these exceptional groups: its fundamental representation is the adjoint one (dim 248) (a unique case among all simple Lie groups), it has no center, nor outer automorphisms; it has 5-torsion, also unique among Lie groups. It appears in modern physics in several disguised forms: the square of it $(E_8)^2$ is the gauge group of the heterotic exceptional strings; it acts also as gauge group in M-theory (D. Freed); it is related to the Hodge diamond of the $K3$ surface (see later), etc.

6.3 Strings and higher dimensions

6.3.1 Higher dimensions in physics: Kaluza–Klein theories

The traditional tool for microphysics has been the *quantum theory of fields*, developed since the 1920s and much improved with the renormalization program of 1947/52 (Schwinger, Feynman, Tomonaga, Dyson, etc.) and for the non-Abelian case ('t Hooft) in 1971/73. Starting around 1930, physicists were frustrated by their inability to quantize gravitation (it is intrinsically non-renormalizable: the coupling constant G_N has dim (length)^2), and new avenues were sought; three deserve our attention here:

1. extra dimensions (of space–time),
2. supersymmetry (mixing bosons with fermions),
3. extended objects (like strings, membranes, etc.).

Extra dimensions

In 1919 T. Kaluza wrote a letter to Einstein, showing that if one sets up general relativity in five dimensions, and somehow disregards the fifth dimension as unobservable, in four dimensions three fields appear: the usual gravitation field $h = h_{\mu\nu}$, a vector field $A = A_\mu$ and a scalar ϕ. Kaluza went on to show that the usual Einstein–Hilbert (EH) Lagrangian in five dimensions,

$$S[h_5] = (\text{const.}) \int \sqrt{g} R_{\text{sc}}(h_5) \mathrm{d}^5 x, \tag{6.66}$$

decomposed in four dimensions as the usual EH action, plus the Lagrangian for the electromagnetic. field $F_{\mu\nu}^2$ (plus an extra piece for the field ϕ, called the *dilaton* field). Later O. Klein interpreted the unobservability of the fifth dimension as due to compactification in a very small sircle S^1, which incidentally proved to quantize the electric charge in the quantum version of the model: the two signs of the charge were the two ways to run through the circle.

Kaluza's and Klein's (KK's) ideas were clearly premature, if very exciting; they were reconsidered in the 1980s, sixty years later, when the ideas of Grand Unified Theories (GUT) appeared, and also supersymmetry.

Let us here only observe why gravitation in five dimensions generates extra matter in four: if M, N run ≤ 5, and $\mu\nu \leq 4$, then

$$g_{MN} \to g_{\mu\nu} + A_{\mu 5} + \Phi_{55}. \tag{6.67}$$

The idea of unifying electromagnetism with gravitation kept Einstein busy until his death, in 1955 (but not necessarily only in the KK approach).

For *Bose* fields it is not difficult to generalize (6.67): besides the graviton, the other Bose fields which appear in modern theories are just p-forms; call them $\phi, A, B, C, D, \ldots$ for 0-, 1-, 2-, 3- and 4-forms respectively. Then in one-dimension descent we have

$$g_{d+1} \rightarrow g_d + A_d + \phi_d; \quad A_{d+1} = A_d + \phi_d;$$
$$B_{d+1} \rightarrow A_d + B_d; \quad C_{d+1} \rightarrow B_d + C_d; \text{ etc.}$$

(6.68)

The split of Fermi fields (gravitinos and ordinary, spin $\frac{1}{2}$ fermions) is not so regular, as spinors have dimension powers of 2; it will be indicated in each case later.

Supersymmetry

There are no compelling arguments to unify bosons and fermions, and several for not doing so: different statistics (in principle!) and different transformation laws under the Lorentz group. Nevertheless, since 1974 (with some Russian antecedents) theorists have looked at enlarged symmetry schemes, mixing up fermions and bosons; it is an ample subject, and we offer only some simple examples.

1. *The Wess–Zumino model* (1973/1974). This is constructed out of a scalar field A, a pseudoscalar field B and a Majorana spinor χ living in Minkowski space. The Lagrangian is free massless at first instance:

$$\mathcal{L} = -\frac{1}{2}(\partial_\mu A)^2 - \frac{1}{2}(\partial_\mu B)^2 - \frac{1}{2}\chi \gamma \partial \chi.$$

(6.69)

Although unmixed for the three massless fields, the action $\int \mathcal{L} d^4 x$ has a Bose–Fermi symmetry. Define

$$\delta A = \bar{e}\chi, \quad \delta B = i\bar{e}\gamma_5\chi, \quad \delta\chi = \partial\gamma(A + i\gamma B)e,$$

(6.70)

where e is a fermionic parameter ($e^2 = 0$); these transformations leave $\mathcal{L}$ invariant. The supersymmetry generator is called Q, so the transformations should be understood roughly as

$$e^{-eQ}Ae^{+eQ} = (1 - eQ)A(1 + eQ) \equiv A + \delta A = A + [Q, A]e$$

(6.71)

as $\{Q, e\} = 0$. Q and e are fermionic (spinorial) objects in some precise sense. One can go on and add (equal) masses and (some) interactions preserving this (super)symmetry (see e.g. [11]): supersymmetry (Susy) can be mantained.

2. *The super Yang–Mills model in 10 dimensions.* Massless fields transform with the (compact) little group of the light cone, here $O(8)$ (it corresponds to $O(2)$,

helicity label, in four dimensions). The three primordial representations of the Spin(8) group have dimension 8:

$$\text{dimension vector, 8-dim chiral } \Delta_{L,R} = 2^{8/2-1} = 8, \text{ real type}: \quad (6.72)$$

which makes it easy to write down a Susy Yang–Mills action in $10 = (1,1) + (8,0)$ dimensions:

$$S = \int \mathcal{L}dx = \int dx \left(-\frac{1}{4}F^2 + \frac{i}{2}\bar{\psi}\Gamma D\psi\right), \quad (6.73)$$

where $F \cdot F = F^2$ with the explicit form

$$F = F^a_{\mu\nu} = \partial_\mu A^a_\nu - \partial_\nu A^a_\mu + g f^a_{bc} A^b_\mu A^c_\nu \quad (6.74)$$

is the field strength, and where "a" labels an index of the gauge group G, $a = 1, 2, \ldots,$ dim G. Notice that ψ in (6.73) that is a chiral field, so S is parity-violating. By dimensional reduction, we get an $\mathcal{N} = 4$ theory in 4D: A in 8D gives A plus 6 ϕs in 4D; ψ in 8D gives 4 ψs in 4D; the Susy pairing is: $A \leftrightarrow \psi_1$, and 3ψs $\leftrightarrow$ 6ϕs, etc. This $\mathcal{N} = 4$ theory enjoys wonderful convergence properties (it is finite; Mandelstam).

3. *Supergravity theory in 11 dimensions.* The particle content is

$$h_{44} + \Psi_{128} + C_{84} \text{ graviton, gravitino and 3-form.} \quad (6.75)$$

This is a remarkable theory: first, 11 is the highest dimension one can set up supergravity, Sugra (Cremmer-Scherk); second, it turns out the 3-form C is coupled to a *membrane* (Townsend); third, the dimensional reduction to 10 dimensions reproduces the particle content of the IIA string theory (see below). Recently there has been some progress in dealing with the divergencies of this theory.

Extended objects

As in so many things, P. A. M. Dirac was a pioneer: the first study of quantum physics of extended objects was a paper by Dirac (1962), trying to describe muons as an *excited membrane* of the electron; it led nowhere. String theory emerged around 1970, and it has an interesting story; milestones were the set-up of fermions in 1971 (P. Ramond), the ambition to cover gravitation in 1975 (J. Scherk–J. Schwarz), a first hint of a theory "of everything"; superstrings appear in 1978, with Susy also on the target space, and no tachyons (GSO projection); uniqueness and claims for a "Theory of Everything (TOE)" in 1984 (Green–Schwarz), when anomaly cancellation pointed to select gauge groups (besides the dimension); the advent of M-theory 1995 (below), and its general dismissal by the community of physicists (even today).

6.3.2 Physics in 10, 11 and 12 dimensions: strings, membranes, M-theory and F-theory

We just mention briefly the five superstring theories established around 1985 as all five (i) are viable theories, (ii) include gravitation, (iii) include fermions with Susy and Sugra. Three of the five include gauge groups, potentially covering the group of the standard model "3, 2, 1", and they are potentially renormalizable.

The five supersymmetric theories make up a pentagon:

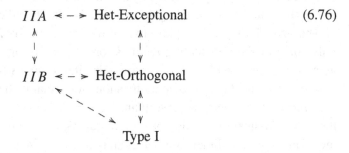

$$IIA \longleftrightarrow \text{Het-Exceptional} \qquad (6.76)$$

A brief description follows:

- IIA describes closed, nonchiral, $\mathcal{N} = 2$ Susy strings with no gauge groups.
- IIB describes closed, chiral, $\mathcal{N} = 2$ Susy strings with no gauge groups.
- Het-Exceptional describes closed, $\mathcal{N} = 1$ Susy strings with E_8^2 gauge group.
- Het-Orthogonal describes closed, $\mathcal{N} = 1$ Susy strings, $O(32)$.
- Type I describes open (and closed) strings, $\mathcal{N} = 1$, $O(32)$ gauge group.

The particle content of superstrings is easy to write: the fundamental supersymmetry is in 8 *Euclidean* dimensions, as mentioned (triality!)

$$8_v \Leftrightarrow \Delta_L \Leftrightarrow \Delta_R. \qquad (6.77)$$

For instance, for the IIA theory, the particle content is

$$(8_v - \Delta_L) \times (8_v - \Delta_R) = (h + B + \phi, A + C; \Psi_{1,2} + \psi_{1,2}), \qquad (6.78)$$

i.e. graviton h, 2-form B and dilaton ϕ appear always (Neveu–Schwarz NS sector); in the RR (Ramond–Ramond) sector multiply two fermions, and one gets A (a 1-form) and C (a 3-form); and then two gravitinos $\Psi(\mathcal{N} = 2)$ and two fermions ψ.

For the IIB theory the chiralities are the same:

$$(8_v - \Delta_L) \times (8_v - \Delta_L) = (h + B + \Phi, B' + D^+ + \Phi'; \Psi_{1,2} + \psi_{1,2}), \qquad (6.79)$$

where $D^\pm$ is a (anti-)selfdual 4-form.

For the Type I theory, with open and closed strings, we get

$$(8_v - \Delta_L) \wedge (8_v - \Delta_L) = (h + \Phi, B; \Psi_1 + \psi_1) \qquad (6.80)$$

plus the gauge group $O(32)$. The wedge $\wedge$ just means: take the symmetric part of the Bose product, and the antisymmetric one of the Fermi, etc.

The content of the heterotic string is the same, but the gauge group is E_8^2 in the first (exceptional) heterotic string. The name "heterotic" comes from the fact that closed $\mathcal{N} = 1$ Susy is performed by substituting the $24 + 2 - (8 + 2)$ dimensions of the "bosonic" string with a gauge group of rank 16, either $O(32)$ or E_8^2.

This is obviously not the place to deal with string theory *in extenso*. We mention only some of the outstanding features: one writes a Lagrangian minimizing the area of the *worldsheet*, and tries to quantize it. *Gravitons* follow at once from the closed sector; the theory is potentially anomalous (mainly because of the endurance of *conformal* symmetry against quantization), that is, some of the symmetries of the naïve classical theory do not survive quantization, unless some conditions are met. This fixes the dimensions of the target space ($24 + 2 = 26$ for the original, bosonic string, and $8 + 2 = 10$ for the superstring) and also the *gauge groups* (as stated above). This critical anomaly cancellation was the big advance in 1984–85 (Green–Schwarz).

Incidentally, we are at odds when quantizing *higher extended* objects; at the moment there is no accepted scheme for quantizing membranes ($p = 2$) or higher p-objects (it was for this reason that for a long time Ed Witten was reluctant to accept e.g. the membrane naturally appearing in 11-dimensional supergravity).

The most appealing theory was the heterotic exceptional string, in part because there was only $\mathcal{N} = 1$ Susy (16 supercharges, down to $\mathcal{N} = 1$ or four in dimension 4), and also because one hoped the E_8^2 group would naturally give rise to some of the GUT groups, like E_6, $SO(10)$ or $SU(5)$.

6.3.3 M-theory

In 1995 three fairly independent advances made the five superstring theories merge in a so-called "M-theory", living in eleven dimensions and including also other extended objects besides strings, in particular membranes:

1. Townsend proved (in January 1995) that 11-dimensional Sugra (see above, particle content) contained a membrane (or $p = 2$ extended object), coupled to the 3-form C (as "old" strings were coupled to the NS 2-form B), becoming the fundamental string upon circle compactification to ten dimensions.
2. Witten proved (in March) that the *strong limit* of the IIA theory developed an 11th dimension, with particle content that of 11-dimensional Sugra.

3. Polchinski proved (in October) that there were some "$D - p$ Branes" as endings of open strings, a kind of "solitonic objects" he himself had discovered earlier. These p-Branes "radiate" ($p + 1$)-forms, in the way that charged particles ($p = 0$) "radiate" the potential A_μ, a 1-form. One finds odd-dimensional Branes in the IIB theory, but even dim-branes in the IIA. We elaborate on membranes below.

The name "M-theory" was coined by Witten; M stands for "membrane", "mother", "mystery", etc., according to taste. To have some idea why the five theories fuse, we remark: IIA and IIB theories are the same from nine dimensions down (no chirality in odd dimensions), and the same for the two heterotic strings; the responsible symmetry was called T-duality. From M-theory in 11 dimensions one goes back to Het-Exceptional by compactification *in a segment* D^1, with the two E_8 groups appearing miraculously to cancel anomalies in the boundary of the segment (we do not go into this). Finally, Polchinski and Witten proved explicitly that in some "strong"/"weak" limit the two (open and closed) theories with the same group, $O(32)$, coincide: a case of the so-called S-duality.

M-theory has not lived up to expectations; 15 years after inception it has not explained anything, nor made a precise testable prediction (overlooking some black-hole entropy calculations of Strominger and Vafa). We shall only say a few words about the physical compactification problem, that is, how do we get from the phantasmal 10 dimensions to our mundane, open, quasi-flat Minkowski space in (apparently) four dimensions!

One of the lessons we have learned, however, is that particles, strings, membranes, and in general p-Branes, are related to the ($p + 1$)-forms they are supposed to radiate, that is to say, membranes are charged as particles are charged, but where particles emit electromagnetic radiation, with 1-form potentials and 2-forms field strength, p-Branes radiate ($p + 1$)-potentials and there are also ($p + 2$) field intensities and even $D - (p + 2)$ duals, ending in $10 - p - 4$ "dual" or *magnetic* Branes; this interesting interplay is quite likely there to stay! For example, in 11-dimensional supergravity, there is a dual $M5$ Brane (Güven), which can be seen as the magnetic dual of the better known $M2$ Brane.

F-theory. In 1996 Cumrum Vafa came up with an extension of M-theory to 12 dimensions, which he called F-theory (F for father, or fundamental?); the idea is that the IIB theory, known to be *selfdual* under the strong limit (S-duality) contains a complex scalar field, $z = \chi + i \exp(\phi)$, where χ is the axion and ϕ the dilaton: this can be seen as a torus fibration from 12 dimensions:

$$T^2 \to \text{F-theory} \to \text{IIB theory} \tag{6.81}$$

as the *moduli space* of the torus T^2 is $\mathbb{C}$, the set of complex numbers.

There are several other arguments in favour of F-theory [1]. The matter content of F-theory is unclear; another confusing feature is that, at face value, F-theory works in $12 = (10, 2)$ dimensions, with all the associated problems related to causality, etc., for having two time-coordinates. For many people (including, at the beginning, Vafa himself!) F-theory was only a way to "track down" the complex field varying over the "surface" of the 10-dimensional IIB theory.

It is amusing that one can relate the particle content of 11-dimensional Sugra (and therefore the low-energy limit of M-theory) to the Moufang plane $\mathbb{O}P^2$ we mentioned before P. Ramond has shown that the three particles (h, Ψ and C) are related to the Euler number of $\mathbb{O}P^2$ being 3, and in fact some representations of F_4 give rise to a triplet of representations of Spin (9): in particular, the above triplet is related to the Id *irrep* of F_4. In fact, one of the putative particle contents of F-theory is related to the "complexification" $\mathbb{O}P_C^2 = E_6/[\text{Spin}(10) \times U(1)]$, but then there is a 27-plet, as $\chi(\mathbb{O}P_C^2) = 27$, and it is related to the fourth power of the primordial Susy doublet,

$$|8_v - \Delta_L|^4 \Rightarrow \text{ a 27-plet of particles.} \tag{6.82}$$

The particles are taken as representations of $O(10)$.

Recently (since 2008) Heckmann [10] has extended this theory considerably, where at the price of forgetting about gravitation one gets closer, it is hoped, to the standard model of particles and forces (see Section 6.4); in particular GUTs, chiral matter, Yukawa couplings, not to mention selected gauge groups, might appear possible, in principle. The development of this approach still goes on.

6.3.4 The Standard Model and its minimal supersymmetric extension

We remind the reader that all known forces are today described as gauge theories; in particular, the gauge group of the Standard Model (SM) is the product $SU(3)(\text{color}) \times SU(2)(\text{weak}) \times U(1)(\approx \text{electromagnetism})$. Besides the $8 + 3 + 1 = 12$ gauge bosons (three "swallowed" by Higgs), the "matter" lies in the bi-fundamental representations: quarks q and $\bar{q}$ and leptons, etc. Besides, there is hope for *God's particle* (a left-over Higgs) supposed (!) to be discovered in the next few years at the Large Hadron Collider at CERN.

We think we understand the gross features of the consequent picture: asymptotic freedom, confinement, radiative corrections in the electroweak sector, etc., but have no idea, for example, why the masses are what they are. We know (in this twenty-first century) that neutrinos have masses, but their type (Majorana?) is still unclear. In fact, the factor from the neutrino masses to the top quark mass approaches $10^{-14} \approx m_v/m_{\text{top}}$, beyond any reasonable calculation.

In spite of these shortcomings, people look beyond the SM. For some, super-symmetry is irresistible, and one looks for clues as to where and why. The most quoted argument (which I like, but it has still many detractors), is that the natural extrapolation of the three coupling constants g_s, g_{wk} and α_{em} by the renormalization group running (the Callan–Symanzik equation), make them cross at very high energies, $\approx 10^{16}$ GeV. It turns out that the "unique" crossing point of the three is much improved if one completes the SM model with the so-called Minimal Supersymmetric Standard Model, MSSM.

In this model there are supersymmetric partners for each particle, with strange names (gauginos: photino, gluino, Wino and Zino for the fermionic partners of gauge bosons, and s-quarks and s-leptons for the bosonic partners of the matter multiplets). Of course, in a way all this is "science fiction", as there is not the slightest experimental evidence favouring any of these, and the threshold for Susy partners approaches the TeV regime.

If supersymmetry is true, where is the scale to see the Susy partners? We do not know for sure, but they cannot be very high in energy, otherwise there are no arguments for them. Most people expect them (!) in the TeV range, which will probably be observed experimentally soon. Still, we would have to understand not only why Susy exists in first place, but also why it is broken badly, at (perhaps) the TeV scale.

6.4 Some issues on compactification

6.4.1 The general problem of compactification

We seem to live in three-plus-one dimensions (perhaps unbounded), whereas strings, M-theory, etc. live in higher dimensions, which we are not aware of. How do we cope with the "fact" of extra, unseen dimensions?

The simplest approach is through *compactification*, i.e. assuming the extra, unseen dimensions are "curled up" in a compact manifold, with dimensions well below, say, the nuclear dimensions 10^{-15} m. In principle one takes the direct product of spaces, for example, if we take flat 4-dimensional space–time,

$$M \approx K \times \text{(Minkowski space, } M = \mathbb{R}^{3,1}), \tag{6.83}$$

where K is a compact manifold with 6, 7 or 8 dimensions for strings, M- and F-theories.

What general properties do we expect K to satisfy, besides dimension(s) and compactness? We shall take it *oriented*, as we shall have to perform integrations, *spinable*, as we do see fermions, and whatever restrictions will make physics back in four dimensional acceptable, either the Standard Model, some GUT approximation, or just MSSM. That will depend, of course, on what physics we start from in the higher dimension, but also, as we have learned lately, the new physics generated by the compactification process: we shall see, for example, that new gauge groups might appear through potential singularities of the K space, or by wrapping on it some extended objects appearing in the total space, M. To maintain $\mathcal{N} = 1$ supersymmetry down to four dimensions will be a condition that recurs often.

6.4.2 Compactification from strings: Calabi–Yau 3-folds

This was historically the first realistic (but unsuccessful) attempt to rescue the SM from the (closed, $\mathcal{N} = 1$) Heterotic Exceptional (E_8^2) string (Gross–Strominger–Candelas–Witten, 1985). The two E_8 groups were very tempting as a starting point to get the GUT group: one E_8 remained hidden, the other spat a $SU(3)$ factor to become E_6, one of the few possible GUT groups (two others were the original $SU(5)$ (Georgi–Glashow, 1974) and $SO(10)$ (Georgi–Fritz–Minkowski, 1976)); E_6 was proposed by F. Gürsey and P. Ramond (1976). In addition, we want to preserve $\mathcal{N} = 1$ Susy in four dimensions (explained below).

In ten dimensions the supercharge $\mathcal{N} = 1$ algebra was fixed by

$$\{Q, Q\} = P_M + \text{"Rest"}, \tag{6.84}$$

$$\dim Q = 2^{10/2-1} = 16, \quad 16 \cdot 17/2 = 136 = 10 + 126. \tag{6.85}$$

We impose the constraint of only $\mathcal{N} = 1$ Susy in four dimensions, which corresponds to four supercharges Q. This is *phenomenological*: we hope that there is some Susy, as we expect supersymmetric breaking to occur close to the 1 TeV scale, and $\mathcal{N} > 1$ "down to earth" means no parity violation ($+1/2 - 1/2 - 1/2 = -1/2$ in helicities, so *both* chiralities would appear for $\mathcal{N} = 2$); but parity violation is a conspicuous feature of nature, which we want to preserve, even with supersymmetry. That means we should break $3/4 = (16_{ini} - 4_{fin})/16$ supercharges. How do we achieve this?

It turns out that $\mathcal{N} = 1$ Susy down to four dimensions is maintained if the compactifying manifold has $SU(3)$ holonomy, so it is, by definition, a CY_3 or Calabi–Yau 3-fold. To see this, recall the identity $SU(4) = \text{Spin}(6)$ (stated in Section 6.2). Now under $SU(3) \subset SU(4)$, the fundamental spinor splits obviously as $4 = 3 + 1$. Hence, the trivial "1" survives, and guarantees four supercharges (four Qs) in four dimensions, so $\mathcal{N} = 1$, as desired, as

$$\dim Q(\text{4-dim}) = 2^{4/2-1} = 2 \text{ complex or 1 Majorana spinor.} \tag{6.86}$$

One can show directly that compactification in a CY_k preserves $1/2^{k-1}$ super-symmetries, so here we have $16/4 = 4$. Here 16 means $\mathcal{N} = 1$ in 10 dimensions, whereas 4 in four dimensions means also $\mathcal{N} = 1$.

The search for CY_3 was an active industry in Austin, Texas, during 1986–90, led by Philip Candelas (who now holds the Rouse Ball Professorship at Oxford). As a by-product, his group discovered mirror symmetry! We shall explain this.

The Hodge diamond for a CY_3 manifold is

$$
\begin{array}{ccccccc}
& & & 1 & & & \\
& & 0 & & 0 & & \\
& h_{2,0} & & h_{1,1} & & h_{0,1} & \\
h_{3,0} & & h_{2,1} & & h_{1,2} & & h_{0,3}
\end{array}
\tag{6.87}
$$

plus Poincaré dual.

Mirror symmetry in this CY_3 context is the exchange $h_{1,1} \leftrightarrow h_{2,1}$. It turns out that the physics in mirror is the same; the two Hodge numbers interchange some complex structure with some Kähler structures, and this has been a big advance in algebraic geometry (we do not elaborate on this here). In an arbitrary CY–k-fold, it is the exchange $h_{1,1} \leftrightarrow h_{k-1,1}$; hence, for example, the $K3$ manifold is self-mirror, as $h_{1,1} \equiv h_{2-1,1}$. Again, this has some consequences.

6.4.3 Compactification from M-theory: manifolds of G_2 holonomy

After the 1995 revolution brought about by the discovery of (still incompletely known) M-theory (Witten, Townsend, Polchinski) it was obvious that we have to face 11 dimensions. The supercharges and particle content (in the low-energy limit) were

$$
\dim Q : 2^{(11-1)/2} = 32, \quad \{Q, Q\} = P_\mu + M2 + M5
$$
$$
32 \cdot 33/2 = 528 = 11 + 55 + 462,
\tag{6.88}
$$

which make even more sense than 10 dimensions: besides the linear momentum (P_μ), the theory describes the $p = 2$ membrane $M2$ (already alluded to; the viewpoint of Townsend and Duff vs. Witten), and a dual, the magnetic 5-brane ($2 \rightarrow 3 \rightarrow 4 \rightarrow 11 - 4 = 7 \rightarrow 6 \rightarrow 5$).

The seven-dimensional compactifying manifold has to be oriented and spinable, of course, and should also leave a spinor covariant constant, so as to have, again, $\mathcal{N} = 1$ down to our mundane four dimensions; the natural spin group would be Spin (7), with a real *irrep* of dim $8 = 2^{(7-1)/2}$. But, happily, it has the subgroup G_2, with the natural *irrep* of dim 7 (also explained above), so G_2 acting in eight dimensions splits as $8 = 7 + 1$, just as we want! This was first emphasized by Townsend and Papadopoulos in 1996. It reinforces the idea that octonions and related objects should play a role in our description of nature.

The search for manifolds with G_2 holonomy is a hard one, as the dimension is odd so we have no resort to complex analysis; some examples were worked out in [5].

As M-theory is incompletely known, there has been not much progress with this G_2 holonomy idea; in any case, in a way M-theory was superseded, I believe, by the F-theory of Cumrum Vafa, after the original work (ca. 1996) was reincarnated in his 2008 version; to this we now turn.

6.4.4 Compactification from F-theory: Vafa's theory (1996, 2008)

We have already discussed F-theory in Section 6.3. With $12 = (10, 2)$ signature, the type of spinor (or supercharge) is still real ($10 - 2 \equiv 0$ mod 8). The number of supercharges is

$$\#Q(\text{12-dim}) = 2^{12/2-1} = 32, \ \text{real type}, \tag{6.89}$$

with the superalgebra

$$\{Q, Q\} = J_{MN} + F^{\pm}$$
$$528 = \binom{12}{2} + \binom{12}{6}/2 = 66 + 462. \tag{6.90}$$

In particular, *no conventional supergravity is possible*, as P_M does not appear in the anticommutator (recall we already said 11 was the maximum dimension for (conventional) supergravity); however, some interpretation can be given. F-theory has other worries: for example, the two times are at odds with conventional causality; some exit avenues have been proposed (e.g. I. Bars proposed to "gauge" one of the times). We shall just address the compactification problem, following the "reincarnation" of the theory by Vafa's group since 2008 [10].

The first idea is to forget about gravitation, and try to understand some of the features of the Standard Model, like GUTs, presence of chiral matter, obtaining the Yukawa couplings, etc., and leave for the future the connection with the more conventional high-dimension theories in relation to gravitation and a "TOE" theory; Vafa calls this the "bottom-up approach".

The most promising compactification approach has two steps, the first in some sense divided into another two:

$$\text{first step: 12 to 8 dimensions, in a } K3 \text{ manifold,} \qquad (6.91)$$

$$\text{second step: 8 to 4 dimensions in a del Pezzo surface.} \qquad (6.92)$$

We explain first the new real four-dimensional manifolds.

$K3$ *complex surface.* The complex surface $K3$ is, topologicaly speaking, the only nontrivial Calabi–Yau 2-fold, i.e. with holonomy $SU(2)$. K stands for the Karakorum high sierra in the Himalayas, and "3" for Kummer, Kähler and Kodaira; the name is due to André Weil, 1954, at the time of the first Mount Everest climb (May 1953). As a complex surface, it has a Hodge diamond as follows:

$$
\begin{array}{ccccc}
 & & 1 & & \\
 & 0 & & 0 & \\
1 & & 20 & & 1
\end{array}
\qquad (6.93)
$$

plus Poincaré dual.

So it is simply connected, the Euler number is $4 \cdot 1 + 20 = 24$, and signature is $22 = (19, 3)$. [Digression: any four-dimensional manifold, if compact and oriented, has another number besides Euler's χ: it is called Hirzebruch's signture τ, and it is the Sylvester signature of the quadratic form given by the *wedge* (although commutative) product of 2-forms, as $\omega_1^{(2)} \wedge \omega_2^{(2)} = (\text{real\#})\text{volume}$.]

Thus $\tau = 16 = 19 - 3$, and miraculously it happens, as $16 = 2 \cdot 8$, that two E_8 groups (or more precisely, singularities giving rise to groups, by the Thom–Arnold procedure; we do not elaborate here) lurk behind $K3$ ($K3$ is easily obtained from the 4-torus $T^4 \approx (S^1)^4$) by "orbifolding" and blow-up, that is, by quotienting by some fix-point group and curing the singularities (see e.g. Aspinwall, arXiv hep-th 94 94 151).

del Pezzo surfaces. The complex projective space $\mathbb{C}P^n$ is the set of (complex) lines in $\mathbb{C}^{n+1}$. We have $\mathbb{C}P^1 \approx S^2$, and $\mathbb{C}P^2$ is the symmetric space $SU(3)/U(2)$; it is a complex, simply connected Kähler manifold, with Hodge diamond

$$
\begin{array}{ccccc}
 & & 1 & & \\
 & 0 & & 0 & \\
 & 0 & 1 & 0 &
\end{array}
\qquad (6.94)
$$

plus Poincaré dual.

Hence $\chi = 3$, and it does *not* admit a spin structure. It is perhaps the simplest four-dimensional real manifold, after S^4 (which is *not* complex). del Pezzo surfaces are obtained from $\mathbb{C}P^2$ after blowing up k points where $0 \leq k \leq 8$.

Descent from 12 to 8: Gauge groups appear! Recall F-theory as an elliptic fibration over IIB theory by the 2-torus.

Exercises

6.1 Consider the flow in $\mathbb{R}^2$ given by the vector field

$$X = x\frac{\partial}{\partial y} - y\frac{\partial}{\partial x}.$$

Find the integral curves. Is there any fixed point? If so, why?

6.2 Find the de Rham cohomology of the circle S^1.

6.3 Prove the following theorem:

If $\xi : P(M, G)$ is a principal bundle $G \to P \to M$ and $H \subset G$ a subgroup, with $X := G/H$ as a coset space, then X *reduces* to $\xi' : P'(M, H)$ if and only if the associated bundle $E(M, X)$ is trivial (i.e. direct product).

6.4 Consider a g-genus surface $S = \sum_g$ with Gauss curvature K. Find expressions for the Ricci and the Riemann curvatures as functions of K.

6.5 The metric g of a general surface can be given in general $(x^1 = u, x^2 = v)$ as:

$$ds^2 = g_{ij}dx^i dx^j = E(u, v)du^2 + 2F(u, v)dudv + G(u, v)dv^2.$$

Find: (i) a geometric construction to express g in so-called *geodesic* coordinates, so that $ds^2 = dx^2 + H(x, y)dy^2$, and (ii) an algebraic way to go from these geodesic coordinates to *isothermal* coordinates $ds^2 = e^{2\sigma(u,v)}(du^2 + dv^2)$.

6.6 Find the expression for Gauss' curvature for the geodesic and for the isothermal form of the metric.

6.7 Find the curvature for the Poincaré plane, that is, the upper-half complex plane with metric

$$ds^2 = \frac{dx^2 + dy^2}{y^2}.$$

6.8 Consider the bundles over the 2-*sphere* S^2 with group $SO(2)$. Show that they are classified by the integers $n \in \mathbb{Z}$. Identify the first three classes, $n = 0, 1, 2$.

6.9 Prove that $\mathrm{Spin}(3) = SU(2)$, $\mathrm{Spin}(4) = [\mathrm{Spin}(3)]^2$, $\mathrm{Spin}(5) = \mathrm{Sq}(2)$ and $\mathrm{Spin}(6) = SU(4)$ from the form of the Dynkin diagrams.

6.10 The Dynkin diagram D_4 exhibits an ostensible triality, that is, the group S_3 permuting the three external balls. Prove that this symmetry is clear already in the center of the Spin group.

6.11 From the "exact homotopy sequence" find the condition: a manifold is a Spin manifold if and only if $w_2 = 0$.

6.12 The Cotton tensor is a rank 3 covariant tensor with the symmetry type $(2, 1)$ and it is traceless. Find its dimension for a manifold of dimension n.

References

[1] L. J. Boya, Arguments for F-theory, *Mod. Phys. Lett.* A **21**, 1–18, 2006.

[2] J. H. Conway, D. A. Smith, *On Quaternions and Octonions*, A. K. Peters, 2003.

[3] M. B. Green, J. H. Schwarz, E. Witten, *Superstring Theory*, Cambridge University Press, 1985.

[4] S. Helgason, *Differential Geometry and Symmetric Spaces*, 1st edn, Academic Press, 1962.

[5] D. Joyce, *Compact Manifolds of Special Holonomy*, Oxford University Press, 2002.

[6] S. Kobayashi, K. Nomizu, *Foundations of Differential Geometry*, J. Wiley, 1963 (Vol. 1), 1969 (Vol. 2).

[7] J. W. Milnor, J. D. Stasheff, *Characteristic Classes*, Princeton University Press, 1974.

[8] M. Nakahara, *Geometry, Topology and Physics*, Institute of Physics, Bristol, 2005.

[9] N. Steenrod, *Topology of Fiber Bundles*, Princeton University Press, 1951.

[10] J. J. Heckmann, C. Vafa, F-Theory and GUTs, arXiv:0809.1098(hep-th).

[11] P. West, *Introduction to Supersymmetry and Supergravity*, World Scientific, 1990.

7

Geometric aspects of the Standard Model and the mysteries of matter

FLORIAN SCHECK

Abstract

The basic structure of gauge theories of fundamental interactions distinguishes *radiation* from *matter.* Radiation is described by Yang–Mills theories, matter particles (i.e. quarks and leptons) are described by a Dirac operator which contains the full complexity of their classification and state mixing. In quantum field theory the two categories intermingle. While the construction of Yang–Mills theories, in essence, is a classical one, the phenomenon of spontaneous symmetry breaking exhibits facets which are linked to the quantum symmetries of gauge theories. Furthermore, noncommutative geometry offers new routes to the standard model of electroweak interactions and reveals some of its otherwise mysterious structure.

We review these matters with regard to both their phenomenology and their theoretical and geometric background. Many examples are given and exercises are provided which illustrate some of the main results.

7.1 Radiation and matter in gauge theories and General Relativity

The basic structure of gauge theories seems to distinguish *radiation* from *matter* as two categories of different origin. The massive and massless vector or tensor bosons, the photon, the $W^{\pm}$- and Z^0-bosons, the gluons, and the graviton, respectively, which are the carriers of the fundamental forces, belong to what may be termed *radiation.* Here we allude to the analogy to (quantum) electrodynamics described by Maxwell's equations and to Einstein's equations for General Relativity (GR). They are described by geometric theories, i.e. Yang–Mills (YM) theories or, in the

Geometric and Topological Methods for Quantum Field Theory, ed. Alexander Cardona, Iván Contreras and Andrés F. Reyes-Lega. Published by Cambridge University Press.

case of GR, by semi-Riemannian geometry in dimension 4. To a large extent, they are *classical* theories.

Matter, i.e. quarks and leptons and composites thereof, *a priori*, seems to belong to a different kind of physics which, at first sight, does not exhibit an underlying geometrical structure. While the equations of motion of YM theories and of GR, taken in isolation, describe nontrivial physics, matter cannot "live on its own" without the gauge bosons that mediate the fundamental interactions – unless one is satisfied with a theory of free particles which remains untestable in experiment.

As soon as one enters the quantum world, however, the two categories, like two rivers, start mingling their waters:

(A) The Higgs particle, as a most prominent example of current interest in particle physics, plays a rather enigmatic role. Its phenomenological role in providing mass terms for some of the vector bosons and for the fermions of the theory suggests that it be another form of "matter", beyond ordinary matter made up of quarks and leptons. Models based on noncommutative geometry, in turn, classify the Higgs field in the generalized YM connection, besides the gauge bosons, and hence declare it to be part of "radiation". As such, it generates parallel transport between universes which are separated by a discrete distance.

(B) The requirement of *renormalizability* of quantum gauge theories with massive vector bosons entails the introduction of so-called Stückelberg (scalar) fields whose place in the classification needs to be clarified.

(C) Quarks and leptons are described by a *Dirac operator* which in its mass sector exhibits a significant, though mysterious, structure. Dirac operators, in turn, are the driving vehicles in constructing noncommutative geometries designed to generalize YM theory – and to describe the Standard Model! They act on the Hilbert space which is spanned by the myriad of quark and lepton states.

(D) Quantization of YM theories is possible only in the absence of *anomalies*. These, in turn, depend on the classification of the matter particles with respect to the structure group. In the minimal electroweak Standard Model, for instance, it needs a conspiracy between the three generations of quarks and leptons to render it renormalizable.

(E) As emphasized particularly by Scharf and his group, the requirement of BRST (Becchi, Rouet, Stora, Tyutin) invariance of the underlying Lagrangian fixes much of its structure and, thereby, intertwines the radiation sector and the matter sector.

(F) Last but not least, the currents and charges of matter or its energy-momentum tensor act as the sources in the equations of motion of YM and gravitational fields, respectively.

In these lecture notes we work out several of the themes alluded to above, both by way of construction and by means of instructive examples. We start with a schematic description of YM theories including spontaneous symmetry breaking (SSB) within the classical geometric framework, and including matter particles. In a first excursion to quantum field theory we describe the stratification of the space of connections and its relevance to anomalies. In order to clarify the phenomenological basis on which YM theories of fundamental interactions are built, we describe some of the most pertinent phenomenological features of leptons and of quarks. Via the Dirac operator describing leptons and quarks, we turn to constructions of the Standard Model in the framework of noncommutative geometry. This, in turn, leads us to a closer analysis of the mass sector and state mixing phenomena of fermions. The intricacies of quantization are illustrated by a semi-realistic model for massive and massless vector bosons.

7.1.1 Schematic construction of gauge theories

The backbone of a gauge model is a *structure group* G which is taken to be simple or semi-simple. For physical reasons *compactness* of G is essential, but why? (Exercise 7.1). For instance, the electroweak sector of the minimal Standard Model (SM) is based on

$$G_{ew} = U(2) = \left\{ \mathbf{U} \in M_2(\mathbb{C}) |\, \mathbf{U}^\dagger \mathbf{U} = \mathbb{1} \right\},$$

which splits into the SU(2) of what we call weak isospin, and the U(1) of weak hypercharge. As is well known, the full SM is built on the structure group

$$G = U(2) \times SU(3),$$

with the SU(3) of colour interactions included. The model is formulated on a principal fibre bundle

$$\mathcal{P} = \left(P \xrightarrow{\pi} M, G \right),$$

where, barring relativity for the moment, $M = \mathbb{R}^{(1,3)}$ is flat four-dimensional Minkowski space. The structure group G is then replaced by the *gauge group* $\mathfrak{G}$. In geometrical terms, this is the group of automorphisms of the principal bundle $\mathcal{P}$ which commutes with the right action R_g of G and which maps every fibre onto itself, namely

$$\Psi \in \mathfrak{G}, \quad \Psi : \mathcal{P} \to \mathcal{P}, \tag{7.1a}$$

$$\pi\left(\Psi(z)\right) = \pi(z), \quad \Psi(z \cdot g) = \Psi(z) \cdot g. \tag{7.1b}$$

As Ψ acts on fibres only, one has

$$\Psi(z) = z\gamma(z), \quad z \in P, \ \Psi \in \mathfrak{G}, \tag{7.2a}$$

$$\gamma : P \to G : z \mapsto \gamma(z), \quad \gamma(z \cdot g) = \mathrm{Ad}\, g^{-1}(\gamma(z)) = g^{-1}\gamma(z)g. \tag{7.2b}$$

Thus, the gauge group $\mathfrak{G}$ can be identified with the set of maps $\gamma : P \to G$, so that $(\gamma\gamma')(z) = \gamma(z)\gamma'(z)$.

The connection form $A \in \Omega^1(P, \mathfrak{g})$ takes values in the Lie algebra Lie $G = \mathfrak{g}$. Its relation to what is called *gauge potential* in physics is affected by local sections

$$\sigma_i : U_i \subset M \to P,$$

such that

$$A^{(\sigma_i)} = \sigma_i^*\omega, \quad A^{(\sigma_i)} \in \Omega^1(U_i, \mathfrak{g}), \ U_i \subset M.$$

Suppose $\{U_i\}$ is a covering of space–time M, and $\{(\varphi_i, U_i)\}$ are the charts of an atlas describing M. The connection as a whole is then obtained in the usual manner by joining the chart representations over a complete atlas.

Radiation

On Minkowski space which is flat and simply connected, the construction of the connection form reduces to[1] the definition

$$A := iq \sum_{k=1}^{N} A^{(k)}\mathbf{T}_k, \quad N = \dim \mathfrak{g}, \tag{7.3a}$$

where the operators $\mathbf{T}_k$ are the generators of G, q is a generalized "charge", and $A^{(k)}$ are one-forms on $M = \mathbb{R}^{(1,3)}$,

$$A^{(k)} = A^{(k)}_\mu(x)\, dx^\mu, \quad k = 1, 2, \ldots, N. \tag{7.3b}$$

The functions $A^{(k)}_\mu(x)$ are components of the gauge fields, linear combinations of which will describe the massive or massless vector fields of the SM. For example, if $A^{(1)}_\mu(x)$ and $A^{(2)}_\mu(x)$ denote the coefficients of the generators $\mathbf{T}_1$ and $\mathbf{T}_2$ of SU(2), the physical W-fields that mediate weak charged current interactions are given by

$$W^{(\pm)}_\mu(x) = \tfrac{1}{\sqrt{2}}\left(A^{(1)}_\mu(x) \pm iA^{(2)}_\mu(x)\right).$$

In Exercise 7.2 one is invited to verify that A is indeed the vehicle which is needed to perform *parallel transport* on the principal bundle.

---

[1] The Lie algebra valued one-forms $A^{(k)}$ are real forms. The optional factor i in the definition (7.3a) renders the operator A Hermitian. This is useful in view of the hermiticity of Lagrangians or actions in physics.

From here on the construction is standard: local gauge transformations on A are

$$A \longmapsto A' = gAg^{-1} + g\mathrm{d}(g^{-1}), \quad g \in \mathfrak{G}. \tag{7.4}$$

The two pieces are seen to be a transformation by "conjugation", well-known, for example, from quantum mechanics, and a genuine local gauge transformation familiar from a U(1) theory such as Maxwell theory. Formally speaking, (7.4) is inhomogeneous and looks like a (generalized) affine transformation. The one-form A serves to construct the covariant derivative (cf. Exercise 7.2)

$$D_A = \mathrm{d} + A, \tag{7.5}$$

which reads, when applied to some (multiplet of) scalar field(s),

$$\partial_\mu \Phi(x) \to \left\{ \mathbb{1} \partial_\mu + iq \sum_{k=1}^{N} A_\mu^{(k)}(x) \mathbf{T}_k \right\} \Phi(x).$$

With respect to gauge transformations the covariant derivative transforms by conjugation only, $D_{A'} = g D_A g^{-1}$, there is no inhomogeneous term as for A. This is important for physics: a term such as $(D_A \Phi, D_A \Phi)$, where the brackets denote a G-invariant scalar product, is automatically invariant under *local* gauge transformations as well. In other terms, the brackets ensure *global* gauge invariance with respect to the structure group G, the covariant derivatives guarantee *local* gauge invariance with respect to the gauge group $\mathfrak{G}$.

The curvature two-form pertaining to the connection A is given by

$$F := D_A^2 = (\mathrm{d}A) + A \wedge A. \tag{7.6}$$

In contrast to D_A itself the action of D_A^2 is *linear* (Exercise 7.3). The analogues of the field strength tensor of Maxwell theory (familiar to physicists) are unveiled in the decomposition

$$F := iq \sum_{k=1}^{N} \mathbf{T}_k \sum_{\mu < \nu} F_{\mu\nu}^{(k)}(x) \, \mathrm{d}x^\mu \wedge \mathrm{d}x^\nu \tag{7.7a}$$

in terms of ordinary, antisymmetric tensor fields $F_{\mu\nu}^{(k)}(x)$ and the base two-forms $\mathrm{d}x^\mu \wedge \mathrm{d}x^\nu$ on Minkowski space. If these tensor fields are decomposed in terms of the component fields $A_\mu^{(k)}(x)$ a new feature appears compared with Maxwell: the field strengths are no longer linear in the A_μ fields,

$$F_{\mu\nu}^{(k)}(x) = \partial_\mu A_\nu^{(k)}(x) - \partial_\nu A_\mu^{(k)}(x) - q \sum_{m,n=1}^{N} C_{kmn} A_\mu^{(m)}(x) A_\nu^{(n)}(x). \tag{7.7b}$$

Obviously, under a local gauge transformation F, Eqn (7.6), transforms by conjugation, $F' = gFg^{-1}$. The quadratic terms in (7.7b) which contain the structure constants C_{kmn} of the structure group, upon squaring F, lead to cubic and quartic interactions among the gauge fields. Indeed, the YM Lagrangian, which must be a local invariant with respect to $\mathfrak{G}$, has the form

$$\mathcal{L}_{YM} = -\frac{1}{4q^2\kappa^{(ad)}} \, \text{tr} \left(F_{\mu\nu} F^{\mu\nu}\right), \tag{7.8}$$

where $\kappa^{(ad)}$ is the normalization of the trace

$$\text{tr} \left(\mathbf{T}_i \mathbf{T}_j\right) = \kappa \, \delta_{ij} \tag{7.9}$$

in the adjoint representation (Exercise 7.4).

The result (7.8) shows that a world containing vector bosons only contains nontrivial physics. Suppose, for example, that the Lagrangian (7.8) describes the photon γ, the Z^0-boson and the charged $W^\pm$-bosons. (Of course, a mechanism is needed, in addition, which renders some of these massive and thereby defines the specific linear combinations of neutral fields which describe the photon and the Z^0.) The Lagrangian (7.8) produces well-defined coupling terms between these particles. For instance, the $WW\gamma$-vertex fixes the anomalous magnetic moment of the W^+ and W^- which, at least in principle, can be measured in scattering experiments. Other, perhaps more prominent and better known examples are provided by the three- and four-gluon vertices of quantum chromodynamics (QCD) which are tested in jet dynamics at e^+e^--colliders.

Matter

While the principle of local gauge invariance based on a group G fixes the structure of the *vector boson sector* to a large extent, the introduction of scalar or fermionic *matter fields* leaves much more freedom of choice. As a first and well-known example, consider the original Higgs mechanism applied to the electroweak sector of the SM, based on the structure group $G = \text{U}(2) \approx \text{U}(1) \times \text{SU}(2)$. The aim is to hide the original symmetry group G in favour of the residual symmetry H of Maxwell theory

$$G = \text{U}_Y(1) \times \text{SU}(2) \longrightarrow H = \text{U}_{em}(1) \tag{7.10}$$

by spontaneous symmetry breaking (SSB), where the $\text{U}_{em}(1)$ is an appropriate linear combination of the original $\text{U}_Y(1)$ and the one-parameter subgroup generated by the operator $\mathbf{T}_3$ of SU(2). For that purpose one introduces a multiplet of scalar fields Φ, classified by quantum numbers y and (t, t_3) of weak hypercharge $\text{U}(1)_Y$ and weak isospin SU(2), respectively, as well as a G-invariant potential $V(\Phi)$ which exhibits a degenerate minimum at some nonvanishing value ϕ^0. The freedom of choice is reflected by the fact that the multiplet Φ can sit in almost any multiplet of weak isospin $t \geq \frac{1}{2}$ provided it contains one substate $(y, t, t_3^{(H)})$ which is electrically neutral such that it develops a vacuum expectation value $v = \sqrt{\langle\phi^0, \phi^0\rangle} \neq 0$. The weak hypercharge of "the" Higgs scalar is $y = 1$, its weak isospin is $(t, t_3^{(H)}) = (\frac{1}{2}, -\frac{1}{2})$ so that its electric charge $Q = t_3 + \frac{1}{2}y$ vanishes. However, there is nothing up to this point that would tell us that the Higgs lives in a doublet with respect to SU(2). All one can deduce from the construction of the minimal electroweak SM is the relation

$$\frac{m_W^2}{m_Z^2 \cos^2 \theta_W} = \frac{t(t+1) - t_3^2}{2t_3^2}, \tag{7.11a}$$

which contains three experimental numbers on its left-hand side:[2] the masses of W and Z, and the squared cosine of the Weinberg angle θ_W. The derivation of the ratio (7.11a) is given in the appendix to this chapter. Strangely enough, it is experiment that tells us that the ratio on the left-hand side of (7.11a) is equal to 1 within a very small error bar,

$$\left.\frac{m_W^2}{m_Z^2 \cos^2 \theta_W}\right|_{\exp} = 1.0004 \, {}^{+0.0008}_{-0.0004}, \tag{7.11b}$$

a value which singles out the doublet case. This reminds one of earlier days of nuclear spectroscopy when people determined spins and parities of nuclear excited states by angular correlations. Clearly, it would be more satisfactory if this assignment were a prediction! In fact, as we shall discuss below, extensions of YM theories within noncommutative geometry put the Higgs into the radiation sector and fix its quantum numbers to the values favoured by (7.11b).

The introduction of fermionic fields such as those describing quarks and leptons also follows standard rules. As in the example discussed above, there is much, and in fact too much, freedom of choice for the corresponding Dirac operator(s). Let $\Psi(x)$ be a set of fermionic fields classified by a reducible or irreducible representation of G. For definiteness, think of the weak isospin and QCD part of the SM, $G = $ SU(2) × SU(3). The U(1) part is always more problematic than the non-Abelian factors because it escapes universality of couplings (Exercise 7.6). A (still classical) Lagrangian including a scalar multiplet of fields Φ and a fermionic multiplet Ψ which is both globally and locally gauge invariant will have the form

$$\mathcal{L} = -\frac{1}{4q^2\kappa^{(\mathrm{ad})}} \, \mathrm{tr} \, \left(F_{\mu\nu}F^{\mu\nu}\right) + \tfrac{1}{2}\left(D_A\Phi, D_A\Phi\right) - V(\Phi)$$

$$+ \tfrac{i}{2}\left(\overline{\Psi}, \gamma \cdot \overset{\leftrightarrow}{D}_A \Psi\right) - \left(\overline{\Psi}, (\mathbf{M} + \varrho\Phi)\Psi\right). \tag{7.12}$$

---

[2] I am skipping a discussion of radiative corrections which appear in the left-hand expression of (7.11a) if one uses bare values for the input parameters. This is a standard topic in the discussion of the minimal SM and the rules for including them are well known; see, for example, the review on the electroweak model [1], p. 125.

Here, as before, D_A denotes the covariant derivative so that, for example, $\gamma \cdot D_A$ stands for $\gamma^\mu D_\mu(A)$, and $V(\Phi)$ is the Higgs potential,

$$V(\Phi) = \tfrac{\lambda}{4}\left\{(\Phi, \Phi) - v^2\right\}^2 + \text{const.} \tag{7.13}$$

(The left–right action of the derivative means $f \overset{\leftrightarrow}{\partial}_\mu g = f(\partial_\mu g) - (\partial_\mu f)g$.)

The last term in (7.12) is not as innocent as it looks at a first glance. The term **M** contains the mass matrices of quarks and of leptons, of unknown origin, which are certainly not diagonal in the base states of the representation Ψ of G. The term proportional to the real number ϱ is a Yukawa coupling of the fermions to the Higgs field which will contribute to the fermion masses through the one component ϕ^0 which develops a vacuum expectation value. Of course, either of these terms can be present only if the factors composing them join in a G-invariant manner. For instance, in the minimal electroweak SM and with one generation of leptons one has

$$\Psi(x) = \begin{bmatrix} L(x) \\ R(x) \end{bmatrix}, \quad \text{where } L(x) = \begin{bmatrix} \nu_L(x) \\ e_L(x) \end{bmatrix} \text{ and } R(x) = \begin{bmatrix} e_R(x) \end{bmatrix}$$

are a left-handed doublet with $y = -1$ and a right-handed singlet with $y = -2$. (Verify the electric charges!) As there is no way of constructing a mass term $(\bar{\Psi}, \mathbf{M}\Psi)$ invariant with respect to G, one *must* rely on the Yukawa coupling for giving the electron a mass. This implies that the parameter ϱ, for every fermion, must be tuned such as to yield the empirical mass.

7.1.2 Mysteries of leptonic interactions

Both the world of quarks and the world of leptons contain a great deal of inner structure which is accessible in experiment but is not understood by any means. Among a longer list of mysteries in this realm we discuss two aspects of special relevance to geometric theories of particle physics.

Chiralities in weak charged-current interactions

One of the great mysteries of lepton physics is the observation that charged weak interactions couple to purely *left-chiral* states only. As long as the three neutrinos were thought to be strictly massless, it was more or less plausible that only neutrino states with *negative* helicity, and antineutrino states with *positive* helicity participated in weak interactions, while states with the opposite helicity did not couple to anything. Helicity plus or minus one-half is an invariant characterization only if the fermion is massless. Furthermore, the occurrence of one state of helicity only is linked to the observed *maximal parity violation* in weak interactions.

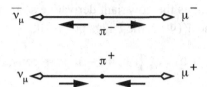

Figure 7.1 Weak decays of π^- (π^+) into μ^- (μ^+) and $\overline{\nu_\mu}$ (ν_μ).

For comparison, consider the interaction of photons with charged particles. A good example is the production process

$$e^+ + e^- \longrightarrow \mu^+ + \mu^-$$

via annihilation of the e^+e^--pair into a virtual photon, and the creation of the $\mu^+\mu^-$-pair by annihilation of the same photon. Suppose the colliding electron and positron beams are unpolarized but the orientation of the spins of μ^+ and μ^- along the momenta or opposite to them are recorded. The spin selection rules (Exercise 7.7) tell us that the chiralities of the positive and the negative muon are correlated: if the μ^+ is right-chiral then the μ^- is left-chiral, but if the μ^+ is left, the μ^- is right. Electromagnetic interactions are strictly parity-conserving, that is to say the two chirality constellations couple with the same strength to the intermediate photon state. As a result, the emerging $\mu^+\mu^-$-pair, like the incident e^+e^--pair, will be found to be unpolarized. In fact, the analogous process with τ-leptons,

$$e^+ + e^- \longrightarrow \tau^+ + \tau^-,$$

is used to produce beams of polarized τ^- by selecting one definite chirality state of its partner τ^+.

In the meantime we have learnt that at least some of the neutrinos do have nonvanishing masses, and, therefore, one might expect that handedness no longer plays a fundamental role. In other terms, models which contain some right-handed weak interactions, besides the dominant left-handed ones, such as left–right symmetric models with appropriate SSB, might describe reality. That this is not so is demonstrated by a beautiful determination of the chirality of the $\overline{\nu_\mu}$. This striking example goes as follows [2]. Muon beams usually stem from charged pion decay, $\pi^- \to \mu^- + \overline{\nu_\mu}$, as illustrated by Fig. 7.1 (see also Exercise 7.8). As this decay is mediated by weak charged interactions which are parity violating, the muon is expected to be longitudinally polarized. Let P_μ be its degree of longitudinal polarization. The negatively charged muon decays predominantly via the process $\mu^- \to e^- \nu_\mu \overline{\nu_e}$. The decay asymmetry of the electron with respect to the muon spin and close to the upper end of the spectrum is calculated to be

$$\left(\frac{d^2\Gamma}{dx\,d\cos\theta}\right)\bigg|_{x\to1} = \frac{m_\mu^5 G_F^2}{144\pi^3}\varrho\left\{1 - P_\mu\frac{\xi\delta}{\varrho}\cos\theta\right\}, \tag{7.14a}$$

where $x = E/E_{\max}$, θ is the angle between the muon spin and the electron momentum, while ϱ, ξ and δ are real parameters which are calculated from the couplings in a general weak interaction Lagrangian [3] but whose explicit form is of no relevance here. Obviously, the absolute value $|P_\mu\xi\delta/\varrho|$ cannot exceed the value 1. The experimental result of Jodidio *et al.* [4]

$$P_\mu\frac{\xi\delta}{\varrho} = 0.9989 \pm 0.0023 \tag{7.14b}$$

is found to be very close to its maximal value 1. By definition, the longitudinal polarization P_μ cannot be larger than 1. Furthermore, from the definitions in terms of coupling constants, the quantity $(\xi\delta/\varrho)$ is smaller than or equal to 1. One concludes that both P_μ and the combination $\xi\delta/\varrho$ must each lie very close to 1. The result for the former has an immediate consequence for the chirality of the $\overline{\nu_\mu}$ of the antineutrino emitted in pion decay, by conservation of angular momentum. As worked out by Fetscher [2], the experimental result (7.14b) implies

$$1 - |h(\overline{\nu_\mu})| < 0.0032 \quad \text{at 90\% CL,} \tag{7.14c}$$

where CL is the confidence level. The signs of $h(\overline{\nu_\mu})$ and $h(\nu_\mu)$, which are opposite of each other, are known from experiment. Therefore, the result (7.14c) is convertible to the information

$$h(\overline{\nu_\mu}) = +1 \quad \text{and} \quad h(\nu_\mu) = -1, \tag{7.15}$$

within very small error bars. This is, by far, the most accurate determination of a neutrino chirality.

Family numbers and selection rules

Another mystery is the separate additive conservation of individual lepton family numbers L_e, L_μ and L_τ. Soon after the minimal SM of electroweak interactions was developed one realized that it could easily accommodate any number of copies of the (e, ν_e) family as well as arbitrary amounts of state mixing between members of different families. The same statement applies to the quark sector (see below). Here again, the geometry of the SM provides almost no constraint on how fermion multiplets should be added to the bosonic sector, except for the well-known conspiracy in the electric charges and numbers of generations and colours needed to cancel chiral anomalies (see Section 7.3.1).

There is a wealth of data supporting the conservation of individual family numbers; see, for example, the compilation in the review Tests of Conservation Laws in [1]. I quote here three prominent examples which show the impressive degree of accuracy to which these conservation laws are known, namely

$$\frac{\Gamma(\mu^- \to e^- + \gamma)}{\Gamma_{\text{total}}} < 1.2 \times 10^{-11} \quad \text{at 90\% CL}, \tag{7.16a}$$

$$\frac{\Gamma(\mu^- \to e^- + e^+ + e^-)}{\Gamma_{\text{total}}} < 1.0 \times 10^{-12} \quad \text{at 90\% CL}, \tag{7.16b}$$

$$\frac{\sigma(\mu^- \text{Ti} \to e^- \text{Ti})}{\sigma(\mu^- \text{-capture on Ti})} < 4.3 \times 10^{-12} \quad \text{at 90\% CL}. \tag{7.16c}$$

The first of these compares the rate of a hypothetical "radiative decay" from the muon to the electron with the total decay rate, where the latter is dominated by the allowed process $\mu^- \to e^- + \nu_\mu + \overline{\nu_e}$. Indeed, the eigenvalues of L_e and L_μ in the decay $\mu^- \to e^- + \nu_\mu + \overline{\nu_e}$ and their additive conservation give $(L_e = 0, L_\mu = 1)$ for the left side, and $(L_e = 1 - 1, L_\mu = 1)$ for the right side. The decay $\mu^- \to e^- + \gamma$, in turn, would lead from $(L_e = 0, L_\mu = 1)$ to $(L_e = 1, L_\mu = 0)$, which is forbidden. The decay (7.16b) differs from the one of (7.16a) by the photon of (7.16a) moving off its mass shell and dissociating into an electron–positron pair. The process (7.16c), finally, compares the cross-section for neutrinoless capture of a muon on a nucleus, yielding an electron and the same nucleus in some excited state, with ordinary muon capture where one has a muon neutrino in the final state.

7.2 Mass matrices and state mixing

In the minimal SM the fermionic states which participate in weak charged-current interactions do not coincide with the eigenstates of mass which couple to the electromagnetic and, in the case of quarks, strong interactions. Therefore, a certain amount of state mixing becomes observable in weak interaction processes. This phenomenon has been studied in great detail and over many years for the three generations of quarks. Qualitatively the same phenomenon is known in neutrino physics through oscillations of neutrino states, although on a quantitative level things are very different.

7.2.1 The case of quarks

The quark states with electric charge $+\frac{2}{3}$ and the states with electric charge $-\frac{1}{3}$ which couple to (charge changing) weak interaction vertices are not identical with the states that occur in strong interactions. The states which couple to weak vertices are "rotated" compared with the mass eigenstates, which are the ones coupling to gluons in QCD. The *mixing matrix*, called CKM-matrix (after its discoverers N. Cabibbo, M. Kobayashi and K. Maskawa), is a unitary 3×3 matrix, four parameters of which are observables. The mixing matrix, in the case of quarks, is close to diagonal. In the analogous case of leptons it seems to be far from diagonal. There are few restrictions on the admissible mass matrices in a given charge sector which are to be inserted in their Dirac operator. In particular, they need not be Hermitian. We shall assume that they are nonsingular. Once the mass matrices are given, their diagonalization yields the mixing matrix and, of course, the mass eigenvalues,

$$\text{mass matrices of } up \text{ and } down \text{ quarks} \Longrightarrow \text{CKM-matrix.}$$

That part of the analysis is trivial. What about the converse?

Suppose the mass eigenvalues and the empirical mixing matrix are given. *What is the space of mass matrices which are compatible with these data?* Can one define parameters that help to sweep the space of admissible mass matrices?

The information on the quark masses is the following [1]: the set of *up*-like quarks u_i, charge $q^{(u)} = +\frac{2}{3}$, have the masses

$$m_u = 1.5 \text{ to } 3.3 \text{MeV}, \quad m_c = 1.27^{+0.07}_{-0.11} \text{ GeV}, \quad m_t = 171.3 \pm 1.1 \pm 1.2 \text{ GeV}. \tag{7.17a}$$

The *down*-like quarks d_i, with charge $q^{(d)} = -\frac{1}{3}$, are known to have the masses

$$m_d = 3.5 \text{ to } 6.0 \text{MeV}, \quad m_s = 105^{+25}_{-35} \text{MeV}, \quad m_b = 4.20^{+0.17}_{-0.07} \text{ GeV}. \tag{7.17b}$$

Regarding the CKM-matrix different conventions for its representation are possible. What is relevant for the problem posed above is the fact that it depends on four real physical parameters, say three mixing angles and one phase (describing CP-violation), all of which are known from experiment.

The key to the inverse analysis is the observation that weak charged-current interactions couple to *left*-chiral fields only. The *right*-chiral fields remain unobservable. In the reconstruction of all mass matrices which are compatible with a given set of data (masses and mixing) one makes use of the complete freedom of choice of right-chiral fields. Furthermore, a simultaneous unitary transformation of the left-chiral fields of charges $+\frac{2}{3}$ and $-\frac{1}{3}$ leaves the observables unchanged. In more detail: let $M^{(q)}$, $q = u$ and $q = d$, be arbitrary, nonsingular mass matrices for the group of *up*-like and *down*-like quarks, respectively. They are diagonalized by bi-unitary transformations of left- and right-chiral fields,

$$U_L^{(q)} M^{(q)} U_R^{(q)\dagger} = \Delta^{(q)}, \quad \Delta^{(q)} = \text{diag}\,(m_1^{(q)}, m_2^{(q)}, m_3^{(q)}), \quad q = u, d. \tag{7.18}$$

As is well known, the unitary matrices $U_L^{(q)}$ diagonalize the Hermitian, "squared" mass matrices $M^{(q)}M^{(q)\dagger}$, while the unitaries $U_R^{(q)}$ diagonalize $M^{(q)\dagger}M^{(q)}$. The CKM-matrix is given by the product of the left unitaries in (7.18), i.e.

$$V_{\text{CKM}} = U_L^{(u)}U_L^{(d)\dagger}. \tag{7.19}$$

The most general transformation of the mass matrices which leaves this matrix invariant reads

$$U^\dagger M^{(q)} V^{(q)} = \hat{M}^{(q)}, \quad q = u, d, \tag{7.20}$$

where U, $V^{(u)}$ and $V^{(d)}$ are arbitrary unitary matrices. Note that the unitary matrices $V^{(u)}$ and $V^{(d)}$ act on *right*-chiral fields and, hence, are independent of each other. The unitary matrix U acts on the *left*-chiral fields and, hence, must be the same in the two charge sectors.

The most economic reconstruction procedure makes use of the polar decomposition theorem for nonsingular matrices [5] (Exercise 7.9).

Any nonsingular M can be written as a product of a lower- (or upper-) triangular matrix T and a unitary matrix W,

$$M = TW \quad \text{with } T \text{ lower triangular, } W \text{ unitary.} \tag{7.21}$$

The decomposition is unique up to multiplication of W from the left by a diagonal unitary matrix diag $(\exp\{i\omega_1\}, \ldots, \exp\{i\omega_n\})$. Closer inspection shows the intimate relation of this theorem to Schmidt's orthogonalization procedure, well-known for example from quantum mechanics.

Since W acts on right-chiral fields and, hence, is unobservable, the essential information on a given mass matrix is contained in the triangular factor, namely

$$T^{(u)} \text{ or } T^{(d)} = \begin{pmatrix} * & 0 & 0 \\ * & * & 0 \\ * & * & * \end{pmatrix},$$

where asterisks denote possibly nonvanishing entries. This form still covers the most general case. Contact to more conventional representations is made by taking the Hermitian "squares", $\hat{H}^{(q)} = T^{(q)}T^{(q)\dagger}$ whose eigenvalues are the squared masses,

$$\hat{H}^{(q)} = M^{(q)}M^{(q)\dagger} = T^{(q)}T^{(q)\dagger} = U^\dagger \hat{D}^{(q)}U, \tag{7.22a}$$

$$\hat{D}^{(q)} = \text{diag}\left(m_1^{(q)2}, m_1^{(q)2}, m_1^{(q)2}\right).$$

More specifically, by an appropriate choice of basis, one obtains the representations in the *up-* and *down*-sectors

$$\hat{H}^{(q+1)} = U^\dagger \hat{D}^{(q+1)} U \qquad \text{(up sector)}, \qquad (7.22b)$$

$$\hat{H}^{(q)} = U^\dagger V_{\text{CKM}} \hat{D}^{(q)} V^\dagger_{\text{CKM}} U \quad \text{(down sector)}. \qquad (7.22c)$$

As we showed earlier [6] the matrix U is known *analytically*. Given the masses and the mixing matrix the remaining freedom in choice of the unitary U is contained in two complex parameters which, in turn, are constrained by a *quadratic* equation [7]. That is to say, the freedom eventually reduces to *one* complex parameter (or two real parameters). While this parameter runs through its domain of definition (bounded by a circle with radius $R = \sqrt{(m_t^2 - m_u^2)/(m_t^2 - m_c^2)}$ about the origin) one sweeps the space of admissible quark mass matrices.

Remark   Earlier analyses such as [8] made use of what was called the nearest-neighbour interaction (NNI) by assuming, from the start, the mass matrices to have the form

$$\hat{M} = \begin{pmatrix} 0 & * & 0 \\ * & 0 & * \\ 0 & * & * \end{pmatrix}.$$

There were two intuitive physical ideas behind this ansatz:

(i) initially, before interactions are switched on, only the third generation has a nonvanishing mass;
(ii) only immediate neighbours are allowed to interact.

Unfortunately intuition was misled; this picture is ill-defined because *any* set of matrices $M^{(q+1)}$, $M^{(q)}$ can be brought to NNI form, just by choosing the bases of chiral states appropriately. In other words, the NNI representation still covers the most general case. Furthermore, if the same ansatz is converted to the triangular representation by the decomposition eqn (7.21) then one sees that now it is the $i = 2, k = 1$-element that vanishes, $\hat{T}^{(q)}_{21} = 0$. This certainly is counter-intuitive because it might suggest that there is no direct interaction between the first and the second generations, while in the first ansatz they seemed to interact strongly.

Since then, Häussling has found an alternative procedure [9] which is technically much simpler than our earlier analysis [7]. He shows that the unobservable right-chiral fields can be chosen such that one arrives at the following representation: with $\Delta^{(u)}$ and $\Delta^{(d)}$ denoting the diagonal matrices of (linear) masses in the *up-* and *down*-sectors, respectively,

$$\Delta^{(u)} V_{\text{CKM}} \Delta^{(d)} = \hat{M}^{(u)\dagger} \hat{M}^{(d)}, \tag{7.23a}$$

$$\hat{M}^{(u)} = V \Delta^{(u)}, \tag{7.23b}$$

$$\hat{M}^{(d)} = \hat{M}^{(u)} \Delta^{(u)-1} V_{\text{CKM}} \Delta^{(d)}. \tag{7.23c}$$

Here the mass matrices $\hat{M}^{(q)}$, $q = u, d$, have rows whose squared norm is $m_{u_i}^2$ or $m_{d_i}^2$, respectively, while any two different rows are orthogonal to each other. The matrix V is an arbitrary unitary. Note that the left-hand side of (7.23a) contains the experimental input. Thus the product $\hat{M}^{(u)\dagger} \hat{M}^{(d)}$ can be determined from experiment. By varying the unitary V one reaches all mass matrices compatible with experiment. Equation (7.23c), finally, yields a linear relationship between the *up* and the *down* sector. The "constant of proportionality" is the experimental input $\Delta^{(u)-1} V_{\text{CKM}} \Delta^{(d)}$.

A few comments on these results follow.

1. All formulae in the analysis [7] are explicit, fairly simple, and can be studied in a transparent manner as a function of the one complex parameter on which they depend. This appears to be the best one can do in reconstructing the mass matrices from the data (mass eigenvalues and observed mixings).
2. That parameter represents, so to speak, "the heart of the matter". The NNI representations are the most rational representations. Therefore, any specific model that is proposed for the quark mass matrices can be tested by converting them to that class of bases and checking for compatibility.
3. The more recent analysis by Häussling [9], of course, is compatible with the NNI representation but presents the advantage of reconstructing *all* mass matrices, up to equivalence due to allowed but unobservable unitary transformations. To quote an analogy: the one-parameter NNI setting is like singling out a specific representative state ψ of a quantum mechanical ray, while the general method yields the whole ray $\{\exp i\alpha \psi\}$. This is true as long as right-chiral fermion fields remain unobservable in weak interactions.
4. Matters would change immediately if one discovered additional interactions which were sensitive to left- and right-chiralities in a physically relevant way. Then the "freedom of phases" decribed in the preceding remark is lost.
5. It is a matter of convention whether one takes the mass matrix of *up*-type quarks to be diagonal and assumes that the mixing occurs in the *down*-sector only. Our analysis above shows that the mixing can be shifted to either one of the two charge sectors, or be distributed over both.

7.2.2 The case of leptons

Everything that was said about quarks in Section 7.2.1 also holds for the three generations of leptons

$$\begin{pmatrix} \nu_e \\ e^- \end{pmatrix}, \quad \begin{pmatrix} \nu_\mu \\ \mu^- \end{pmatrix}, \quad \text{and} \begin{pmatrix} \nu_\tau \\ \tau^- \end{pmatrix}. \tag{7.24}$$

In defining the charge -1 and charge 0 states that participate in (charge changing) weak interactions, with reference to the mass eigenstates, one is free to assume neutrino states to be mixed, or electron-like states to be mixed, or even a combined configuration where both charge sectors mix. The electromagnetic and weak *neutral* interactions are diagonal in the flavours and, therefore, are insensitive to mixing described by unitary matrices. Unfortunately, although the masses are better defined than for quarks the experimental information on neutrino masses and mixing is much more scarce. The present state of knowledge is as follows [1] (for a more detailed discussion see the review [10]):

$$\Delta m_{21}^2 = (7.59 \pm 0.20) \times 10^{-5} \text{eV}^2,$$
$$\sin^2(2\theta_{12}) = 0.87 \pm 0.03,$$
$$\Delta m_{32}^2 = (2.43 \pm 0.13) \times 10^{-3} \text{eV}^2, \tag{7.25}$$
$$\sin^2(2\theta_{23}) > 0.92,$$
$$\sin^2(2\theta_{13}) < 0.15, \quad \text{at CL } 90\%.$$

As such, these data are not sufficient for an analysis of the kind discussed above. Nevertheless, there are certain patterns as well as specific models that were proposed by various people (including the author) and it might be worthwhile to check them for internal consistency in the general framework developed for quarks.

7.3 The space of connections and the action functional

As we emphasized previously in Section 7.1.1, although meant to describe *quantum field theories*, the construction of non-Abelian gauge theories seems to be a purely *classical* construction. The remaining three sections will show that this is not really true for several reasons. First, one can show that not every classical YM theory, after quantization, becomes a viable theory. Second, it may be that YM theories are embedded in the more restrictive framework of noncommutative geometry so that, again, not every structure group can be "gauged" and be converted to an acceptable quantum YM theory.

7.3.1 The axial anomaly, a reminder

A class of obstructions that were known already very early through the work of Adler, Bell, Jackiw, and others, concern local and global *anomalies*. For example, the renormalizability of the minimal electroweak SM is threatened by the triangle anomaly involving an axial current. The axial vector part of the fermionic current

$$a_\mu = \sum_{f=e,\mu,\tau} \overline{\Psi_f(x)}\gamma_\mu\gamma_5 U(Y)\Psi_f(x) + \sum_{q=1}^{3}\sum_{c=1}^{3} \overline{\Psi_{q,c}(x)}\gamma_\mu\gamma_5 U(Y)\Psi_{q,c}(x), \quad (7.26)$$

which couples to the gauge field $A_\mu^{(0)}(x)$, produces an anomaly in its divergence which is proportional to

$$S = S_{\text{leptons}} + S_{\text{quarks}} \quad \text{with}$$

$$S_{\text{leptons}} = \sum_{e,\mu,\tau} \text{tr}\,\{U(T_m T_m Y)\} \quad \text{and} \quad S_{\text{quarks}} = \sum_{q,c} \text{tr}\,\{U(T_m T_m Y)\}, \quad (7.27)$$

where T_m is a component of weak isospin and U denotes the respective fermion representations. The sums run over the three generations of leptons and quarks, and, in the case of the quarks, over the colour quantum number. Note that only isospin doublets contribute and that it is sufficient to consider the component $m = 3$ only. Then, marking the factors 3 from flavour and from colour, one finds

$$S_{\text{leptons}} = 3_{\text{f}} \cdot \frac{1}{4} \cdot (-2) = -\frac{3}{2}, \quad S_{\text{quarks}} = 3_{\text{f}} \cdot \frac{1}{4} \cdot 3_{\text{c}} \cdot \frac{2}{3} = \frac{3}{2}, \quad (7.28)$$

whose sum indeed vanishes. Thus, the electroweak SM is safe only if there is this conspiracy between the lepton families and the quark multiplets. We note in passing that the factor 3, which stems from the colour degrees of freedom, is essential in explaining the absolute magnitude of the amplitude for $\pi^0 \to \gamma\gamma$ decay.

7.3.2 Geometric route to anomalies

It is well known that anomalies occur at the order $\hbar$, i.e. at the level of one-loop diagrams. This may be the reason why they can be identified also within the geometric approach [11, 12], beyond the algebraic analysis sketched above. Without going into details let me describe the essence of this approach by the following somewhat sketchy remarks. A more detailed account can be found in the papers quoted above and in [13].

A given YM theory is formulated on a principal fibre bundle

$$\mathcal{P} = \left(P \xrightarrow{\pi} M, G\right). \quad (7.29)$$

Both the gauge group $\mathfrak{G}$ which is the group of vertical automorphisms on $\mathcal{P}$, and the space $\mathfrak{A}$ of connections on $\mathcal{P}$ are infinite dimensional. The space of connections is an affine space and, hence, is mathematically simple. From the point of view of physics, the space $\mathfrak{A}$ contains far too much freedom. Physics can only depend on gauge potentials which are not gauge equivalent. So, roughly speaking, the space of connections should be divided into classes of gauge-equivalent connections and only these classes should appear in the action. Now, the action of the gauge group $\mathfrak{G}$ on $\mathfrak{A}$, in general, is highly nontrivial so that

$$M := \mathfrak{A}/\mathfrak{G} \tag{7.30}$$

is a rather complicated object which, in general, is not a manifold. In other words, brute-force division by the gauge group is dangerous, and perhaps not even possible. The action functional, integrated over the fermions, is formulated on $\mathfrak{A}$. On the other hand, the effective functional must depend only on the gauge-inequivalent connections.

A more gentle way of performing this division is to do it, if possible, stepwise. For that purpose one studies the stratification of $\mathfrak{A}$ by the action of the gauge group, so that it is decomposed as follows

$$\mathfrak{A} = \mathfrak{A}^{(J_0)} \cup \mathfrak{A}^{(J_1)} \cup \cdots \mathfrak{A}^{(J_k)}, \tag{7.31a}$$

where the individual stratum is characterized by the stability group $\mathfrak{G}_A$ of its elements A being conjugate to J_i, $i = 0, \ldots, k$. That is to say

$$\mathfrak{A}^{(J_i)} = \left\{ A \in \mathfrak{A} \,|\, \mathfrak{G}_A = \psi J_i \psi^{-1}, \psi \in \mathfrak{G} \right\}. \tag{7.31b}$$

Here J_i is isomorphic to a subgroup of the structure group G. The number of strata is countable. The main stratum (index J_0) is characterized by J_0 isomorphic to the centre C of G. The stratification (7.31a) is unique and natural once the framework is defined by giving $\mathcal{P}(M, G)$. The formal definition

$$\mathcal{M}_i = \mathfrak{A}^{(J_i)}/\mathfrak{G}$$

yields an orbit bundle decomposition which bears some similarity to the compactification procedure in Kaluza–Klein theories. The spaces $\mathcal{M}_i$ are parts of the space of physical, gauge-inequivalent connections. The trouble is that, in general, they cannot be joined to a smooth manifold.

Physics comes in via an action functional $S(A, \bar{\psi}, \psi)$ which is a classical functional and is strictly gauge-invariant. At the quantum level the central quantity is the generating function

$$Z(A) = \int [\mathcal{D}\chi][\mathcal{D}\bar{\chi}] \exp\{-(S + \bar{\chi}\,\partial_A\,\chi)\} \tag{7.32}$$

obtained while integrating the fermionic degrees of freedom. There are two possibilities. Either $Z(A)$ is strictly invariant under a gauge transformation ψ, $Z(\psi A) = Z(A)$, or it is equivariant but not strictly invariant, $Z(\psi A) = \varrho^{-1}(A, \psi)Z(A)$, where ϱ is the action of the gauge transformation. In the first case one can safely divide by the gauge group to obtain a perfectly acceptable functional. The theory has no anomaly and can be reduced to the space $\mathcal{M}$ of gauge-inequivalent connections. In the second case, in contrast, there must be anomalies, the reduction is not possible. Therefore, in the geometric framework anomalies are obstructions to the reduction procedure which are due to quantization.

Without going into further, mostly technical details, let me sketch a constructive way of identifying anomalies in this geometric setting. One well-known ansatz is to make use of what is called the *pointed gauge group* $\mathfrak{G}^*$. This is the subgroup of $\mathfrak{G}$ which is the stability group of an arbitrary but fixed point p_0 of the principal fibre bundle $\mathcal{P}(M, G)$,

$$\mathfrak{G}^* = \mathfrak{G}_{p_0} = \{\psi \in \mathfrak{G}|\, \psi(p_0) = p_0\}. \tag{7.33}$$

In other words the pointed gauge group acts like the identity in the fibre over p_0. Singer showed that the action of $\mathfrak{G}^*$ on $\mathfrak{A}$ is free [14]. Therefore,

$$\begin{array}{ccc} \mathfrak{G}^* & \longrightarrow & \mathfrak{A} \\ & & \downarrow \\ & & \mathcal{M}^* = \mathfrak{A}/\mathfrak{G}^* \end{array}$$

is a principal fibre bundle. The functional $Z(A)$ is a trivial section

$$Z : \mathfrak{A} \longrightarrow \mathrm{Det} := \mathfrak{A} \times \mathbb{C} \tag{7.34a}$$

in the determinant bundle. If one divides by the pointed gauge group one obtains the reduced section

$$Z^* : \mathfrak{A}/\mathfrak{G}^* \longrightarrow (\mathfrak{A} \times \mathbb{C})/\mathfrak{G}^* =: \mathrm{Det}^*. \tag{7.34b}$$

If $Z(A)$ is strictly invariant then the action of $\mathfrak{G}^*$ on $\mathbb{C}$ is trivial so that (7.34b) reduces to

$$Z^* : \mathcal{M}^* \longrightarrow \mathcal{M}^* \times \mathbb{C}, \quad \mathcal{M}^* = \mathfrak{A}/\mathfrak{G}^*. \tag{7.34c}$$

In turn, if $Z(A)$ is equivariant but not strictly invariant then the action of the pointed gauge group on $\mathbb{C}$ is not trivial. In this situation Det^* has a twist; the integration over $[\mathcal{D}A]$ is not possible. Geometrically speaking there is a topological anomaly.

Even if no such anomaly is encountered, the story is not finished. There remains the "division" by the remainder $\mathfrak{G}/\mathfrak{G}^*$ which is isomorphic to the structure group G. This last step is particularly important because it is the structure group which defines the conserved charges of the theory. Again, if Z^* is strictly invariant the final division poses no problem. If it is not, but is (only) equivariant, one obtains

$$Z^{**} : \mathcal{M} \longrightarrow \left(\mathcal{M}^* \times \mathbb{C}\right) / G =: \mathrm{Det}^{**}; \qquad (7.35)$$

the functional Z^{**} is nontrivial, and one has found an anomaly.

In summary, by following this geometrical method one identifies all topological as well as possible global, nonperturbative anomalies. More on this can be found in the references given above.

7.4 Constructions within noncommutative geometry

The reconstruction of the minimal SM as well as of more general gauge and gravitational theories by means of noncommutative geometry is of geometrical origin but goes far beyond the classical framework of local gauge theories. The class of admissible gauge groups is restricted, spontaneous symmetry breaking (SSB) occurs as a rather natural phenomenon, and at least part of what we observe in the matter sector obtains a geometrical backbone. There are essentially three lines of proceeding that were explored extensively:

(i) the construction of the action by means of Dirac operators as advocated by A. Connes and his collaborators,

(ii) the somewhat more empirical construction within the Mainz–Marseille model, and

(iii) the numerous programs of formulating quantum field theory on noncommutative spaces as pioneered by Madore, Grosse, Wulkenhaar, and others.

I will not discuss the third group, which, in fact, would cover a series of lectures of its own. Instead, I briefly highlight the constructions (i) and (ii) but without going into much detail.

7.4.1 Spectral triples and all that

The original construction of the Standard Model by A. Connes and J. Lott [15] led rather naturally to SSB with a Higgs potential whose parameters were functions of the quark masses. Thus, for a while it seemed as though the essential parameters of the minimal SM (Weinberg angle, Higgs mass) could be predicted. This, however, was not successful [16]. Furthermore, the model struggled with the correct assignments regarding additive quantum numbers. A much more ambitious approach was proposed later by A. Chamseddine and A. Connes [17] by postulating the spectral action principle. In both approaches the Dirac operator D plays the central role. The spectral action principle asserts that the operator D is all that is needed to define the bosonic part of the action. Since the disjoint union of spaces corresponds to direct sums of Dirac operators, the action functional - which is determined by D – must be additive and, hence, must have the form

$$S = \mathrm{tr}\,(f(D/\Lambda)),\tag{7.36}$$

where f is an even function of its real variable, and Λ is a parameter which fixes the mass scale. The theory is determined by a *spectral triple* $(\mathcal{A}, \mathcal{H}, D)$ containing a $*$-algebra $\mathcal{A}$, a Hilbert space $\mathcal{H}$ and a Dirac operator D, A and D being represented on Hilbert space. Of course, matters are not as simple as that. In fact the spectral *triple* is rather at least a *quintet* because further data are needed to define it. The simplest realistic example, in terms of physics, is

$$(\mathrm{C}^\infty(M), L_2(M, S), \eth),\tag{7.37}$$

with M a compact oriented spin manifold. Here the algebra is the algebra of smooth functions on M and is commutative. The Dirac operator reduces to the ordinary partial derivatives. In view of the SM, in turn, one chooses the data to be

$$\mathcal{A} = \mathrm{C}^\infty(M) \otimes \mathcal{A}_F,\tag{7.38a}$$

$$\mathcal{H} = L_2(M, S) \otimes \mathcal{H}_F,\tag{7.38b}$$

$$D = \eth \otimes \mathbb{1}_F + \gamma^5 \otimes D_F.\tag{7.38c}$$

The algebra $\mathcal{A}_F$ is finite and is chosen to be [18]

$$\mathcal{A}_F = \mathbb{C} \oplus \mathbb{H}_L \oplus \mathbb{H}_R \oplus M_3(\mathbb{C}).\tag{7.39}$$

A connection is introduced into this model by replacing D by the covariant operator

$$D \longrightarrow D_A = D + A + JAJ^{-1}.\tag{7.40}$$

Implementing charge conjugation properly – and this applies to all NC models – needs special consideration as was first noted in [19]. The operation J, called reality structure, which was introduced later, does this job in Connes' framework. With the choices (7.38a)–(7.38c) the connection is

$$A = \gamma^5 \otimes \Phi - i\gamma^\mu \otimes A_\mu,\tag{7.41}$$

where Φ is a scalar field on M which takes values in $\mathcal{A}_F$, whereas A_μ pertains to a one-form which takes its values in the Lie algebra Lie $(U(\mathcal{A}_F))$. (Note that the structure group must be found in a unitary subalgebra of $\mathcal{A}_F$. This limits the class of structure groups which one can reproduce in this way.)

Of course, there is much more to be said about this fascinating theoretical framework. It provides an interesting ansatz for combining YM theories with gravity and there are branches of it exploring various directions. Its weakness, from a physicist's point of view, is the Euclidean framework. Although there were attempts to generalize spectral triples to Minkowski signature, there are still no satisfactory answers. Our short excursion may be sufficient to illustrate our main assertion: noncommutative geometry brings in *more* structure into gauge theories of fundamental interactions.

7.4.2 The bosonic sector à la Mainz–Marseille

The Mainz–Marseille model is a more heuristic construction but, I claim, is closer to phenomenology because it is formulated on Minkowski space with the right causal signature from the start and because it contains less freedom than other models. Instead of a detailed exposition of the model and of what it can do and what it cannot, I illustrate its salient features by three items.

The model is based on a bi-graded differential structure by composing the exterior algebra on M^4 (Minkowski space) and a graded Lie algebra akin to the electroweak structure group U(2) [19], [20]. In the minimal case, the graded algebra is chosen to be

$$SU(2|1) = \left\{ \mathbf{M} \,|\, \mathbf{M}^\dagger = -\mathbf{M}, \, \text{Str} \, \mathbf{M} = 0 \right\}, \tag{7.42}$$

the symbol Str denoting the super-trace. In the defining representation these matrices have the form

$$\mathbf{M} = \begin{pmatrix} * & *| & * \\ * & *| & * \\ \hline * & *| & * \end{pmatrix} \equiv \begin{pmatrix} \mathbf{A}_{2\times2}| & \mathbf{C}_{1\times2} \\ \mathbf{D}_{2\times1}| & \mathbf{B}_{1\times1} \end{pmatrix}. \tag{7.43}$$

The blocks along the diagonal are *even* with regard to the algebra grading in (7.42), the blocks which sit off the main diagonal are *odd*. (In this representation the super-trace is Str $\mathbf{M} = \text{tr}\,\mathbf{A} - \mathbf{B}$.) In terms of generators of the graded Lie algebra (7.42) the ones in the even part must be the T_k of SU(2) and T_0 (or Y) of the U(1) factor. The odd generators, in turn, sit in the off-diagonal blocks.

A rather natural way of constructing a connection for this model is to put ordinary gauge fields into the diagonal blocks of (7.43) along with the even generators of SU(2|1). If one takes the bi-grading seriously then the connection should have total grade 1 where the total grade is the sum of the exterior form grade and the internal (matrix) grade. This suggestive structure invites one to fill the odd blocks in the connection by fields which must be zero-forms (with regard to the exterior algebra), one-forms within the algebra and, last but not least, doublets with respect to weak isospin. In other words, one obtains an isospin doublet, scalar field just the way the empirical SM had imposed on us. Conversely, if the Higgs field were not a doublet, it would not fit into the connection, and this whole picture would fail.

If one then sits down and works out the Lagrangian of this model [20] one finds the SM Lagrangian with a Higgs doublet which sits in the right (shifted) SSB phase and a potential $V(\Phi)$ which has the correct shape. Thus, SSB is an unavoidable consequence of the model! Schematically and without repeating a detailed calculation, this may be understood as follows.

Generally, representations of SU(2|1) are characterized by two quantum numbers, I_0 (denoted such because it is parent of weak isospin) and Y_0 (parent of weak hypercharge). As shown by Marcu [21] and described in a book by Scheunert [22], the *adjoint representation* of SU(2|1) has the quantum numbers $[I_0 = 1, Y_0 = 0]$, and decomposes in terms of SU(2) × U(1) as follows

$$[I_0 = 1, Y_0 = 0] \longrightarrow (I = 1, Y = 0) \oplus (I = 0, Y = 0)$$
$$\oplus (I = \tfrac{1}{2}, Y = 1) \oplus (I = \tfrac{1}{2}, Y = -1). \quad (7.44)$$

Note that this is precisely what one needs: a triplet of gauge bosons with vanishing weak hypercharge, a singlet with vanishing hypercharge and two doublets with $Y = \pm 1$ for the "standard" Higgs fields!

The model also suggests more structure in the *fermionic* sector of the SM. In the case of quarks, each of the three *quark* generations is classified by the simplest *typical representation* of SU(2|1) which reads, together with its decomposition in terms of SU(2) × U(1),

$$\left[I_0 = \tfrac{1}{2}, Y_0 = \tfrac{1}{3}\right] \longrightarrow$$
$$(I = \tfrac{1}{2}, Y = \tfrac{1}{3}) \oplus (I = 0, Y = \tfrac{4}{3}) \oplus (I = 0, Y = -\tfrac{2}{3}). \quad (7.45)$$

Furthermore, SU(2|1) possesses reducible but indecomposable representations where these generations are joined by semi-sums in the following way:

$$\left[I_0 = \tfrac{1}{2}, Y_0 = \tfrac{1}{3}\right] \oplus\!\!\!\!\!\!\!\!\!\!\!\!\!\!- \left[I_0 = \tfrac{1}{2}, Y_0 = \tfrac{1}{3}\right] \oplus\!\!\!\!\!\!\!\!\!\!\!\!\!\!- \left[I_0 = \tfrac{1}{2}, Y_0 = \tfrac{1}{3}\right]. \quad (7.46)$$

In this representation the generators have block triangular form, just like the (lower) triangular mass matrices considered earlier; see eqn (7.21).

Likewise, one generation of *leptons* fits into the *fundamental* representation of SU(2|1),

$$\left[I_0 = \tfrac{1}{2}, Y_0 = -1\right] \rightarrow (I = \tfrac{1}{2}, Y = -1) \oplus (I = 0, Y = -2). \quad (7.47)$$

Here again, identical representations of this kind may be joined in a semi-direct sum analogous to (7.46) so as to describe the three lepton families.

It is tempting to use these reducible but indecomposable representations of SU(2|1) for classifying leptons and quarks [19], [23]. If one does so, one discovers a scheme which does not fix absolute parameters but reveals certain textures in the mass matrices which are in agreement with experiment.

Finally, the graded structure of the Mainz–Marseille model also fits very well with the quantum number assignment of quarks and leptons, with the absence of anomalies, and with charge quantization [24].

7.5 Further routes to quantization via BRST symmetry

Spontaneous symmetry breaking within the SM and its extensions remains a puzzle. Although the *classical* geometric setting for describing SSB from the original symmetry group G to the residual symmetry H,

$$G \longrightarrow H,$$

in relation to Goldstone's theorem, looks convincing, it is not at all clear whether nature has chosen this route. Here again, and in view of the forthcoming searches at the Large Hadron Collider (LHC) it might be instructive to first consider the present experimental situation.

7.5.1 General remarks on SSB and experimental information

The standard model is a renormalizable quantum field theory. Thus, it allows us to calculate radiative corrections to an impressive accuracy. However, radiative corrections applied to a specific observable at a given energy scale always receive contributions from other parts of the theory, including contributions from constituents which, for kinematical reasons, cannot be seen (yet) at the scales in question. A striking example is the *top*-quark t whose mass was deduced from radiative corrections, to a fair accuracy, before it was actually discovered. To get a feeling for analyses of this kind let me quote a global fit to all observables within the SM published recently by the Gfitter Group [25]. Excluding the directly measured and by now well-known mass of the top from the fit, radiative corrections alone yield $m_t = 178.2\,^{+9.8}_{-4.2}$ GeV, which is not far from the value $m_t = 172.4 \pm 1.2$ GeV obtained from experiment.

The same fit when used to predict the Higgs mass m_H, is less conclusive. There is an experimental lower limit of the order of 114 GeV from LEP and Tevatron experiments. If one excludes that constraint one obtains

$$M_\mathrm{H} = 80\,^{+30}_{-23} \text{ GeV}. \tag{7.48}$$

The complete fit, including the mass limits, yields

$$M_\mathrm{H} = 116.4\,^{+18.3}_{-1.3} \text{ GeV}. \tag{7.49}$$

Figure 7.2 shows the *pull values* for the complete fit, the pull value of an observable being defined by

$$\frac{1}{\sigma|_\mathrm{meas}}\left(\mathcal{O}|_\mathrm{fit} - \mathcal{O}|_\mathrm{meas}\right),$$

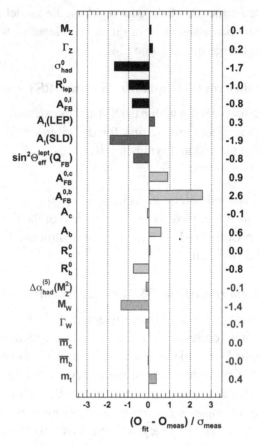

Figure 7.2 Pull values of observables in a *complete* fit to electroweak observables (taken from [25]).

with $\sigma|_{\text{meas}}$ the error in the measurement. The hadronic asymmetry into *b*-quarks yields a tendency to rather high values of the Higgs mass while the leptonic asymmetries either agree with the overall fit or would prefer an even lower value than that. Obviously, the situation is much less clear than it was for the top before its discovery. This is further illustrated by the next two figures. Figure 7.3 shows the results for the Higgs mass that one obtains if all sensitive observables are excluded from the standard fit, except the one indicated. This information is complementary to the information given in Fig. 7.4, which shows the results for M_{H} obtained from the standard fit but excluding the respective measurements.

Various alternatives were discussed in the literature including variations of the minimal SM (two doublet Higgs models, or more, as well as other scenarios).

Note added in proof (December 2012): In the meantime a particle with mass of about 126 MeV was discovered by experiments at the Large Hadron Collider (LHC) which might be the expected Higgs particle. However, it will need further studies to identify its detailed properties.

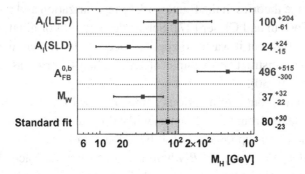

Figure 7.3 Results for M_H from standard fit excluding the respective measurements (taken from [25]).

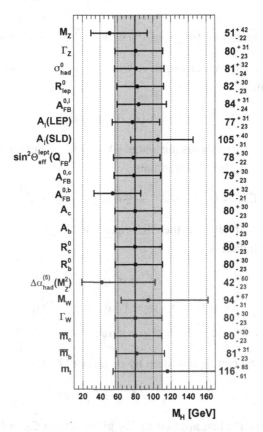

Figure 7.4 Results for M_H from standard fit excluding the respective measurements (taken from [25]).

7.5.2 A simple model

Scharf was among the first to point out that SSB need not be based on the traditional Higgs mechanism but could be derived as a consequence of causal gauge invariance (CGI) [26]. Causal gauge invariance is a systematic method of treating perturbative gauge theories in the framework of regularization and renormalization developed by Epstein and Glaser [27]. Scharf developed it originally for quantum electrodynamics [28] but it was also applied successfully to massive non-Abelian theories [29]. The instructive example of an Abelian theory was worked out in quite some detail in [30].

We examined this route from various points of view [31]. The following simple model, which is taken from this reference, will help to understand some of the ideas which guide one in this approach. Let $\vec{\Phi}$ be a doublet of scalar fields, regarded as a complex singlet $\Phi = v + \varphi + iB$ where v is a constant (a "vacuum expectation value" in the standard picture), B is a Stückelberg field whose role is to give mass to vector bosons of the model, and φ is a scalar field. The model contains a Higgs Lagrangian,

$$\mathcal{L}_\Phi = \tfrac{1}{2} \left(\partial_\mu + ig A_\mu \right) \Phi^\dagger \left(\partial^\mu + ig A^\mu \right) \Phi + \tfrac{1}{2}\mu^2 \Phi^\dagger \Phi - \tfrac{1}{4}\lambda(\Phi^\dagger \Phi)^2. \tag{7.50a}$$

As it contains a massive vector boson A whose mass is $m = gv$, the model needs a gauge fixing term,

$$\mathcal{L}_{\text{gf}} = -\tfrac{1}{2}\Lambda \left(\partial_\mu A^\mu + \tfrac{m}{\Lambda} B \right)^2, \tag{7.50b}$$

where the constant Λ determines the gauge ($\Lambda = 1$ is the Feynman gauge). Finally, it is accompanied by a ghost Lagrangian

$$\mathcal{L}_{\text{gh}} = \left(\partial_\mu \tilde{u} \right) \cdot s A^\mu - \tfrac{m}{\Lambda} \tilde{u} \, s B \equiv \mathcal{L}_{\text{gh}}^{(0)} + \mathcal{L}_{\text{gh}}^{(1)}, \tag{7.50c}$$

where $\tilde{u}$ is the antighost field, s is the BRS-operator. The action of the BRS-operator on the fields of the model is

$$s A^\mu = \partial^\mu u, \tag{7.51a}$$

$$su = 0, \quad s\tilde{u} = - \left(\Lambda \partial_\mu A^\mu + mB \right), \tag{7.51b}$$

$$s\Phi = igu\Phi \quad \text{or} \quad sB = mu + g\,u\varphi, \quad s\varphi = -g\,Bu. \tag{7.51c}$$

The gauge fixing term can be rewritten by means of (7.51b),

$$\mathcal{L}_{\text{gf}} = \tfrac{1}{2} \left(\partial_\mu A^\mu + \tfrac{m}{\Lambda} B \right) (s\tilde{u}).$$

One then shows that s when applied to $\mathcal{L}_\Phi$ gives zero, $s\mathcal{L}_\Phi = 0$ (Exercise 7.10). Regarding the other two terms, (7.50b) and (7.50c), one shows that

$$s \left(\mathcal{L}_{\text{gf}} + \mathcal{L}_{\text{gh}} \right) = \partial_\mu \left((s\tilde{u})(s A^\mu) \right)$$

is a total divergence (see Exercise 7.10). The nilpotency of s is easy to check. One has $(s \circ s)\,\Phi = 0$ while $(s \circ s)\,\tilde{u}$ vanishes for solutions of the field equations. The field equations of the vector field A and of the Stückelberg field are seen to be

$$\left(\Box + m^2\right) A^\mu = (1 - \Lambda)\partial^\mu(\partial_\nu A^\nu), \tag{7.52}$$

$$\left(\Box + \tfrac{m^2}{\Lambda}\right) B = 0. \tag{7.53}$$

If a gauge other than the Feynman gauge is chosen, i.e. if $\Lambda \neq 1$, the mass of the Stückelberg field is $\frac{m}{\sqrt{\Lambda}}$. Working out the Lagrangian in terms of the field degrees of freedom one obtains

$$\mathcal{L}^{(0)} = \mathcal{L}_{\text{kin}}^{\text{YM}}(A) + \tfrac{1}{2}m^2 A_\mu A^\mu - \Lambda\tfrac{1}{2}\left(\partial_\mu A^\mu\right)^2 + \partial_\mu \bar{u}\, \partial^\mu u - \tfrac{m^2}{\Lambda}\bar{u}u$$

$$+ \tfrac{1}{2}\left(\partial_\mu B \partial^\mu B\right) - \tfrac{m^2}{2\Lambda}B^2 + \tfrac{1}{2}\left(\partial_\mu \varphi\, \partial^\mu \varphi\right) - \tfrac{1}{2}m_H^2 \varphi^2 - m\partial_\mu \left(A^\mu B\right). \tag{7.54}$$

The individual contributions to this Lagrangian are easy to identify. The first term is the YM "kinetic" term, the second term is the mass term of the vector boson. The third comes from the gauge fixing (7.50b), the fourth and fifth terms come from the ghost Lagrangian (7.50c). The two contributions that follow pertain to the B-field, followed by kinetic and mass terms of the scalar field φ. In a similar way one works out the interaction terms at order g and order g^2 (see [31]) and verifies that the total Lagrangian is consistent with BRST symmetry, $s(\mathcal{L}^{(0)} + \mathcal{L}_{\text{int}}) = \partial_\mu I^\mu$. What the model shows is this: the CGI approach yields structures which look very much alike the ad-hoc Higgs mechanism without assuming a potential with a degenerate minimum *ab initio*. The BRS symmetry is instrumental in this construction.

7.6 Some conclusions and outlook

Starting from classical field theory, the basic structure of gauge theories seems to distinguish radiation from matter as two categories which, *a priori*, have little to do with each other. As soon as one enters the quantum world, however, the distinction can no longer be maintained. Quarks and leptons are described by a Dirac operator which, in turn, is the driving force in the construction of noncommutative geometries designed to generalize Yang–Mills theory. It is not known to what extent the realistic Dirac operator, with regard to the complexity of its mass sectors, determines the (NC) geometry on which quantum field theories for the fundamental interactions should be built.

The Higgs particle plays a rather enigmatic role. Its phenomenology from radiative corrections and global fits to all observables within the Standard Model is not in a satisfactory state. Its model role of providing mass terms for some of the vector bosons and for the fermions of the theory suggests that it is another form of "matter". Models based on noncommutative geometry, on the other hand, classify the Higgs field in the generalized Yang–Mills connection, in addition to the gauge bosons, and hence declare it to be part of "radiation". So, what is it?

We worked out some of these themes by way of construction and by means of instructive examples. We started with a schematic description of Yang–Mills theories including spontaneous symmetry breaking (SSB) within the classical geo- metric framework, and including matter particles. In a first excursion into quantum field theory we described the stratification of the space of connections and its relevance to anomalies. In order to clarify the phenomenological basis on which Yang–Mills theories of fundamental interactions are built, we reviewed some of the most pertinent phenomenological features of leptons and of quarks. Constructions of the Standard Model in the framework of noncommutative geometry were briefly summarized. This, in turn led us to a closer analysis of the mass sector and state mixing phenomena of fermions. The intricacies of quantization were illustrated by a semi-realistic model for massive and massless vector bosons.

Acknowledgements

It is a pleasure to thank the organizers (in alphabetical order) Sergio Adarve, Alexander Cardona, Hernán Ocampo, Sylvie Paycha and Andrés Reyes for having organized the wonderful summer school in the old city of Villa de Leyva. Marta Kovacsics deserves special thanks for the excellent administrative coordination and her perfect organization of all aspects of the meeting.

The discussions with the other speakers on various topics of interest, physics, mathematics, life in general, were very stimulating and it was a pleasure to spend quite some time with them.

Last, but certainly not least, I wish to thank the students and other participants at the school for their vivid interest, their questions and, at times, their patience in more difficult topics (including the art of dancing Salsa).

Exercises

7.1 What does *compactness* of a Lie group imply for its Killing metric? Why must the structure group of a Yang–Mills theory be compact?

7.2 Within the framework of a Yang–Mills theory over Minkowski space, study the parallel transport of a scalar field by means of the connection.

7.3 Verify that, while the action of a covariant derivative D_A is not *linear*, the action of D_A^2 is.

7.4 In a unitary representation of a compact simple Lie group one has

$$\mathrm{tr}\left[U(T_i), U(T_j)\right] = \kappa\, \delta_{ij}.$$

Show that the constant κ depends on the representation but does not depend on either i or j. Study the examples of spinor and triplet representations of SU(2).

7.5 Let the structure group be $G = SO(3)$. Construct a Lagrangian for the local gauge theory whose structure group is G, and add a triplet of scalar fields to this Lagrangian.

7.6 When a YM theory is based on a reductive Lie algebra one says that couplings of matter particles to gauge fields are *universal* in the sense that ratios of physical couplings within a multiplet are fixed. The aim of this exercise is to clarify this statement. Does it hold for $\mathfrak{u}(1)$? If so, when are couplings universal?

7.7 Chirality selection rules: consider a fermion–fermion vertex coupling to a *scalar* field or to a *vector* field, respectively,

$$\overline{\psi^{(k)}(x)}\,(a\mathbb{1} + ib\gamma_5)\,\psi^{(i)}(x),$$

$$\overline{\psi^{(k)}(x)}\gamma_\mu\,(a\mathbb{1} + ib\gamma_5)\,\psi^{(i)}(x).$$

Work out the chirality selection rules at these vertices.

7.8 Experiment tells us that the decay $\pi^+ \to e^+ \nu_e$ has a probability which is about 10^{-4} smaller than for the decay $\pi^+ \to \mu^+ \nu_\mu$, even though the electron is 207 times lighter than the muon. As the available phase space in the electronic decay is much *larger* than in the muonic channel, the decay rate for $\pi^+ \to e^+ \nu_e$ should be about five times larger than for $\pi^+ \to \mu^+ \nu_\mu$. Making use of the result of the previous exercise can you explain this discrepancy?

7.9 The decomposition theorem says that any nonsingular matrix M can be written as the product of a (lower) triangular matrix T and a unitary matrix W, $M = TW$. Prove this theorem by induction. Establish its relation to the Schmidt's orthogonalization procedure. Show that the decomposition is unique up to multiplication of W by a diagonal unitary matrix from the left.

7.10 Show that the Lagrangian (7.54) of the toy model is BRST-invariant.

Appendix: Proof of relation (7.11a)

Let $\Phi = \{\phi^{(i)}\}$ be a multiplet of scalar fields (irreducible representation of SU(2)) such that

$$U^\Phi(T^2)\Phi = t(t + 1)\Phi, \quad U^\Phi(T_3)\phi^{(i)} = t_3^{(i)}\phi^{(i)}, \quad t_3^{(i)} = -t, -t + 1, \ldots, t.$$

All components of Φ have the same eigenvalue t_0 of the generator T_0 of the U(1) factor, $U^\Phi(T_0)\phi^{(i)} = t_0\phi^{(i)}$. Let $V(\Phi)$ be a quartic potential with a degenerate, absolute minimum at some $\Phi_0 = (\phi_0^{(1)}, \ldots, \phi_0^{(i)}, \ldots)$ which is not identically zero.

In the bosonic sector make the ansatz

$$A_\mu^{(0)}(x) = A_\mu^{(\gamma)}(x) \cos\theta_W + A_\mu^{(Z)}(x) \sin\theta_W, \tag{7.55a}$$

$$bA_\mu^{(3)}(x) = -A_\mu^{(\gamma)}(x) \sin\theta_W + A_\mu^{(Z)}(x) \cos\theta_W. \tag{7.55b}$$

Here $A_\mu^{(0)}(x)$ and $A_\mu^{(3)}(x)$ are the companions of the generators T_0 and T_3, respectively, while $A_\mu^{(\gamma)}(x)$ and $A_\mu^{(Z)}(x)$ are supposed to become the photon and the Z^0 fields, respectively. The angle θ_W is called the Weinberg angle. In this minimal version it remains a free parameter and has to be determined from experiment. The action of the connection (7.3a) on the scalar multiplet is

$$U^\Phi(A_\mu)\Phi = iq \sum_{k=0}^{3} A_\mu^{(k)}(x) U^\Phi(T_k)\Phi \tag{7.56}$$

$$= iq \left\{ \tfrac{1}{\sqrt{2}} \left[W_\mu^- U^\Phi(T_+) + W_\mu^+ U^\Phi(T_-) \right] \right.$$

$$\times A_\mu^{(Z)} U^\Phi(T_3 \cos\theta_W + T_0 \sin\theta_W) \tag{7.57}$$

$$\left. + A_\mu^{(\gamma)} U^\Phi(-T_3 \sin\theta_W + T_0 \cos\theta_W) \right\} \Phi.$$

Here we have replaced T_1 and T_2 by the ladder operators $T_\pm = T_1 \pm iT_2$.

Obviously, the factor $(-T_3 \sin\theta_W + T_0 \cos\theta_W)$ which multiplies the photon field must be proportional to the electric charge operator since only charged particles couple to photons. Now, if Φ_0 has the form $\Phi_0 = (0, 0, \ldots, \phi_0^{(i)} = v \neq 0, 0, \ldots)$ this means that the component $\phi^{(i)}$ of Φ must be electrically neutral. The quantum numbers of the nonvanishing component must be related by the condition

$$t_0^{(i)} = t_3^{(i)} \tan\theta_W. \tag{7.58}$$

Another way of expressing this is: Φ_0 pertains to the ground state of the theory, i.e. the vacuum. Only electrically neutral fields can develop a nonvanishing vacuum expectation value $v \neq 0$.

Possible mass terms for the vector bosons originate from the scalar product

$$\left(U^\Phi(A_\mu)\Phi_0, U^\Phi(A^\mu)\Phi_0 \right). \tag{7.59}$$

Form here on it is straightforward to compute the term (7.59) and to isolate the mass terms of $W^\pm$ and Z^0. Making use of the identity $T_+T_- + T_-T_+ = 2(\vec{T}^2 - T_3^2)$ and inserting (7.58) one obtains

$$\left(U^\Phi(A_\mu)\Phi_0, U^\Phi(A^\mu)\Phi_0 \right)$$

$$= q^2 v^2 \left\{ \left[t(t+1) - (t_3^{(i)})^2 \right] W_\mu^- W^{+\mu} + \frac{1}{\cos^2\theta_W} (t_3^{(i)})^2 A_\mu^{(Z)} A^{(Z)\mu} \right\}. \tag{7.60}$$

One reads off the masses of $W^{\pm}$ and Z^0 from (7.60), obtaining

$$m_W^2 \propto \frac{1}{2} q^2 v^2 \left[t(t+1) - (t_3^{(i)})^2 \right], \tag{7.61a}$$

$$m_Z^2 \propto q^2 v^2 \cos^{-2} \theta_W (t_3^{(i)})^2. \tag{7.61b}$$

As the constant of proportionality is the same for both, the ratio (7.11a) follows from these equations.

From this derivation one sees very clearly that there is little that restricts the choice of the multiplet for Φ. The only condition is that Φ have one component which is electrically neutral.

References

[1]   C. Amsler *et al.* (Particle Data Group), Review of Particle Physics, *Phys. Lett.* **667** (2008) 1–1340
[2]   W. Fetscher, *Phys. Lett.* **140B** (1984) 117
[3]   F. Scheck, *Electroweak and Strong Interactions – Phenomenology, Concepts, Models,* 3rd edn. Heidelberg, New York: Springer Verlag, 2012
[4]   A. Jodidio *et al., Phys. Rev.* **D34** (1986) 1967; **D37** (1988) 237
[5]   F. D. Murnaghan, *The Theory of Group Representations.* New York: Dover, 1963; *The Unitary and Rotation Groups.* Washington, DC: Spartan, 1962
[6]   R. Häussling and F. Scheck, *Phys. Lett,* **B336** (1994) 477
[7]   S. Falk, R. Häussling and F. Scheck, *Phys. Rev.* **D65** (2002) 093011-1
[8]   G. C. Branco, L. Lavoura and F. Mota, *Phys. Rev.* **D39** (1989) 3443
[9]   R. Häussling, private communication and to be published
[10] B. Kayser, Neutrino mass, mixing, and flavor change, in [1]
[11] A. Heil, A. Kersch, N. Papadopoulos, B. Reifenhäuser and F. Scheck, *J. Geom. Phys.* **7** (1990) 489
[12] A. Heil, A. Kersch, N. Papadopoulos, B. Reifenhäuser, F. Scheck and H. Vogel, *J. Geom. Phys.* **6** (1989) 237
[13] F. Scheck, in *Rigorous Methods in Particle Physics,* S. Ciulli, F. Scheck, W. Thirring (Eds.), Springer Tracts in Modern Physics **119**. Berlin: Springer, 1990, p. 202
[14] I. M. Singer, *Soc. Mathématique de France, Astérisque* (1985) 323
[15] A. Connes and J. Lott, *Nucl. Phys.* **B18** (1990) 29
[16] M. Paschke, *Phys. Lett.* **B414** (1997) 323
[17] A. Chamseddine and A. Connes, *Comm. Math. Phys.* **186** (1997) 731
[18] A. Connes, Noncommutative geometry and the standard model with neutrino mixing, (2006) hep-th/0608226
[19] R. Coquereaux, G. Esposito-Farèse and F. Scheck, *Int. J. Mod. Phys.* **A7** (1992) 6555
[20] R. Coquereaux, R. Häussling, N.A. Papadopoulos and F. Scheck, *Int. J. Mod. Phys.* **A7** (1992) 2809
[21] M. Marcu, *J. Math. Phys.* **21** (1980) 1277
[22] M. Scheunert, *The Theory of Lie-Superalgebras,* Lecture Notes in Mathematics. Berlin: Springer, 1979
[23] R. Häussling and F. Scheck, *Phys. Lett.* **B336** (1994) 477; R. Häussling, M. Paschke, and F. Scheck, *Phys. Lett.* **B417** (1997) 312

[24] F. Scheck, *Phys. Lett.* **B284** (1992) 303
[25] The Gfitter Group, H. Flächer, M. Goebel, J. Haller, A. Hoecker, K. Mönig and J. Stelzer, CERN-OPEN-2008-024, DESY-08-160, arXiv: 0811.0009
[26] G. Scharf, *Quantum Gauge Theories: A True Ghost Story.* New York: John Wiley & Sons, 2001
[27] H. Epstein and V. Glaser, *Ann. Inst. H. Poincaré* **A19** (1973) 211
[28] G. Scharf, *Finite Quantum Electrodynamics: the Causal Approach.* Berlin: Springer, 1995
[29] M. Dütsch, F. Krahe, T. Hurth and G. Scharf, *Nuovo Cim.* **A106** 1029 (1993); **107** (1994) 375
[30] J.M. Gracia-Bondía, On the causal gauge principle, in *Proceedings of Workshop on Renormalization*, MPIM Bonn; hep-th/0809.0160
[31] M. Dütsch, J. Gracia-Bondía, F. Scheck and J. C. Varilly, Quantum gauge models without (classical) Higgs mechanism, *Eur. Phys. J.* **C69** (2010) 599–621

8

Absence of singular continuous spectrum for some geometric Laplacians

LEONARDO A. CANO GARCÍA

Abstract

We provide two examples of spectral analysis techniques of Schrödinger operators applied to geometric Laplacians. In particular we show how to adapt the method of analytic dilation to Laplacians on complete manifolds with corners of codimension 2 finding the absence of singular continuous spectrum for these operators, a description of the behavior of its pure point spectrum in terms of the underlying geometry, and a theory of quantum resonances.

Introduction

Spectral geometry studies the interactions between the geometry of a Riemannian manifold and the spectral analysis of its associated Laplacian. These interactions have been deeply studied in the case in which the manifold is closed (see [Cha84] [Ros97] for basic introductions). In the closed case the Laplacian is a self-adjoint operator with compact resolvent and hence its spectrum is purely discrete. This contrasts with the general case since, if (M, g) is a geodesically complete Riemannian manifold, not necessarily closed, it is known that its Laplacian $\Delta_g : C_c^\infty(M) \subset L^2(M, dvol_g) \to L^2(M, dvol_g)$ is essentially self-adjoint but its spectral resolution is not purely discrete in general.

In order to show more clearly the new spectral phenomena for Δ_g that the lack of compactness of M brings, let us recall the spectral theorem for self-adjoint operators.

Theorem 8.0.1 [RS80] Let A: $\mathcal{H} \to \mathcal{H}$ be a self-adjoint operator acting on a separable Hilbert space $\mathcal{H}$. Then there exists a collection of Borel measures on

Geometric and Topological Methods for Quantum Field Theory, ed. Alexander Cardona, Iván Contreras and Andrés F. Reyes-Lega. Published by Cambridge University Press. © Cambridge University Press 2013.

$\mathbb{R}$, $\{\mu_i\}_{i \in I}$ where $I \subseteq \mathbb{N}$, and there exists a unitary operator $\mathcal{F}$ such that the following diagram commutes

$$
\begin{array}{ccc}
\mathcal{H} & \xrightarrow{\ A\ } & \mathcal{H} \\
\mathcal{F} \uparrow & & \downarrow \mathcal{F}^* \\
\bigoplus_{i \in I} L^2(\mathbb{R}, \mu_i) & \xrightarrow{\oplus_{i \in I} x_i} & \bigoplus_{i \in I} L^2(\mathbb{R}, \mu_i),
\end{array}
$$

where the operator $\oplus_{i \in I} x_i$ is the direct sum of the multiplication operators $x_i : L^2(\mathbb{R}, \mu_i) \to L^2(\mathbb{R}, \mu_i)$ that, specifically, send a function $f \in L^2(\mathbb{R}, \mu_i)$ to the function $x_i f$.

Given a self-adjoint operator A, as in the theorem above, the Lebesgue decomposition theorem for measures induces a decomposition of the Hilbert space $\mathcal{H}$ into three important A-invariant subspaces $\mathcal{H}_{ac}$, $\mathcal{H}_{pp}$ and $\mathcal{H}_{sing}$ that, using theorem 8.0.1, are the associated to the three Hilbert spaces

$$
\bigoplus_{\mu_i \text{ is ac}} L^2(\mathbb{R}, \mu_i), \quad \bigoplus_{\mu_i \text{ is pp}} L^2(\mathbb{R}, \mu_i) \quad \text{and} \quad \bigoplus_{\mu_i \text{ is sing}} L^2(\mathbb{R}, \mu_i),
$$

where ac means absolutely continuous, pp means pure point and sing means singular continuous measures (all of them defined with respect to the Lebesgue measure in $\mathbb{R}$). We have already said that when (M, g) is a closed Riemannian manifold, its Laplacian has purely discrete spectrum, in other words the invariant subspaces $L^2_{ac}(M, dvol_g)$ and $L^2_{sing}(M, dvol_g)$, corresponding to the self-adjoint operator $A := \Delta_g$, are empty. In these lecture notes we will show examples of manifolds (M, g) for which $L^2_{ac}(M, dvol_g)$ is not empty and $L^2_{sing}(M, dvol_g)$ is empty.

The former description of the spectral analysis of self-adjoint operators justifies the question about how to prove that a geometric Laplacian associated to a geodesically complete manifold does not have singular continuous spectrum. These lecture notes provide, through the friendly environment of precise examples and without pretending to give a complete answer, an illustration of how to work around this question. A common feature of these examples is the control on the Riemannian metric of the open manifold at infinity.

Let us describe the contents more carefully. For Laplacians acting on manifolds with cylindrical or cusp ends and their natural generalizations it is well known in the literature (see [Gui89] [Hus05] [Mül83]) how to find a meromorphic extension of their resolvent and how, using it, to prove the absence of singular continuous spectrum. In Section 8.1 we illustrate the classical method of obtaining a meromorphic extension of the resolvent for manifolds with cylindrical and cusp ends, and we show the relation between it and the absence of singular continuous spectrum, that is, fundamentally, the limit absorption principle. In Section 8.2 we explain the notion of a complete manifold with corner of codimension 2 and how to prove absence of singular continuous spectrum for Laplacians on this kind of manifold. Following [Can11] we describe the way to apply the technique of analytic dilation to such Laplacians, a context in which the techniques explained in Section 8.1 do not apply.

The method of analytic dilation was originally applied to N-particle Schrödinger operators; a classic reference in that setting is [Gér93]. It has also been applied to the black-box perturbations of the Euclidean Laplacian in the series of papers [SZ93a], [SZ93b], [SZ94], [SZ95]. In [Bal97] it is used to study Laplacians on hyperbolic manifolds. The analytic dilation has also been applied to the study of the spectral and scattering theory of quantum wave guides and Dirichlet boundary domains; see e.g. [DEM98], [KS07]. It has also been applied to symmetric spaces of non-compact type in [MV02], [MV04], [MV07]. In [Kal10] it is applied to manifolds with analytic asymptotically cylindrical end. In each of these settings new ideas and new methods are applied. In [Can12] the method of Mourre estimates, also coming from the spectral analysis of Schrödinger operators and with strong relations to the analytic dilation method, is adapted to work out the case of compatible Laplacians on complete manifolds with corners of codimension 2; this method proves the absence of continuous spectrum too.

8.1 Meromorphic extension of the resolvent and singular continuous spectrum

In this section we give the main ideas of a method for meromorphically extending the resolvent of Laplacians on manifolds with cylindrical and cusp ends and show why such extension is enough to have absence of singular continuous spectrum. The results of this section were obtained in [Gui89], [Mül83] but we base our exposition on [Hus05] since we consider it is easier to understand for non-experts.

Definition 8.1 Let M be an open manifold with a decomposition $M = M_0 \cup M_\infty$ in a compact manifold M_0 with boundary $Y := \partial M_0$ and a non-compact manifold M_∞ with boundary and suppose that $\partial M_\infty = M_0 \cap M_\infty = \partial M_0$. If M is endowed with a complete Riemannian metric such that

(i) M_∞ is isometric to $Y \times \mathbb{R}_+$ with the natural product metric $g_y + du \otimes du$, we say that M **is a manifold with cylindrical end**;
(ii) M_∞ is isometric to $Y \times \mathbb{R}_+$ with the Riemannian metric $\frac{du \otimes du}{u^2} + \frac{g_y}{u^2}$, we say that M **is a a manifold with cusp end**.

Let Δ_{cyl} be the Laplacian associated to a manifold (M, g_M) with cylindrical end. We recall that on $M_\infty = Y \times \mathbb{R}_+$ the Laplacian Δ_{cyl} has the form

$$\Delta_{0,M} := -\frac{\partial^2}{\partial u^2} + \Delta_Y$$

where Δ_Y is the Laplacian on Y associated to the Riemannian metric g_Y. Let (N^n, g_N) be a manifold with cusp, with $N_\infty = Y \times \mathbb{R}_+$, and let Δ_{cusp} be the Laplacian associated to (N^n, g_N). Δ_{cusp} has the form

$$\Delta_{0,N} := -u^2\frac{\partial^2}{\partial u^2} + (n-1)u\frac{\partial}{\partial u} + u^2\Delta_Y$$

on N_∞. An important approach, in quantum scattering theory, is to consider the Laplacians Δ_{cyl} and Δ_{cusp} as perturbations of the operators $\Delta_{0,M}$ and $\Delta_{0,N}$ respectively. Roughly speaking, this idea is formalized in the following way. Using Dirichlet boundary conditions, we consider the operators $\Delta_{0,M}$ and $\Delta_{0,N}$ as self-adjoint operators in $L^2(Y \times \mathbb{R}_+, dvol_{g_M})$ and $L^2(Y \times \mathbb{R}_+, dvol_{g_N})$. Observe that $L^2(Y \times \mathbb{R}_+, dvol_{g_M})$ is naturally isomorphic to $L^2(Y, dvol_{g_Y}) \otimes L^2(\mathbb{R}_+, du)$; furthermore, if $\phi_i \in C^\infty(Y)$ is an orthonormal basis of eigenvectors of Δ_Y with eigenvalues μ_i, then it provides an isomorphism of $L^2(Y, dvol_{g_Y}) \otimes L^2(\mathbb{R}_+, du)$ with $\oplus_{i \in \mathbb{N}} L^2(\mathbb{R}_+, du)$, hence $L^2(Y \times \mathbb{R}_+, dvol_{g_M}) \cong \oplus_{i \in \mathbb{N}} L^2(\mathbb{R}_+, du)$. Modulo this last isomorphism, we have:

$$\Delta_{0,M} = \bigoplus_{i \in \mathbb{N}} \left(-\frac{\partial^2}{\partial u^2} + \mu_i \right). \tag{8.1}$$

The analogue of (8.1) in the context of manifolds with cusps is

$$\Delta_{0,M} = \bigoplus_{i \in \mathbb{N}} \left(-u^2\frac{\partial^2}{\partial u^2} + (n-1)u\frac{\partial}{\partial u} + u^2\mu_j \right). \tag{8.2}$$

From this point we continue our exposition over the manifold with cylindrical end M, indicating how analogous methods apply in the case of manifolds with cusps and referring to [Mül83] for details.

Fourier transform and (8.2) give us a formula for the resolvent of the operators $-\frac{\partial^2}{\partial u^2} + \mu_i$ and hence for the resolvent of $\Delta_{0,M}$. The analogue of such a formula in the case of the manifold N with cusp is technically harder to obtain and can be found in [Mül83, Lemma 2.68]. Define the double of the compact manifold with boundary M_0 as $\tilde{M} := M_0 \cup_Y M_0$, where we are identifying the boundary Y of two disjoint copies of M_0 and we endow $\tilde{M}$ with the natural differential and Riemannian structure. We have also a nice formula for the resolvent of the Laplacian $\Delta_{\tilde{M}}$ of $\tilde{M}$ using a spectral resolution of $\Delta_{\tilde{M}}$. In order to apply this knowledge about the resolvents in M_∞ and $\tilde{M}$, we construct a *parametrix* for the resolvent of Δ_M, i.e.

an operator $P(\lambda)$ such that $R(\lambda) - P(\lambda)$ is compact in some weighted L^2-space, where $\lambda \in \mathbb{C} - \mathbb{R}_+$ and $R(\lambda) := (\Delta_M - \lambda)^{-1}$ denotes the resolvent of Δ_M. We proceed as in [Hus05]: for $0 \geq a \geq b$, let $\rho(a, b) \in C^\infty(M, [0, 1])$ be such that

$$\rho(a, b)(x) = \begin{cases} 0 & \text{for } x \in M \cup (Y \times [0, a]); \\ 1 & \text{for } x \in Y \times [b, \infty). \end{cases}$$

We define the functions:

$$\Phi_1 := 1 - \rho\left(\tfrac{4}{5}, 1\right), \quad \Psi_1 := 1 - \rho\left(\tfrac{2}{5}, \tfrac{3}{5}\right)$$
$$\Phi_2 := \rho\left(0, \tfrac{1}{5}\right), \quad \Psi_2 := 1 - \Psi_1,$$

for which $\Psi_1 + \Psi_2 = 1$,

$$\Phi_j(x) = 1 \text{ for } x \in supp(\Psi_j), \quad \text{and dist}(supp \nabla \Phi_j, supp \Psi_j) \geq \tfrac{1}{5}.$$

We define the operator $S(\lambda)$ with Schwartz kernel

$$S(x_1, x_2, \lambda) := \sum_{j=1}^{2} \Psi_j(x_1) R_j(x_1, x_2, \lambda) \Phi_j(x_2), \tag{8.3}$$

where $R_1(x_1, x_2, \lambda)$ is the Schwartz kernel of the resolvent of the Laplacian $\Delta_{\tilde{M}}$, on $\tilde{M}$ the double of the manifold with boundary M_0, and $R_2(x_1, x_2, \lambda)$ is the Schwartz kernel of the resolvent of $\Delta_{0,M}$. Using the explicit expressions of the Schwartz kernels $R_1(x_1, x_2, \lambda)$ and $R_2(x_1, x_2, \lambda)$ and (8.1), it is possible to prove the following.

Lemma 8.1.1 [Hus05, Lemma 3.8] For $\lambda \in \mathbb{C} - \mathbb{R}_+$, the operator $S(\lambda)$ is a parametrix of $R(\lambda)$ in the sense that $R(\lambda) - S(\lambda)$ is L^2-compact.

The meromorphic extension of the resolvent will have a domain contained in the following surface, which we call a *spectral surface*:

$$\Sigma_s := \{ \Lambda := (\Lambda_i) \in \mathbb{C}^{\mathbb{N}} : \forall i, j \in \mathbb{N}, \ \Lambda_i^2 + \mu_i = \Lambda_j^2 + \mu_j \}.$$

Σ_s is a covering of $\mathbb{C}$ with projection $\pi_s(\Lambda) := \Lambda_i^2 + \mu_i$.

As we said previously, we will consider the resolvent acting on weighted L^2-spaces that we shall define now:

$$L_\delta^2(M) := \left\{ f : M \stackrel{\text{meas}}{\to} \mathbb{C} : \int_0^\infty \int_Y e^{2\delta u} |f(y, u)|^2 dvol_{g_Y} du < \infty \right\}. \tag{8.4}$$

For all $\delta > 0$, we have the inclusions

$$L_\delta^2(M) \subset L^2(M) \subset L_{-\delta}^2(M).$$

If we define the *physical domain FD* by

$$FD := \{\Lambda \in \Sigma_s : \Lambda_i \geq 0\},$$

then we can identify FD with $\mathbb{C} - \mathbb{R}_+$. We denote by Σ_s^μ the connected component of $\pi_s^{-1}(\mathbb{C} - [\mu, \infty))$ and, for $\epsilon > 0$, $\mu(\epsilon) := \min\{\mu \in \sigma(\Delta_Y) : \mu \leq \epsilon\}$. Now we can define the domains Ω_ϵ, for $\epsilon > 0$, where we will extend the resolvent of Δ_M,

$$\Omega_\epsilon := \left(FD \cup \pi_s^{-1}(\{z \in \mathbb{C} : |z| \leq \epsilon\})\right) \cap \Sigma_s^{\mu(\epsilon)}.$$

From the explicit formulas of the resolvents $R_1(\lambda)$ and $R_2(\lambda)$ in equation (8.3), and the definition of the weigthed L^2-spaces in (8.4), we deduce the following.

Lemma 8.1.2 [Hus05, Lemma 3.20]   For all $\delta \geq \epsilon > 0$ the function $\lambda \mapsto S(\lambda)$ has a meromorphic extension to Ω_ϵ as a continuous operator from $L_\delta^2(M)$ to $L_{-\delta}^2(M)$.

For $\Lambda \in \Omega_\epsilon$, let us define the operators

$$G(\Lambda) := S(\Lambda)(\Delta_M - \pi_s(\Lambda)) - Id.$$

We will use the following important tool of functional analysis to meromorphically extend the resolvent.

Theorem 8.1.3 [Hus05, Appendix] (Analytic Fredholm theorem)   Let $U \subset \mathbb{C}$ be an open and connected set and let $T(z)$, for $z \in U$, be an analytic family of compact operators of a Hilbert space $\mathcal{H}$. Suppose that, for $z_0 \in U$, $(Id - T(z_0))^{-1}$ exists, then the family $(Id - T(z))^{-1}$ is meromorphic in U with values in the bounded linear operators of $\mathcal{H}$ and poles contained in the set $\{z \in U : 1 \in \sigma(T(z))\}$.

As for Lemma 8.1.1, it is possible to prove that $G(\Lambda)$, as a bounded operator of $L_{-\delta}^2(M)$, is compact. This fact, Lemma 8.1.2 and Theorem 8.1.3 imply the following.

Theorem 8.1.4 [Hus05, Theorem 3.24]   The resolvent $R(\lambda)$ has a meromorphic extension from FD to Ω_ϵ as a continuous operator from $L_\delta^2(M)$ to $L_{-\delta}^2(M)$.

The next theorem provides the connection between the meromorphic extension of the resolvent and the absence of singular continuous spectrum.

Theorem 8.1.5 [RS79]   Let H be a self-adjoint operator with resolvent $R(\lambda) := (H - \lambda)^{-1}$.

(i) Let (a, b) be a bounded interval and $\varphi \in \mathcal{H}$. Suppose that there exists $p > 1$ for which:

$$\sup_{0 < \epsilon < 1} \int_a^b |\text{Im}(\varphi, R(x + i\epsilon)\varphi)|^p dx < \infty. \tag{8.5}$$

Then $E_{(a,b)}\varphi \in \mathcal{H}_{\text{ac}}.$

(ii) Let (a, b) be a bounded interval. Suppose that there is a dense subset D in $\mathcal{H}$ such that for $\varphi \in D$ the inequality (8.5) holds for some $p > 1$. Then H has purely absolutely continuous spectrum on (a, b).

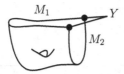

Figure 8.1 Compact manifold with corner of codimension 2.

8.2 Analytic dilation on complete manifolds with corners of codimension 2

In [Mül96] it has been explained how to meromorphically extend the resolvent of a generalized Laplacian on a complete manifold with corner of codimension 2, using the method outlined in Section 8.1 under the hypothesis that the Laplacian on the corner has kernel $\{0\}$. It turns out that to weaken this hypothesis and try to use the methods of Section 8.1 is not easy and new methods should be used to prove absence of singular continuous spectrum. In this section we survey the method of analytic dilation applied to compatible Laplacians on complete manifolds with corners of codimension 2. This method appeared originally in the context of Schrödinger operators and was adapted in [Can11] to this geometric context.

Following [Mül96], we explain the notions of *compact and complete manifolds with corner of codimension 2*. Let X_0 be a compact oriented Riemannian manifold with boundary M and suppose that there exists a hypersurface Y of M that divides M into two manifolds with boundary M_1 and M_2, i.e. $M = M_1 \cup M_2$ and $Y = M_1 \cap M_2$ (see Fig. 8.1). Assume also that a neighborhood of Y in M is diffeomorphic to $Y \times (-\varepsilon, \varepsilon)$. We say that the manifold X_0 **has a corner of codimension 2** if X_0 is endowed with a Riemannian metric g that is a product metric on small neighborhoods, $M_i \times (-\varepsilon, 0]$ of the M_is and on a small neighborhood, $Y \times (-\varepsilon, 0]^2$, of the corner Y. If X_0 has a corner of codimension 2, we say that X_0 is a **compact manifold with corner of codimension 2**.

Example 8.2 For $i = 1, 2$, let M_i be a compact oriented Riemannian manifold with boundary $\partial M_i := Y_i$. Suppose that on a neighborhood $Y_i \times (-\varepsilon, 0]$ of Y_i the Riemannian metric g_i of M_i is a product metric, i.e. $g_i := g_{Y_i} + du \otimes du$ where u is the coordinate associated to $(-\varepsilon, 0]$ in $Y_i \times (-\varepsilon, 0]$. Then $M_1 \times M_2$ is a compact manifold with corner of codimension 2.

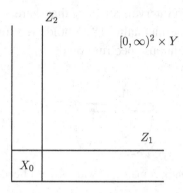

Figure 8.2 Sketch of a complete manifold with corner of codimension 2.

From the compact manifold with corner X_0 we construct a complete manifold X. Let

$$Z_i := M_i \cup_Y (\mathbb{R}^+ \times Y), \quad i = 1, 2,$$

where the bottom $\{0\} \times Y$ of the half-cylinder is identified with $\partial M_i = Y$. Then Z_i is a manifold with cylindrical end (see Definition 8.1). Define the manifolds

$$W_1 := X_0 \cup_{M_2} (\mathbb{R}_+ \times M_2) \quad \text{and} \quad W_2 := X_0 \cup_{M_1} (\mathbb{R}_+ \times M_1).$$

Observe that W_i is an n-dimensional manifold with boundary Z_i that can be equipped with a Riemannian metric compatible with the product metric of $\mathbb{R}_+ \times M_2$ and the Riemannian metric of X_0. Set:

$$X := W_1 \cup_{Z_1} (\mathbb{R}_+ \times Z_1) = W_2 \cup_{Z_2} (\mathbb{R}_+ \times Z_2),$$

where we identify $\{0\} \times Z_i$ with Z_i, the boundary of W_i. Figure 8.2 is a sketch, in particular the lines that enclose the picture should not be thought of as boundaries.
Let $T \geq 0$ be given and set

$$Z_{i,T} := M_i \cup_Y ([0, T] \times Y), \quad \text{for } i = 1, 2, \tag{8.6}$$

where $\{0\} \times Y$ is identified with Y, the boundary of M_i. $Z_{i,T}$ is a family of manifolds with boundary which exhausts Z_i. Next we attach to X_0 the manifold $[0, T] \times M_1$ by identifying $\{0\} \times M_1$ with M_1. The resulting manifold $W_{2,T}$ is a compact manifold with corner of codimension 2, whose boundary is the union of M_1 and $Z_{2,T}$ (Fig. 8.3). The manifold X has associated a natural exhaustion given by

$$X_T := W_{2,T} \cup_{Z_{2,T}} ([0, T] \times Z_{2,T}), \quad T \geq 0, \tag{8.7}$$

where we identify $Z_{2,T}$ with $\{0\} \times Z_{2,T}$.

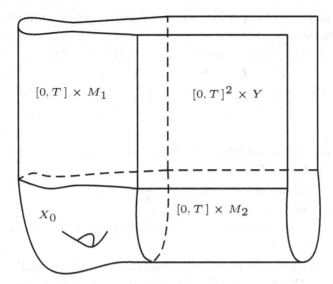

Figure 8.3 X_T, element of the exhaustion of X.

For each $T \in [0, \infty)$, X has two submanifolds with cylindrical ends, namely $M_i \times \{T\} \cup (Y \times \{T\}) \times [0, \infty)$, for $i = 1, 2$.

Let E be a Hermitian vector bundle over a complete manifold with corner of codimension 2, X. Let Δ be a generalized Laplacian acting on $C^\infty(X, E)$. The operator Δ is a **compatible Laplacian** over X if it satisfies the following properties:

(i) There exists a Hermitian vector bundle E_i over Z_i such that $E|_{\mathbb{R}_+ \times Z_i}$ is the pullback of E_i under the projection $\pi : \mathbb{R}_+ \times Z_i \to Z_i$, for $i = 1, 2$. We suppose also that the Hermitian metric of E is the pullback of the Hermitian metric of E_i. On $\mathbb{R}_+ \times Z_i$, we have

$$\Delta = -\frac{\partial^2}{\partial u_i^2} + \Delta_{Z_i},$$

where Δ_{Z_i} is a compatible Laplacian acting on $C^\infty(Z_i, E_i)$.

(ii) There exists a Hermitian vector bundle S over Y such that $E|_{\mathbb{R}_+^2 \times Y}$ is the pullback of S under the projection $\pi : \mathbb{R}_+^2 \times Y \to Y$. We assume also that the Hermitian product on $E|_{\mathbb{R}_+^2 \times Y}$ is the pullback of the Hermitian product on S. Finally we suppose that the operator Δ restricted to $\mathbb{R}_+^2 \times Y$ satisfies

$$\Delta = -\frac{\partial^2}{\partial u_1^2} - \frac{\partial^2}{\partial u_2^2} + \Delta_Y,$$

where Δ_Y is a generalized Laplacian acting on $C^\infty(Y, S)$.

Examples of compatible Laplacians are the Laplacian acting on forms and Laplacians associated to compatible Dirac operators (see [Mül96]). Since X is a manifold with bounded geometry and the vector bundle E has bounded Hermitian metric, the operator $\Delta : C_c^\infty(X, E) \to L^2(X, E)$ is essentially self-adjoint (see [Shu91, Corollary 4.2]). Similarly $\Delta_{Z_i} : C_c^\infty(Z_i, E_i) \to L^2(Z_i, E_i)$ is also essentially self-adjoint for $i = 1, 2$.

Definition 8.3
- Let H and $H^{(i)}$ be the self-adjoint extensions of $\Delta : C_c^\infty(X, E) \to L^2(X, E)$ and $\Delta_{Z_i} : C_c^\infty(Z_i, E_i) \to L^2(Z_i, E_i)$.
- Let b_i be the self-adjoint extension of $-\frac{d^2}{du_i^2} : C_c^\infty(\mathbb{R}_+) \to L^2(\mathbb{R}_+)$ obtained by imposing Dirichlet boundary conditions at 0.
- Let H_i be the self-adjoint operator $b_i \otimes \mathrm{Id} + \mathrm{Id} \otimes H^{(i)}$ acting on $L^2(\mathbb{R}_+) \otimes L^2(Z_i, E_i)$.
- Let $H^{(3)}$ be the self-adjoint operator associated to the essentially self-adjoint operator $\Delta_Y : C^\infty(Y, S) \to L^2(Y, S)$ and let H_3 be the self-adjoint operator $H_3 := b_1 \otimes \mathrm{Id} \otimes \mathrm{Id} + \mathrm{Id} \otimes b_2 \otimes \mathrm{Id} + \mathrm{Id} \otimes \mathrm{Id} \otimes H^{(3)}$ acting on $L^2(\mathbb{R}_+) \otimes L^2(\mathbb{R}_+) \otimes L^2(Y)$.

The operators H_i are called **channel operators** for $i = 1, 2, 3$.

The self-adjoint operators H_1 and H_2 have a free channel of dimension 1 (associated to b_1 and b_2, respectively); the operator H_3 has a free channel of dimension 2 (associated to $b_1 \otimes \mathrm{Id} \otimes \mathrm{Id} + \mathrm{Id} \otimes b_2 \otimes \mathrm{Id}$). In some parts of this text we abuse the notation and denote by H, H_i and $H^{(i)}$ the Laplacians acting on distributions and the self-adjoint operators previously defined.

Throughout the next section we will use the following notation. Let H be a self-adjoint operator acting on a Hilbert space $\mathscr{H}$. We define the Banach space $\mathscr{H}_2(H)$ as the domain of H with the norm $||\varphi||_2 := ||(|H| + i)\varphi||$. Similarly, we define the Banach space $\mathscr{H}_1(H)$ as the completion of $\mathscr{H}_2(H)$ with the norm $||\varphi||_1 := ||(|H| + i)^{1/2}\varphi||$, and $\mathscr{H}_{-1}(H)$ and $\mathscr{H}_{-2}(H)$ as the dual spaces associated to $\mathscr{H}_1(H)$ and $\mathscr{H}_2(H)$.

8.2.1 Analytic dilation

In this section we explain the method of analytic dilation for the Laplacian acting on functions that we will denote H. However, the method generalizes to the compatible Laplacians explained in the earlier sections. The analytic dilation of a many-body Schrödinger operator depends on the analytic dilation of their subsystem Hamiltonians (see [HS00]). In a similar way the analytic dilation of H is described in terms of the spectral theory of the operators $H^{(1)}$, $H^{(2)}$ and $H^{(3)}$, explained above. For $\theta > 0$, we define the operator $U_{i,\theta} : L^2(Z_i) \to L^2(Z_i)$ that essentially is the dilation operator by $\theta + 1$ up to a compact set. More precisely:

$$U_{i,\theta} f(x) = \begin{cases} f(x) & \text{for } x \in M_i, \\ (\theta + 1)^{1/2} f((\theta + 1)u, y) & \text{for } x = (u, y) \in [0, \infty) \times Y \\ & \text{and for } u \text{ big enough}, \end{cases}$$

and $U_{i,\theta} f$ is extended to the whole manifold Z_i in such a way that it sends $C_c^\infty(Z_i)$ into $C_c^\infty(Z_i)$, and it becomes a unitary operator on $L^2(Z_i)$. We refer the reader to [Can11] for the technical details. Similarly, the operators $U_\theta : L^2(X) \to L^2(X)$ are defined by

$$U_\theta f(x) = \begin{cases} f(x) & \text{for } x \in X_0, \\ (\theta + 1)^{1/2} U_{i,\theta} f((\theta + 1)u_i, z_i) & \text{for } x = (u_i, z_i) \in [0, \infty) \times Z_i \\ & \text{and for } u_i \text{ big enough}. \end{cases}$$

Again $U_\theta f$ is extended to the whole X in such a way that, for $f \in C_c^\infty(X)$, $U_\theta f \in C_c^\infty(X)$, and U_θ becomes a unitary operator in $L^2(X)$.

For $\theta \in [0, \infty)$, define $H_\theta := U_\theta H U_\theta^{-1}$, a closed operator with domain $\mathcal{H}_2(H)$. We have that

$$\mathcal{H}_2(H) = \{f \in L^2(X) : \Delta_{\text{dist}} f \in L^2(X)\}$$

is the second Sobolev space associated to (X, g). Consider the set:

$$\Gamma := \{\theta := \theta_0 + i\theta_1 \in \mathbb{C} : \theta_0 > 0, \theta_0 > |\theta_1| \text{ and } \text{Im}(\theta)^2 < 1/2\}. \tag{8.8}$$

We will extend the family H_θ from $[0, \infty)$ to Γ (Fig. 8.4).

In [Can11], the next theorem is proved.

Theorem 8.2.1 [Can11]   The family $(H_\theta)_{\theta \in [0,\infty)}$ extends to an holomorphic family for $\theta \in \Gamma$, which satisfies:

(1) H_θ is a closed operator with domain $\mathcal{H}_2(H)$ for $\theta \in \Gamma$,
(2) for $\varphi \in \mathcal{H}_2(H)$ the map $\theta \mapsto H_\theta \varphi$ is holomorphic in Γ.

A holomorphic family of operators satisfying (1) and (2) is called **a holomorphic family of type A**. This theorem is proved using the analogous result that the family $\{H^{(i),\theta}\}_{\theta \in [0,\infty)}$ extends to a holomorphic family of type A in Γ, where $H^{(i),\theta}$ denotes the closed operator associated to $U_{i,\theta} \Delta_{Z_i} U_{i,\theta}^{-1}$ with domain

$$\mathcal{H}_2(H^i) = \{f \in L^2(Z_i) : \Delta_{\text{dist}}(f) \in L^2(Z_i)\},$$

the second Sobolev space associated to (Z_i, g_i).

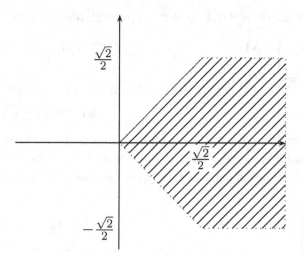

Figure 8.4 Sketch of the region Γ.

The families H_θ and $H_\theta^{(i)}$ extend to domains larger than Γ, but Γ is enough for our goals. In particular, Γ is chosen because for $\theta \in \Gamma$ it is easy to prove that $H_\theta^{(i)}$ is m-sectorial (see [Can11, Section 2.7]), a fact that will be important for the proof of Theorem 8.2.2 where the Ichinose lemma is a main tool. We define

$$\theta' := \frac{1}{(\theta + 1)^2}.$$

The parameter θ' is very important in the description of the essential spectrum of H_θ, as we can see in the next theorem.

Theorem 8.2.2 [Can11] For $\theta \in \Gamma$,

$$\sigma_{ess}(H_\theta) = \bigcup_{\mu \in \sigma(H^{(3)})} (\mu + \theta'[0, \infty))$$

$$\cup \bigcup_{\lambda_1 \in \sigma_{pp}(H^{(1),\theta})} \left(\lambda_1 + \theta'[0, \infty) \right)$$

$$\cup \bigcup_{\lambda_2 \in \sigma_{pp}(H^{(2,\theta)})} \left(\lambda_2 + \theta'[0, \infty) \right).$$

It is possible to associate to $(U_\theta)_{\theta \in [0,\infty)}$ a set of functions $\mathcal{V} \subset \mathcal{H}_2(H)$ that satisfies:

(i) $\mathcal{V}$ is dense in $L^2(X)$,
(ii) for $\varphi \in \mathcal{V}$, $U_\theta \varphi$ is defined for all $\theta \in \Gamma$,
(iii) $U_\theta \mathcal{V}$ is dense in $L^2(X)$ for all $\theta \in \Gamma$.

The elements of a subset of $\mathscr{H}_2(H)$ which satisfy (i) and (ii) will be called **analytic vectors**. We denote by Λ the left-hand plane, more explicitly:

$$\Lambda := \{(x, y) \in \mathbb{C} : x < 0\}. \tag{8.9}$$

We denote by $R(\lambda)$ the resolvent of H and by $R(\lambda, \theta)$ the resolvent of H_θ. Using the general analytic dilation theory of Aguilar–Balslev–Combes (see [Bal97]) we describe the nature of the spectrum of H. This theory is based on:

  (i) knowledge of the essential spectrum of H_θ, provided by Theorem 8.2.2,
  (ii) the following equation, which is a consequence of the unitarity of U_θ,

$$\langle R(\lambda)f, g \rangle_{L^2(X)} = \langle R(\lambda, \theta)U_\theta f, U_\theta g \rangle_{L^2(X)}, \tag{8.10}$$

  for $f, g \in \mathscr{V}$ and $\theta \in [0, \infty)$.

Since the right-hand side of (8.10) is defined for $\lambda \in \Lambda$ and $\theta \in \Gamma$, (8.10) provides a meromorphic extension of the functions $\lambda \mapsto \langle R(\lambda)f, g \rangle_{L^2(X)}$ from Λ to $\mathbb{C} - \sigma(H_\theta)$. From this, using Aguilar–Balslev–Combes, we deduce the following theorem.

Theorem 8.2.3 [Can11]

(1) For $f, g \in \mathscr{A}$ the function $\lambda \mapsto \langle R(\lambda)f, g \rangle_{L^2(X)}$ extends from Λ to $\mathbb{C} - \sigma(H_\theta)$.
(2) For all $\theta \in \Gamma$, H_θ has no singular continuous spectrum.
(3) The accumulation points of $\sigma_{\mathrm{pp}}(H)$ are contained in $\{\infty\} \cup \sigma(H^{(3)}) \cup \cup_{i=1}^2 \sigma_{\mathrm{pp}}(H^{(i)})$.

In the case of manifolds with cylindrical ends we have shown in Section 8.1 the absence of singular continuous spectrum for their Laplacians; in [Don84], by giving a polynomial bound to the increase in the number of L^2-eigenvalues, it is proved that the unique possible accumulation point of the pure point spectrum of a Laplacian on a manifold with cylindrical end is ∞.

Acknowledgments

The author is very grateful to the organizers of the summer school in Villa de Leyva for giving him the opportunity to take part in this nice school and to write this paper for its proceedings. Special thanks are due to Alexander Cardona who helped to improve the presentation of this chapter. The author also wants to thank Professor Werner Müller at Bonn University for introducing him to these topics, and for his constant advice during the writing of the PhD thesis on which Section 8.2 and the paper [Can11] are based.

References

[Bal97] E. Balslev. Spectral deformation of Laplacians on hyperbolic manifolds. *Comm. Anal. Geom.*, **5**(2): 213–247, 1997.

[Can11] L. Cano. Analytic dilation on complete manifolds with corners of codimension 2. PhD thesis, University of Bonn, 2011.

[Can12] L. Cano. Mourre estimates for compatible laplacians on complete manifolds with corners of codimension 2. *Ann. Global Anal. Geom.*, 2012.

[Cha84] I. Chavel. *Eigenvalues in Riemannian geometry*, Pure and Applied Mathematics 115. Orlando, FL: Academic Press, 1984. Including a chapter by Burton Randol, with an appendix by Jozef Dodziuk.

[DEM98] P. Duclos, P. Exner and B. Meller. Exponential bounds on curvature-induced resonances in a two-dimensional Dirichlet tube. *Helv. Phys. Acta*, **71**(2): 133–162, 1998.

[Don84] H. Donnelly. Eigenvalue estimates for certain noncompact manifolds. *Michigan Math. J.*, **31**(3): 349–357, 1984.

[Gér93] C. Gérard. Distortion analyticity for N-particle Hamiltonians. *Helv. Phys. Acta*, **66**(2): 216–225, 1993.

[Gui89] L. Guillopé. Théorie spectrale de quelques variétés à bouts. *Ann. Sci. École Norm. Sup. (4)*, **22**(1): 137–160, 1989.

[HS00] W. Hunziker and I. M. Sigal. The quantum N-body problem. *J. Math. Phys.*, **41**(6): 3448–3510, 2000.

[Hus05] R. Husseini. Zur spektraltheorie verallgemeinerter laplace-operatoren auf mannigfaltigkeiten mit zylindrischen enden. Diplomarbeit, Rheinischen Friedrich-Wilhelms-Universität Bonn, 2005.

[Kal10] V. Kalvin. The aguilar-baslev-combes theorem for the laplacian on a manifold with an axial analytic asymptotically cylindrical end. arXiv:10032538v2, 2010.

[KS07] H. Kovařík and A. Sacchetti. Resonances in twisted quantum waveguides. *J. Phys. A*, **40**(29): 8371–8384, 2007.

[Mül83] W. Müller. Spectral theory for Riemannian manifolds with cusps and a related trace formula. *Math. Nachr.*, **111**: 197–288, 1983.

[Mül96] Werner Müller. On the L^2-index of Dirac operators on manifolds with corners of codimension two. I. *J. Diff. Geom.*, **44**(1): 97–177, 1996.

[MV02] R. Mazzeo and A. Vasy. Resolvents and Martin boundaries of product spaces. *Geom. Funct. Anal.*, **12**(5): 1018–1079, 2002.

[MV04] R. Mazzeo and A. Vasy. Analytic continuation of the resolvent of the Laplacian on SL(3)/SO(3). *Amer. J. Math.*, **126**(4): 821–844, 2004.

[MV07] R. Mazzeo and A. Vasy. Scattering theory on SL(3)/SO(3): connections with quantum 3-body scattering. *Proc. Lond. Math. Soc. (3)*, **94**(3): 545–593, 2007.

[Ros97] Steven Rosenberg. *The Laplacian on a Riemannian manifold*, London Mathematical Society Student Texts 31. Cambridge: Cambridge University Press, 1997.

[RS79] M. Reed and B. Simon. *Methods of modern mathematical physics. III*. New York: Academic Press, 1979.

[RS80] M. Reed and B. Simon. *Methods of modern mathematical physics. I*, 2nd edn. New York: Academic Press, 1980.

[Shu91] M. A. Shubin. Spectral theory of elliptic operators on non-compact manifolds. Paper presented at the Summer School on Semiclassical Methods, Nantes, 1991.

[SZ93a] J. Sjöstrand and M. Zworski. Estimates on the number of scattering poles near the real axis for strictly convex obstacles. *Ann. Inst. Fourier (Grenoble)*, **43**(3): 769–790, 1993.

[SZ93b] J. Sjöstrand and M. Zworski. Lower bounds on the number of scattering poles. *Comm. Partial Diff. Equations*, **18**(5–6): 847–857, 1993.

[SZ94] J. Sjöstrand and M. Zworski. Lower bounds on the number of scattering poles. II. *J. Funct. Anal.*, **123**(2): 336–367, 1994.

[SZ95] J. Sjöstrand and M. Zworski. The complex scaling method for scattering by strictly convex obstacles. *Ark. Mat.*, **33**(1): 135–172, 1995.

9
Models for formal groupoids

IVÁN CONTRERAS

Abstract

This chapter is a brief overview of some semiclassical objects with
particular relevance in Poisson geometry and deformation quantization:
formal groupoids. We give a categorical description of the object, study
its associated algebraic structure (Hopf algebroid), mentioning its relevance in Poisson geometry as formal realizations of Poisson manifolds.

9.1 Motivation and plan

The relation between smooth manifolds and their algebras of smooth functions
has been studied deeply and the problem of connecting geometric information and
algebraic data appears frequently in Lie theory and deformation theory, among
others.

   In particular, the notion of a groupoid appears naturally as a generalization of
the structure of a group and it helps to understand geometric spaces. Its study in
differential geometry allows us to link, for example, Lie groupoids and foliations
of Poisson manifolds. In a more general setting, the notion of a groupoid object
in a category $\mathcal{C}$ [1] can be introduced, and this generalized version of groupoid
appears as a solution of what is called the *Integrability problem* or the generalized
Lie Third Theorem for Lie algebroids [5] and in particular for Poisson manifolds
[6], [4]. The main objective in this overview is to discuss different approaches
to describe *formal groupoids*, which can be defined categorically as a groupoid
object in a certain category, in which the properties of the object are encoded in
the spaces of infinite jets associated to smooth manifolds. The first part of these
lecture notes is devoted to describing groupoids in general, with some particular

Geometric and Topological Methods for Quantum Field Theory, ed. Alexander Cardona, Iván Contreras and
Andrés F. Reyes-Lega. Published by Cambridge University Press. © Cambridge University Press 2013.

examples. Specifically, the formal groupoid will appear in a special category where the morphisms are infinite jets of functions over manifolds.

Alternatively, the formal groupoid described by Karabegov in (see [7]), is written in terms of a formal neighborhood $C^\infty(P, M)$ with respect to a local pair manifold (P, M). Now, this object lives in the category of commutative algebras and Theorem 9.19 relates both definitions, where the formal groupoid defined in [7] appears as the contravariant version of the J_∞-**Mic** formal groupoid; the pullback construction in the second one corresponds to the push-out construction in the first one. In Section 9.3 the construction given in [7] is outlined. In addition, formal groupoids have a natural structure of *Hopf algebroids* (Theorem 9.25).

In Poisson geometry, the notion of symplectic groupoid appears frequently, especially in the context of symplectic realizations and integration of Poisson manifolds. The purpose of Section 9.4 is to discuss the particular case of formal symplectic groupoids and the connection with Poisson geometry. Some of the perspectives given by this approach are related to the explicit examples of formal symplectic groupoids as formal integration of Poisson manifolds, in particular, the formal deformation of the canonical generating functions associated to the trivial symplectic groupoid $T^*\mathbb{R}^n \rightrightarrows \mathbb{R}^n$ ([3]) as integration of the Lie algebroid T^*M associated to a Poisson manifold (M, Π).

9.2 Definitions and examples

A groupoid is, by definition, a small category[1] with invertible morphisms. The following definition allows us to talk about groupoid objects in categories.

Definition 9.1 Let $\mathcal{C}$ be a category. A groupoid object in $\mathcal{C}$, denoted by $\Sigma \rightrightarrows B$, corresponds to the following data:

1. Two objects Σ, B in $\mathcal{C}$.
2. Two morphisms $s, t : \Sigma \rightrightarrows B$ called the *source* and *target*, respectively, such that the pullbacks

$$
\begin{array}{ccc}
\Sigma \times_{(s,t)} \Sigma & \xrightarrow{\pi_2} & \Sigma \\
\downarrow{\scriptstyle \pi_1} & & \downarrow{\scriptstyle t} \\
\Sigma & \xrightarrow{s} & B
\end{array}
$$

and the associated pullbacks $\Sigma \times_{(s,t)} (\Sigma \times_{(s,t)} \Sigma)$ and $(\Sigma \times_{(s,t)} \Sigma) \times_{(s,t)} \Sigma$ exist.

---

[1] This means that the objects and morphisms are sets and not proper classes.

3. The morphisms

$$m : \Sigma \times_{(s,t)} \Sigma \longrightarrow \Sigma, \quad i : \Sigma \longrightarrow \Sigma, \quad \varepsilon : B \longrightarrow \Sigma$$

called *multiplication, inversion* and *unit* maps, respectively, satisfying certain compatibility axioms (associativity, unitality, compatibility of the inversion with source and target; for details see e.g. [1]). Therefore, we can think about a groupoid as the following set of arrows and objects

$$\Sigma \times_{(s,t)} \Sigma \xrightarrow{\ m\ } \Sigma \xrightarrow{\ i\ } \Sigma \underset{t}{\overset{s}{\rightrightarrows}} B \xrightarrow{\ \varepsilon\ } \Sigma.$$

The different versions of groupoid objects in categories are subject to the nature of the objects and the morphisms, together with the existence of the corresponding pullbacks.

In the category **Set**, with object sets and morphisms maps between them, a groupoid object is what is usually called a *groupoid* $\Sigma \rightrightarrows B$ with the objects and maps described by:

$$\Sigma \times_{(s,t)} \Sigma := \{(a, b) \in \Sigma \times \Sigma \mid t(a) = s(b)\},$$

$$\Sigma_3 := \{(a, b, c) \in \Sigma \times \Sigma \times \Sigma \mid (a, b) \in \Sigma \times_{(s,t)} \Sigma, (b, c) \in \Sigma \times_{(s,t)} \Sigma\}$$

and the following axioms for the maps:

(A.1) $s(m(a, b)) = s(a), \forall a, b \in \Sigma \times_{(s,t)} \Sigma.$
(A.2) $t(m(a, b)) = t(b), \forall a, b \in \Sigma \times_{(s,t)} \Sigma.$
(A.3) $s \circ \varepsilon = \mathrm{id}_M.$
(A.4) $t \circ \varepsilon = \mathrm{id}_M.$
(A.5) $s(i(a)) = t(a), \forall a \in \Sigma.$
(A.6) $m(a, i(a)) = \varepsilon(s(a)) \forall a \in \Sigma.$
(A.7) $m(i(a), a) = \varepsilon(t(a)) \forall a \in \Sigma.$
(A.8) $m(a, m(b, c)) = m(m(a, b), c) \forall (a, b, c) \in \Sigma_3.$

As examples of groupoid objects in **Set** we have the following:

Example 9.2 The arbitrary disjoint union of groups

$$G = \sqcup_{\lambda \in \Lambda} G_\lambda$$

is a groupoid over Λ, where the product of a and b is defined if and only if $a, b \in G_\lambda$, for some $\lambda \in \Lambda$. The source and target map corresponds to

$$s = t : G \longrightarrow \{\lambda\}.$$

The unit map sends 1_λ to λ and the inversion map is the inverse in G_λ. As we can see, a group can be understood as a groupoid over a one-element base.

Example 9.3 Given an equivalence relation R on a set X, the groupoid structure over X is given by

$$s = \pi_1 R, \quad t = \pi_2 R, \quad \varepsilon(x) = (x, x), \quad i(x, y) = (y, x)$$

and the product is defined by $m((x, y), (y, z)) = (x, z)$. When $R = X \times X$ we call this the *coarse groupoid* and when $R = \Delta_X$, denoting the diagonal relation, this is called the *fine groupoid*.

Some interesting examples appear as generalizations of Lie groups. Let **Man** be the category whose objects are smooth manifolds and morphisms are smooth maps. A groupoid object in this category is usually called a *Lie groupoid*, with the condition that the source map s is a submersion.[2]

Example 9.4 If $\pi : E \longrightarrow M$ is a vector bundle over a manifold M we can define the *fiber groupoid* over M, where $s = t = \pi$, the unit map corresponds to the zero section and the inverse and multiplication map correspond to the fiberwise inverse and multiplication, respectively. As a particular case, the cotangent bundle $T^*M \rightrightarrows M$ is a groupoid over M.

9.2.1 Groupoid objects in special categories

We call a *local manifold pair* a pair (P, M) of smooth manifolds such that M is a closed submanifold of P (see [10]).

Definition 9.5 We define the category **S** (called the *strict category*) whose objects are local manifold pairs and a morphism from (P, M) to (Q, N) is a smooth map $f : P \longrightarrow Q$ such that $f(M) \subseteq N$. A groupoid object in **S** over M corresponds to a groupoid object with $B = (M, M)$, M a smooth manifold.

Remark Given a Lie groupoid $G \rightrightarrows M$ over M, we can identify it with a groupoid object in the category **S** over M: the maps are the same and the objects correspond to:

$$G \longmapsto (G, M); \quad M \longmapsto (M, M);$$

---

[2] The fact that s is a submersion implies that t is a submersion as well and that the pullback construction for $\Sigma \times_{(s,t)} \Sigma$ has the structure of a smooth manifold.

and associated to them the pullbacks:

$$G \times_{(s,t)} G \longmapsto (G \times_{(s,t)} G, M),$$

$$G_3 \longmapsto (G_3, M).$$

Definition 9.6 We define the category **G** (called the *Germ category*) whose objects are local manifold pairs and a morphism from (P, M) to (Q, N) is an equivalence class $[f]_G$ of smooth maps $f : P \longrightarrow Q$ such that $f(M) \subseteq N$ is defined by the following equivalence relation: $f \sim g$ iff there exists an open set U, $M \subseteq U \subseteq P$, such that $f \mid_U = g \mid_U$. Equivalently we can define the germ of a function f at a point p as an equivalence class $[f]_p$ given by the relation: $f \sim_p g$ iff there exists an open set $U \ni p$ such that $f \mid_U = g \mid_U$. Therefore we can describe the equivalence classes $[f]_G$ using the following equivalence: $f \sim_G g$ iff the germs $[f]_p$ and $[g]_p$ of the functions f and g, with $p \in M$ coincide. This description suggests the name for this category. A groupoid object in **G** over M corresponds to a groupoid object in this category with $B = (M, M)$. This object is usually called a *local groupoid* and its study is relevant, for example, in problems related to symplectic realizations of Poisson manifolds (see [10]).

Definition 9.7 Consider the category **S**. We define the following equivalence relation of objects in **S**: $(P_1, M_1) \sim (P_2, M_2)$ if and only if they satisfy the following conditions:

1. $M_1 = M_2 = M$.
2. There exists a local manifold pair (P, M) such that P is an open subset of P_1 and P_2 simultaneously. The equivalence class $[P, M]$ is called a *microfold pair* (see [3]).

Definition 9.8 The *infinite jet space* $Jet_p^\infty(f)$ of a map $f : P \longrightarrow Q$ at a point $p \in P$ is defined as the equivalence class $[f]$ given by the following relation: $f \sim g$ if and only if their Taylor expansions at the point p coincide.

Definition 9.9 Two morphisms $f_1 : (P_1, M) \longrightarrow (Q_1, N)$, $f_2 : (P_2, M) \longrightarrow (Q_2, N)$ are equivalent if and only if they satisfy the following:

1. There exists an open submanifold P of P_1 and P_2 (i.e. $[P_1, M] = [P_2, M]$).
2. There exists an open submanifold Q of Q_1 and Q_2 (i.e. $[Q_1, N] = [Q_2, N]$).
3. For the functions $f_1|_P$, $f_2|_P : (P, M) \longrightarrow (Q, N)$ we have that

$$Jet_m^\infty(f_1|_P) = Jet_m^\infty(f_2|_P)$$

for all $m \in M$.

Definition 9.10 We define the category J_∞-**Mic** (which will be called the *infinite jet microcategory*) whose objects are microfold pairs and a morphism from $[P, M]$ to $[Q, N]$ corresponds to the equivalence class $[f]_{J_\infty}$ given by the relation descibed in Definition 9.9.

The following lemma ensures the existence of the corresponding pullback for the groupoid object in J_∞-**Mic**.

Lemma 9.11 *Let* $[s]_{J_\infty}$, $[t]_{J_\infty} : [P, M] \longrightarrow [M, M]$ *in* J_∞ *be such that, for all* $s \in [s]_{J_\infty}$, $t \in [t]_{J_\infty}$, $s(m) = t(m) = m$, *for all* $m \in M$. *Then, with respect to these maps, there exists a pullback in* J_∞-**Mic**, *denoted by* $[P, M] \times_{([s],[t])} [P, M]$.

Proof This lemma follows from the following fact: Given that $s(m) = t(m) = m$, $\forall m \in M$, this implies that $ds : T_m P \longrightarrow T_{s(m)} M$ and $dt : T_m P \longrightarrow T_{t(m)} M$ have maximal rank. By a continuity argument, there exists a neighborhood U, such that $M \subseteq U \subseteq P$ and ds and dt have maximal rank at points of U. Hence, the pullback construction $U \times_{(s,t)} U$ with respect to the restrictions of s and t in U:

$$
\begin{array}{ccc}
U \times_{(s,t)} U & \xrightarrow{\;\pi_1\;} & U \\
\downarrow{\scriptstyle \pi_2} & & \downarrow{\scriptstyle t} \\
U & \xrightarrow{\;s\;} & M
\end{array}
$$

as a set has the structure of a manifold.

Using the following inclusions:

$$\Delta_M \subset U \times_{(s,t)} U, \; ; s|_U(M) \subset M, t|_U(M) \subset M, \pi_1(\Delta_M) \subset M, \pi_2(\Delta_M) \subset M$$

we can define the microfold pair $(U \times_{(s,t)} U, \Delta_M)$ and taking the infinite jets of $s|_u$ and $t|_u$ we have the following commutative diagram:

$$
\begin{array}{ccc}
[U \times_{(s,t)} U, \Delta_M] & \xrightarrow{\;\pi_1\;} & [U, M] = [P, M] \\
\downarrow{\scriptstyle [\pi_2]} & & \downarrow{\scriptstyle [t]} \\
[U, M] = [P, M] & \xrightarrow{\;[s]\;} & [M, M]
\end{array}
$$

and the microfold pair $(U \times_{(s,t)} U, \Delta_M)$ inherits the universal property from the pullback $U \times_{(s,t)} U$. $\qquad\square$

This suggests the following description for a groupoid object in this category.

Definition 9.12 Let $[s], [t] : [P, M] \longrightarrow [M, M]$. A J_∞-*formal groupoid* $[P, M] \rightrightarrows [M, M]$ over M is a groupoid object in the category J_∞-**Mic** over (M, M) with the following condition for the source and target morphims

$[s]_{J_\infty}$ and $[t]_{J_\infty}$:

$$\forall s \in [s]_{J_\infty}, \ t \in [t]_{J_\infty} : \quad s(m) = t(m) = m, \ \forall m \in M.$$

where $[\varepsilon] : [M, M] \longrightarrow [G, M]$ corresponds to the infinite jet of the inclusion of M in G.

9.3 Algebraic structure for formal groupoids

9.3.1 Formal groupoid as cogroupoid object

Let (P, M) be a local manifold pair. We define the following commutative algebra:

$$C^\infty(P, M) := C^\infty(P)/I_M^\infty$$

where $I_M^\infty = \bigcap_{k=1}^\infty I_M^k$ and I_M^k corresponds to the ideal of functions in $C^\infty(P)$ such that their Taylor polynomials of order k are identically zero at points on M. Given two algebra morphisms $S, T : C^\infty(M) \longrightarrow C^\infty(P, M)$, we define the following quotient algebra

$$C^\infty(P, M) \underset{(S,T)}{*} C^\infty(P, M) := C^\infty(P \times P, M \times M)/\langle 1 \otimes Tf - Sf \otimes 1 \rangle,$$

for $f \in C^\infty(M)$. Here $f_1 \otimes f_2 \in C^\infty(P \times P, M \times M)$, $f_1, f_2 \in C^\infty(P, M)$, is defined by

$$(f_1 \otimes f_2)(x, y) := f_1(x) \cdot f_2(y).$$

Then, we have the following definition of a formal groupoid according to [7].

Definition 9.13 A formal groupoid over a manifold M corresponds to the following data:

1. a local manifold pair (P, M),
2. two algebra morphisms: $S, T : C^\infty(M) \longrightarrow C^\infty(P, M)$,
3. the following algebra morphisms:
   $E : C^\infty(P, M) \longrightarrow C^\infty(M)$
   $I : C^\infty(P, M) \longrightarrow C^\infty(P, M)$
   $\Delta : C^\infty(P, M) \longrightarrow C^\infty(P, M) \underset{(S,T)}{*} C^\infty(P, M)$ satisfying the contravariant
   versions of the compatibility axioms for a groupoid object.

We have the following propositions:

Proposition 9.14 Let (P, M) and (Q, N) local manifold pairs, $f : P \longrightarrow Q$, $f(M) \subseteq N$. Consider the dual map $f^* : C^\infty(Q) \longrightarrow C^\infty(P)$ described by

$$f^* : q \longmapsto q \circ f, \forall q \in C^\infty(Q).$$

Then f^* induces a morphism $\{f^*\} : C^\infty(Q, N) \longrightarrow C^\infty(P, M)$ in the quotient algebras and if $[f]_{J_\infty} = [g]_{J_\infty}$ then $\{f^*\} = \{g^*\}$.

Proof Using that $f(M) \subseteq N$ we have that $f^*(I_N^\infty) \subseteq I_M^\infty$ and therefore f^* passes through the quotients $C^\infty(Q)/I_N^\infty$ and $C^\infty(P)/I_M^\infty$ and hence induces a map $\{f^*\} : C^\infty(Q, N) \longrightarrow C^\infty(P, M)$. Now, consider two maps $f, g \in [f]_{J_\infty}$ and their duals $f^*, g^* : C^\infty(Q) \longrightarrow C^\infty(P)$. Then, for all $q \in C^\infty(Q)$, $f^*(q) = q \circ f$ and $g^*(q) = q \circ g$ in $C^\infty(P)$ coincide in their Taylor expansions at points of M, hence $(f^* - g^*)(q) \in I_M^\infty$ and $\{f^*\} = \{g^*\}$ as we wanted. $\qquad \square$

Proposition 9.15 Let $(P, M) \rightrightarrows (M, M)$ a J_∞- formal groupoid over M. Let $S = [s^*]$ and $T = [t^*]$ be algebra morphisms from $C^\infty(M)$ to $C^\infty(P, M)$. The we have the following algebra isomorphim:

$$C^\infty(P, M) \underset{(S,T)}{*} C^\infty(P, M) \cong C^\infty((P, M) \times_{(s,t)} (P, M)).$$

Proof Using Lemma 9.11 we have a natural inclusion $i : (P, M) \times_{(s,t)} (P, M) \longrightarrow (P \times P, M \times M)$ inducing the dual map

$$[i]^* : C^\infty(P \times P, M \times M) \longrightarrow C^\infty((P, M) \times_{(s,t)} (P, M)),$$

that is surjective. Using the inclusion map

$$j : \langle 1 \otimes Tf - Sf \otimes 1 \rangle \longrightarrow C^\infty(P \times P, M \times M)$$

we have the following short exact sequence of alegbras:

$$0 \to \langle 1 \otimes Tf - Sf \otimes 1 \rangle \overset{j}{\to} C^\infty(P \times P, M \times M)$$
$$\overset{[i]^*}{\to} C^\infty((P, M) \times_{(s,t)} (P, M)) \to 0.$$

(The fact that $\langle 1 \otimes Tf - Sf \otimes 1 \rangle \subseteq Ker([i]^*)$ follows from the definition of S and T. The other inclusion is a consequence of a more general result, involving the local Hadamard's lemma; for a complete proof see [2].)

Therefore we have a canonical algebra isomorphism

$$C^\infty((P, M) \times_{(s,t)} (P, M)) \cong C^\infty(P \times P, M \times M)/\langle 1 \otimes Tf - Sf \otimes 1 \rangle. \quad \square$$

Proposition 9.16 Let $(P, M) \rightrightarrows (M, M)$ a J_∞-formal groupoid over M. Let $S = [s^*]$ and $T = [t^*]$ be algebra morphisms from $C^\infty(M)$ to $C^\infty(P, M)$.

We define the following quotient

$$C^\infty(P, M) \otimes_{(S,T)} C^\infty(P, M) := C^\infty(P, M) \otimes C^\infty(P, M)/\langle 1 \otimes Tf - Sf \otimes 1 \rangle.$$

Then there exists and algebra isomorphism (see [9]):

$$C^\infty(P, M) \otimes_{(S,T)} C^\infty(P, M) \cong C^\infty((P, M) \times_{(s,t)} (P, M)).$$

Proof Consider the following morphism defined pointwise in $(P, M) \times_{(s,t)}$ (P, M):

$$\phi : C^\infty(P, M) \otimes C^\infty(P, M) \longrightarrow C^\infty((P, M) \times_{(s,t)} (P, M)),$$

$$(f_1 \otimes f_2) \longrightarrow f|_{(g_1, g_2)} := f_1|_{g_1} f_2|_{g_2},$$

with $(g_1, g_2) \in (P, M) \times_{(s,t)} (P, M)$. In order to prove that this map is well defined observe that:

$$\phi(Sf \otimes 1) = [s^*]f|_{g_1} = f \circ s(g_1)$$

and

$$\phi(1 \otimes_T f) = [t^*]f|_{g_2} = f \circ t(g_2).$$

Therefore, the image of the ideal generated by $Sf \otimes 1 - 1 \otimes Tf$ vanishes on points of M, hence, this map is well defined on the tensor product $C^\infty(P, M) \otimes C^\infty(P, M)$. In order to prove that this map is an isomorphism we will use the following fact that describes C^∞-functions in local coordinates:

Lemma 9.17 *Let* $\dim(M) = k$ *and* $\dim(G) = n$. *Using that M is a submanifold of G, there exists an open set U, $M \subseteq U \subseteq G$, and local coordinates*

$$(x_1, x_2 \ldots x_k, y_{k+1}, y_{k+2}, \ldots, y_n)$$

such that $f \in C^\infty(U, M)$ *can be written as (using multiindex notation):*

$$f(x, y) = \sum_{|\alpha|} \partial^{|\alpha|}/\partial y_{k+1}^{|\alpha_{k+1}|} \cdots \partial y_n^{|\alpha_n|} f(x, 0) \frac{y^\alpha}{\alpha!}.$$

Using this lemma for the functions $f_1|_{g_1}$ and $f_2|_{g_2}$ we have the injectivity condition, and describing a function $f \in C^\infty((P, M) \times_{(s,t)} (P, M))$, with $(P, M) \times_{(s,t)} (P, M)$ naturally included in $(P \times P, M \times M)$ and using local coordinates for $P \times P$, Lemma 9.17 gives us the expansion of a function $f \in C^\infty((P, M) \times_{(s,t)} (P, M))$ as a product of two power series, each one with respect to (P, M), therefore we have the surjectivity condition, as desired.

Now, the following proposition allows us to relate the pullback construction for the groupoid objects to a push-out construction in commutative algebras in a formal groupoid according to Karabegov, and hence justify the contravariant version of the groupoid axioms. □

Proposition 9.18 $C^\infty(P, M) \otimes_{(S,T)} C^\infty(P, M)$ *has the structure of a push-out in commutative algebras.*

Proof Consider the tensor product $C^\infty(P, M) \otimes C^\infty(P, M)$ and its canonical inclusions Π_1, $\Pi_2 : C^\infty(P, M) \longrightarrow C^\infty(P, M) \otimes C^\infty(P, M)$. By definition we have the following commutative diagram:

$$
\begin{array}{ccc}
C^\infty(P, M) \otimes_{(S,T)} C^\infty(P, M) & \xleftarrow{\Pi_1} & C^\infty(P, M) \\
\uparrow{\scriptstyle \Pi_2} & & \uparrow{\scriptstyle S} \\
C^\infty(P, M) & \xleftarrow{\quad T \quad} & C^\infty(M)
\end{array}
$$

The universal property is given by the canonical map γ given by the following commutative diagram:

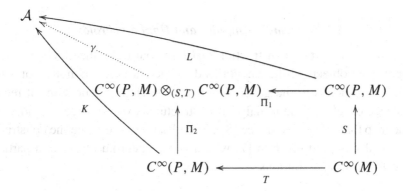

where γ is defined by $\gamma[f \otimes g] = Kf \cdot Lg, \forall f, g \in C^\infty(P, M)$. $\qquad \square$

In order to describe the relation between the J_∞ category and the Karabegov construction, using a result proved in, we get

Theorem 9.19 *Let $C^\infty(\mathbf{Man})$ be the category with objects corresponding to algebras of C^∞-functions defined over smooth manifolds, where the morphisms correspond to algebra morphisms. Then the functor*

$$
C^\infty : M \longrightarrow C^\infty(M), \quad f \longrightarrow f^*
$$

induces an equivalence of categories between **Man** *and* $C^\infty(\mathbf{Man})$ *(is fully faithful).*

Therefore, we are able to state the following theorem:

Theorem 9.20 *Let* **Comm Alg** *be the category whose objects are commutative algebras, where the morphisms correspond to algebra morphisms. The* *functor*

$$F_{J_\infty} : J_\infty \longrightarrow \text{Comm Alg},$$

$$(P, M) \longmapsto C^\infty(P, M),$$

$$[f] \in \text{Hom}\,((P, M), (Q, N)) \longmapsto \{f^*\}$$

is faithful and the image of a J_∞-formal groupoid under F_{J_∞} is a formal groupoid according to [7] and it corresponds to a cogroupoid object in Comm Alg.

Proof This follows from the previous propositions. $\square$

9.3.2 Formal groupoids and Hopf algebroids

There exists an important construction that turns out to be the algebraic version of the groupoid objects in the categories described above. The notion of a Hopf algebroid corresponds to the generalization of the Hopf algebras and it requires the existence of left and right bialgebroid structures over a k-algebra and also an antipode map that relates them (see [8]). The idea now is to relate the construction of the formal groupoid given in [7] with a Hopf algebroid object in a particular category. First, some definitions:

Definition 9.21 An *A-algebra* U corresponds to a triple (U, m_u, η) with $U \in (A^e\text{-Mod})$, $m_u : U \oplus_A U \longrightarrow U$, given by $u \otimes v \longrightarrow uv$ $\eta : A \longrightarrow U$ in such way that they are (A, A)-bimodule maps and we have unitalitity and associativity conditions:

$$m_u(\eta \otimes \text{id}_u) = m_u(\text{id}_u \otimes \eta) \quad \text{(unitality condition)},$$

$$m_u(m_u \otimes \text{id}_u) = m_u(\text{id}_u \otimes m_u) \quad \text{(associativity condition)}.$$

We define a morphism of A-algebras as an (A, A) bimodule morphism $f : U \longrightarrow V$ such that:

$$m_V(f \otimes f) = fm_U,$$

$$f\eta_U = \eta_V.$$

We can define the category A-Alg with objects corresponding to A-algebras and morphims corresponding to A-algebra morphisms.

Similarly, in the category A^e-Mod we can define the comonoid object, that is, a triple (V, Δ, ε), with $V \in (A^e\text{-Mod})$, equipped with left and right actions L_A

and R_A, and $\Delta : V \longrightarrow V \otimes_A V$, $\varepsilon : V \longrightarrow A$ in such way that they are (A, A)-bimodule maps and we have counitatitity and coassociativity conditions:

$$L_A(\varepsilon \otimes \mathrm{id}_V) = R_A(\mathrm{id}_V \otimes \varepsilon) = \mathrm{id}_V \quad (\textit{counitality condition}),$$

$$(\Delta \otimes \mathrm{id}_V)\Delta = (\mathrm{id}_V \otimes \Delta)\Delta \quad (\textit{coassociativity condition}).$$

The next step is to consider a structure that relates the notion of an A^e-algebra and an A-coalgebra, but, before that, we need an important of description for an A-algebra U. We will use a theorem (see [8]) that describes the A-algebras in terms of k-algebra morphisms and it implies that the bimodule structure on U is characterized by the map η:

Theorem 9.22 *There is a bijective correspondence between the A^e-algebras U and the k-algebra homomorphisms*

$$\eta : A \longrightarrow U,$$

and the (A, A)-bimodule structure on U is described by:

$$aub := \eta(a)u\eta(b), \quad a, b \in A, \ u \in U.$$

Using this description, we define the source and target of U as the restrictions of η:

$$s := \eta(\cdot \otimes_k A),$$

$$t := \eta(A \otimes_k \cdot),$$

and the left module structure on U as

$$(a \otimes b, u) \longrightarrow \eta(a \otimes b)u, \quad a, b \in A, \ u \in U.$$

With respect to this action we define the following tensor product:

$$U \otimes_A U := U \otimes_k U / \mathrm{span}\{\eta(1 \otimes_k a)u \otimes_k u' - u \otimes_k \eta(a \otimes_k 1)u' \,|\, a \in A, u, u' \in U\}.$$

These tensor product has the structure of an A^e-left module, given by

$$(a \otimes_k b) \cdot (u \otimes_A u') := s(a)u \otimes_A t(b)u'.$$

The problem that appears here is that this tensor product does not have the structure of a A^e-algebra, because the product in each one of the components is not well defined in general for this quotient, therefore it is necessary to restrict the tensor product to the following k-submodule

$$U \times_A U := \left\{ \sum_i u_i \otimes_A u_i' \in U \otimes_A U \,\bigg|\, \sum_i u_i t(a) \otimes_A u_i' = \sum_i u_i t(a) \otimes_A u_i' s(a) \right\},$$

for all $a \in A$, this will be called the *Takeuchi product* of U with itself. Using the k-ring morphism $\eta_{U \times_A U}(a \otimes b) := s(a) \otimes_A t(b)$ we can verify that the Takeuchi product is an A^e-algebra.

A structure that relates the notions of A^e-algebra and A^e-coalgebra will be called a *left A-bialgebroid* and is defined as follows.

Definition 9.23 A left A-bialgebroid consists of a k-module U with the structure of an A^e-algebra (U, s, t) and with the structure of an A^e-coalgebra (U, Δ, ε) related in the following way:

1. The (A, A)-bimodule structure of the A^e-colagebra is given by:
$$a \cdot u \cdot b := s(a)t(b)u, \quad a, b \in A, \ u \in U.$$

2. With this structure, the coproduct Δ is a k-algebra morphism between U and $U \times_A U$ (the Takeuchi product).

3. (*Counital properties*)
$$\varepsilon(a \cdot u \cdot b) = a\varepsilon(u)b,$$
$$\varepsilon(uu') = \varepsilon(us(\varepsilon u')) = \varepsilon(ut(\varepsilon u')),$$

$\forall a, b \in A, \ u, u' \in U$.

In an analogous way we can define a right A-bialgebroid and then we are able to define a Hopf algebroid.

Definition 9.24 Let (U_l, s_l, t_l) be a left A-bialgebroid and (U_r, s_r, t_r) a right A-bialgebroid. A Hopf algebroid U has the structure of a left A-bialgebroid and right A-bialgebroid with the following compatibility axioms:

1. (*Counitality compatibility*)
$$s_l \varepsilon_l t_r = t_r, \quad t_l \varepsilon_l s_r = s_r, \quad s_r \varepsilon_r t_l = t_l, \quad t_r \varepsilon_r s_l = s_l.$$

2. (*Coassociativity compatibility*)
$$(\Delta_l \otimes U)\Delta_r = (\mathrm{id}_U \otimes \Delta_r)\Delta_l,$$
$$(\Delta_r \otimes U)\Delta_l = (\mathrm{id}_U \otimes \Delta_l)\Delta_r.$$

3. (*Antipode axioms*)
$$S(t_l(a)ht_r(b)) = s_r(a)S(u)s_l(b),$$
$$m_U(S \otimes_A \mathrm{id}_U)\Delta_l = s_r \varepsilon_r, \quad m_U(\mathrm{id}_U \otimes_A S)\Delta_r = s_l \varepsilon_l.$$

Here, we can state the following theorem, proved in a more general version by Kowalzig, which gives the claimed algebraic structure for the formal groupoid.

Theorem 9.25 *Let $A = C^\infty(M)$ be the $\mathbb{R}$-algebra of C^∞-functions over a manifold M and let $A^e = C^{\infty(M)} \otimes C^{\infty(M)}$. Consider the category A^e-**Alg** defined above. Let $[P, M] \rightrightarrows [M, M]$ be a groupoid object defined in J^∞-**Mic** and let $F_{J_\infty}([P, M] \rightrightarrows [M, M])$ be its corresponding K-formal groupoid. Then $F_{J_\infty}([P, M] \rightrightarrows [M, M])$ has the structure of a Hopf algebroid over A.*

Proof Using the groupoid structure defined on J^∞-**Mic**, we want to prove that this implies the structure of a Hopf algebroid. The source and target maps in

$$[G, M] \rightrightarrows [M, M]$$

induce the algebra morphisms:

$$s^*, t^* : C^\infty(M, M) \longrightarrow C^\infty(G, M)$$

and, according to Theorem 9.22, defining $\eta := s^* \otimes t^*$ we can describe $C^\infty(G, M)$ as an A^e-algebra (the associativity and unitality conditions here come from the corresponding conditions associated to the algebra of C^∞-functions).

We need to check that we have the structure of a left (right) bialgebroid. Therefore, we must construct the structure of the A-coalgebra. Now, if G_2 corresponds to the object of composable pairs, we associate the algebra $C^\infty(G_2, M)$.

Using the isomorphism described in Proposition 9.16 we can define the comultiplication map $\Delta : C^\infty(G, M) \otimes_{C^\infty(M)} C^\infty(G, M) \longrightarrow C^\infty(G, M)$ given by

$$\Delta := \phi \circ [m]^*$$

where ϕ is the isomorphism in Proposition 9.16 and $[m]^*$ is the dual map corresponding to the multiplication map

$$[m] : [P, M] \times_{s,t} [P, M] \longrightarrow [M, M]$$

defined in the groupoid object $[P, M] \rightrightarrows [M, M]$.

The coassociativity condition for Δ comes from the dual version of the associativity condition. Similarly, the map $\varepsilon^* : C^\infty(G, M) \longrightarrow C^\infty(M, M)$ inherits the counitality condition from ε. Hence, we obtain that $C^\infty(G, M)$ has the structure of an A-coalgebra, with the (A, A)-bimodule structure given by

$$f \cdot h \cdot g := s^*(f) h t^*(g).$$

To prove that $C^\infty(G, M)$ corresponds to a left A-bialgebroid we refer to Definition 9.23 and the verification of the axioms is straightforward. $\qquad\square$

9.4 The symplectic case

In this section we outline the formal version of the symplectic groupoids, introduced independently by Karasev, Weinstein and Zakrzewsky (see e.g. [10]).

9.4.1 Definitions

Definition 9.26 A *symplectic groupoid* $G \rightrightarrows M$ consists of a groupoid object G over M equipped with a nondegenerate closed form ω (G is a symplectic manifold) in such way that the graph of the multiplication map $m : G_2 \longrightarrow G$ is a Lagrangian (i.e. maximal isotropic) submanifold of $G \times \overline{G} \times \overline{G}$ with respect to the symplectic form $\omega \oplus \omega \ominus \omega$.

This Lagrangian property for the product is equivalent to the maximality condition $(\mathrm{Dim}(Gr(m)) = 3/2\, \mathrm{Dim}(G))$ and a compatibility of the symplectic form ω with the product m as follows:

Proposition 9.27 $Gr(m) \subseteq G \times G \times G$ is Lagrangian if and only if

$$m^*\omega = pr_1^*\omega + pr_2^*\omega,$$

where $pr_1, pr_2 : G_2 \longrightarrow G$ denote the canonical projections.

Proof

$$
\begin{array}{ccc}
G_2 \xrightarrow{pr_2} G & \quad & TG_2 \xrightarrow{d(pr_2)} TG \\
\downarrow{\scriptstyle pr_1} \quad \downarrow{\scriptstyle s} & & \downarrow{\scriptstyle d(pr_1)} \quad \downarrow{\scriptstyle ds} \\
G \xrightarrow{t} M & & TG \xrightarrow{dt} TM
\end{array}
$$

We have that

$$T Gr(m) = \mathrm{Graph}(dm : TG_2 \longrightarrow TG)$$

and that

$$T(G_2)_{(g,h)} = \{(X, Y) \in T_g G \times T_h G \,|\, ds(X) = dt(Y)\}.$$

Therefore we have that

$$T Gr_{(g,h,m(g,h))} = \{(X, Y, m_*(X, Y)) \in T_g G \times T_h G \times T_{m(g,h)} G \,|\, ds(X) = dt(Y)\}.$$

Hence,

$$\omega \oplus \omega \ominus \omega((X, Y, m_*(X, Y)), (X', Y', m_*(X', Y')))$$

corresponds to

$$\omega(X, X') + \omega(Y, Y') - \omega(m_*(X, Y), m_*(X', y')),$$

$$pr_1^*\omega + pr_2^\omega - m^*\omega,$$

where the isotropic condition is equivalent to having the last expression equal to zero and by a dimensional argument the proof is complete. □

9.4.2 Examples of symplectic groupoids

Acording to the construction of examples of groupoid objects in the category **Man** considered before we have the following symplectic versions:

Example 9.28 If (M, ω) is a symplectic manifold, the coarse groupoid $M \times \bar{M} \rightrightarrows M$, equipped with the symplectic form $\omega \ominus \omega$, is a symplectic groupoid. It can be proved using the multiplication defined in Example 9.3.

Example 9.29 Let M be a manifold. The fiber groupoid $T^*M \rightrightarrows M$, equipped with the canonical symplectic form ω_{can} on the cotangent bundle, is a symplectic groupoid.

9.4.3 The formal version

Definition 9.30 Let (P, M) a local manifold pair. We define the space of k-J_∞ forms with respect to (P, M) as:

$$\Omega_{\mathrm{form}}^k(P, M) = \Omega^k(P)/I_M^\infty(\Omega^k(P))$$

where $I_M^\infty(\Omega^k(P))$ denotes the ideal of k-forms on P such that, locally, their coefficients (C^∞-functions on M) vanish at points of M.

Therefore we have the following definition of a J_∞-*symplectic form*:

Definition 9.31 A 2-J_∞ form $[\omega] \in \Omega_{\mathrm{form}}^2(P, M)$ is called *symplectic* if there exists an open neighborhood U with $M \subseteq U \subseteq P$ such that $\omega|_U$ is a symplectic 2-form.

Definition 9.32 A *formal symplectic groupoid* object in the category J^∞-**Mic** corresponds to the following data:

1. a local manifold pair (P, M),
2. a J^∞-groupoid object $(P, M) \rightrightarrows (M, M)$ over M,
3. a 2-J_∞ form $[\omega]$ with the following compatibility property:

$$m^*[\omega] = pr_1^*[\omega] + pr_2^*[\omega].$$

Remark The notion of formal symplectic groupoid given by Weinstein *et al.* is compatible with the definition given above, since the Lagrangian properties for the graph of the multiplication Δ can be restated as the compatibility condition between m^* and $[\omega]$. In fact, Karabegov describes a *formal symplectic groupoid* as a formal groupoid over a Poisson manifold M. More precisely, we have the following.

Definition 9.33 A *formal symplectic groupoid* over a Poisson manifold M corresponds to the following data:

1. a local manifold pair (P, M), with P a symplectic manifold and M a Poisson Lagrangian submanifold of P,
2. a K-formal groupoid, with S a Poisson morphism $S : C^\infty(M) \longrightarrow C^\infty(P, M)$ and T an anti-Poisson morphism $T : C^\infty(M) \longrightarrow C^\infty(P, M)$ such that Sf and Tg Poisson commute, for all $f, g \in C^\infty(M)$,
3. the involutive inverse map $I : C^\infty(P, M) \longrightarrow C^\infty(P, M)$ is an antisymplectic morphism,
4. the comultiplication map $\Delta : C^\infty(P, M) \longrightarrow C^\infty(P, M) *_{S,T} C^\infty(P, M)$ has the compatibility condition described in Definition 9.32.

Given this definition of formal symplectic groupoid, Karabegov proves the existence of a unique realization of a Poisson manifold M. It is interesting to notice the relation of this construction to some other construction, given by Cattaneo–Dherin–Felder in the context of deformation quantization (see [3]), where the existence of the symplectic groupoid structure comes from a formal deformation of the groupoid structure of the cotangent bundle $T^*\mathbb{R}^n$ and $\mathbb{R}^n$ equipped with the zero Poisson structure.

These results show that any Poisson manifold (M, Π) can be formally integrated in the sense that it is always possible to find a formal symplectic groupoid in such a way that the space of objects corresponds to the Poisson manifold M and the structure maps of the groupoid are compatible with the Poisson structure.

In this context it would be interesting to understand the algebraic structure (the Hopf algebroid) of the formal symplectic groupoid, that is conjecturally associated to the Lie algebroid structure associated to a Lie groupoid. In the case of Poisson manifolds, formal deformations of the Poisson structure would conjecturally control the deformations. Further research can be done in this direction.

Acknowledgements

I am very grateful to Marius Crainic and Benoit Dherin for their guidance during my Master's thesis project in the Master Class programme *Aspects of Calabi–Yau geometries*, held by Utrecht University; this document is an overview of such work. I thank Universidad de los Andes for the financial support to attend the summer school in Villa de Leyva and Zürich University, where this document was completed.

References

[1] Baues, H. J., Quintero, A. *Infinite Homotopy Theory*. Kluwer Academic, 2000.

[2] Bunge, M., Dubuc, E. Archimedian local C^∞ rings and models of synthetic differential geometry. *Cahiers Topol. Geom. Diff. Categ.* **27**(3), 1–22, 1986.

[3] Cattaneo, A., Dherin, B., Weinstein, A. Cotangent microbundle category, I. arXiv:0712.1385, 2007.

[4] Cattaneo, A., Felder, G. Poisson sigma models and symplectic groupoids. In *Quantization of singular symplectic quotients*, Progress in Mathematics 198. Birkhäuser, 2001, pp. 61–63.

[5] Crainic, M., Loja, R. Integrability of Lie brackets. *Ann. Math.* **157**(2), 575–620, 2003.

[6] Crainic, M., Loja, R. Integrability of Poisson brackets. *J. Diff. Geom.* **66**, 71–137, 2004.

[7] Karabegov, A. Formal symplectic groupoid of a deformation quantization. *Commun. Math. Phys.* **258**, 223–256, 2005.

[8] Kowalzig, N. Hopf algebroids and their cyclic theory. PhD thesis, Utrecht University, 2009.

[9] Kowalzig, N., Posthuma, H. The cyclic theory of Hopf algebroids. arXiv:0904.4736v1, 2009.

[10] Weinstein, A. Symplectic groupoids and Poisson manifolds. *Bull. Amer. Math. Soc.* **16**(1), 101–104, 1987.

10

Elliptic PDEs and smoothness of weakly Einstein metrics of Hölder regularity

ANDRÉS VARGAS

Abstract

This chapter is broadly divided into two parts. In the first, a brief but self-contained review of the interior regularity theory of elliptic PDEs is presented, including relevant preliminaries on function spaces. In the second, as an application of the tools introduced in the first part, a detailed study of the Einstein condition on Riemannian manifolds with metrics of Hölder regularity is undertaken, introducing important techniques such as the use of harmonic coordinates and giving some consideration to the smoothness of the differentiable structure of the underlying manifold.

10.1 Introduction

Partial differential equations (PDEs) play a fundamental role in many areas of pure and applied sciences. Particularly known is their ubiquitous presence in the description of physical systems, but it is less obvious to students how their application to geometry is not only useful but often decisive. As an introduction to the world of PDEs in geometry, the aim of this chapter is two-fold. First, to give a brief (and rather condensed) overview of the minimum number of basic preliminaries that allow us to write down precise and complete statements of several important theorems from the regularity theory of elliptic PDEs, which is one of the most needed aspects for applications. Second, to present a detailed treatment of some useful techniques required to study geometric equations involving the Ricci tensor and, in particular, to put into use the definitions and tools introduced in the first part to study the smoothness of weak solutions to the Einstein condition on Riemannian manifolds with metrics of Hölder regularity $C^{1,\alpha}$.

Geometric and Topological Methods for Quantum Field Theory, ed. Alexander Cardona, Iván Contreras and Andrés F. Reyes-Lega. Published by Cambridge University Press. © Cambridge University Press 2013.

From the analytical perspective, the reader is assumed to be familiar with preliminaries in functional analysis and to have been exposed to basic material on PDEs. On the geometric side, an introductory course in Riemannian geometry and/or general relativity is sufficient to be acquainted with the concepts employed here. It must be clearly stated that, even in the geometric part, most of the discussion is heavily inclined towards the analytic side of the problem.

Owing to their extension, there is no space here to enter into the finer details and proofs of the results mentioned in the review on function spaces and elliptic PDEs that corresponds to the first two sections of this chapter. Even simple examples have been left out in this part. As a partial remedy, in the geometric context of the second part, consisting of the last two sections, the use of these analytic tools is shown with detail and their power is made clear with the proof of the smoothness of weakly Einstein metrics (in harmonic coordinates). Apart from this, relevant references to required preliminaries and to detailed material on the topics covered is given along the way.

Concerning the result on regularity of weak solutions to the Einstein condition for $C^{1,\alpha}$ metrics, a few comments are in order. Although the author is not aware of its origin, its relevance seems to be associated with the developments in the convergence theory of Riemannian manifolds during the 1980s and 1990s, and particularly with the convergence and compactness theorems in the $C^{0,\alpha}$ topology under Ricci curvature bounds studied in [An90] and [AC92]. The result, which is well known, is explicitly mentioned in [Pe98] and is a particular consequence of regularity results for quasilinear equations treated in [Ta11] and [Ta00]. Nevertheless, a complete, accessible and self-contained presentation of the problem may be of interest, especially to people learning the analytic machinery needed for geometric applications of the regularity theory of PDEs, to whom this chapter is mainly addressed.

10.2 Basics on function spaces

This section collects some definitions and useful results from analysis of function spaces that constitute the basic prerequisites for the study of solutions to PDEs. The generalization of these notions to Riemannian manifolds is given some attention at the end of Section 10.3, after a brief overview of elliptic PDEs and regularity. No attempt is made to present a fully detailed account on the subject, but relevant points are indicated.

10.2.1 Notational conventions

For simplicity, a partial derivative of order $k \in \mathbb{N}_0$ with respect to $x \in \mathbb{R}$ is denoted by $\partial_x^k := \partial/\partial x$ with the convention $\partial^0 f(x) := f(x)$. As usual, the norm of a multi-index $\ell = (l_1, \ldots, l_n) \in \mathbb{N}_0^n$ is defined as the integer $|\ell| := \sum_{i=1}^n l_i$, the notation $\partial^\ell := \partial_{x_1}^{l_1} \cdots \partial_{x_n}^{l_n} \equiv \partial_1^{l_1} \cdots \partial_n^{l_n}$ stands for any composition of partial derivatives of total order $|\ell|$ and $x^\ell := x_1^{l_1} \cdots x_n^{l_n}$, where $x = (x_1, \ldots, x_n) \in \mathbb{R}^n$. The standard Euclidean inner product in $\mathbb{R}^n$ is denoted by $\langle \cdot, \cdot \rangle_{\mathbb{R}^n}$ and its corresponding norm $|\cdot|_{\mathbb{R}^n}$ will often be abbreviated by $|\cdot|$ when no confusion arises. Moreover, this norm should be distinguished from norms on function spaces for which the notation $\|\cdot\|$ is used.

If $\Omega \subset \mathbb{R}^n$, its closure is denoted by $\bar{\Omega}$ and $\Omega' \Subset \Omega$ means that Ω' is compactly contained in Ω, i.e. $\Omega' \subset \bar{\Omega}' \subset \Omega$ and $\bar{\Omega}'$ is compact. A function $f : \Omega \to \mathbb{R}^m$ has compact support in Ω if $\operatorname{supp}(f) \Subset \Omega$.

10.2.2 Differentiable functions

Let Ω be an arbitrary subset of $\mathbb{R}^n$. The set of bounded continuous functions $f : \Omega \to \mathbb{R}^m$ with all its partial derivatives (in Ω) continuous and bounded up to order k is denoted by $C^k(\Omega, \mathbb{R}^m)$, or simply by $C^k(\Omega)$. This space comes endowed with the sup-norm

$$\|f\|_{C^k(\Omega)} := \sum_{|\ell| \le k} \sup_{x \in \Omega} |\partial^\ell f(x)|.$$

Some authors (see [AF03]) write this set as $C_B^k(\Omega)$, to make a distinction with the convention where functions in $C^k(\Omega)$ are not required to have bounded derivatives (including order zero). The subscript "B" is omitted here and, for the larger space of k-times differentiable functions whose derivatives are finite but not necessarily bounded in Ω, the notation $C_{\mathrm{loc}}^k(\Omega)$ is used, given that those derivatives are still bounded locally.

Notice that if $\Omega = \mathbb{R}$, then $x^n \notin C^k(\mathbb{R})$ for any $n \in \mathbb{Z}_+$, but still $x^n \in C^k(\Omega')$ for each bounded $\Omega' \subset \mathbb{R}$. Similarly, for open and bounded subsets like $I^\circ :=$ $(0, 1) \subset \mathbb{R}$, it holds $x^{-n} \notin C^k(I^\circ)$, but $x^{-n} \in C^k(K)$ for any compact $K \subset I^\circ$. In general, for a domain $\Omega \subset \mathbb{R}^n$, the presence of a supremum in the norm implies $C^k(\Omega) = C^k(\bar{\Omega})$. Then, if Ω is open, a distinction with functions on $\bar{\Omega}$ only arises by studying their behavior on compact subsets. This requires the use of local function spaces C_{loc}^k mentioned in the previous paragraph (see also Remark 1 below).

Although C^k spaces are complete with the sup-norm and satisfy inclusions $C^{k+1}(\Omega) \hookrightarrow C^k(\Omega)$, the images are not closed in their targets. This "bad" behavior makes them inadequate for analytic purposes (in particular, for studying PDEs) and justifies the introduction of Hölder spaces.

Notation $C_c^k(\Omega)$ refers to the set of compactly supported functions in $C^k(\Omega)$ and, for the rest of this section, unless otherwise explicitly stated, Ω stands for an open and bounded subset of $\mathbb{R}^n$.

10.2.3 Hölder spaces

Let $k \in \mathbb{Z}_+$ and $\alpha \in (0, 1]$. To the Hölder space $C^{k,\alpha}(\Omega)$ belong the functions $f : \Omega \to \mathbb{R}^m$ with bounded continuous derivatives of all orders $|\ell| \le k$ for which the norm

$$\|f\|_{C^{k,\alpha}(\Omega)} := \|f\|_{C^k(\Omega)} + \sum_{|\ell|=k} \sup_{\substack{x,y \in \Omega \\ x \neq y}} \frac{|\partial^\ell f(x) - \partial^\ell f(y)|}{|x - y|^\alpha} < \infty.$$

With this norm Hölder spaces are complete, hence Banach, spaces and there are compact inclusions $C^{k,\beta}(\Omega) \hookrightarrow C^{k,\alpha}(\Omega)$ whenever $0 < \alpha < \beta \le 1$. Functions like $|x|^\alpha \in C^{0,\alpha}(\mathbb{R})$ are called α-Hölder continuous, but notice that $|x|^\alpha \notin C^1(\mathbb{R})$. In general, there are proper inclusions $C^{k+1}(\Omega) \subsetneq C^{k,\alpha}(\Omega) \subsetneq C^k(\Omega)$, even for $\alpha = 1$, that corresponds to a Lipschitz condition on the derivatives. So beware, with the notation used here, $C^{k,1}(\Omega) \neq C^{k+1}(\Omega)$. Nevertheless $C^{k,0}(\Omega) = C^k(\Omega)$, showing that $\alpha = 0$ is redundant.

10.2.4 Zygmund spaces

Let $k \in \mathbb{Z}_+$ and $\alpha \in (0, 1]$. To the Zygmund space $C^{k+\alpha}_*(\Omega)$ belong the functions $f : \Omega \to \mathbb{R}^m$ with bounded continuous derivatives of all orders $|\ell| \le k$ for which the norm

$$\|f\|_{C^{k+\alpha}_*(\Omega)} := \|f\|_{C^k(\Omega)}$$

$$+ \sum_{|\ell|=k} \sup_{\substack{x,h \in \Omega \\ h \neq 0}} \frac{|\partial^\ell f(x + h) - 2\partial^\ell f(x) + \partial^\ell f(x - h)|}{|h|^\alpha} < \infty.$$

Since $\alpha = 0$ is not allowed in this description, for a positive integer l the space $C^l_*(\Omega)$ has to be understood as $C^{(l-1)+1}_*(\Omega)$. In other words, requiring $l = k + \alpha \in \mathbb{F}_+$ it is necessary to choose $k = l - 1$ and $\alpha = 1$ to get the correct definition of the norm. In this case there are proper inclusions $C^{k+1}(\Omega) \subsetneq C^{k+1}_*(\Omega) \subsetneq C^{k,\alpha}(\Omega)$, for all $\alpha \in (0, 1]$.

On the other hand, when $k + \alpha \notin \mathbb{Z}_+$, Hölder and Zygmund spaces actually coincide, i.e. if $\alpha \in (0, 1)$ then $C^{k,\alpha}(\Omega) = C^{k+\alpha}_*(\Omega)$ holds. For this reason they are also known as *Hölder–Zygmund spaces*.

Intuitively speaking, the inclusions $C^{k+\alpha}_*(\Omega) \subsetneq C^k(\Omega)$ allow us to regard $\alpha \in (0, 1)$ as a kind of fractional differentiability. The case $\alpha = 1$ for Hölder spaces does not behave as nicely, in the regularity theory of elliptic PDEs, as $\alpha = 1$ for Zygmund spaces. Therefore, it is common to avoid $\alpha = 1$ in Hölder spaces and work instead with Zygmund spaces for any $\alpha \in (0, 1]$.

There is an alternative characterization of Zygmund spaces by means of *Littlewood–Paley theory*, which is a very useful tool to study regularity of PDEs, but it is not treated here. Details can be found, for example, in [Tr92] or [Ta11].

10.2.5 Lebesgue spaces

Let $p \in (0, \infty)$. To the Lebesgue space $L^p(\Omega)$ belong the measurable functions $f : \Omega \to \mathbb{R}^m$ with norm

$$\|f\|_{L^p(\Omega)} := \left(\int_\Omega |f(x)|^p dx \right)^{1/p} < \infty.$$

Actually, it is only a quasi-norm for $p \in (0, 1)$ since a constant C would be needed in the triangle inequality: $\|f + g\|_{L^p} \le C(\|f\|_{L^p} + \|g\|_{L^p})$. Functions in $L^p(\Omega)$ are called p-integrable and two of them are identified when they differ only on a set of measure-zero. This means that they are equivalence classes and need to be defined, merely, almost everywhere in Ω. For simplicity, the same standard notation for functions is used. In the case $p = \infty$, the space $L^\infty(\Omega)$ consists of essentially bounded measurable functions, i.e. the norm

$$\|f\|_{L^\infty(\Omega)} := \operatorname*{ess\,sup}_{x \in \Omega} |f(x)| < \infty.$$

L^p spaces are Banach spaces for $p \in [1, \infty)$ and this will be the case considered from now on. Furthermore, if $\Omega \subset \mathbb{R}^n$ has finite measure and $1 \le p \le q \le \infty$, there is a continuous embedding $L^q(\Omega) \hookrightarrow L^p(\Omega)$.

Finally, recall that $L^2(\Omega, \mathbb{R}^m)$ carries a canonical inner product

$$\langle f, g \rangle_{L^2(\Omega, \mathbb{R}^m)} := \int_\Omega \langle f(x), g(x) \rangle_{\mathbb{R}^m} \, dx$$

that induces its norm. Therefore, $L^2(\Omega, \mathbb{R}^m)$ is a Hilbert space. Moreover, a function $f \in L^2(\Omega, \mathbb{R}^m)$ is said to be orthogonal to some subspace $\mathcal{F} \subset L^2(\Omega, \mathbb{R}^m)$ if $\langle f, g \rangle_{L^2} = 0$ for all $g \in \mathcal{F}$. This is denoted by $f \perp \mathcal{F}$.

10.2.6 Weak derivatives

Let $f, g \in L^1(\Omega, \mathbb{R}^m)$. The function g is called an ℓ-weak (partial) derivative of f if for all *test* functions $\eta \in C_c^{|\ell|}(\Omega, \mathbb{R}^m)$ the following holds

$$\int_\Omega \langle g, \eta \rangle_{\mathbb{R}^m} dx = (-1)^{|\ell|} \int_\Omega \langle f, \partial^\ell \eta \rangle_{\mathbb{R}^m} dx. \tag{10.1}$$

In this case, it is also said that $g = \partial^\ell f$ in the *weak sense*. Weak derivatives (if they exist) are unique up to sets of measure zero and, if $f \in C^{|\ell|}(\Omega, \mathbb{R}^m)$, any classical derivative $\partial^\ell f$ is already a weak derivative.

10.2.7 Sobolev spaces

Let $k \in \mathbb{N}_0$ and $p \in [1, \infty)$. To the Sobolev space $W^{k,p}(\Omega)$ belong the functions $f \in L^p(\Omega)$ with weak derivatives $\partial^\ell f \in L^p(\Omega)$ for all orders $|\ell| \leq k$. The corresponding norm is

$$\|f\|_{W^{k,p}(\Omega)} := \sum_{|\ell| \leq k} \|\partial^\ell f\|_{L^p(\Omega)}.$$

By definition, for $k = 0$, it holds that $W^{0,p}(\Omega) = L^p(\Omega)$. In addition, spaces $H^{k,p}(\Omega)$ and $H_c^{k,p}(\Omega)$ are defined as the closures of $C^\infty(\Omega)$ and $C_c^\infty(\Omega)$ in the norm $\| \cdot \|_{W^{k,p}(\Omega)}$, respectively. By a classical result of Meyers–Serrin [MS64], for any $\Omega \subset \mathbb{R}^n$ it holds that $W^{k,p}(\Omega) = H^{k,p}(\Omega)$.

Remark 1 All the previous function spaces, denoted generically by $\mathcal{F}(\Omega)$, admit local versions $\mathcal{F}_{\mathrm{loc}}(\Omega)$ consisting of functions whose norms are only required to be bounded locally, i.e. on some open neighborhood of every point of Ω. This implies (because the norms are defined through sums of sup-norms or through integrals) that they will be bounded on compact subsets of Ω. Hence, it is usual to define

$$\mathcal{F}_{\mathrm{loc}}(\Omega) := \{f : \Omega \to \mathbb{R}^m \mid f \in \mathcal{F}(\Omega') \text{ for all open } \Omega' \Subset \Omega\}.$$

Clearly $\mathcal{F}(\Omega) \subset \mathcal{F}_{\mathrm{loc}}(\Omega)$. Notice also that the notion of ℓ-weak derivatives introduced above is well defined on the larger space of locally integrable functions $L_{\mathrm{loc}}^1(\Omega, \mathbb{R}^m)$.

10.2.8 Embeddings

Let $\Omega \subset \mathbb{R}^n$ be open and bounded. The *Sobolev scaling exponent* $w \in \mathbb{R}$ of a Sobolev space $W \equiv W^{k,p}(\Omega)$ is defined as the real number

$$w \equiv w(k, p) := k - \frac{n}{p}.$$

Its relevance is explained as follows. After scaling down the Euclidean norm on $\mathbb{R}^n$ by a factor $\lambda \in (1, \infty)$ such that $| \cdot |_{\mathbb{R}^n} \mapsto \lambda^{-1} | \cdot |_{\mathbb{R}^n}$, the boundedness of $\Omega \subset \mathbb{R}^n$ is preserved and $\mathrm{Vol}(K) \mapsto \lambda^{-n} \mathrm{Vol}(K)$ for all compact $K \subset \Omega$. Additionally, because of the decrease of distances in $\mathbb{R}^n$, each derivative of a function $f \in C^k(\Omega)$ increases by a factor λ, so that $\partial^\ell f(x) \mapsto \lambda^{|\ell|} \partial^\ell f(x)$ for all $|\ell| \leq k$. The same will be true for weak derivatives. In conclusion, the rescaling sends

$$\|f\|_W \longmapsto \|f\|_{W,\lambda} := \sum_{|\ell| \leq k} \lambda^{|\ell| - (n/p)} \|\partial^\ell f\|_{L^p}$$

and these norms are equivalent for each $\lambda > 1$ because

$$\lambda^{-(n/p)} \|f\|_W \leq \|f\|_{W,\lambda} \leq \lambda^w \|f\|_W.$$

Moreover, since

$$[f]_{k,p} := \sum_{|\ell| = k} \|\partial^\ell f\|_{L^p} \leq \|f\|_W$$

it holds also that

$$\lambda^w [f]_{k,p} \leq \|f\|_{W,\lambda} \leq \lambda^w \|f\|_W.$$

Now, let W_1 and W_2 be two Sobolev spaces on Ω with corresponding scaling exponents (say) $w_1 := w(k_1, p_1) < w_2 := w(k_2, p_2)$, such that $k_1 \leq k_2$. Then, given any compactly supported function $f \in C_c^\infty(\Omega)$, there has to exist a constant $c \equiv c(w_1, w_2) > 0$ independent of f such that $\|f\|_{W_1} \leq c\|f\|_{W_2}$. Indeed, suppose to the contrary that for any such constant there is always an f for which $c\|f\|_{W_2} < \|f\|_{W_1}$. If by rescaling both sides we get also $c\|f\|_{W_2,\lambda} \leq \|f\|_{W_1,\lambda}$, for any contraction of the Euclidean norm of $\mathbb{R}^n$ by a $\lambda > 1$, then

$$c\lambda^{w_2}[f]_{k_2,p_2} \leq c\|f\|_{W_2,\lambda} > \|f\|_{W_1,\lambda} \leq \lambda^{w_1}\|f\|_{W_1},$$

therefore

$$c\lambda^{w_2 - w_1}[f]_{k_2,p_2} > \|f\|_{W_1}.$$

Since $w_2 - w_1 > 0$, by choosing a λ sufficiently large, it is clear that for all $f \in C_c^\infty(\Omega)$ the left-hand side can always be made larger than the right-hand side, yielding a contradiction. This argument indicates that $\|f\|_{W_1} \leq c'\|f\|_{W_2}$ holds for some constant $c' > 0$.

Similar considerations are possible for Hölder spaces $C^{k,\alpha}(\Omega)$ with $\alpha \in (0, 1)$ using the *Hölder scaling exponent* $h(k, \alpha) := k + \alpha$.

The previous argument suggests that there should exist embeddings between the spaces involved. A precise formulation is given by the following theorem.

Theorem 10.1 *Let k_1, k_2 be integers with $0 \leq k_1 \leq k_2$ and p_1, p_2, α reals with p, p_1, $p_2 \geq 1$ and $\alpha \in (0, 1)$. Suppose $w(k, p) := k - (n/p)$ and $h(k, \alpha) := k + \alpha$ denote the Sobolev and Hölder scaling exponents introduced above and assume Ω is a fixed open bounded subset of $\mathbb{R}^n$.*

- **Sobolev embeddings**

  For exponents $w(k_1, p_1) \leq w(k_2, p_2)$ there are continuous embeddings $W^{k_2,p_2}(\Omega) \hookrightarrow W^{k_1,p_1}(\Omega)$.

- **Sobolev into Hölder/Zygmund embeddings**

  For exponents $h(k_1, \alpha) \leq w(k_2, p_2)$ there are continuous embeddings $W^{k_2,p_2}(\Omega) \hookrightarrow C^{k_1,\alpha}(\Omega)$. Moreover, if $p_2 > 1$ and $r := w(k_2, p_2)$ then $W^{k_2,p_2}(\Omega) \hookrightarrow C^r_*(\Omega) \hookrightarrow C^{k_1,\alpha}(\Omega)$.

- **Rellich–Kondrakov compactness**

  When $w(k_1, p_1) < w(k_2, p_2)$, the embedding $W^{k_2,p_2}(\Omega) \hookrightarrow W^{k_1,p_1}(\Omega)$ is compact, i.e. the closure of the image is compact in $W^{k_1,p_1}(\Omega)$.

  When $h(k_1, \alpha) < w(k_2, p_2)$, the embedding $W^{k_2,p_2}(\Omega) \hookrightarrow C^{k_1,\alpha}(\Omega)$ is compact, i.e. the closure of the image is compact in $C^{k_1,\alpha}(\Omega)$.

Remark 2 To easily analyze the validity of an embedding, note that, for Sobolev spaces, if $k_2 \geq k_1 \geq 0$ and $p_2 \geq p_1 \geq 1$ the inclusions are always guaranteed and, if $p_1 > p_2$, it is only necessary to check whether $(k_2 - k_1) \geq (p_2^{-1} - p_1^{-1})n$ or not. Similarly, for the Sobolev into Hölder embeddings to hold, it suffices that $k_2 - k_1 \geq \alpha + (n/p_2)$. In this last situation almost everywhere identical functions are identified in the target Hölder space.

10.3 Elliptic operators and PDEs

Results on existence and uniqueness of solutions to PDEs are as essential for applications in geometry as they are in other areas. Since many expressions involving curvature and other geometric identities produce elliptic equations, what follows is a brief summary of standard terminology and important theorems on their behavior. Detailed treatments of most of the material in this section can be found in many texts on elliptic PDEs. Recommended modern references are [Jo07], [Eva10], [GT01] and [Ta11]. Classical texts such as [Mo66], [LU68] and [BJS79] contain earlier accounts of the interior elliptic regularity theory presented in different degrees of generality. For a nice exposition of these tools in the context of applications to geometry see [Ka93]. Also [Au98] contains a useful part on preliminaries (from where the different notions of weak solutions were taken) as well as more advanced geometric applications. Finally, it should be pointed out that in what follows, and given their importance to geometry, elliptic equations and operators are presented from the general perspective of (real) elliptic systems, i.e. functions on which operators act are assumed to be vector-valued most of the time. This is necessary for the general point of view of operators acting on sections of vector bundles, although this particular development is not explained, nor strictly required here.

10.3.1 Generalities

Let $\mathcal{F}$, $\mathcal{F}'$ denote generic function spaces that may vary depending on the specific situation. A (possibly nonlinear) partial differential operator of order $k \in \mathbb{N}_0$, defined on a domain $\Omega \subset \mathbb{R}^n$, is a map $P : \mathcal{F}(\Omega, \mathbb{R}^m) \to \mathcal{F}'(\Omega, \mathbb{R}^{m'})$ such that for any $u \in \mathcal{F}(\Omega, \mathbb{R}^m)$ its image under P has the form

$$(Pu)(x) \equiv P(x, (\partial^\ell u)_{|\ell| \le k}).$$

On the right, with an obvious abuse of notation, the expression $P(-, -)$ is generally assumed to be at least a continuous function of its arguments and, when it actually depends smoothly on them, the operator P is called *smooth*. If $P(x, (\partial^\ell u)_{|\ell| \le k})$ depends linearly on u, the operator is called *linear* and when it is linear only in the higher-order derivatives of u, i.e. all $\partial^\ell u$ for $|\ell| = k$, the operator is called *quasilinear*. In other cases the operator is called *fully* or *completely nonlinear*.

To make this definition more concrete, any k-order linear differential operator P can be written in the form $P = \sum_{|\ell| \le k} a_\ell(x) \partial^\ell$, where the coefficient functions $a_\ell(x)$ are linear maps from $\mathbb{R}^m$ to $\mathbb{R}^{m'}$ at each $x \in \Omega$ and, as functions $a_\ell : \Omega \to \mathrm{Hom}(\mathbb{R}^m, \mathbb{R}^{m'})$, can be bounded, continuous, Hölder, smooth, etc., according to the context. Similarly, any k-order quasilinear differential operator P acting on a function $u \in \mathcal{F}(\Omega, \mathbb{R}^m)$ can be written as

$$Pu = \sum_{|\ell| \le k} a_\ell(x, (\partial^\lambda u)_{|\lambda| < k}) \partial^\ell u + b(x, (\partial^\lambda u)_{|\lambda| < k}),$$

where, of course, it is assumed that the maximal order of derivatives of u cannot exceed k. For many regularity results it is necessary to assume that the functions $a_\ell(-, -)$ and $b_\ell(-, -)$ depend smoothly on their arguments, or at least that they are more regular than the derivatives $\partial^\lambda u$ of order $|\lambda| = k - 1$.

The *formal adjoint* $P^* : \mathcal{F}'(\Omega, \mathbb{R}^{m'}) \to \mathcal{F}(\Omega, \mathbb{R}^m)$ of a linear partial differential operator $P : \mathcal{F}(\Omega, \mathbb{R}^m) \to \mathcal{F}'(\Omega, \mathbb{R}^{m'})$ is (densely) defined by the condition

$$\langle u, P^* v \rangle_{L^2(\Omega, \mathbb{R}^m)} = \langle Pu, v \rangle_{L^2(\Omega, \mathbb{R}^{m'})}$$

for all $u \in C_c^\infty(\Omega, \mathbb{R}^m)$, $v \in C_c^\infty(\Omega, \mathbb{R}^{m'})$. To extend this definition to $\mathcal{F}$ and $\mathcal{F}'$, it is generally assumed that $\mathcal{F}(\Omega, \mathbb{R}^m) \subset L^2(\Omega, \mathbb{R}^m)$ and $\mathcal{F}'(\Omega, \mathbb{R}^{m'}) \subset L^2(\Omega, \mathbb{R}^{m'})$.

Explicitly, if P is linear of order k, then P^* is also and

$$Pu = \sum_{|\ell| \le k} a_\ell(x) \partial^\ell u \quad \Longrightarrow \quad P^* v = \sum_{|\ell| \le k} (-1)^{|\ell|} \partial^\ell (a_\ell^*(x) v),$$

where $a_\ell^*(x) \in \mathrm{Hom}(\mathbb{R}^{m'}, \mathbb{R}^m)$ denotes the adjoint of the linear map $a_\ell(x) \in \mathrm{Hom}(\mathbb{R}^m, \mathbb{R}^{m'})$.

A linear differential operator $P : \mathcal{F}(\Omega, \mathbb{R}^m) \to \mathcal{F}'(\Omega, \mathbb{R}^{m'})$, of even order $2k$, is said to be given in *divergence form* if acting on some $u \in \mathcal{F}(\Omega, \mathbb{R}^m)$ it can be written as

$$Pu = \sum_{|\ell|, |\ell'| \leq k} \partial^\ell (a_{\ell, \ell'}(x) \partial^{\ell'} u) + \sum_{|\ell| \leq k} b_\ell(x) \partial^\ell u,$$

where $a_{\ell, \ell'}(x), b_\ell(x) \in \mathrm{Hom}(\mathbb{R}^m, \mathbb{R}^{m'})$. Clearly, if the coefficients $a_{\ell, \ell'}(-)$ are of class $C^{|\ell|}$, an operator given in divergence form can be written in standard form and vice versa. The simplest example occurs for second-order operators acting on scalar functions. In this case, an operator written in divergence form reads

$$Pu = \sum \partial_i (a_{ij}(x) \partial_j u) + \sum b_i(x) \partial_i u + c(x) u.$$

10.3.2 Generalized solutions to PDEs

Let $P : \mathcal{F}(\Omega, \mathbb{R}^m) \to \mathcal{F}'(\Omega, \mathbb{R}^{m'})$ be a partial differential operator of order k. By this notation it will be implicitly understood that in the equation $Pu = f$ the functions are, initially, $u \in \mathcal{F}(\Omega, \mathbb{R}^m)$ and $f \in \mathcal{F}'(\Omega, \mathbb{R}^{m'})$. A *classical solution* to this equation is, in general, assumed to belong to $C^k_{\mathrm{loc}}(\Omega, \mathbb{R}^m)$. However, it is possible to define different kinds of generalized solutions to $Pu = f$ according to the properties of P.

Definition 10.2 Let $k, l \in \mathbb{N}_0$, $p \in [1, \infty)$ and P a partial differential operator of order k as above. The equation $Pu = f$ admits a generalized solution u in the following senses:

- **Strong solutions** If $P : W^{k+l,p}(\Omega) \to W^{l,p}(\Omega)$ is linear with $L^\infty_{\mathrm{loc}}(\Omega)$ coefficients, a solution is called *strong* if there is a sequence (u_i) of classical solutions such that $u_i \to u$ in $W^{k+l,p}$ and $Pu_i \to f$ in $W^{l,p}$. Therefore, weak derivatives $\partial^\lambda u \in W^{l,p}$ for all $|\lambda| \leq k$ and $Pu = f$ hold almost everywhere.
- **$W^{l,1}$-weak solutions** If $P : W^{l,1}(\Omega, \mathbb{R}^m) \to L^1(\Omega, \mathbb{R}^{m'})$ is linear with coefficients $a_\ell \in C^{|\ell|+l}(\Omega)$ for each $|\ell| \leq k$, a $W^{l,1}$-*weak solution* exists if for all $v \in C^\infty_c(\Omega, \mathbb{R}^{m'})$ the following holds:

$$\int_\Omega \langle u, P^* v \rangle_{\mathbb{R}^m} \, dx = \int_\Omega \langle f, v \rangle_{\mathbb{R}^{m'}} \, dx.$$

- **$W^{k,p}$-weak solutions** If $P : W^{k,p}(\Omega, \mathbb{R}^m) \to L^1(\Omega, \mathbb{R}^{m'})$ is linear of even order $2k$ with $L^\infty_{\mathrm{loc}}(\Omega)$ coefficients and admits a divergence form, a $W^{k,p}$-*weak solution* exists if for all $v \in C_c^\infty(\Omega, \mathbb{R}^{m'})$ the following holds

$$\sum_{|\ell|,|\ell'|\leq k} (-1)^{|\ell|} \int_\Omega \langle a_{\ell,\ell'} \partial^{\ell'} u, \partial^\ell v \rangle_{\mathbb{R}^{m'}} \, dx + \sum_{|\ell|\leq k} \int_\Omega \langle b_\ell \, \partial^\ell u, v \rangle_{\mathbb{R}^{m'}} \, dx$$

$$= \int_\Omega \langle f, v \rangle_{\mathbb{R}^{m'}} \, dx.$$

This condition is well defined for any $p \in [1, \infty)$.

10.3.3 *Elliptic partial differential operators*

Let $\Omega \subset \mathbb{R}^n$. A k-order linear differential operator

$$P = \sum_{|\ell|\leq k} a_\ell(x)\partial^\ell : \mathcal{F}(\Omega, \mathbb{R}^m) \longrightarrow \mathcal{F}'(\Omega, \mathbb{R}^{m'})$$

is *elliptic at* $x_0 \in \Omega$ if its *principal symbol* $\sigma_P : \Omega \times \mathbb{R}^n \to \operatorname{Hom}(\mathbb{R}^m, \mathbb{R}^{m'})$, defined by

$$\sigma_P(x, \xi) := \sum_{|\ell|=k} a_\ell(x)\xi^\ell,$$

is a linear isomorphism as an element of $\operatorname{Hom}(\mathbb{R}^m, \mathbb{R}^{m'})$ at $x_0 \in \Omega$, for all $\xi \in \mathbb{R}^n \setminus \{0\}$. When this holds for any $x \in \Omega$ the operator is said to be *elliptic in* Ω. In addition, P is called *underdetermined elliptic* if the application $\sigma_P(x, \xi) : \mathbb{R}^m \to \mathbb{R}^{m'}$ is surjective, and *overdetermined elliptic* if it is injective, for those same $x \in \Omega$ and $\xi \in \mathbb{R}^n \setminus \{0\}$. Clearly, ellipticity requires $m = m'$, underdetermined ellipticity $m \leq m'$ and overdetermined ellipticity $m \geq m'$. Moreover, if P is underdetermined (or overdetermined) elliptic at some $x \in \Omega$, then PP^* (or P^*P) is elliptic at these points.

It is common to complement the ellipticity condition at a point $x \in \Omega$ with the stronger requirement that the quadratic form

$$B_\xi(v, w) := \langle \sigma_P(x, \xi)v, w \rangle_{\mathbb{R}^m},$$

where $v, w \in \mathbb{R}^m$, is always positive (or negative) definite for every $\xi \in \mathbb{R}^n \setminus \{0\}$. This is called *strong ellipticity* and, by changing ξ to $-\xi$ in the definition, it is easily seen that it can only hold for operators of even order.

Remark 3 For complex-valued operators (and functions) strong ellipticity is introduced, more generally, with a quadratic form B that is the real part of an appropriate Hermitian product, but this case is not considered here (see, for example, [BJS79]).

In the real scalar case, i.e. when $m = 1$, ellipticity and strong ellipticity are equivalent since in this case the principal symbol is just a real number that is required to be non-zero.

When $m = m'$, ellipticity on Ω can be strengthened to *uniform ellipticity* by requiring the existence of a constant $C > 0$ such that

$$|\langle \sigma_P(x, \xi)v, v \rangle_{\mathbb{R}^m}| \geq C|\xi|^k|v|^2$$

for any $v \in \mathbb{R}^m$ and all $x \in \Omega, \xi \in \mathbb{R}^n \setminus \{0\}$.

Remark 4 The condition of ellipticity on Ω may only be needed almost everywhere, for example, if the operator is assumed to act on functions in some $L^p(\Omega)$. A general linear, second-order, partial differential operator $P : \mathcal{F}(\Omega, \mathbb{R}) \to \mathcal{F}'(\Omega, \mathbb{R})$ can be explicitly written as

$$P := \sum_{i,j=1}^{n} a_{ij}(x)\partial_i\partial_j + \sum_{i=1}^{n} b_i(x)\partial_i + c(x).$$

In this case, and assuming P has (essentially) bounded coefficients, uniform ellipticity implies that there are constants $C_1, C_2 > 0$ such that for (almost) every $x \in \Omega \subset \mathbb{R}^n$ and all $\xi = (\xi_1, \ldots, \xi_n) \in \mathbb{R}^n \setminus \{0\}$,

$$C_1|\xi|^2 \leq \sum a_{ij}(x)\xi_i\xi_j \leq C_2|\xi|^2.$$

This condition is frequently taken as the definition of uniform ellipticity for second-order operators and in some contexts can be simply called ellipticity.

Finally, a general nonlinear k-order partial differential operator

$$P \equiv P(x, (\partial^\ell u)_{|\ell| \leq k})$$

is *elliptic* at $v \in C^k(\Omega, \mathbb{R}^m)$ if the ellipticity condition holds for its *linearization* $L_v P$ at v, defined by

$$(L_v P)u := \frac{d}{dt}\Big|_{t=0} P(x, (\partial^\ell(v + tu))_{|\ell| \leq k}).$$

10.3.4 Elliptic regularity of solutions

There are different theorems on existence and regularity of solutions to PDEs defined through linear elliptic operators acting on function spaces in $\mathbb{R}^n$. Extensions of these results are available to quasilinear and also to fully nonlinear operators under appropriate conditions. Furthermore, such theorems can be carried locally to the manifold setting, and sometimes even globally, using additional hypotheses (like compactness of the manifold).

Assuming existence of solutions for elliptic equations, in what follows a quick summary of important results on interior regularity of elliptic equations is presented. First, the following interior estimates hold for general linear elliptic differential operators in $\mathbb{R}^n$ with Hölder coefficients.

Theorem 10.3 (Interior elliptic regularity) *Let $k, l \in \mathbb{N}_0$, $\alpha \in (0, 1)$ and $p \in (1, \infty)$. Suppose P is a linear, elliptic, partial differential operator of order k in a bounded open set $\Omega \subset \mathbb{R}^n$ and assume $\Omega' \Subset \Omega$.*

- *L^p-estimates If $P : W^{k,p}(\Omega, \mathbb{R}^m) \to W^{l,p}(\Omega, \mathbb{R}^m)$ has C^l coefficients and $Pu = f$ weakly in $L^1(\Omega, \mathbb{R}^m)$, then $u \in W^{k+l,p}(\Omega', \mathbb{R}^m)$ and for some constant $C > 0$ the following holds*

$$\|u\|_{W^{k+l,p}(\Omega')} \leq C\big(\|f\|_{W^{l,p}(\Omega)} + \|u\|_{L^1(\Omega)}\big).$$

- ***Schauder estimates** If $P : C^{k,\alpha}(\Omega, \mathbb{R}^m) \to C^{l,\alpha}(\Omega, \mathbb{R}^m)$ has $C^{l,\alpha}$ coefficients and $Pu = f$, then $u \in C^{k+l,\alpha}(\Omega', \mathbb{R}^m)$ and for some constant $C > 0$ the following holds*

$$\|u\|_{C^{k+l,\alpha}(\Omega')} \leq C\big(\|f\|_{C^{l,\alpha}(\Omega)} + \|u\|_{C^0(\Omega)}\big).$$

Remark 5 The conclusion $u \in C^{k+l,\alpha}(\Omega', \mathbb{R}^m)$ in the Schauder estimates part also holds assuming only that $P : W^{k,p}(\Omega, \mathbb{R}^m) \to C^{l,\alpha}(\Omega, \mathbb{R}^m)$ has $C^{l,\alpha}$ coefficients and $Pu = f$ almost everywhere, i.e. it is enough to suppose that initially u belongs to $W^{k,p}(\Omega, \mathbb{R}^m)$ rather than $C^{k,\alpha}(\Omega, \mathbb{R}^m)$.

As an example of possible generalizations of regularity theorems for linear operators, the following theorem provides elliptic regularity results for quasilinear operators. The statement combines particular cases of more general results considered in Theorems 4.4 and 4.5, Chapter 14 of [Ta11].

Theorem 10.4 (Quasilinear elliptic regularity) *Assume $\Omega' \Subset \Omega$ for a bounded open set $\Omega \subset \mathbb{R}^n$ and $p \in (1, \infty)$. Let P be a k-order quasilinear elliptic operator of the form*

$$Pu = \sum_{|\ell| \leq k} a_\ell(x, (\partial^\lambda u)_{|\lambda| < k}) \partial^\ell u,$$

whose coefficients $a_\ell(-, -)$ depend smoothly on their arguments.

- *If $r > \alpha \in [0, 1)$ and $P : C^{k,\alpha}(\Omega, \mathbb{R}) \to C^r_*(\Omega, \mathbb{R})$ then $u \in C^{k+r}_*(\Omega', \mathbb{R})$.*
- *If $r > 0$, $\alpha \in [0, 1)$ and $P : C^{k-1,\alpha}(\Omega, \mathbb{R}) \cap W^{k,p}(\Omega, \mathbb{R}) \to C^r_*(\Omega, \mathbb{R})$ then $u \in C^{k+r}_*(\Omega', \mathbb{R})$.*

Useful regularity theorems for second-order elliptic PDEs that can be written in divergence form are as follows.

Theorem 10.5 (Ellipticity in divergence form) *Assume $\Omega' \Subset \Omega$ for $\Omega \subset \mathbb{R}^n$ open and bounded. Let P be a second-order, linear, elliptic operator such that the equation $Pu = f$ is given in divergence form.*

- **Sobolev spaces** *If $n < q < p < \infty$ and $P : W^{1,q}(\Omega, \mathbb{R}) \to W^{k,p}(\Omega, \mathbb{R})$ has $W^{k,p}$ coefficients then $u \in W^{k+2,p}(\Omega, \mathbb{R})$.*
- **Zygmund spaces** *If $r \geq s \geq 1$ and $P : C^1_*(\Omega, \mathbb{R}^m) \to C^{s-1}_*(\Omega, \mathbb{R}^m)$ has C^r_* coefficients, then $u \in C^{s+1}_*(\Omega', \mathbb{R}^m)$.*

The Sobolev space part follows from the (more general) assertions in Proposition 2.2.I and equation (2.2.56) of [Ta91]. Furthermore, the Zygmund space part is also valid for any $s > 0$, but that would require negative-order Zygmund spaces not introduced here. This result is a consequence of Proposition 1.1 in [Ta00], where broader hypotheses and conclusions are given, in terms of the modulus of continuity of the functions involved.

Remark 6   In Theorem 10.5, the Sobolev part works in a more general setting namely, the statement holds for any second-order, quasilinear, elliptic PDE in divergence form

$$Pu = \sum_{i,j=1}^{n} \partial_i(a_{ij}(x, u)\partial_j u) + Q(x, u, \partial u) = f,$$

where $Q(x, u, \partial u)$ is a quadratic form in ∂g and the functions $a_{ij}(-, -)$, $Q(-, -, \partial g)$ are smooth in their dashed $(-)$ arguments x and u.

Theorems 10.4 and 10.5 are just examples of many possible generalizations of standard elliptic regularity. Variations include the use of boundary conditions, different function spaces, results for completely nonlinear differential operators, etc. Interested readers are referred to detailed texts as [GT01] and [Ta11].

10.3.5 Existence of solutions

A bounded and linear partial differential operator $P : \mathcal{F}(\Omega, \mathbb{R}^m) \to \mathcal{F}'(\Omega, \mathbb{R}^{m'})$ between Banach spaces is called *Fredholm* if $\operatorname{Ker} P \subset \mathcal{F}(\Omega, \mathbb{R}^m)$ and $\operatorname{Ker} P^* \subset \mathcal{F}'(\Omega, \mathbb{R}^{m'})$ are finite-dimensional subspaces. These operators have a well-defined *Fredholm index*

$$\operatorname{Ind} P := \dim(\operatorname{Ker} P) - \dim(\operatorname{Ker} P^*).$$

An application of the so-called Fredholm alternative for elliptic differential operators with Hölder coefficients ensures the existence of solutions to their associated elliptic PDEs.

Theorem 10.6 (Fredholm alternative) *In addition to the hypotheses of Theorem 10.3, assume $l \geq k$ and that*

- $P : W^{k+l,p}(\Omega, \mathbb{R}^m) \to W^{l,p}(\Omega, \mathbb{R}^m)$ *has C^l coefficients, or*
- $P : C^{k+l,\alpha}(\Omega, \mathbb{R}^m) \to C^{l,\alpha}(\Omega, \mathbb{R}^m)$ *has $C^{l,\alpha}$ coefficients.*

Then P is a Fredholm operator whose adjoint is an elliptic operator

- $P^* : W^{l,p}(\Omega, \mathbb{R}^m) \to W^{l-k,p}(\Omega, \mathbb{R}^m)$ *with C^{l-k} coefficients, or*
- $P^* : C^{l,\alpha}(\Omega, \mathbb{R}^m) \to C^{l-k,\alpha}(\Omega, \mathbb{R}^m)$ *with $C^{l-k,\alpha}$ coefficients,*

respectively. Moreover, a solution to $Pu = f$ exists if and only if $f \perp \text{Ker } P^*$. The solution is unique if in addition $u \perp \text{Ker } P$.

More general existence results are available, covering variations of these and other cases, but precise statements are left to the specialized literature on the subject mentioned at the beginning of this section.

10.3.6 *Function spaces and PDEs on manifolds*

To conclude this condensed review of preliminaries, some consideration must be given to the definition on Riemannian manifolds of the previously introduced function spaces.

Let (M, g) be an n-dimensional Riemannian manifold. Note first that local versions of function spaces are readily well defined. In fact, given a local coordinate chart $\phi : U \subset M \to \mathbb{R}^n$ of the maximal atlas $\mathcal{A}_M$ of M and any $f : M \to \mathbb{R}^m$, define the pull-back $\phi^* f := f \circ \phi^{-1} : \phi(U) \to \mathbb{R}^m$. Then

$$\mathcal{F}(U, \mathbb{R}^m) := \{f : U \to \mathbb{R}^m \mid \phi^* f \in \mathcal{F}(\phi(U), \mathbb{R}^m)\} \quad \text{and}$$

$$\mathcal{F}_{\text{loc}}(M, \mathbb{R}^m) := \{f : M \to \mathbb{R}^m \mid \phi^* f \in \mathcal{F}_{\text{loc}}(\phi(U), \mathbb{R}^m);$$

$$\text{for all } (\phi, U) \in \mathcal{A}_M\}.$$

To rigorously introduce function spaces $\mathcal{F}(M, \mathbb{R}^m)$ in the Hölder, Zygmund, Lebesgue and Sobolev cases, a partition-of-unity argument suffices. Namely, with the aid of a partition of unity $\Psi = \{\psi_\gamma\}$ whose compact supports are subordinated to some (not necessarily maximal) atlas $\mathcal{A} = \{(U_\gamma, \phi_\gamma)\}$ of M, any function $f : M \to \mathbb{R}^m$ can be written as $f(x) = \sum_\gamma \psi_\gamma(x) f(x)$. Therefore, it is enough to require that each compactly supported function $\psi_\gamma \cdot f$ belongs to $\mathcal{F}(U_\gamma, \mathbb{R}^m)$. It is not difficult to see that this property does not really depend on the atlas or the partition of unity Ψ chosen. Explicitly,

$$\mathcal{F}(M, \mathbb{R}^m) := \{f : M \to \mathbb{R}^m \mid \phi_\gamma^*(\psi_\gamma \cdot f) \in \mathcal{F}(\phi_\gamma(U_\gamma), \mathbb{R}^m);$$

$$\text{for all } (\phi_\gamma, U_\gamma) \in \mathcal{A} \text{ and } \psi_\gamma \in \Psi\}.$$

Recall that local integration of real-valued functions on a Riemannian manifold (M, g) is well defined and canonical, by use of the Riemannian volume element $dv_g := \sqrt{|g|}\, dx_1 \cdots dx_n$. Explicitly, on a coordinate chart $(U, \phi) \in \mathcal{A}$ and for $f \in C_c(U, \mathbb{R})$ integration on U is defined by

$$\int_U f \, dv_g \equiv \int_U \hat{f} \, dx := \int_{\phi(U)} \phi^* \hat{f} \, dx_1 \cdots dx_n,$$

where $\hat{f} := \sqrt{|g|}\, f$. If M is compact this integration is well defined in a coordinate independent way on the whole of M by using, once again, a partition-of-unity argument. This makes it possible to introduce Lebesgue and Sobolev spaces on M with the norms defined directly by integration of functions on the manifold. In general, this procedure will depend on the Riemannian metric g chosen, but the resulting norms turn out to be equivalent. Moreover, the Lebesgue and Sobolev spaces obtained will be the same as the ones defined above.

Concerning partial differential operators and PDEs, the definitions in this section are easily carried over to the manifold setting. Also, local properties, including interior regularity theorems, continue to hold. Nevertheless, care should be taken in how they are used. In general, results such as embedding theorems (of function spaces) and regularity estimates of PDEs are only applicable on local neighborhoods of points of M and not globally, or, at best, need modifications to be extended to the whole manifold. One simple but important case where the embeddings of Theorem 10.1 are globally valid is when M is assumed to be closed, i.e. compact without boundary. With this assumption, interior estimates also hold globally, replacing Ω and Ω' in Theorem 10.3 by M.

10.4 Riemannian regularity and harmonic coordinates

It is common to assume that the underlying differentiable structure of a manifold M is smooth. Moreover, owing to an important theorem of H. Whitney, any C^r-differentiable manifold has a C^∞ atlas contained in the C^r maximal atlas of M. Nevertheless, when working with Riemannian metrics of low regularity it is usually enough, and sometimes important, to consider less regular differentiable structures on M. The aim of this section is to deal with some details of non-smooth differentiable structures on manifolds admitting $C^{k,\alpha}$ Riemannian metrics on them, and to introduce a particularly useful system of coordinates with respect to which tensor fields have the best regularity possible on a given manifold.

10.4.1 Regularity of differentiable and Riemannian structures

Notation In what follows $\mathcal{A}_M = \{(U_\gamma, \phi_\gamma)\}$ will always denote the maximal atlas that provides the differentiable structure of an n-dimensional manifold M, and its regularity will be given explicitly in non-smooth situations. General coordinate charts (U, ϕ) on M are of the form $\phi = (x^1, \ldots, x^n) : U \subset M \to \mathbb{R}^n$ and may or may not belong to $\mathcal{A}_M$ depending on their compatibility with charts in $\mathcal{A}_M$. Components will be denoted by indices a, b, c, d, i, j running from 1 to n and, for simplicity, the Einstein sum convention will be implicitly used in expressions where an index appears repeated in upper and lower positions.

To consider the regularity of a Riemannian structure (given by a metric g) on M, recall that there is a general notion of regularity for tensor fields, which is defined componentwise. In the concrete case of Hölder regularity, a tensor field T is said to be of class $C^{k,\alpha}$ on an arbitrary coordinate chart (U, ϕ) of M if all its components are $C^{k,\alpha}$ functions on U, and it is said to be of class $C^{k,\alpha}$ globally on M, if it is $C^{k,\alpha}$ on every chart $(U_\gamma, \phi_\gamma) \in \mathcal{A}_M$.

Now, to have a well-defined function f of class $C^{k,\alpha}$ with respect (only) to an individual chart (U, ϕ) that does not necessarily belong to $\mathcal{A}_M$, no conditions on the smoothness of the manifold are required. This is because the differentiability is given by $f \circ \phi^{-1}$ and is independent of $\mathcal{A}_M$. Not that even constant functions can only be as smooth as the chart allows. Nevertheless, when the compatibility with charts of $\mathcal{A}_M$ is important, as is the case with the components g_{ab} of a metric that is supposed to be defined globally on M, it is required (and enough) to suppose that M carries a $C^{k+1,\alpha}$ differentiable structure. In fact, from a $C^{k+1,\alpha}$ atlas for M it is possible to construct $C^{k,\alpha}$ local trivializations of the tangent bundle TM and, hence, a $C^{k,\alpha}$ bundle atlas for TM. In this way, sections of TM on which the metric g acts will be able to have well-defined $C^{k,\alpha}$ regularity with respect to (U, ϕ), compatible with the regularity intended for g.

Indeed, let (U, ϕ) be an arbitrary chart on M. Then, coordinate vector fields ∂_a associated to (U, ϕ) act upon $f \in C^{k+1,\alpha}(M)$, defining new functions $\partial_a f$ whose values at a point $p = \phi^{-1}(x) \in U \cap U_\gamma$ are given by

$$(\partial_a f)(p) := \frac{\partial}{\partial x^a}(f \circ \phi^{-1})(x) = \frac{\partial}{\partial x^a}[(f \circ \phi_\gamma^{-1}) \circ (\phi_\gamma \circ \phi^{-1})](x). \qquad (10.2)$$

If the chart (U, ϕ) is compatible with $\mathcal{A}_M$ (and hence of class $C^{k+1,\alpha}$ with respect to it), all compositions $\phi_\gamma \circ \phi^{-1}$ are at least $C^{k+1,\alpha}$ and according to (10.2) the functions $\partial_a f$ will have at least $C^{k,\alpha}$ regularity on each $U \cap U_y$. This implies that all coordinate vector fields ∂_a have that same regularity on U, even if seen from different charts of $\mathcal{A}_M$ that intersect it (and for which a transformation of components is required). This allows the components of the metric to be of class $C^{k,\alpha}$ with respect to (U, ϕ) and, simultaneously, with respect to other charts of $\mathcal{A}_M$ that intersect it.

Concerning the Christoffel symbols Γ^a_{bc} of the Levi-Civita (or Riemannian) connection associated to g, the relation

$$\Gamma^a_{bc} = \frac{1}{2} g^{ad} (\partial_b g_{cd} + \partial_c g_{bd} - \partial_d g_{bc}) \tag{10.3}$$

implies that they will be of class $C^{k-1,\alpha}$ in the same chart (U, ϕ). Conversely, the following holds.

Theorem 10.7 (DeTurck–Kazdan) *Let g be a Riemannian metric of class C^1 in some local coordinates (U, ϕ). If the Christoffel symbols Γ^c_{ab} in (U, ϕ) are of class $C^{k,\alpha}$ for $k \geq 0$, then the metric g is actually of class $C^{k+1,\alpha}$ in these coordinates.*

Proof The proof follows from the observation that, for some given components Γ^c_{ab} of class $C^{k,\alpha}$, the equation

$$P(g) := \partial_b g_{ci} + \partial_c g_{bi} - \partial_i g_{bc} - 2g_{ij} \Gamma^j_{bc} = 0,$$

obtained from (10.3), is linear and overdetermined elliptic. Hence, it can be converted into a second-order, linear (determined) elliptic equation by applying the formal adjoint P^* at both sides $(P^* P)g = 0$.

Since the metric g is C^1 in (U, ϕ) it is necessary to look for weak solutions. Asume first that $k \geq 1$, then coefficients of $P^* P$ in (U, ϕ) are derivatives of Γ^c_{ab} of class $C^{k-1,\alpha}$. As a consequence, the operator $P^* P : C^1(U) \cap W^{2,2}(U) \to C^{k-1,\alpha}(U)$ and $P^* P g = 0$ a.e. (almost everywhere) in U, hence Schauder estimates (with Remark 5) imply $g \in C^{k+1,\alpha}(U)$.

For $k = 0$, it is important to note that weak solutions can be obtained by requiring that for any $\eta \in C^\infty_c(U)$ the following holds

$$0 = \langle P^*(Pg), \eta \rangle_{L^2(U)} = -\langle Pg, P\eta \rangle_{L^2(U)} = - \int_U (P\eta)(Pg)\, dx,$$

where on the right-hand side $P\eta \in C^{0,\alpha}(U)$ because it has coefficients with Γ^c_{ab}. This means that $(P\eta)Pg = 0$ must hold a.e. in U and the operator $(P\eta)P : C^1(U) \cap W^{2,2}(U) \to C^{0,\alpha}(U)$ has $C^{0,\alpha}$ coefficients. Again, by interior estimates for weak solutions, this yields the conclusion $g \in C^{1,\alpha}(U)$. □

A local formula for the Riemann curvature tensor, in terms of the Christoffel symbols, is given by

$$R^d_{abc} = \partial_b \Gamma^d_{ac} - \partial_c \Gamma^d_{ab} + (\Gamma^i_{ac}\Gamma^d_{bi} - \Gamma^i_{ab}\Gamma^d_{ci}). \tag{10.4}$$

It shows that the Riemann tensor (and all curvatures derived from it) can be explicitly written in terms of the metric and its first and second partial derivatives. If g is of Hölder class $C^{k,\alpha}$, this expression and the previous discussion indicate that the curvatures would be, *a priori*, of class $C^{k-2,\alpha}$. Nevertheless, the converse does not hold unless appropriate coordinates are used. See Theorem 10.12.

10.4.2 Harmonic coordinates

On any Riemannian manifold with differentiable structure of class at least C^2 and metric of Hölder class $C^{1,\alpha}$, there exists a special system of coordinates on which the components of the metric tensor (and of tensor fields in general) have optimal regularity. They are called harmonic coordinates and are very useful for dealing with "geometric" PDEs on manifolds. Their regularity properties were first studied by Sabitov–Shefel' in [SS76], DeTurck–Kazdan in [DK81] and Jost–Karcher in [JK82]. Proofs and additional results related to the theorems that follow are left to these references. More refined statements for metrics of lower regularity can be found in Chapter 3 of [Ta00].

Let (M, g) denote an n-dimensional Riemannian manifold whose differentiable structure is given by a maximal atlas $\mathcal{A}_M$ of class at least C^2. Recall that the *Laplace–Beltrami operator* Δ of (M, g), or *Laplacian* on functions for short, depends on g and acts on a function $u \in C^2(M)$ through a local expression

$$\Delta u := \mathrm{div}(\nabla u) = \frac{1}{\sqrt{|g|}}\partial_i\big(g^{ij}\sqrt{|g|}\,\partial_j u\big) \tag{10.5}$$

$$= g^{ij}\partial_i\partial_j u + \frac{1}{\sqrt{|g|}}\partial_i\big(\sqrt{|g|}g^{ij}\big)\partial_j u,$$

where $|g| := \det(g_{ij})$. Defining $\Gamma^i := g^{ab}\Gamma^i_{ab}$, it can be written in terms of the Christoffel symbols as

$$\Delta u = g^{ij}\partial_i\partial_j u - \Gamma^i\partial_i u.$$

The Laplacian is the fundamental example of a second-order, linear, elliptic operator.

Definition 10.8 A *harmonic coordinate chart* (U, ϕ) on (M, g) is a local chart for which all coordinate functions x^i are harmonic. Namely, $\Delta x^i = 0$.

Acting on coordinate functions x^i, the Laplacian reads in shortened form as $\Delta x^i = -\Gamma^i$ and, therefore, a coordinate system $\phi = (x^1, \ldots, x^n)$ will be harmonic if and only if all $\Gamma^i = 0$. In such harmonic charts the Laplacian, acting on a generic function $u \in C^2(M)$, takes the simpler form

$$\Delta u = g^{ij} \partial_i \partial_j u. \tag{10.6}$$

Existence of harmonic coordinates on neighborhoods of every point of a Riemannian manifold is asserted in the next theorem. See [SS76], [DK81] and [JK82] for more details and proofs of the following two theorems.

Theorem 10.9 (DeTurck–Kazdan) *Let $k \geq 1$ and suppose M is a $C^{k+1,\alpha}$ manifold endowed with a metric g of class $C^{k,\alpha}$ with respect to a local coordinate chart (U, ϕ). Then any point $p \in U$ has a (possibly smaller) neighborhood admitting harmonic coordinates that are $C^{k+1,\alpha}$ functions with respect to (U, ϕ). Furthermore,*

1. *these harmonic coordinates can be chosen such that all components $g_{ij} = \delta_{ij}$,*
2. *all harmonic charts about p have the same $C^{k+1,\alpha}$ regularity,*
3. *the metric g is of class $C^{k,\alpha}$ in any harmonic chart, while it is at least of class $C^{k-2,\alpha}$ in geodesic normal coordinates, and*
4. *any tensor which in (U, ϕ) is of class $C^{l,\beta}$, with $l \geq k$, $\beta \geq \alpha$, is at least of class $C^{k,\alpha}$ in harmonic coordinates.*

Optimal regularity of the metric (or any other tensor) in harmonic coordinates follows from property 4 in Theorem 10.9. Briefly, the consequences are that, in changing from arbitrary to harmonic coordinates, the regularity of the metric is, at worst, preserved, while changing to normal coordinates involves always the loss of at least two derivatives.

On compact manifolds the situation is even better owing to the existence of uniform Hölder bounds on the harmonic structure.

Theorem 10.10 (Jost–Karcher) *Let (M, g) be a compact Riemannian manifold and $\alpha \in (0, 1)$. About any point $p \in M$ there exists a ball $B_r(p)$ of some fixed radius $r > 0$, admitting harmonic coordinates and such that*

1. *transition functions have a uniform $C^{2,\alpha}$ bound,*
2. *the metric g has a uniform $C^{1,\alpha}$ bound, and*
3. *its associated Christoffel symbols have a uniform $C^{0,\alpha}$ bound.*

The radius r is bounded from below by a quantity that depends on the dimension n, the injectivity radius inj_M and upper bounds on the absolute value of the sectional curvature $|\sec|$. All Hölder bounds also depend on these quantities.

In conclusion, it is always possible to find harmonic coordinates on balls of some fixed radius about any point of the manifold, and to construct with them a harmonic atlas that provides a uniformly bounded differentiable structure. Moreover, in this atlas the Riemannian structure will also carry uniform Hölder bounds. These bounds are useful, for example, to prove one version of the compactness theorems for Riemannian manifolds (see [Pt87], [HH97], [Pe97]).

Remark 7 The proof of Theorem 10.10 assumes that M has at least a C^3 differentiable structure. This ensures that the metric can be of class C^2, allowing for Christoffel symbols and Riemann curvature to be well defined in a classical way through (10.3) and (10.4), respectively.

10.5 Ricci curvature and the Einstein condition

A useful local expression for the Ricci tensor in terms of the Christoffel symbols is obtained by a contraction of equation (10.4) and, using (10.3), it can be made to depend, exclusively, on the metric. This expression is conveniently simplified by the use of harmonic coordinates, making the Ricci tensor linear in the second-order derivatives of the metric. The following classical result gives the details.

Theorem 10.11 (Lanczos, [La22]) *In arbitrary local coordinates the components of the Ricci tensor can be written, in terms of the metric g and the contractions $\Gamma^c = g^{ij}\Gamma^c_{ij}$, as*

$$\text{Ric}_{ab} = -\frac{1}{2}g^{ij}\partial_i\partial_j g_{ab} + \frac{1}{2}\left(g_{ac}\partial_b\Gamma^c + g_{bc}\partial_a\Gamma^c\right) + Q_{ab}(g,\partial g), \qquad (10.7)$$

where $Q(g,\partial g)$ is a quadratic form depending only (and smoothly) on g and its first partial derivatives ∂g. Moreover, in harmonic coordinates these components take the simpler form,

$$\text{Ric}_{ab} = -\frac{1}{2}g^{ij}\partial_i\partial_j g_{ab} + Q_{ab}(g,\partial g). \qquad (10.8)$$

Proof Since $\text{Ric}_{ab} = R^i_{aib}$, replacing (10.3) in the first two terms of (10.4) and simplifying through symmetries produces

$$\text{Ric}_{ab} = \tfrac{1}{2}g^{ij}(\partial_b\partial_i g_{aj} + \partial_a\partial_j g_{bi} - \partial_i\partial_j g_{ab} - \partial_a\partial_b g_{ij}) + \widetilde{Q}_{ab}(g,\partial g), \qquad (10.9)$$

where $\widetilde{Q}_{ab}(g,\partial g)$ is a function depending only on g and its first partial derivatives ∂g. Furthermore, it is homogeneous of degree 2 in ∂g. By (10.3), the following holds

$$g_{ac}\Gamma^c = g_{ac}g^{ij}\Gamma^c_{ij} = \tfrac{1}{2}g_{ac}g^{ij}g^{cd}(\partial_i g_{jd} + \partial_j g_{id} - \partial_d g_{ij})$$

$$= g^{ij}(\partial_i g_{aj} - \tfrac{1}{2}\partial_a g_{ij}).$$

Taking the partial derivative ∂_b at both sides of the last equality gives

$$\partial_b(g_{ac}\Gamma^c) = \partial_b g^{ij}(\partial_i g_{aj} - \tfrac{1}{2}\partial_a g_{ij}) + g^{ij}(\partial_b\partial_i g_{aj} - \tfrac{1}{2}\partial_b\partial_a g_{ij}).$$

Adding this and the same expression with a and b interchanged produces, after reordering,

$$g^{ij}(\partial_b\partial_i g_{aj} + \partial_a\partial_i g_{bj} - \partial_a\partial_b g_{ij})$$
$$= \partial_b(g_{ac}\Gamma^c) + \partial_a(g_{bc}\Gamma^c) - \partial_b g^{ij}\partial_i g_{aj} - \partial_a g^{ij}\partial_i g_{bj}$$
$$+ \tfrac{1}{2}(\partial_b g^{ij}\partial_a g_{ij} + \partial_a g^{ij}\partial_b g_{ij}).$$

Finally, replacing all this in formula (10.9) yields, after additional rearrangements,

$$\text{Ric}_{ab} = -\tfrac{1}{2}g^{ij}\partial_i\partial_j g_{ab} + \tfrac{1}{2}(g_{ac}\partial_b\Gamma^c + g_{bc}\partial_a\Gamma^c)$$
$$+ \tfrac{1}{2}(\partial_b g_{ac} + \partial_a g_{bc})\Gamma^c - \tfrac{1}{2}(\partial_b g^{ij}\partial_i g_{aj} - \partial_a g^{ij}\partial_i g_{bj})$$
$$+ \tfrac{1}{4}(\partial_b g^{ij}\partial_a g_{ij} + \partial_a g^{ij}\partial_b g_{ij}) + \tilde{Q}_{ab}(g, \partial g).$$

The last two rows of this equality depend only on g and its first derivatives. They can be abbreviated by $Q_{ab}(g, \partial g)$, and this proves (10.7). Since $\Gamma^c = 0$ in harmonic coordinates, the whole expression simplifies to (10.8) as claimed. □

In the previous section, the Laplace–Beltrami operator on (M, g) was shown to become $\Delta = g^{ij}\partial_i\partial_j$ in any harmonic coordinate chart, therefore

$$\text{Ric}_{ab} = -\frac{1}{2}\Delta g_{ab} + Q_{ab}(g, \partial g) \tag{10.10}$$

in such coordinates. This means that any equation (in the metric g) of the form $\text{Ric}_{ab} = F_{ab}(g, \partial g)$, where the right-hand side is a function that does not depend on higher derivatives of g, becomes a second-order quasilinear elliptic equation when written in harmonic charts. Here resides the importance of harmonic coordinates in the study of Ricci curvature.

Concerning regularity, in harmonic coordinates the metric always has two degrees of differentiability more than the Ricci tensor. This is not necessarily true in an arbitrary chart as one could naïvely expect. In fact, a smooth Ricci tensor can even have a non-smooth metric in certain coordinates. An example is easily constructed by taking the pull-back of a Ricci-flat metric by an appropriate non-smooth diffeomorphism of the manifold.

Theorem 10.12 *Let g be of class C^2 in some local chart containing $p \in M$ and suppose that in harmonic coordinates about p the Ricci tensor Ric is of class $C^{k,\alpha}$ for some $k \geq 0$, then in these coordinates g is of class $C^{k+2,\alpha}$.*

For the proof of the previous theorem and additional related statements see [DK81, Theorem 4.5].

10.5.1 Einstein condition and $C^{1,\alpha}$ metrics

Recall that the *Einstein condition* holds in a Riemannian manifold (M, g) of dimension n, when there is a constant $\kappa \in \mathbb{R}$ such that the Ricci tensor satisfies

$$\text{Ric} = \kappa g.$$

Taking the trace on both sides of this equality yields the well-known fact that $\kappa = \text{scal}_g / n$, where scal_g denotes the scalar curvature of (M, g). In local coordinates the Einstein condition reads $\text{Ric}_{ab} = \kappa g_{ab}$ and therefore, using (10.10), a quasilinear elliptic equation is obtained for g_{ab} in a harmonic chart:

$$-\frac{1}{2} g^{ij} \partial_i \partial_j g_{ab} + Q_{ab}(g, \partial g) = \kappa g_{ab}. \tag{10.11}$$

Notice that for some appropriate function $Q'_{ab}(g, \partial g)$, that is still a quadratic form in ∂g, equation (10.11) can be written also in a divergence-like form

$$-\frac{1}{2} \partial_i (g^{ij} \partial_j g_{ab}) + Q'_{ab}(g, \partial g) = \kappa g_{ab}. \tag{10.12}$$

The resulting expression is useful because it suggests introducing the following definition, in the same spirit of the generalized solutions to linear PDEs given in Definition 10.2.

Definition 10.13 Let (U, ϕ) be a harmonic coordinate chart on a manifold M. A Riemannian metric g of class $C^{1,\alpha}(U) \cap L^2_{\text{loc}}(U)$ is a *weak solution* to the Einstein condition in (U, ϕ) if for any test function $\eta \in C^1_c(U)$ the following holds

$$\frac{1}{2} \int_U (g^{ij} \partial_j g_{ab}) \, \partial_i \hat{\eta} \, dx + \int_U Q'_{ab}(g, \partial g) \hat{\eta} \, dx = \int_U \kappa g_{ab} \hat{\eta} \, dx,$$

where $\hat{\eta} := \sqrt{|g|} \, \eta$ and $dx := dx_1 \cdots dx_n$.

If this condition is satisfied on any chart of a harmonic atlas $\mathcal{A}_h$ of M, the Riemannian metric g will be called *weakly Einstein* with respect to $\mathcal{A}_h$.

The previous definition still makes sense for Riemannian metrics with lower regularity but, since in harmonic coordinates the metric is (at least) of class $C^{1,\alpha}$, more generality is not necessary here. Similarly, it is clear that a definition of weak solutions to the Einstein condition can be made in arbitrary coordinate charts, but it is in harmonic ones that the best regularity for the metric and a quasilinear expression for the Ricci tensor are achieved, hence a general definition is not better to study the regularity of these solutions.

An improvement to Theorem 10.12 for weakly Einstein metrics in a harmonic chart is given by the following.

Theorem 10.14 *Let M be an n-dimensional $C^{2,\alpha}$-manifold with a Riemannian metric g of class $C^{1,\alpha}$. If the components of g are weak solutions to the Einstein condition in a harmonic coordinate chart (U, ϕ), then they are smooth with respect to (U, ϕ).*

Proof First, note that g is of class $C^{1,\alpha}$ with respect to the differentiable structure of M and by Theorem 10.9 it is also at least of class $C^{1,\alpha}$ with respect to (U, ϕ). Now, using (10.12) the Einstein condition in harmonic coordinates reads explicitly

$$-\partial_i(g^{ij}\partial_j g_{ab}) = 2\kappa g_{ab} - 2Q'_{ab}(g, \partial g) =: F_{ab}(g, \partial g). \qquad (10.13)$$

Here $Q'_{ab}(g, \partial g)$, and hence $F_{ab}(g, \partial g)$ also, are smooth on their arguments and depend at most on the first derivatives of g. Thus, they are at least of class $C^{0,\alpha}$ in (U, ϕ).

By assumption, the metric is a weak solution to (10.13) in (U, ϕ), therefore $g_{ab} \in C^{1,\alpha}(U) \cap L^2_{\text{loc}}(U) \subset C^{1,\alpha}(U) \cap W^{1,2}_{\text{loc}}(U)$ and satisfies

$$\int_U (g^{ij}\partial_j g_{ab})\, \partial_i \hat\eta \, dx = \int_U F_{ab}(g, \partial g)\hat\eta \, dx, \qquad (10.14)$$

where $g^{ij}\partial_j g_{ab}$ and $F_{ab}(g, \partial g)$ belong to $C^{0,\alpha}(U) \cap L^1_{\text{loc}}(U)$. This means that F_{ab} is a sum of weak derivatives of $g^{ij}\partial_j g_{ab}$, i.e. those derivatives, denoted by $-\partial_i(g^{ij}\partial_j g_{ab})$, exist and, by uniqueness, coincide with F_{ab} a.e. in U. In conclusion, the quasilinear equation

$$-\partial_i(g^{ij}\partial_j g_{ab}) = F_{ab}(g, \partial g) \qquad (10.15)$$

does hold a.e. in the harmonic chart (U, ϕ). To show that this implies $g_{ab} \in C^{2,\alpha}(U)$ it is possible to invoke the Schauder estimates of Theorem 10.3 with Remark 5. The process can be repeated again to raise the regularity once more and, by consecutive iteration of this argument (called a bootstrap), prove the smoothness of g_{ab} with respect to the harmonic chart (U, ϕ). But rigorously, this procedure has the drawback that F_{ab} depends on g and ∂g, unlike f in the Schauder estimates, where it is assumed to depend on x only.

Instead of going into the proof of these estimates for the situation in (10.15), it is easier to invoke the second part of Theorem 10.4. Indeed, writing the equation as the action of a quasilinear operator P gives

$$P g_{ab} := \partial_i(g^{ij}\partial_j g_{ab}) + F_{ab}(g, \partial g) = 0. \qquad (10.16)$$

Now, combining $g_{ab} \in C^{1,\alpha}(U) \cap W^{1,2}_{\text{loc}}(U)$ with the fact that $g^{ij}\partial_i g_{ab}$ admits weak derivatives means that $g_{ab} \in C^{1,\alpha}(U) \cap W^{2,2}_{\text{loc}}(U)$. In addition, the function $f(x) = 0$ on the right-hand side of (10.16) is as smooth as the chart (U, ϕ) allows, which exclusively depends on the differentiability of $\phi^* f := f \circ \phi^{-1}$, but it is clearly at least $C^{0,\alpha}(U)$ because of (10.15).

Assume, in general, that f is of class $C^{k,\alpha}$ for some $k \geq 0$, then $P : C^{1,\alpha}(U) \cap W^{2,2}_{\text{loc}}(U) \to C^{k,\alpha}(U)$ and Theorem 10.4 implies that the metric $g_{ab} \in C^{k+1,\alpha}_{\text{loc}}(U)$. Then, g_{ab} has one degree of differentiability more than the constant function $f(x) = 0$. Since $g_{ab}(x) - g_{ab}(x) = f(x)$, this is impossible, unless both g_{ab} and f are smooth with respect to the chart (U, ϕ). □

Now, if some atlas $\mathcal{A}_h$ on a smooth manifold M consists of harmonic coordinate charts, assertion (1) of Theorem 10.10 implies that all its transition functions are (at least) of class $C^{2,\alpha}$. This and the previous result imply the following.

Corollary 1 *If g is weakly Einstein with respect to all charts in a harmonic coordinate atlas $\mathcal{A}_h = \{(U_\gamma, \phi_\gamma)\}$ of M, then $\mathcal{A}_h$ is a smooth atlas, and g is a smooth Riemannian metric with respect to it.*

Remark 8 Notice that a harmonic coordinate atlas $\mathcal{A}_h$ might define a different differentiable structure on M than the one given by its underlying maximal atlas $\mathcal{A}_M$. In particular, it is possible that the overall regularity of $\mathcal{A}_M$ may not be able to be increased owing to the presence of badly behaved charts. Nevertheless, a different atlas of M (not necesarilly a subatlas of $\mathcal{A}_M$) may have better regularity. This means that, in principle, $\mathcal{A}_h$ could be smooth without $\mathcal{A}_M$ being so. But, if the differentiable structure of these atlases is compatible, transition functions between charts of $\mathcal{A}_h$ and $\mathcal{A}_M$ are smooth, therefore $\mathcal{A}_h \subset \mathcal{A}_M$ by the maximality of the latter and a $C^{1,\alpha}$ metric g on M that is weakly Einstein with respect to $\mathcal{A}_h$ will be smooth in the original atlas of M.

Finally, as a consequence of the analyticity of smooth Einstein metrics, the following immediately holds.

Corollary 2 *On a connected manifold M, with $\dim M \geq 3$, any $C^{1,\alpha}$ metric which is weakly Einstein in harmonic coordinates (in the sense of Theorem 10.14) is actually real analytic in these coordinates.*

Details of related and more refined results in a broader context follow from theorems in [Ta11] and [Ta00] as was mentioned previously.

Acknowledgements

The author wishes to express his gratitude to the organizers of the Summer School "Geometric and Topological Methods for Quantum Field Theory" (Villa de Leyva, Colombia, 2009) for the kind invitation to contribute to this volume, to the Department of Mathematics of Universidad de los Andes (Bogotá) for financial support during the school, and to the Department of Mathematics of Pontificia Universidad Javeriana (Bogotá) where these notes were written.

References

[AF03] R. A. Adams and J. F. Fournier. *Sobolev Spaces*, 2nd edition. Pure and Applied Mathematics (Amsterdam) 140. Amsterdam: Elsevier/Academic Press, 2003.

[An90] M. T. Anderson. Convergence and rigidity of manifolds under Ricci curvature bounds. *Invent. Math.* **102** (1990), 429–445.

[AC92] M. T. Anderson and J. Cheeger. C^α compactness for manifolds with Ricci curvature and injectivity radius bounded below. *J. Diff. Geom.* **35** (1992), 265–281.

[Au98] T. Aubin. *Some Nonlinear Problems in Riemannian Geometry*. Springer Monographs in Mathematics. Berlin: Springer-Verlag, 1998.

[BJS79] L. Bers, F. John and M. Schechter. *Partial Differential Equations*. Reprint of the 1964 original. Lectures in Applied Mathematics 3A. Providence, RI: American Mathematical Society, 1979.

[DK81] D. DeTurck and J. Kazdan. Some regularity theorems in Riemannian geometry. *Ann. Sc. Éc. Norm. Supér.* (4) **14** (1981), 249–260.

[Eva10] L. C. Evans. *Partial Differential Equations*, 2nd edition. Graduate Studies in Mathematics 19. Providence, RI: American Mathematical Society, 2010.

[GT01] D. Gilbarg and N. S. Trudinger. *Elliptic Partial Differential Equations of Second Order*. Reprint of the 1998 edition. Classics in Mathematics. Berlin: Springer-Verlag, 2001.

[HH97] E. Hebey and M. Herzlich. Harmonic coordinates, harmonic radius and convergence of Riemannian manifolds. *Rend. Mat. Appl.* Ser. 7, **17**(1997) 569–605.

[Jo07] J. Jost, *Partial Differential Equations*, 2nd edition. Graduate Texts in Mathematics 214. New York: Springer, 2007.

[JK82] J. Jost and H. Karcher. Geometrische Methoden zur Gewinnung von a-priori-Schranken für harmonische Abbildungen. *Man. Math.* **40** (1982), 27–77.

[Ka93] J. Kazdan. Applications of partial differential equations to problems in geometry. Lecture Notes, 1983, 1993. Available at http://www.math.upenn.edu/~kazdan.

[La22] C. Lanczos. Ein Vereinfachendes Koordinatensystem für die Einsteinschen Gravitationsgleichungen. *Phys. Z.* **23** (1922), 537–539.

[LU68] O. Ladyzenskaya and N. Ural'tseva. *Linear and Quasilinear Elliptic Partial Differential Equations*. New York: Academic Press, 1968.

[MS64] N. Meyers and J. Serrin. $H = W$, *Proc. Nat. Acad. Sci. USA* **51** (1964), 1055–1056.

[Mo66] C. B. Morrey, Jr. *Multiple Integrals in the Calculus of Variations*. Reprint of the 1966 edition. Classics in Mathematics. Berlin: Springer-Verlag, 2008.

[Pt87] S. Peters. Convergence of Riemannian manifolds. *Compos. Math.* **62** (1987), 1–6.

[Pe97] P. Peterson. Convergence theorems in Riemannian geometry. In K. Grove and P. Peterson (eds) *Comparison Geometry*. MSRI Publications 30. Cambridge: Cambridge University Press, 1997, pp. 167–202.

[Pe98] P. Petersen. *Riemannian Geometry*, 1st edition. Graduate Texts in Mathematics 171. New York: Springer-Verlag, 1998.

[SS76] I. Kh. Sabitov and S. Z. Shefel'. Connections between the order of smoothness of a surface and its metric. (Russian) *Sibirsk. Mat. Ž.* **17**: 4 (1976), 916–925. English translation: *Siberian Math. J.* **17**: 4 (1977), 687–694.

[Ta91] M. E. Taylor. *Pseudodifferential Operators and Nonlinear PDE*. Progress in Mathematics 100. Boston, MA: Birkhäuser, 1991.

[Ta00] M. E. Taylor. *Tools for PDE: Pseudodifferential Operators, Paradifferential Operators, and Layer Potentials*. Mathematical Surveys and Monographs 81. Providence, RI: American Mathematical Society, 2000.

[Ta11] M. E. Taylor. *Partial Differential Equations III. Nonlinear Equations*, 2nd edition. Applied Mathematical Sciences 117. New York: Springer, 2011.

[Tr92] H. Triebel. *Theory of Function Spaces II*. Monographs in Mathematics 84. Basel: Birkhäuser Verlag, 1992.

11

Regularized traces and the index formula for manifolds with boundary

ALEXANDER CARDONA AND CÉSAR DEL CORRAL

Abstract

Let D be a first order differential operator acting on the space of section of a finite rank vector bundle over a smooth manifold M with boundary X. In this chapter we show that the index of D, associated to Atiyah–Patodi–Singer type boundary conditions, can be expressed as a weighted (super-)trace of the identity operator, generalizing the corresponding result in the case of closed manifolds obtained in [19]. We also show that the reduced eta-invariant can be expressed as a weighted (super-)trace of an identity operator so that, actually, the index of D can be expressed as a sum of two weighted super-traces of identity operators, one giving rise to the integral term in the Atiyah–Patodi–Singer theorem and the other one corresponding to the η-term.

Introduction

Let D be a positive order differential operator acting on the space of section of a finite rank vector bundle $E \to M$ over a smooth manifold M. When the manifold is closed, it is a well-known result that the index of such an operator can be written as a weighted (super-)trace of the identity, i.e. as a regularization of the trace of the identity with respect to some positive differential operator, usually of Laplacian type (see e.g. [19], [20]). These weighted traces (or pseudo-traces) are neither independent of the reference operator used to define them, nor traces on the algebra of classical pseudo-differential operators, but the obstructions associated to these

---

Geometric and Topological Methods for Quantum Field Theory, ed. Alexander Cardona, Iván Contreras and Andrés F. Reyes-Lega. Published by Cambridge University Press. © Cambridge University Press 2013.

anomalies can be computed explicitly in terms of Wodzicki residues, giving rise to *local* terms, i.e. terms given by integrals of smooth densities on the manifold M. In particular, the index of a positive order differential operator of this type can be shown to be a super-residue in the case of M closed, and therefore local (see [5], [6] or [19] for a self-contained review). In the case of manifolds *with* boundary similar results about the weighted traces can be obtained (see [13], [14] and references therein) but, in contrast with the closed case, the index of a Dirac-type operator is known to be given by a sum of local and non-local terms [1].

Let us be more precise. Let M be a compact oriented Riemannian manifold with boundary $\partial M = X$, such that $M = X \times [0, 1]$ near the boundary. Let D be a first order elliptic differential operator of Dirac type acting on sections of a vector bundle E over M, and let us assume (following [1]; see also [17]) that in the neighborhood on which the boundary of M looks like $M = X \times [0, 1]$, D takes the form

$$D = \gamma \left(\frac{\partial}{\partial u} + A \right), \tag{11.1}$$

where u denotes the inward normal coordinate, $\gamma : E|_X \to E|_X$ is a bundle isomorphism and $A : \Gamma(X, E|_X) \to \Gamma(X, E|_X)$ is elliptic and satisfies

$$\gamma^2 = -I, \quad \gamma^* = -\gamma, \quad A\gamma = -\gamma A, \quad A^* = A.$$

Let us now consider the projection

$$P_+ : L^2(X, E|_X) \to \Gamma_A^+(X, E|_X) \tag{11.2}$$

onto the space $\Gamma_A^+(X, E|_X) = \mathrm{Sp}\{\varphi | A\varphi = \lambda\varphi, \ \lambda \geq 0\}$, then P_+ is a pseudo-differential operator and it defines the space of boundary conditions $\Gamma_{P_+}(M, E) = \{\varphi | P_+\varphi = 0\}$. In [1] Atiyah, Patodi and Singer proved the following.

Theorem 11.1 *Let* $D_{P_+} : \Gamma_{P_+}(M, E) \to \Gamma(M, E)$ *be the restriction of D to* $\Gamma_{P_+}(M, E)$. *Then D_{P_+} is a Fredholm operator with index given by*

$$\mathrm{Ind}\,(D_{P_+}) = \int_M \omega_D - \frac{1}{2}\,(\eta(A) + \dim \ker A)\,, \tag{11.3}$$

where the local density ω_D is given by the constant term in the local asymptotic expansion of $\mathrm{str}(e^{-t\Delta_D}(x, x))$ *as* $t \to 0$.

This theorem has been widely generalized to the case of any first-order differential operator D with much more general boundary conditions by Gerd Grubb in [10]. In this chapter we use some results about the asymptotic behavior of traces of holomorphic families of pseudo-differential operators to show that the index of a Dirac-type operator on a manifold with boundary can be still be expressed as a weighted super-trace of the identity operator, which is *not* a super-residue of Wodzicki type and, as a consequence, non-local. More results on the asymptotic analysis of traces of operator families for manifolds with boundary where carried out in [9], [10], [11], further analysis on the relation between weighted traces and Wodzicki residues in the case of manifolds with boundary can found in [10], [13], [14].

We will finally use a trace defect, or *anomaly* – associated to the dependence of a weighted trace on the reference operator – to show that the index of a differential operator of Dirac type, satisfying Atiyah–Patodi–Singer boundary conditions, can be computed as a sum of weighted super-traces of two different "identities", one coming from the usual asymptotic analysis of the heat kernel and its relation with weighted traces in the closed case (giving rise to a *local* term) and the other given by the decomposition induced by the boundary conditions on the space of sections on which the operator acts (giving rise to the *non-local* η-term). This result follows from some observations about the invariance of weighted traces of particular classical pseudo-differential operators under certain operations on the weight (such us taking absolute values, Laplacians, etc).

The chapter is organized as follows: in Section 11.1 we prove a technical result on the behavior near zero of the meromorphic function defined by the Mellin transform of a smooth function decaying exponentially at infinity. In Section 11.2 we review the basic definitions and results relating the index of an elliptic positive order classical pseudo-differential operator with regularized trace anomalies and regularized determinant anomalies, in the cases of manifolds both with and without boundary. In particular, we show that the index (11.3) of a Dirac operator can be expressed as a weighted super-trace of the identity. In Section 11.3 we consider the eta-invariant and we show that the reduced eta-invariant can also be expressed as a weighted super-trace of an identity defined by the Atoyah–Patodi–Singer boundary conditions (11.2). We prove finally Theorem 11.7, giving a formula for the index of a Dirac-type operator as a sum of weighted super-traces of the identity.

11.1 General heat kernel expansions and zeta functions

Recall that, given a smooth function $f \in C^\infty(\mathbb{R}^+)$ decaying exponentially at infinity, the *Mellin transform* of f is the function defined by

$$\mathrm{M}[f](z) = \frac{1}{\Gamma(z)} \int_0^\infty f(t) t^{z-1} \, dt. \tag{11.4}$$

The meromorphic structure of the Mellin transform of a smooth function f is determined by its asymptotic behavior near zero [19]. The next result gives the finite part of $M[f]$ at $z = 0$ for the general positive functions we will work with in what follows.

Proposition 11.2 Consider, for a real number r, a smooth function $f \in C^\infty(\mathbb{R}^+)$ decaying exponentially at infinity, with asymptotic behavior around $t = 0$ of the form

$$f(t) \sim_{t=0} \sum_{j=0}^\infty a_j t^{\frac{j-r}{q}} + \sum_{l=1}^k \sum_{\frac{j-r}{q} \notin \mathbb{Z}} b_{j,l} t^{\frac{j-r}{q}} \log^l t + \sum_{l=1}^{k+1} \sum_{j=0}^\infty c_{j,l} t^j \log^l t, \quad (11.5)$$

where q is positive, $j \in \mathbb{N}$, $l = 0, \ldots, k$ and $r, a_j, b_{j,l}, c_{j,l} \in \mathbb{R}$ depend on f. Then the Mellin transform of f is a meromorphic function on $\mathbb{C}$, with poles of order $\leq k+1$ at $z = 0$. Moreover,

$$\text{f.p.}|_{z=0} M[f](z) = \text{f.p.}_{t=0} f(t) + \sum_{i=1}^{k+1} \frac{(\frac{1}{\Gamma})^{(i+1)}(0)}{(i+1)!} c_{0,i}, \quad (11.6)$$

where f.p. $|_{z=a}$ denotes the finite part of $f(z)$ at $z = a$. Thus, the difference f.p.$|_{z=0} M[f] - $ f.p.$_{t=0} f(t)$ is given by the logarithmic coefficients $c_{0,l}$.

Proof Integration by parts implies the identity

$$\int_0^1 t^\alpha \log^l t \, dt = \sum_{i=0}^l \frac{l!(-1)^i}{(l-i)!} \left[\frac{t^{\alpha+1}}{(\alpha+1)^{i+1}} \log^{l-i} t \right]_{t=0}^1, \quad (11.7)$$

from which it follows that $\frac{1}{\Gamma(z)} \int_0^1 t^{z-1+a} \log^l t \, dt = o(z)$. Now, for N large enough,

$$M[f](z) = \frac{1}{\Gamma(z)} \int_0^1 t^{z-1} f(t) dt + \frac{1}{\Gamma(z)} \int_1^\infty t^{z-1} f(t) \, dt$$

$$= \frac{1}{\Gamma(z)} \left(\sum_{j=0}^\infty a_j \int_0^1 t^{z-1} t^{\frac{j-r}{q}} \, dt + \sum_{l=0}^k \sum_{\frac{j-r}{q} \notin \mathbb{Z}, j \leq N} b_{j,l} \int_0^1 t^{z-1} t^{\frac{j-r}{q}} \log^l t \, dt \right.$$

$$\left. + \sum_{l=0}^{k+1} \sum_{j=0}^N c_{j,l} \int_0^1 t^{z-1} t^j \log^l t \, dt \right) + \frac{1}{\Gamma(z)} \int_1^\infty t^{z-1} f(t) dt + o(z)$$

and, using (11.7),

$$M[f](z) = \frac{1}{\Gamma(z)} \left(\sum_{j=0}^{\infty} a_j \left[\frac{t^{z+\frac{j-r}{q}}}{z + \frac{j-r}{q}} \right]_{t=0}^{1} \right.$$

$$+ \sum_{l=0}^{k} \sum_{i=0}^{l} \frac{l!(-1)^i}{(l-i)!} \sum_{\frac{j-r}{q} \notin \mathbb{Z}, j \leq N} b_{j,l} \left[\frac{t^{z+\frac{j-r}{q}}}{(z+\frac{j-r}{q})^{i+1}} \log^{l-i} t \right]_{t=0}^{1}$$

$$\left. + \sum_{l=0}^{k+1} \sum_{i=0}^{l} \frac{l!(-1)^i}{(l-i)!} \sum_{j=0}^{N} c_{j,l} \left[\frac{t^{z+j}}{(z+j)^{i+1}} \log^{l-i} t \right]_{t=0}^{1} \right)$$

$$+ \frac{1}{\Gamma(z)} \int_{1}^{\infty} t^{z-1} f(t) dt + o(z),$$

which shows that a meromorphic extension of $M[f]$ exists in a neighborhood of $z = 0$.

Since $\frac{1}{\Gamma(z)} = \sum_{m=1}^{M} \frac{(\Gamma^{-1})^{(m)}(0)}{m!} z^m + o(z^M)$, $\frac{1}{\Gamma(0)} = 0$ and $(\frac{1}{\Gamma})'(0) = 1$, the finite part of $M[f]$ will be given by the terms in which $l = i$, i.e.

$$\frac{a_j}{\Gamma(z)} \left[\frac{t^{z+\frac{i+r}{q}}}{z + \frac{j-r}{q}} \right]_{t=0}^{t=1}$$

and

$$\sum_{l=0}^{k+1} \frac{l!(-1)^l}{\Gamma(z)} \sum_{j=0}^{\infty} c_{j,l} \left[\frac{t^{z+j}}{(z+j)^{l+1}} \right]_{t=0}^{t=1},$$

so that

$$f.p.|_{z=0} M[f](z) = f.p._{t=0} f(t) + \sum_{l=1}^{k} \frac{(-1)^l (\frac{1}{\Gamma})^{(l+1)}(0)}{(l+1)!} c_{0,l}. \qquad \square$$

In particular, if f has no logarithmic divergences its Mellin transform is holomorphic at $z = 0$, and the order of poles there is given by the power of the logarithmic divergence of f at $t = 0$. As we shall see in the next section, when applied to the trace of the heat kernel of a Laplacian, equation (11.6) relates different regularized traces and pseudo-traces on pseudo-differential operators, in the case of manifolds both with and without boundary, with index terms (see [5], [12], [14], [19]).

11.2 Weighted traces, weighted trace anomalies and index terms

11.2.1 The Wodzicki residue and weighted traces for closed manifolds

Let E be a vector bundle above a smooth n-dimensional closed Riemannian manifold M, and let $Cl(E)$ denote the algebra of classical pseudo-differential operators acting on smooth sections of E. Let us denote by $Ell(E)$, $Ell^*(E)$ and $Ell^*_{\mathrm{ord}>0}(E)$ the set of elliptic, invertible elliptic and invertible elliptic with positive order operators, respectively, acting on sections of E, and $Ad(E)$ the subset of $Ell^*_{\mathrm{ord}>0}(E)$ containing the invertible admissible elliptic classical pseudo-differential operators which have positive order, i.e. pseudo-differential operators for which complex powers can be defined in the sense of Seeley [21]. For $Q \in Ad(E)$ and $A \in Cl(E)$, the map $z \mapsto \mathrm{tr}(AQ^{-z})$ is meromorphic with a simple pole at zero. Given $Q \in Ad(E)$, the *Wodzicki residue* of $A \in Cl(E)$ is defined by [22]

$$\mathrm{res}(A) = q \, \mathrm{Res}_{z=0} \left(\mathrm{tr}(AQ^{-z}) \right), \tag{11.8}$$

where q denotes the order of Q. The definition of $\mathrm{res}(A)$ is independent of the choice of Q, and it is (up to a constant) the only *trace* on the algebra $Cl(E)$, i.e.

$$\mathrm{res}([A, B]) = 0, \tag{11.9}$$

for any $A, B \in Cl(E)$ and it is given by a local expression involving the symbol of the operator

$$\mathrm{res}(A) = \frac{1}{(2\pi)^n} \int_M \int_{|\xi|=1} \mathrm{tr}_x \left(\sigma_{-n}(x, \xi) \right) d\xi \, d\mu_M(x), \tag{11.10}$$

where n is the dimension of M, μ_M the volume measure on M, tr_x the trace on the fiber above x and σ_{-n} the homogeneous component of order $-n$ of the symbol of A. Finally, if A is of finite rank, or if its order is less than $-n$, then $\mathrm{res}(A) = 0$.

The Wodzicki residue is *not* an extension of the finite-dimensional trace, and extensions to the algebra $Cl(E)$ of the ordinary trace on trace class operators do not exist. An alternative to the Wodzicki residue trace is a *weighted trace* that, even if it is no longer tracial on $Cl(E)$, extends the usual trace on finite-rank operators (see [5], [6], [19]).

In what follows, by a *weight* we shall mean an element of $Ad(E)$, often denoted by Q, and by q we shall denote its order. We shall take complex powers Q^{-z} of operators $Q \in Ad(E)$, which involves a choice of spectral cut for the operator Q. However, in order to simplify notation, we shall drop the explicit mention of the spectral cut. For $A \in Cl(E)$ and $Q \in Ad(E)$ the map $z \mapsto \mathrm{tr}(AQ^{-z})$ is a meromorphic function with a simple pole at $z = 0$ and we can set the following (see [5]).

Definition 11.2.1 Let Q be a weight and A in $Cl(E)$. We call a Q-weighted trace of A the expression

$$\operatorname{tr}^Q(A) = \text{f.p.}|_{z=0}\left(\operatorname{tr}(AQ^{-z})\right), \qquad (11.11)$$

where f.p. denotes the finite part.

Notice that, when Q has positive leading symbol, we can recover the ζ-regularized trace (11.11) using a heat-kernel expansion. It follows from Proposition 11.2 that, if f has an asymptotic expansion for small t of the form

$$f(t) \sim \sum_{k \geq -n} f_k t^{\frac{k}{q}} + c \log t, \qquad (11.12)$$

then its Mellin transform $M[f]$ is a meromorphic function with poles contained in the set $\frac{n}{q} - \frac{\mathbb{N}}{q}$, and with a Laurent series around zero of the form $-c\,z^{-1} + (f_0 - \gamma c) + O(z)$, where γ is the Euler constant. For $A \in Cl(E)$ let $f(t) = \operatorname{tr}(Ae^{-tQ})$, then $f(t)$ behaves as in (11.12), where $q = \operatorname{ord}Q$ (see [3], [8]) and (11.6) yields

$$\text{f.p.}|_{z=0}\left(\operatorname{tr}(AQ^{-z})\right) = \text{f.p.}|_{z=0}M[f](z) = \text{f.p.}|_{t=0}\left(\operatorname{tr}(Ae^{-tQ})\right) - \gamma \cdot \operatorname{res}(A),$$

where γ is the Euler constant. Thus, if $\operatorname{res}(A) = 0$,

$$\operatorname{tr}^Q(A) = \text{f.p.}|_{t=0}\left(\operatorname{tr}(Ae^{-tQ})\right). \qquad (11.13)$$

Remark 11.2.2 Notice that weighted traces extend the usual finite-dimensional traces, i.e.

$$\operatorname{tr}^Q(A) = \operatorname{tr}(A),$$

whenever A is a finite rank operator. The notion of weighted trace can be extended to the case when Q is a non-injective self-adjoint elliptic with positive order. Being elliptic, such an operator has a finite-dimensional kernel and the orthogonal projection P_Q onto this kernel is a pseudo-differential operator of finite rank. Since Q is an elliptic operator so is the operator $\bar{Q} = Q + P_Q$, for the ellipticity is a condition on the leading symbol which remains unchanged when adding P_Q. Moreover, as Q is self-adjoint, the range of Q is given by $R(Q) = (\ker Q^*)^{\perp} = (\ker Q)^{\perp}$ so that $\bar{Q}$ is onto. As $\bar{Q}$ is injective and onto, it is invertible and, being self-adjoint, $\bar{Q} \in Ad(E)$. We set

$$\operatorname{tr}^Q(A) = \text{f.p.}|_{z=0}\left(\operatorname{tr}(A(\bar{Q})^{-z})\right). \qquad (11.14)$$

11.2.2 Weighted trace anomalies

Unlike Wodzicki residues, weighted traces are not tracial and depend on the weight Q. As a matter of fact both $\operatorname{tr}^Q([A, B])$ and $\operatorname{tr}^{Q_1}(A) - \operatorname{tr}^{Q_2}(A)$, for $Q, Q_1, Q_2 \in$

$Ad(E)$ and $A, B \in Cl(E)$, can be expressed in terms of Wodzicki residues (see Proposition 11.3 below), and we call these obstructions *weighted trace anomalies*. We will assign the *first weight anomaly* to the fact that weighted traces depend on the choice of the weight and *coboundary anomaly* to the fact that weighted traces are not traces. Weighted trace anomalies play an important role in the study of the geometry of the determinant line bundle and relate to anomalies quantum field theories [6]. As it was shown in [5] and [6], both of these obstructions can be computed in terms of Wodzicki residues.[1]

Proposition 11.3

1. For $Q_1, Q_2 \in Ad(E)$ with positive orders q_1, q_2 we have

$$\text{tr}^{Q_1}(A) - \text{tr}^{Q_2}(A) = -\text{res}\left(A\left(\frac{\log Q_1}{q_1} - \frac{\log Q_2}{q_2}\right)\right). \tag{11.15}$$

2. Given $A, B \in Cl(E)$, $Q \in Ad(E)$ with positive order q, we have

$$\text{tr}^{Q}([A, B]) = -\frac{1}{q}\text{res}\left(A[\log Q, B]\right). \tag{11.16}$$

Being residues of classical pseudo-differential operators, both the first trace anomaly and the coboundary anomaly are local, and they are actually related to index terms when applied to geometric operators of Dirac type [6].

11.2.3 Weighted traces, regularized determinants and multiplicative anomalies

Given $A, Q \in Ad(E)$ with common spectral cut, the map $z \mapsto \text{tr}\left((\log A)Q^{-z}\right)$ is meromorphic on the complex plane with a simple pole at the origin [15]. In order to define *weighted determinants*, in [7] weighted traces were extended to logarithms of pseudo-differential operators.

Definition 11.2.3 Given $A, Q \in Ad(E)$ we set

$$\text{tr}^{Q}(\log A) := \text{f.p.}|_{z=0}\left(\text{tr}(\log A\, Q^{-z})\right).$$

As before, Q is referred to as the weight and $\text{tr}^{Q}(\log A)$ as the Q-weighted trace of $\log A$, and we shall not make explicit mention in the notation of the determination of the logarithm underlying this definition. Extending our previous observations about the anomalies associated to weighted traces, Ducourtioux studied the corresponding obstructions on logarithms of pseudo-differential operators and applied them to the

---

[1] Recall that, although the logarithm of a classical pseudo-differential operator is not classical, the bracket $[\log Q, A]$ and the difference $\frac{\log Q_1}{q_1} - \frac{\log Q_2}{q_2}$ of two such logarithms lie in $Cl(E)$.

study of the *multiplicative anomaly* of the Q-weighted determinant, which she defined by

$$\det_Q A = \exp \operatorname{tr}^Q(\log A), \tag{11.17}$$

for any admissible operator A.

Notice that, as a particular case when $Q = A$, we retrieve the well-known ζ-regularized determinant

$$\det_\zeta A = \exp \operatorname{tr}^A(\log A) = \exp(-\zeta_A'(0)), \tag{11.18}$$

where $\zeta_A(z) = \sum_{\lambda \in \operatorname{spec} A} \lambda^{-z}$ is the spectral ζ-function of A. In fact the *multiplicative anomaly* [15], defined for any two operators $A, B \in \mathcal{A}d(E)$ of order a and b, respectively, by the expression

$$M_\zeta(A, B) = \frac{\det_\zeta AB}{\det_\zeta A \det_\zeta B},$$

which generally differs from 1, can be computed in terms of Wodzicki residues, namely (see [7])

$$\log M_\zeta(A, B) = \frac{1}{2a}\operatorname{res}\left(\left(\frac{b \log A - a \log B - a\lambda(A, B)}{a + b}\right)^2\right)$$
$$+ \frac{1}{2b}\operatorname{res}\left(\left(\frac{a \log B - b \log A - b\lambda(A, B)}{a + b}\right)^2\right)$$
$$+ \operatorname{tr}^{AB}(\lambda(A, B))$$

where

$$\lambda(A, B) := \log AB - \log A - \log B. \tag{11.19}$$

Now, $\lambda(A, B) = 0$ if A and B commute, so that the multiplicative anomaly reduces to

$$\log M_\zeta(A, B) = \frac{1}{ab(a + b)}\operatorname{res}\left((a \log B + b \log A)^2\right) \tag{11.20}$$

in that case. In particular, even in the case $B = A^*$, with the formal adjoint of A for the L^2-structure induced by a Riemannian metric on M and a Hermitian one on E, we have $F_\zeta(A, A^*) \neq 0$ and $\det_\zeta(A^*A) \neq |\det_\zeta(A)|^2$. However, if A is self-adjoint this multiplicative anomaly vanishes, i.e. $\det_\zeta A^2 = \det_\zeta(A^*A) = |\det_\zeta A|^2$.

As for the ζ-regularized determinant, there is a multiplicative anomaly associated to the weighted determinant $\det_Q$ which comes from the weighted trace anomalies and also from fact that the weighted trace of $\lambda(A, B)$ is not zero in general (see [7], [18], [19]). However, it has been shown in [20] that the Wodzicki residue of

$\lambda(A, B)$ is always zero for classical operators A and B, so that the determinant associated to the Wodzicki trace does not present any multiplicative anomaly.

11.2.4 The index as a super-trace of the identity

Let us turn now to the relation of these traces with the index theory for elliptic operators. Recall that, whenever Q has a positive leading symbol, we can recover the weighted trace (11.11) using a heat-kernel expansion (see [3], [4]), i.e.

$$\text{f.p.}|_{z=0}\text{tr}(AQ^{-z}) = \text{f.p.}|_{t=0}\text{tr}(Ae^{-tQ}) - \gamma \cdot \text{res}(A),$$

where $q = \text{ord}Q$ and γ is the Euler constant, for operators A with vanishing Wodzicki residue we have (11.13). Thus, when Q is a *differential* operator and $A = I$ we have, as $t \to 0$, $\text{tr}(e^{-tQ}) \sim \sum_{k \geq -\frac{\dim M}{q}} a_k t^{\frac{k}{q}}$ (see e.g. [8]), so $\text{tr}^Q(I) = \text{f.p.}|_{t=0}(\text{tr}(e^{-tQ}))$. Thus, together with the McKean–Singer formula [16], this leads to a formula for the index $\text{Ind}(A) = \dim \ker A - \dim \text{coker}A$ of an elliptic operator with positive order as a (super-)trace of the identity [19]:

$$\text{Ind}(A) = \text{str}^{\Delta_A}(I), \tag{11.21}$$

where the weight is the Laplacian $\Delta_A = A^*A + AA^*$ associated to A and the weighted super-trace is the extension of the usual super-trace, defined as $\text{str}^{\Delta_A}(B) = \text{tr}^{A^*A}(B_+) - \text{tr}^{AA^*}(B_-)$ with respect to the grading on $Cl(E)$ induced by the grading $E = E_+ \oplus E_-$ of the bundle E. Actually, the index can also be expressed in terms of differences of Wodzicki residues, or *super-residues*, as (see [19], [20])

$$\text{Ind}(A) = -\frac{1}{2a}\left[\text{res}(\log(A^*A + \pi_{\Delta_A})) - \text{res}(\log(AA^* + \pi_{\Delta_A}))\right]$$

$$= -\frac{1}{2a}s\text{res}(\log \bar{\Delta}_A). \tag{11.22}$$

Notice that the locality of the index in this case follows directly from the locality of the Wodzicki residue observed in (11.10).

Let us turn now to the case of manifolds with boundary and, in particular, to Theorem 11.1 [1]. Let P_+ denote the pseudo-differential projection associated to the Atiyah–Patodi–Singer boundary conditions (11.2). Then it has been shown that

$$\text{tr}(\varphi e^{-t\Delta_{P_+}}) \sim \sum_{-n \leq k \leq 0} a_k(\varphi)t^{\frac{k}{2}} + \sum_{k \geq 0}(a'(\varphi)\ln t)t^{\frac{k}{2}} + a''(\varphi)t^{\frac{k}{2}}, \tag{11.23}$$

where φ is any compactly supported bundle morphism and Δ_{P_+} is the Laplacian operator associated to the Dirac operator D_{P_+} (see [9] and [10] for more general boundary conditions). It follows from Proposition 11.2 and the fact that $a'(I) = 0$

(see Theorem 4.5 in [11]) that

$$\text{f.p.}|_{z=0}\text{tr}(\Delta_{P_+}^{-z}) = \text{f.p.}_{t=0}\text{tr}(e^{-t\Delta_{P_+}}),$$

so that, with the same super-notations used in (11.21), we have proved the following.

Theorem 11.4

$$\text{Ind}\,(D_{P_+}) = str^{\Delta_{P_+}}(I). \tag{11.24}$$

Nevertheless, in the case of manifolds with boundary such a super-trace of the identity *cannot* be expressed as a super-residue and, as is well known, the index of D is not local in general (see [11]).

11.3 Eta-invariant and super-traces

11.3.1 The eta-invariant for self-adjoint operators

Let $A \in Ell^*_{\text{ord}>0}(E)$ be a self-adjoint elliptic (classical) pseudo-differential operator. If A is not positive, its spectrum contains negative eigenvalues but its ζ-function can still be defined using complex powers, taking now λ_k^{-z} to be $|\lambda_k|^{-z}e^{-i\pi z}$ if λ_k is negative. In [1] Atiyah, Patodi and Singer define, for large $\Re(z)$, the η-function of A as the trace of the operator $A|A|^{-z-1}$, i.e.

$$\eta_A(z) = \sum_{k\in\mathbb{Z}}(\text{sign}\lambda_k)\lambda_k^{-z}. \tag{11.25}$$

They showed that this function extends meromorphically to the whole z-plane and, moreover, that $\eta_A(z)$ is *finite* at $z = 0$. Its value at $z = 0$ measures the *asymmetry* of the spectrum of A. Following [1] we define the η-*invariant* of A by

$$\eta(A) = \text{f.p.}|_{z=0}\,\text{tr}\left(\text{Sign}(A)|A|^{-z}\right) = \text{tr}^{|A|}(\text{Sign}(A)), \tag{11.26}$$

where the *sign of* A is the classical pseudo-differential operator defined by $\text{Sign}(A) := A|A|^{-1}$.

The eta-invariant of a positive order self-adjoint elliptic pseudo-differential operator A has been shown to measure a complex phase appearing in the ζ-determinant of the operator. Actually, part of the proof of such a result in [6] involves the fact that, in that context, the ζ-regularized determinant defined by (11.18) is invariant under the operation of taking absolute values in the weight, i.e. $\text{tr}^A(\log A) = \text{tr}^{|A|}(\log A)$. The eta-invariant, which is defined to be a trace of the sign of the operator, weighted by the operator itself, is also insensitive to the operation of changing that weight by its absolute value.

Proposition 11.5 $\eta(A) = \mathrm{tr}^A(\mathrm{Sign}(A))$.

Proof Let us write $A = |A|\mathrm{Sign}(A)$. Then, since $[|A|, \mathrm{Sign}(A)] = 0$, $\log A = \log |A| + \log(\mathrm{Sign}(A))$ and, writing $\mathrm{Sign}(A) = e^{\frac{i\pi}{2}(\mathrm{Sign}(A)-I)}$, it follows that

$$\frac{\log(A)}{a} - \frac{\log |A|}{a} = \frac{i\pi}{2a}(\mathrm{Sign}(A) - I).$$

Let us show that $\mathrm{tr}^A(\mathrm{Sign}(A)) = \mathrm{tr}^{|A|}(\mathrm{Sign}(A))$. It follows from the first weight anomaly formula (11.15) that

$$\mathrm{tr}^A(\mathrm{Sign}(A)) - \mathrm{tr}^{|A|}(\mathrm{Sign}(A)) = -\mathrm{res}\left(\mathrm{Sign}(A)\left(\frac{\log(A)}{a} - \frac{\log |A|}{a}\right)\right)$$

$$= -\mathrm{res}\left(\mathrm{Sign}(A)\left(\frac{i\pi}{2a}(\mathrm{Sign}(A) - I)\right)\right)$$

$$= -\frac{i\pi}{2a}\mathrm{res}\left(\mathrm{Sign}(A)^2 - \mathrm{Sign}(A)\right)$$

$$= -\frac{i\pi}{2a}\mathrm{res}\left(I - \mathrm{Sign}(A)\right)$$

$$= \frac{i\pi}{2a}\mathrm{res}\left(\mathrm{Sign}(A)\right) = 0,$$

since $\mathrm{res}\left(\mathrm{Sign}(A)\right) = 0$ as shown in [2]. □

11.3.2 The reduced eta-invariant as a super-trace of the identity

A comparison with formula (11.21) shows that the first term in this index formula is the (super-)trace (weighted by the Laplacian Δ_D associated to D) of the identity. We want now turn to the *reduced eta-invariant*, defined by

$$\tilde{\eta}_A = \eta_A + \dim \ker A \tag{11.27}$$

to show it is also a (super-)trace of the identity, actually also weighted by a Laplacian-type operator. We start relating the trace weighted by an operator A to the one weighted by its associated Laplacian Δ_A.

Proposition 11.6 For $Q \in \mathcal{A}d(E)$ and $A \in \mathcal{C}l(E)$,

$$\mathrm{tr}^{Q^*Q}(A) = \frac{1}{2}\mathrm{tr}^Q(A) + \frac{1}{2}\mathrm{tr}^{Q^*}(A) - \frac{1}{2q}\mathrm{res}(A\lambda(Q^*, Q)) \tag{11.28}$$

and

$$\mathrm{tr}^{QQ^*}(A) = \frac{1}{2}\mathrm{tr}^Q(A) + \frac{1}{2}\mathrm{tr}^{Q^*}(A) - \frac{1}{2q}\mathrm{res}(A\lambda(Q, Q^*)). \tag{11.29}$$

In particular, if Q is a self-adjoint weight,

$$\mathrm{tr}^Q(A) = \mathrm{tr}^{QQ^*}(A) = \mathrm{tr}^{Q^*Q}(A). \tag{11.30}$$

Proof From (11.19) it follows that

$$\frac{\log(Q^*Q)}{2q} - \frac{\log Q}{q} = \frac{\log Q^*}{2q} - \frac{\log Q}{2q} + \frac{\lambda(Q^*,Q)}{2q},$$

so that

$$\mathrm{tr}^{Q^*Q}(A) - \mathrm{tr}^Q(A) = -\mathrm{res}\left[A\left(\frac{\log Q^*}{2q} - \frac{\log Q}{2q} + \frac{\lambda(Q^*,Q)}{2q}\right)\right]$$

$$= \frac{1}{2}\mathrm{tr}^{Q^*}(A) - \frac{1}{2}\mathrm{tr}^Q(A) - \frac{1}{2q}\mathrm{res}(A\lambda(Q^*,Q))$$

from which the first identity follows. The second identity can be proved on the same lines. Finally, to obtain (11.30) it suffices to remember that, if two operators A and B commute, then $\lambda(A,B) = 0$. $\qquad\square$

As a consequence of the this result we have the following.

Corollary 11.3.1 If A is a self-adjoint operator

$$\eta(A) = \mathrm{tr}^A(\mathrm{Sign}(A)) = \mathrm{tr}^{A^*A}(\mathrm{Sign}(A)) = \mathrm{tr}^{AA^*}(\mathrm{Sign}(A)).$$

Let us now come back to the geometric situation in which the Atiyah–Patodi–Singer theorem (11.3) was formulated. If we write $P_- = I - P_+$, where P_+ denotes the projection onto the space of sections given by boundary conditions (11.2), decomposing the space of sections $\Gamma(X, E|_X)$ as

$$\Gamma(X, E|_X) = \Gamma_+(X, E|_X) \oplus \Gamma_-(X, E|_X)$$

with the obvious notation, we have a $\mathbb{Z}_2$-grading on such a space of sections with respect to which the sign of the operator A can be seen as a $\mathbb{Z}_2$-graded identity, denoted I_X (it has only $+1$-eigenvectors on such a decomposition, but its super-trace is a sum of both ±1-eigenvalues). Thus, the introduction of the boundary conditions realized by the pseudo-differential projection (11.2) is behind the appearance of the sign of the operator A in the computation of the weighted trace of the identity, and appears itself in form of a non-local eta-term in the index formula.

Theorem 11.7 *The index of the Dirac operator (11.1) can be written as*

$$\mathrm{Ind}\,(D_{P_+}) = \mathrm{str}^{\Delta_D}(I) - \frac{1}{2}\mathrm{str}^{\Delta_A}(I_X), \tag{11.31}$$

where I_X denotes the identity induced by the $\mathbb{Z}_2$-grading on $\Gamma(X, E|_X)$.

Proof Since $\Gamma_{\pm}(X, E|_X)$ is the ± 1-eigenspace of $\mathrm{Sign}(A)$, it follows from Corollary 11.3.1 that, if A is non invertible,

$$\eta(A) + \dim \ker A = \mathrm{str}^{\Delta_A}(I_X). \tag{11.32}$$

Combining this with (11.21), the index formula (11.3) reads

$$\mathrm{Ind}\,(D_{P_+}) = \mathrm{str}^{\Delta_D}(I) - \frac{1}{2}\mathrm{str}^{\Delta_A}(I_X),$$

and we have a formula for the index of the operator (11.1) as a sum of weighted (super-)traces of the identity.   □

Acknowledgements

The authors are indebted to Sylvie Paycha, Steven Rosenberg, Simon Scott and Nicolás Martínez for many stimulating discussions on the analysis and geometry of weighted traces and index theory. This research has been supported by the Vicerrectoría de Investigaciones and the Faculty of Sciences of the Universidad de los Andes.

References

[1] Atiyah, M., Patodi, V. and Singer, I. M. Spectral asymmetry and Riemanian geometry. *Bull. Lond. Math. Soc.* **5** (1973), 229–234.

[2] Atiyah, M., Patodi, V. and Singer, I. M. Spectral asymmetry and Riemannian geometry I, II and III. *Math. Proc. Camb. Phil. Soc.* **77** (1975), 43–69; **78** (1975), 405–432, **79** (1976), 71–99.

[3] Berline, N., Getzler, E. and Vergne, M. *Heat Kernels and Dirac Operators.* Grundlehren Math. Wiss. 298. Berlin: Springer-Verlag, 1992.

[4] Bismut, J-M. Index theorem and the heat equation. In *Proceedings of the International Congress of Mathematicians*, Vols. 1, 2 (Berkeley, California, 1986). American Mathematical Society, 1987, Providence, RI, 1987.

[5] Cardona, A., Ducourtioux, C., Magnot, J. P. and Paycha, S. Weighted traces on algebras of pseudo-differential operators and geometry on loop groups. *Infin. Dimen. Anal. Quant. Probab. Relat. Top.* **5**: 4 (2002), 503–540.

[6] Cardona, A., Ducourtioux, C. and Paycha, S. From tracial anomalies to anomalies in quantum field theory. *Comm. Math. Phys.* **242**: 1–2 (2003), 31–65.

[7] Ducourtioux, C. Weighted traces on pseudo-differential operators and associated determinants. PhD thesis, Université Blaise Pascal, 2001.

[8] Gilkey, P. *Invariance Theory, the Heat Equation, and the Atiyah-Singer Index Theorem*, 2nd edn. Boca Raton, FL: CRC Press, 1995.

[9] Grubb, G. Heat operator trace expansions and index for general Atiyah-Patodi-Singer boundary problems. *Comm. Partial Diff. Equations* **17**:11–12 (1992), 2031–2077.

[10] Grubb, G. Trace expansions for pseudodifferential boundary problems for Dirac-type operators and more general systems. *Ark. Mat.* **37**:1 (1999), 45–86.

[11] Grubb, G. Spectral boundary conditions for generalizations of Laplace and Dirac operators. *Comm. Math. Phys.* **240** (2003), 243–280.

[12] Grubb, G. On the logarithm component in trace defect formulas. *Comm. Partial Diff. Equations* **30** (2005), 1671–1716.

[13] Grubb, G. Trace defect formulas and zeta values for boundary problems. In *Traces in Number Theory, Geometry and Quantum Fields*, Aspects of Mathematics E**38**, Wiesbaden: Friedr. Vieweg, 2008, pp. 137–153.

[14] Grubb, G. The local and global parts of the basic zeta coefficient for operators on manifolds with boundary. *Math. Ann.* **341**:4 (2008), 735–788.

[15] Kontsevich, M. and Vishik, S. Determinants of elliptic pseudo-differential operators, Max Planck Institut Preprint, 1994.

[16] McKean, H. P. and Singer, I. M. *Curvature and the eigenvalues of the Laplacian.* *J. Diff. Geom.* **1**:1 (1967), 43–69.

[17] Müller, W. Eta invariants and manifolds with boundary. *J. Diff. Geom.* **40**:2 (1994), 311–377.

[18] Ouedraogo M-F. and Paycha, S. The multiplicative anomaly for determinants revisited: locality. Preprint, 2007.

[19] Paycha, S. *Regularized Integrals, Sums and Traces: Analytic Aspects.* Providence, RI: American Mathematical Society, 2012.

[20] Scott, S. The residue determinant. *Comm. Partial Diff. Equations* **30**:4–6 (2005), 483–507.

[21] Seeley, R. T. Complex powers of an elliptic operator. In *Proceedings of the Symposium on Pure Mathematics*, Vol. 10. Providence, RI: American Mathematical Society, 1967, pp. 288–307.

[22] Wodzicki, M. *Non Commutative Residue.* Lecture Notes in Mathematics **1289**. Berlin: Springer Verlag, 1987.

Index

Printed in the United States
by Baker & Taylor Publisher Services